Filippo Gazzola
Franco Tomarelli
Maurizio Zanotti

ANALYTIC FUNCTIONS INTEGRAL TRANSFORMS DIFFERENTIAL EQUATIONS

theoretical topics and solved exercises

Società Editrice
ESCULAPIO

Filippo Gazzola
Department of Mathematics
Politecnico di Milano
20133 Milano - Italy
filippo.gazzola@polimi.it

Franco Tomarelli
Department of Mathematics
Politecnico di Milano
20133 Milano - Italy
franco.tomarelli@polimi.it

Maurizio Zanotti
Structural and Machine designer
maurizio_zanotti@alice.it

ISBN 978-88-9385-345-3

Società Editrice Esculapio s.r.l.
Via Terracini, 30 - 40131 Bologna
www.editrice-esculapio.com - info@editrice-esculapio.it

Editorial staff: Carlotta Lenzi, Laura Tondelli, Laura Brugnoli

On front cover:
Victor Vasarely, *Ellen-2*,
oil on canvas, 1974
Il Castello Arte Moderna e Contemporanea, Milano

Società Editrice
ESCULAPIO

Preface

Differential equations play a relevant role in many disciplines and provide powerful tools for modeling and analysis in applied sciences.
The book shows several methods for the study of *ordinary differential equations* and of *partial differential equations*. A broad space is reserved to Fourier and Laplace transforms together with their applications to solution of boundary value and/or initial value problems for differential equations.
Basic prerequisites concerning analytic functions of complex variable and L^p spaces are synthetically presented in the first two chapters.
Techniques based on integral transforms and Fourier series are shown in specific chapters, first in the easier framework of integrable functions and later in the general framework of distributions. The less elementary distributional context allows to deal also with differential equations with highly irregular data and pulse signals.
Theory is introduced concisely while learning of miscellaneous methods is achieved step-by-step through the proposal of many exercises of increasing difficulty. Additional recap exercises are collected in dedicated sections.
Several tables for easy reference of main formulas are available at the end of the book.
The introduction of every topic is organized as follows: first a short list of main definitions and properties is given; then the subject is studied by facing many examples and proposing exercises whose detailed solution is eventually shown.
A whole chapter collects self-evaluation trials and examination problems with their solutions.
The book content can be used for a one-semester course, or may be split into two didactic modules: one at undergraduate level, and a follow-up advanced module at graduate level. In this respect detailed information is supplied in the ensuing section, that is labeled Reader's Guide.
The presentation of book matter is based on the long-standing didactic experience of several courses: Analisi Matematica III and Analisi Matematica D for Ingegneria Matematica (Politecnico di Milano), Mathematical Analysis

Tools for Engineering and Metodi Analitici e Statistici for Electrical Engineering (Politecnico di Milano), Advanced Mathematical Analysis for Chemical Engineering (Politecnico di Milano), Differential Equations for Electronic Engineering (Politecnico di Milano), Metodi Analitici e Numerici for Energy Engineering and Mechanical Engineering (Politecnico di Milano), Complementi di Analisi Matematica for Chemical Engineering (Politecnico di Milano), Metodi Matematici per l'Ingegneria (Università di Pavia), Istituzioni di Analisi Superiore (Università del Piemonte Orientale).

We thank Elvise Berchio and Marco Boella for many useful remarks.
We wish to express our warm thanks also to the students of the Laurea and Master in Mathematical Engineering, Electrical Engineering and Chemical Engineering at Politecnico di Milano for their enthusiasm and valuable suggestions.

The present text is the English translation of the last revised and extended edition of the Italian book "Analisi Complessa - Trasformate - Equazioni Differenziali" ([6]) published by Esculapio in 2023, and is available also as an eBook. We would like to acknowledge the collaboration of Michele Di Cristo and Fabio Scardigli in the translation.

Milano, January 30, 2023

Filippo Gazzola Franco Tomarelli Maurizio Zanotti

Reader's Guide

The content is divided in 10 Chapters according to the main subjects. Along the text the reader can find: 308 Exercises together with their worked solutions, 14 self-evaluation Trials, 52 Figures, 27 summarizing Tables.

The Chapters from 1 to 8 are split in 2 Sections: the first Section presents a synopsis of definitions, properties and theoretical tools, the second one is more extensive and contains exercises about the subject of the first Section; their detailed solutions are described in the text. Also theoretical questions and the more significant proofs of main theorems are presented in the second Section. In each Chapter theory and exercises are shown the in increasing order of difficulty.

The first Chapter (devoted to the theory of analytic functions of one complex variable) and the second one (which introduces the L^p spaces) are preliminary to to study of all subsequent ones.

All subsequent chapters (from third to eighth) are self-contained and largely independent from each other, in such a way that several different reading path are possible, accordingly to the interests of the reader. In particular the book content can be organized in independent courses.

Chapter 3 introduces the distributions or generalized functions in a operative way; it can be omitted when reading first, without inconvenient in the comprehension of next chapters if the reader is not interested in tackling differential equations in the distributional framework. Explicitly, the tools from the theory of distributions are exploited only in the subsections 4.1.4, 4.2.2, 4.2.3; 5.1.2, 5.2.2; 6.1.5, 6.1.6, 6.1.7, 6.1.8 6.2.3.

Chapter 4 shows the powerful and flexible tool of Fourier transform together with its applications to the study and solution of differential equations. In the exercises two separate subsections are devoted to ordinary differential and partial differential equations: in particular there are the detailed computations of the fundamental solution for heat equation, wave equation and other important equation of Mathematical Physics.

Chapter 5 introduces the Fourier series and essential tools for their use in the study of differential problems in bounded domains of $\mathbb{R}^n$ and the analysis of periodic signals. In particular the method is focussed on the study of periodic solutions of ordinary differential equations.

Chapter 6 introduces the method of Laplace transform and shows many applications to the solution of differential equations: initial value problems, problems with conditions on the support of the solution and differential equations with delay terms.

Chapter 7 is devoted to differential equations modeling transport phenomena. Several techniques are presented, the attentions is focussed mainly on Duhamel method and separation of variables.

Chapter 8 is devoted to differential equations modeling diffusion phenomena. These equations are tackled in many ways, among them are the one mentioned before. The smoothing effect and the "infinite" speed of propagation associated to the heat operator are discussed.

Chapter 9 proposes self-evaluation survey problems and their solutions: they refer to the content of all previous chapters of the book. This chapter is split in two parts: the first one is a collection of trials built with multiple-choice questions covering all the topics, with complete solutions in separate pages; the second one proposes exercises to be solved in detail, with explanation about answers and computations. Many of these exercises are extracted from written examination proofs in recent years.

Chapter 10 is a collection of Tables that gather all the necessary prerequisites, even the elementary ones, and the main results proven in the text about transforms and differential equations.

Contents

Notations

$\mathbb{R}$	real numbers;
$\mathbb{N}$	natural numbers together with 0;
$\mathbb{Z}$	integer numbers;
$\mathbb{Q}$	rational numbers;
$\mathbb{C}$	complex numbers;
$\mathbb{R}^n$	n–dimensional Euclidian space;
$\{\mathbf{e}_1, \mathbf{e}_2, \cdots, \mathbf{e}_n\}$	canonical basis of $\mathbb{R}^n$: $\mathbf{e}_j = (0, 0, \ldots, 1, \ldots, 0)$;
$\mathbf{x} = (x_1, \ldots, x_n)$	vector $\mathbf{x} \in \mathbb{R}^n$ with components $x_j \in \mathbb{R}$ with respect to the canonical basis of $\mathbb{R}^n$;
$\lvert\mathbf{x}\rvert$	euclidian norm of $\mathbf{x} \in \mathbb{R}^n$;
$B_R(\mathbf{x})$	n-dimensional open ball with radius R centered at $\mathbf{x}$, disk is used as a synonym of ball in dimension $n = 2$;
∂E	boundary of the set $E \subset \mathbb{R}^n$;
$\overline{E}$	closure of the set $E \subset \mathbb{R}^n$;
ω_n	volume of the unit ball $B_1(\mathbf{0}) \subset \mathbb{R}^n$;
σ_{n-1}	surface area of the unit sphere $S^{n-1} = \partial B_1(\mathbf{0}) \subset \mathbb{R}^n$;
x^+	positive part of the real number x;
x^-	negative part of the real number x;
$o(1)$	infinitesimal;
$o(\lvert x\rvert^a)$	"little o" of $\lvert x\rvert^a$: infinitesimal term, faster than $\lvert x\rvert^a$ as $x \to 0,\ a > 0$;
$O(\lvert x\rvert^a)$	"big O" of x^a: a term whose modulus is estimated by $C\lvert x\rvert^a$;
$\asymp$	of the same order;
$\sim$	asymptotic to;
$\operatorname{Re} z$	real part of $z \in \mathbb{C}$;
$\operatorname{Im} z$	imaginary part of $z \in \mathbb{C}$;
$\lvert z\rvert$	modulus of $z \in \mathbb{C}$: $\lvert z\rvert = \left((\operatorname{Re} z)^2 + (\operatorname{Im} z)^2\right)^{1/2}$;
$\arg(z)$	argument of $z \in \mathbb{C} \setminus \{0\}$: oriented angle between $\mathbf{e}_1$ and z ;
$\operatorname{phase}(z)$	phase of $z \in \mathbb{C} \setminus \{0\}$: $\operatorname{phase}(z) = z/\lvert z\rvert$;
$\operatorname{res}(f(z), z_0)$	residue of the function f at the point $z = z_0 \in \mathbb{C}$;
$\operatorname{Ind}(\gamma, z_0)$	winding number of the curve γ with respect to the point $z_0 \in \mathbb{C}$;

$v : \Omega \mapsto E$	map v with domain $\Omega \subset \mathbb{R}^n$ and range in $E \subset \mathbb{R}^m$;
$\mathrm{spt}(v)$	support of the map v;
$v(x_+)$	right-limit at the point x of the map v (dependent on one real variable);
$v(x_-)$	left-limit at the point x of the map v (dependent on one real variable);
$v_k \to v$ in A	sequence of functions v_k convergent in the space A to the function v as $k \to +\infty$;
$L^p(\Omega)$	space of measurable functions with Lebesgue integrable p–th power of their modulus over Ω;
$L^\infty(\Omega)$	space of essentially bounded functions with domain Ω;
$L^2_T(\mathbb{R})$	space of T periodic functions belonging to $L^2_{loc}(\mathbb{R})$;
$\\|v\\|_{C^0}$, $\\|v\\|_{C^k}$	lagrangian norms of the function v;
$\\|v\\|_p$	L^p norm of the function v;
$\\|v\\|_\infty$	L^∞ norm of the function v;
$\mathcal{H}(\Omega)$	space of holomorphic (i.e. analytic) functions with domain Ω;
$H^k(\mathbb{R}^n)$	Sobolev space of $L^2(\mathbb{R}^n)$ functions whose derivatives up to the order k belong to $L^2(\mathbb{R}^n)$;
$H^k_T(\mathbb{R})$	Sobolev space of T periodic functions whose derivatives up to the order k belong to $L^2_{loc}(\mathbb{R})$;
$\mathcal{D}(\Omega)$	test functions in Ω, i.e. the space of c^∞ functions with compact support;
$\mathcal{D}'(\Omega)$	distributions in Ω, i.e. the space of linear continuous functional on $\mathcal{D}(\Omega)$;
$\mathcal{S}(\mathbb{R}^n)$	Schwartz space: smooth functions in $\mathbb{R}^n$ which have fast decay at ∞ together with all derivatives;
$\mathcal{S}'(\mathbb{R}^n)$	tempered distributions in $\mathbb{R}^n$;
$\mathcal{S}'_T(\mathbb{R})$	T periodic tempered distributions in $\mathbb{R}$;
$\mathcal{E}(\Omega)$	$C^\infty(\Omega)$ functions;
$\mathcal{E}'(\Omega)$	compactly supported distributions in Ω;
$BV(I)$	bounded variation functions in I;
$\mathcal{V}$	space of Fourier transforms of $L^1(\mathbb{R}^n)$ functions;
$\boldsymbol{\alpha} = (\alpha_1, \dots, \alpha_n)$	multi-index (element of $\mathbb{N}^n$) with components $\alpha_j \in \mathbb{N}$;
$\|\boldsymbol{\alpha}\|$, $\mathrm{ord}(\boldsymbol{\alpha})$	order of multi-index $\boldsymbol{\alpha}$: $\mathrm{ord}(\boldsymbol{\alpha}) = \|\boldsymbol{\alpha}\| = \alpha_1 + \alpha_2 + \cdots + \alpha_n$;
$D^{\boldsymbol{\alpha}} v$	partial derivative: $D^{\boldsymbol{\alpha}} = \dfrac{\partial^{\alpha_1}}{\partial x_1^{\alpha_1}} \cdots \dfrac{\partial^{\alpha_n}}{\partial x_n^{\alpha_n}}$;

$\nabla_{\mathbf{x}} v, D_{\mathbf{x}} v$	gradient of the function v: $\nabla_{\mathbf{x}} v = D_{\mathbf{x}} v = \left(\frac{\partial v}{\partial x_1}, \ldots, \frac{\partial v}{\partial x_n} \right)$;
$\Delta_{\mathbf{x}} v$	Laplacian of the function v dependent on n variables: $\Delta_{\mathbf{x}} v = \sum_{j=1}^{n} \frac{\partial^2 v}{\partial x_j^2}$;
$\Box_c\, v$	d'Alembertian of the function v dependent on $n+1$ variables (wave operator with speed propagation c): $\Box_c\, v = \left(\partial_t^2 - c^2 \Delta_{\mathbf{x}}\right) v$;
$\Box\, v$	d'Alembertian of the function v dependent on $n+1$ variables (wave operator with speed propagation $c = 1$): $\Box\, v = \left(\partial_t^2 - \Delta_{\mathbf{x}}\right) v$;
$U(\mathbf{x})$	fundamental solution of a differential operator;
$K(\mathbf{x}, \mathbf{y})$	kernel associated to the fundamental solution of a differential operator: $K(\mathbf{x}, \mathbf{y}) = U(\mathbf{x} - \mathbf{y})$;
λ_f	abscissa of convergence for Laplace transform of f;
$\mathcal{L}\{v\}$	Laplace transform of v: $V = \mathcal{L}\{v\}$;
$\mathcal{F}\{v\}$	Fourier transform of v ;
$\widehat{v}$	Fourier transform of v : $\widehat{v} = \mathcal{F}\{v\}$;
$\check{v}$	function v evaluated with sign change of the variable: $\check{v}(\mathbf{x}) = v(-\mathbf{x})$;
$u * v$	convolution of u and v;
$u \otimes v$	tensor product of u e v;
$\delta_{\mathbf{x}_0}(\mathbf{x})$	Dirac delta concentrated at $\mathbf{x}_0 \in \mathbb{R}^n$;
$\mathrm{comb}_T(t)$	comb function with period $T > 0$: $\mathrm{comb}_T(t) = \sum_{k=-\infty}^{+\infty} \delta(t - kT)$;
p	(un-normalized) rectangular pulse: $p(t) = 1$ if $\|t\| < 1$, $p(t) = 0$ if $\|t\| > 1$;
rect	(normalized) rectangular pulse: $\mathrm{rect}(t) = 1$ if $\|t\| < 1/2$, $p(t) = 0$ if $\|t\| > 1/2$;
H	Heaviside function or unit-step function: $H(t) = 1$ if $\|t\| < 0$, $H(t) = 0$ if $\|t\| > 0$;
sign	sign function: $\mathrm{sign}(t) = 1$ if $t > 0$, $\mathrm{sign}(t) = -1$ if $t < 0$;

Sinc	(un-normalized) sampling function: $\mathrm{Sinc}(t) = \dfrac{\sin t}{t}$ if $t \neq 0$, $\mathrm{Sinc}(0) = 1$;
sinc	(normalized) sampling function: $\mathrm{sinc}(t) = \dfrac{\sin(\pi t)}{\pi t}$ if $t \neq 0$, $\mathrm{sinc}(0) = 1$;
Si	sine integral function: $\mathrm{Si}(t) = \int_0^t \dfrac{\sin \tau}{\tau}\, d\tau = \dfrac{\pi}{2} - \int_t^\infty \dfrac{\sin \tau}{\tau}\, d\tau = \int_0^t \mathrm{Sinc}(\tau)\, d\tau$;
erf	error function: $\mathrm{erf}(t) = \dfrac{2}{\sqrt{\pi}} \int_0^t e^{-\tau^2}\, d\tau$;
erfc	complementary error function: $\mathrm{erfc}(t) = \dfrac{2}{\sqrt{\pi}} \int_t^\infty e^{-\tau^2}\, d\tau$;
p.v.	Cauchy principal value;
f.p.	finite part;
iff	if and only if;
s.t.	such that;
a.e.	almost everywhere (with respect to the Lebesgue measure);
$\setminus$	set difference: $A \setminus B = \{a \in A : \ a \notin B\}$.

Chapter 1
Analytic functions of complex variable

The notion of derivative can be extended in a natural way to the case of functions of complex variable: one can apply the same formal definition of the real case (limit of difference quotient) with the only difference that also complex increments of the variable are allowed.
Existence of the derivative in the complex sense is a by far richer property than existence of derivative in the real sense: for instance, if a complex-valued function admits the first derivative in the complex sense in a nonempty open subset of the complex plane $\mathbb{C}$ then this function admits derivatives of every order in the same open set.
The definition of every elementary function can be extended (in a unique way!) in open sets of the complex plane in such a way that the function obtained by this extension admits a derivative in the complex sense.
This chapter shows the main properties of complex-valued functions that admit a derivative in their domain, which is an open subset of $\mathbb{C}$: these functions are called holomorphic or analytic functions of complex variable.
The chapter has two sections: the first one exposes topics from the theory, the second one proposes exercises together with their worked solutions.
The knowledge of analytic functions provides elegant and powerful tools for evaluating integrals over curves in $\mathbb{C}$: for this reason the complex variable techniques are effective in the computation of integral transforms, which are the object of Chapters 4 and 6 about Fourier and Laplace transform. On this subject the reader may refer also to [3], [7], [25].

1.1 Definitions and techniques

Topics from the theory are displayed in several subsections: first there is a reminder of basic properties of complex numbers and curves in the complex plane, then the main tools for dealing with analytic functions are presented together with their application to evaluation of definite integrals.

1.1.1 Algebra of complex numbers

The symbol $\mathbb{C}$ denotes the set of complex numbers, say the set of ordered pairs (x, y) of real numbers (associated to the set of points in the cartesian plane) endowed with the algebraic operations of sum and product, which are defined by formulae (1.1),(1.2) below.
The complex number $(0,1)$ is denoted by i and is called *imaginary unit* . Then every complex number (x, y) can be represented in the standard way by $x + iy$ (short notation for $x\,(1,0) + y\,(0,1)$). Analogously we write 0 in place of $(0,0)$ and 1 in place of $(1,0)$.
The real numbers x and y are called respectively *real part* and *imaginary part* of the complex number $x + iy$ and we write $x = \operatorname{Re} z$ e $y = \operatorname{Im} z$.
If $z_1 = x_1 + iy_1$ and $z_2 = x_2 + iy_2$ are complex numbers, then their sum and product are defined as follows:

$$z_1 + z_2 = (x_1 + x_2) + i\,(y_1 + y_2)\,, \tag{1.1}$$
$$z_1 \cdot z_2 = (x_1 x_2 - y_1 y_2) + i\,(x_1 y_2 + x_2 y_1)\,. \tag{1.2}$$

Notice that $i^2 = i \cdot i = -1$, thanks to the second definition.
Sum and product fulfil the following properties, for any $z_1,\ z_2,\ z_3 \in \mathbb{C}$:

$$z_1 + z_2 = z_2 + z_1 \qquad z_1 z_2 = z_2 z_1$$
$$(z_1 + z_2) + z_3 = z_1 + (z_2 + z_3) \qquad (z_1 z_2)\, z_3 = z_1\,(z_2 z_3)$$
$$z_1 + 0 = z_1 \qquad 1 \cdot z_1 = z_1$$
$$z + (-z) = 0 \qquad z = x + iy \neq 0 \quad \Rightarrow \quad \exists\, z^{-1} = \frac{x - iy}{x^2 + y^2}$$
$$(z_1 + z_2)\, z_3 = z_1 z_3 + z_2 z_3\,.$$

If $z = x + iy$, then the complex number $\overline{z} = x - iy$ is called *conjugate* of z. Notice that $z = \overline{z}$ if and only if z is real (say $y = 0$).
The *modulus* of z is the nonnegative real number $|z| = \sqrt{x^2 + y^2}$, where $\sqrt{}$ denotes the arithmetic square root. Hence $|z| = 0$ if and only if $z = 0$, that is $x = y = 0$. Moreover:

$$|z_1 + z_2| \le |z_1| + |z_2|\,, \qquad |z_1 z_2| = |z_1|\,|z_2|\,, \qquad \forall z_1,\ z_2 \in \mathbb{C}\,.$$

For every $z \in \mathbb{C}$ we get $z\,\overline{z} = |z|^2$. For every $z \in \mathbb{C}\backslash\{0\}$ we get $z^{-1} = \overline{z}/\,|z|^2$.
The *exponential* of a complex number $z = x + iy$ is defined as follows:

$$e^{x+iy} = e^x\,(\cos y + i \sin y)$$

and fulfils $e^{z_1 + z_2} = e^{z_1} e^{z_2}$ for any $z_1, z_2 \in \mathbb{C}$. In particular,

$$\sin\vartheta = \frac{e^{i\vartheta} - e^{-i\vartheta}}{2i} = -\,i\,\sinh(i\vartheta), \quad \cos\vartheta = \frac{e^{i\vartheta} + e^{-i\vartheta}}{2} = \cosh(i\vartheta), \quad \forall \vartheta \in \mathbb{R}.$$

For every $z = x + iy \neq 0$ there exists (unique) $\vartheta \in [0, 2\pi)$, which is called principal value of the *argument* of z, such that if $\rho = |z|$ we have

$$x + iy = \rho\,(\cos\vartheta + i\sin\vartheta) = \rho\, e^{i\vartheta}$$

The expressions in the box above are called respectively *cartesian representation*, *polar representation* (or *trigonometric representation*) and *exponential representation* of the complex number z. The first one is handy for evaluating sums, the second and the third are suitable for evaluating products.
We emphasize that for a given complex number z there are infinitely many choices (values) of its argument. In fact, if $z = \varrho\, e^{i\vartheta}$, then we have also $z = \varrho\, e^{i\vartheta_k}$ for every choice of $\vartheta_k = \vartheta + 2k\pi\,,\ k \in \mathbb{Z}\,.$
More precisely arg is a multi-valued function with domain $\mathbb{C} \setminus \{0\}$ and range in $\mathbb{R}$: $\vartheta \in \arg(z)$ iff ϑ is an oriented angle between $\mathbf{e}_1$ and z.
If $z = \rho\,(\cos\vartheta + i\sin\vartheta)$ and $w = r\,(\cos\varphi + i\sin\varphi\ )$, then the following formula holds true

de Moivre's formula (Abraham de Moivre, 1667-1754).

$$z\,w \;=\; \rho\, r\ (\cos(\vartheta + \varphi) + i\sin(\vartheta + \varphi)) = \rho\, r\, e^{i(\vartheta+\varphi)}\,,$$

$$z^n \;=\; \rho^n\,(\cos(n\vartheta) + i\sin(n\vartheta)) = \rho^n e^{in\vartheta}\,, \qquad \forall n \in \mathbb{N}\,.$$

For any complex number $z = \rho\, e^{i\vartheta}$ different from 0 (say, s.t. $\rho > 0$) and every fixed integer $n > 0$, there exist exactly n complex n-th roots w_1, w_2, ..., w_n of z, which are different one from another, and no more: they are the n complex numbers w_j fulfilling $w_j^n = z\,(j = 1, \ldots, n)$. Their list is the following:

$$w_1 = \rho^{1/n} e^{i\frac{\vartheta}{n}} \quad w_2 = \rho^{1/n} e^{i\frac{\vartheta+2\pi}{n}} \quad w_3 = \rho^{1/n} e^{i\frac{\vartheta+4\pi}{n}} \quad \ldots \quad w_n = \rho^{1/n} e^{i\frac{\vartheta+2(n-1)\pi}{n}}$$

where $\rho^{1/n}$ denotes the arithmetic n-th root of the positive real number ρ.
Every algebraic equation of degree greater then 0 has solutions in the complex field, as stated by the following theorem.

Theorem 1.1. ***Fundamental Theorem of Algebra****. Every non-constant single-variable polynomial with complex coefficients has at least one complex root. More precisely, every polynomial P of degree $n \geq 1$, has exactly n complex roots z_j (if they are counted with their multiplicity in the factorization), that is there exist $z_1, \ldots, z_n \in \mathbb{C}$, not necessarily different from each other, and a coefficient $a \in \mathbb{C} \setminus \{0\}$ such that*

$$P(z) = a\,(z - z_1)\,(z - z_2)\cdots(z - z_n)\;.$$

1.1.2 Curves in the complex plane

Definition 1.2. *(Curves)* *Given a bounded closed interval* $I = [a, b] \subset \mathbb{R}$, *a curve in the complex plane is a continuous map* $\gamma : I \mapsto \mathbb{C}$. *The terms* arc, path, oriented path *and* parametric curve *are commonly used as synonyms of curve.*
Given a curve $\gamma : I \mapsto \mathbb{C}$*, the set* $\gamma(I)$ *is called image of* γ. *Though they are often confused in the spoken language, the distinction between a curve and its image is important, since a given image is not a prescribed route but it is a set which can be run through in many different ways.*
A closed curve γ *, or* loop *is a curve s.t.* $\gamma(a) = \gamma(b)$.
A curve $\gamma : [a, b] \mapsto \mathbb{C}$ *is simple if*

$$\forall t, \tau \in [a, b] , \qquad \gamma(t) = \gamma(\tau) \implies \{ \textit{either } t = \tau \textit{ or } \{t, \tau\} = \{a, b\} \} .$$

If γ *is a* $\mathbb{C}$*-valued* $C^1(I)$ *function we say that* γ *is regular.*
We say that the curve γ *is piecewise* C^1 *if it is continuous in* I *and the interval* I *can be subdivided in a finite collection of of closed sub-intervals* I_j *such that the restriction of* γ *in* I_j *is* $C^1(I_j)$ *for every* j.

Definition 1.3. *Given an open set* $\Omega \subset \mathbb{C}$*, a piecewise* C^1 *curve* $\gamma : I \to \Omega$ *and a continuous function* $f : \Omega \to \mathbb{C}$*, we define the line integral or path integral (also called contour integral when* γ *is a loop) of* f *along* γ *as follows*

$$\int_\gamma f \; := \; \int_\gamma f(z)\, dz \; = \; \int_I f(\gamma(t))\, \gamma'(t)\, dt \, . \tag{1.3}$$

Due to the properties of integrals over an interval of the real line, the line integral of the differential form $f(z)dz = (u + iv)(dx + idy) = (u\, dx - v\, dy) + i(v\, dx + u\, dy)$ (defined by (1.3) above) fulfils, for every choice of the functions f and g (assumed to be integrable over the image of γ):

- $\displaystyle\int_\gamma (f(z) + g(z))\, dz = \int_\gamma f(z)\, dz + \int_\gamma g(z)\, dz \,,$

- $\displaystyle\int_\gamma \alpha\, f(z)\, dz \; = \alpha \int_\gamma f(z)\, dz \,, \qquad \forall \alpha \in \mathbb{C} \,,$

- $\displaystyle\int_\gamma f(z)\, dz \; = \; - \int_{-\gamma} f(z)\, dz \,,$ $\quad$ where $(-\gamma)(t) := \gamma(a + b - t),\ t \in [a, b],$

- $\displaystyle\Big| \int_\gamma f(z)\, dz \Big| \; \le \; \int_\gamma |f(z)| ds \,,$ $\quad$ where $ds \; = \; |\gamma'(t)| dt \, .$

- If γ is a piecewise C^1 curve of length $L(\gamma)$ and the function f is defined and integrable over the image of γ, then the following estimate holds true:

$$\left| \int_\gamma f(z)\, dz \right| \;\leq\; L(\gamma)\; \sup_t |f(\gamma(t))| \,. \tag{1.4}$$

Definition 1.4. *The curves* $\gamma : I \mapsto \mathbb{C}$ *and* $\eta : J \mapsto \mathbb{C}$ *are called* equivalent curves *if there exists* $\varphi : I \mapsto J$ *s.t.* $\gamma = \eta \circ \varphi\,,\ \varphi \in C^1(I),\ \varphi'(t) > 0\ \forall t \in I$.

Theorem 1.5. *Path integrals of* $f(z)\,dz$ *over equivalent curves are the same. Precisely, if* γ *and* η *are equivalent, then for any* f *we have*

$$\int_\gamma f \;=\; \int_\eta f \,.$$

Definition 1.6. ***(Winding number of a curve with respect to a point)*** *Let* γ *be a curve in* $\mathbb{C}$ *whose image does not contain* z_0 , *say* $\gamma : [a,b] \mapsto \mathbb{C} \setminus \{z_0\}$. *Select a choice of the argument depending continuously on the curve parameter* $\vartheta(t) = \arg(\gamma(t) - z_0)$, *such that* $\gamma(t) = z_0 + |\gamma(t) - z_0| e^{i\vartheta(t)},\ \forall t \in [a,b]$, *and set:*

$$\operatorname{Ind}(\gamma, z_0) \;=\; \frac{1}{2\,\pi} \Big(\vartheta(b) - \vartheta(a) \Big) \,. \tag{1.5}$$

The value $\operatorname{Ind}(\gamma, z_0)$ *is called* winding number of γ with respect to z_0, *since it is a counter of the tours and fractions of tour performed by* γ *around* z_0.

Theorem 1.7. *The winding number fulfils the following properties.*

1. *The value of the winding number does not depend on the selected continuous choice of the argument* ϑ .
2. *If* γ *describes an arc of circumference (in such a way that the arc is ran through exactly one time) then its winding number with respect to the center of the circumference is the amplitude of the related center angle, with positive sign if* $t \mapsto \gamma(t)$ *corresponds to a counterclockwise motion and negative sign if the motion is clockwise.*
3. *The winding number of a closed curve is always an integer number.*
4. *Equivalent curves have the same winding number.*
5. *If we fix a closed curve* $\gamma : I \mapsto \mathbb{C}$, *then* $\operatorname{Ind}(\gamma, z)$ *continuously depends on* z, *if* z *varies in the complement of the image* $\gamma(t)$. *Hence* $z \mapsto \operatorname{Ind}(\gamma, z)$ *is a constant function in each connected component of* $\mathbb{C} \setminus \gamma(I)$ *and vanishes in the unique unbounded connected component.*
6. *If we fix the point* $z_0 \in \mathbb{C}$, *then the winding number remains constant under any "continuous deformation" in* $\mathbb{C}\setminus\{z_0\}$ *of the closed curve* γ. *This property is precisely stated as follows: the winding number* $\operatorname{Ind}(\gamma, z_0)$ *does not change under homotopic deformations of* γ *in* $\mathbb{C}\setminus\{z_0\}$ *(see Definition 1.10 below of closed homotopic curves).*

Remark 1.8. Since we are interested in the evaluation of path integrals, we will always implicitly assume that all the curves which are mentioned in the theorems and exercises of this chapter are piecewise C^1, without writing it explicitly.

Remark 1.9. Whenever we deal with an open set $\Omega \subset \mathbb{C}$ whose boundary $\partial\Omega$ is the image of one (or finitely many) piecewise C^1 closed curves, we always assume the following convention: the orientation of $\partial\Omega$ corresponds to walking exactly once along all the boundary while keeping the set Ω always on the left-hand side; such orientation is called *positive orientation of the boundary*. This entails that the tangent vector τ is oriented in such a way to coincide with counterclockwise $\pi/2$ radians rotation of the outward normal to the set, at each boundary point (counterclockwise denotes the shorter rotation which overlaps the x axis onto the y axis).
For instance: on the boundary of the unitary disk $\partial B_1(0) := \{z \in \mathbb{C} : |z| < 1\}$ it is always understood the parametrization $\gamma(t) = e^{it}$, $t \in [0, 2\pi]$, which is related to a counterclockwise uniform motion performing exactly one turn around the origin (or any other parametrization equivalent to it), except when a different choice is prescribed explicitly.

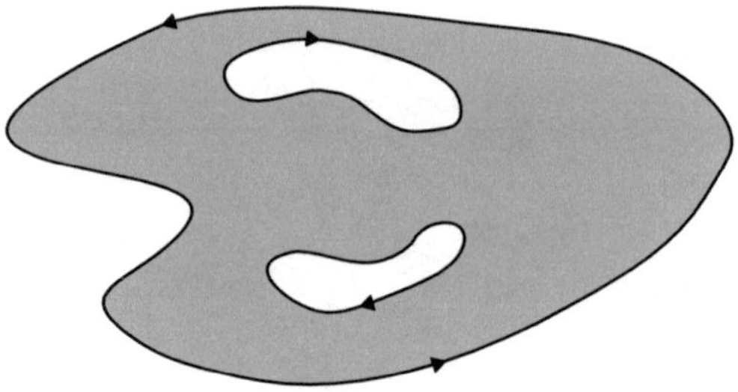

Figure 1.1 - Oriented boundary of an open subset (grey region) of the complex plane.

Beyond equivalence there is another relationship between curves that keeps invariant the path integral of an analytic function (for the definition and properties of analytic functions we refer to Section 1.1.3). This relationship is given by the notion of Ω-homotopy, which expresses the possibility of a continuous deformation of a curve into another one, with all the intermediate steps of the deformation contained in Ω. A mechanical model of this situation is obtained by setting an elastic string in the region Ω of the plane and allowing only deformations of the string which remain in Ω, with no cuts and/or gluing, possibly admitting self-intersections; that is, beginning, ending and intermediate steps are not required to be simple. The formal definition of this relationship is given as follows.

Definition 1.10. ***(Ω-homotopy of closed curves)*** *Two closed curves, whose image is contained in the open set $\Omega \subset \mathbb{C}$, are Ω-homotopic if they are equivalent to a pair of closed curves γ_0 and γ_1, parametrized in $[0,1]$ with* *image in Ω such that there is a continuous function* $H:[0,1]\times[0,1]\mapsto\Omega$ *which fulfils*

$$\begin{cases} t\mapsto H(t,s) \quad \text{is a closed curve} & \forall s\in[0,1]\,, \\ H(t,0)=\gamma_0(t), \quad H(t,1)=\gamma_1(t) & \forall t\in[0,1]\,. \end{cases} \tag{1.6}$$

Ω-homotopy of closed curves is an equivalence relationship (that is, it enjoys symmetry, reflexivity and transitivity). The homothopy relationship allows also to give a formal definition of open sets without holes: see Definition 1.12 below of simply connected open set.

Definition 1.11. *(**Connected open sets**) An open set $\Omega\subset\mathbb{C}$ is* connected *if it is not the union of two nonempty disjoint open sets.*

Definition 1.12. *(**Simply connected open sets**) An open set $\Omega\subset\mathbb{C}$ is* simply connected *if Ω is connected and every loop with image in Ω is Ω-homotopic to a constant.*
An open set $\Omega\subset\mathbb{C}$ is disconnected if it is not connected.

Theorem 1.13. *(**Jordan curve Theorem**, Camille Jordan, 1838 – 1922). The complement in $\mathbb{C}$ of a simple closed curve in $\mathbb{C}$ is disconnected. Precisely this complement is the union of 2 disjoint connected open sets: one is bounded and simply connected (it is called "interior" region bounded by the curve), the other one is unbounded and not simply connected (it is called "exterior" region bounded by the curve).*

The path integral $\displaystyle\int_\gamma f(z)dz$ of the differential form $f(z)dz$ along the closed curve γ, is often denoted by $\displaystyle\oint_\gamma f(z)dz$.

1.1.3 Tools and methods based on analytic functions

In the present subsection, Ω always denotes a nonempty open subset of the complex plane $\mathbb{C}$, namely Ω fulfils

$$\emptyset\neq\Omega\subset\mathbb{C} \qquad \text{and} \qquad \forall z\in\Omega\ \exists r>0:\ \{w\in\mathbb{C}:\ |w-z|<r\}\subset\Omega\,. \tag{1.7}$$

Definition 1.14. *Assume* (1.7) *and that f is a function $f:\Omega\mapsto\mathbb{C}$.*
We say that the function f has complex derivative *at $z_0\in\Omega$ if there exists a complex number, denoted by $f'(z_0)$, s.t.*

$$\lim_{h\to0,\ h\in\mathbb{C}} \frac{f(z_0+h)-f(z_0)}{h} = f'(z_0)\,. \tag{1.8}$$

The number $f'(z_0)$ is called derivative of f and is also denoted by $\dfrac{df}{dz}(z_0)$. In the sequel we will write shortly "f has derivative at z_0", with the meaning: f has complex derivative at z_0.

The equality (1.8) means:

$$\forall \varepsilon > 0 \, \exists \, \delta \in (0, r) \text{ s.t. } \quad \left| \frac{f(z_0 + h) - f(z_0)}{h} - f'(z_0) \right| < \varepsilon \,, \; \forall h \in \mathbb{C}\colon |h| < \delta .$$

Hence a function f has complex derivative at z_0 if and only if f is differentiable at z_0, say:

$$f(z) - f(z_0) \;=\; f'(z_0)\,(z - z_0) \;+\; o(z - z_0)\,.$$

Theorem 1.15. ***(Cauchy-Riemann equations).***
Given an open set $\Omega \subset \mathbb{C}$, $z_0 \in \Omega$ and $f : \Omega \mapsto \mathbb{C}$, we set:

$$f(z) = u(x,y) + iv(x,y), \quad z = x + iy, \quad z_0 = x_0 + iy_0,$$
$$u = \operatorname{Re} f, \; v = \operatorname{Im} f, \quad x = \operatorname{Re} z, \; y = \operatorname{Im} z, \quad x_0 = \operatorname{Re} z_0, \; y_0 = \operatorname{Im} z_0\,.$$

Then f has complex derivative at $z_0 = x_0 + iy_0 \in \Omega$ iff u and v are differentiable at (x_0, y_0) and the following equalities hold true at (x_0, y_0):

Cauchy-Riemann equations

$$\begin{cases} \dfrac{\partial u}{\partial x}(x_0, y_0) = \dfrac{\partial v}{\partial y}(x_0, y_0) \\[2ex] \dfrac{\partial u}{\partial y}(x_0, y_0) = -\dfrac{\partial v}{\partial x}(x_0, y_0)\,. \end{cases} \tag{1.9}$$

Cauchy-Riemann equations hold true at (x_0, y_0) iff the following equations hold true at z_0

holomorphy conditions:
$$\begin{cases} f_x = f' = u_x + i\,v_x = v_y - i\,u_y \\ f_y = i\,f' = i\,u_x - v_x = i\,v_y + u_y \end{cases} \tag{1.10}$$

where:
$$\begin{cases} f' = df/dz, \;\; f_x = \partial f/\partial x, \;\; f_y = \partial f/\partial y, \\ u_x = \partial u/\partial x, \;\; u_y = \partial u/\partial y, \;\; v_x = \partial v/\partial x, \;\; v_y = \partial v/\partial y\,. \end{cases}$$

Definition 1.16. ***(Holomorphic functions).***
Assume that Ω fulfils (1.7). A function $f : \Omega \mapsto \mathbb{C}$ which has complex derivative at every point of Ω is called holomorphic in Ω.
$\mathcal{H}(\Omega)$ denotes the set of holomorphic functions in Ω. So $f \in \mathcal{H}(\Omega)$ means that f is holomorphic in Ω. $\mathcal{H}(\Omega)$ is a complex vector space, endowed with the algebraic sum of functions and the multiplication by complex numbers.

Definition 1.17. *A function $f\colon \Omega \mapsto \mathbb{C}$ is analytic in Ω if it has an expansion in positive powers series (Taylor series of f) around any point of the open set Ω. This amounts to say that:*

$$\forall z_0 \in \Omega \ \exists r > 0,\, c_n \in \mathbb{C} : \quad f(z) = \sum_{n=0}^{+\infty} c_n (z - z_0)^n \quad \forall z \in \Omega : |z - z_0| < r\,,$$

where the equality means that the series is absolutely convergent and its sum coincides with the left-hand side.
In complex analysis: an entire function is a complex-valued function which is analytic in the whole complex plane; a transcendental entire function is a non polynomial entire function.

Typical examples of entire functions are: polynomials, exponential function, trigonometric functions sine and cosine and their hyperbolic counterparts (hyperbolic sine and hyperbolic cosine function), error function and any sum, product and composition of the previous ones. Neither the natural logarithm nor the square root functions (starting from the elementary definition in $\mathbb{R}^+$) can be continued analytically to an entire function.

Theorem 1.18. *$f : \Omega \to \mathbb{C}$ is holomorphic in Ω iff it is analytic in Ω.*

So the terms holomorphic and analytic refer to equivalent properties and they are used as synonyms in the text. We emphasize the difference with respect to differentiability on a real interval: in fact, for any $m \in \mathbb{N}$ we can show examples of $C^m(\mathbb{R})\backslash C^{m+1}(\mathbb{R})$ functions (e.g $(x^+)^{m+1}$) and examples of $C^\infty(\mathbb{R})$ functions that are not real analytic (e.g. $\exp(-1/x^2)$ extended with 0 value at $x = 0$).
We recall the main properties of the *power series*

$$\sum_{0}^{+\infty} c_n\, w^n\,. \tag{1.11}$$

- The series (1.11) has *convergence radius*

$$R \;=\; \left(\limsup_{n\to+\infty} \sqrt[n]{|c_n|}\right)^{-1}\,, \tag{1.12}$$

  this estimate is called *Hadamard criterion* and it holds true also in limit cases, provided we set $1/0 = +\infty$ and $1/+\infty = 0$.
- The series (1.11) is absolutely convergent in its *convergence disk*
  $$B_R(0) \;:=\; \{w \in \mathbb{C} : \ |w| < R\}$$
  while it is not even simply convergent if $|w| > R$.
- The sum of the power series (1.11) is an analytic function of w in the convergence disk.

- The series (1.11) uniformly converges in every closed disk of the kind $\{w \in \mathbb{C} : |w| \le r\}$ with $r < R$.
- The derivative and antiderivative series of (1.11) have the same radius of convergence of (1.11).
- The derivative of the power series coincides with the sum of the derivative series, in the common convergence disk.

Now we look for possible extensions of domains of elementary functions (defined in $\mathbb{R}$ or in a subset of $\mathbb{R}$) to suitable open subsets of $\mathbb{C}$. It is possible to define such extensions and, due to the next Theorem, this extension is uniquely defined. We recall that a segment in $\mathbb{C}$ is a set of the following kind $\{z_0 + t(z_1 - z_0) : \ t \in [0,1]\}$ and a segment is nontrivial when it contains more than a single point (e.g. $z_1 \neq z_0$).

Theorem 1.19. ***(Uniqueness of the analytic extension)*** *If $\Omega \subset \mathbb{C}$ is a connected open set, f and g are two analytic functions in Ω which coincide on a nontrivial segment $E \subset \Omega$ then f and g coincide in whole Ω.*

Fundamental examples are given by entire functions: *exponential, sine* e *cosine* of complex variable. Starting from their definition as functions of real variable, we write their Taylor series with center at 0 and notice that the convergence radius is $R=+\infty$. Hence we get

$$\exp(z) = \sum_{n=0}^{+\infty} \frac{z^n}{n!} \qquad \forall z \in \mathbb{C}, \tag{1.13}$$

$$\sin(z) = \sum_{n=0}^{+\infty} (-1)^n \frac{z^{2n+1}}{(2n+1)!} \qquad \forall z \in \mathbb{C}, \tag{1.14}$$

$$\cos(z) = \sum_{n=0}^{+\infty} (-1)^n \frac{z^{2n}}{(2n)!} \qquad \forall z \in \mathbb{C}. \tag{1.15}$$

These functions of complex variable fulfil the identity

$$\exp(z) = \exp(\operatorname{Re} z) \Big(\cos(\operatorname{Im} z) + i \sin(\operatorname{Im} z) \Big).$$

Nevertheless there are cases where the extension is allowed only in open subsets strictly contained in $\mathbb{C}$ (it is anyway uniquely defined in any open set where it is allowed).
For instance the series $\sum_{n=0}^{+\infty} (-1)^n z^{2n}$ with center at 0 has convergence radius equal to 1, moreover

$$\sum_{n=0}^{+\infty} (-1)^n z^{2n} = \frac{1}{z^2+1} \qquad \forall z \in \mathbb{C} : \ |z| < 1;$$

actually the function $z \mapsto (z^2+1)^{-1}$ is defined for any $z \in \mathbb{C} \setminus \{\pm i\}$ and is the unique extension of the sum of the power series above (convergent in the unitary disk only); there is no analytic extension at $z = \pm i$.
Table 10.2 shows a wide list of Taylor expansions.

Sum, product and composition of elementary functions are analytic and their derivatives are easily deduced as the analytic extension of the corresponding derivatives on the real line. Some examples are listed below:

$$\frac{d}{dz}\exp(z) = \exp(z)\,, \quad \frac{d}{dz}\sin(z) = \cos(z)\,, \quad \frac{d}{dz}\cos(z) = -\sin(z)\,.$$

Uniqueness of the analytic extension (Theorem 1.19) entails several identities for analytic functions of complex variable. For instance:

$$(\sin(z))^2 + (\cos(z))^2 = 1 \qquad \forall z \in \mathbb{C}\,.$$

Nevertheless this kind of extension cannot be exploited for inequalities. For instance: $|\sin(x)| \le 1\ \forall x \in \mathbb{R}$, but $|\sin(z)|$ is unbounded if z varies in $\mathbb{C}$, in fact: $\sin(iy) = i\sinh(y)\,,\ \forall y \in \mathbb{R}$.

Several properties of holomorphic or analytic functions are presented in the sequel. We start from the cornerstone of the theory of analytic functions, say Cauchy Theorem about contour integrals, which easily entails all the subsequent statements of this Section.

Theorem 1.20. ***(Cauchy Theorem**, Augustin-Louis Cauchy, 1789−1857).* *Assume that $f : \Omega \mapsto \mathbb{C}$ is holomorphic. Then*

$$\int_{\gamma_1} f(z)\,dz = \int_{\gamma_2} f(z)\,dz \qquad \text{if } \gamma_1 \text{ and } \gamma_2 \text{ are } \Omega\text{-homotopic loops.} \tag{1.16}$$

In particular (1.16) *entails the following properties.*

- *If f is holomorphic in Ω, then*

$$\int_{\gamma} f(z)\,dz = 0 \qquad \forall \text{ loop } \gamma \text{ which is } \Omega\text{-homotopic to a constant.} \tag{1.17}$$

- *If Ω is simply connected and f is holomorphic in Ω, then*

$$\int_{\gamma} f(z)\,dz = 0 \qquad \forall \text{ loop } \gamma \text{ in } \Omega\,. \tag{1.18}$$

- *Assume U is an open set whose boundary ∂U is the image of finitely many piecewise C^1 non intersecting simple loops and $\overline{U} \subset \Omega$. Then*

$$\int_{\partial U} f(z)\,dz = 0\,, \tag{1.19}$$

where each loop on the boundary ∂U is swept exactly once by path integral with the conventional orientation, as stated in Remark 1.9.

Theorem 1.21. (Cauchy Formula).
Let f be holomorphic in $\Omega \subset \mathbb{C}$ and $\overline{B_R(z_0)} \subset \Omega$. Then

$$f(z_0) = \frac{1}{2\pi i}\oint_{\partial B_r(z_0)} \frac{f(z)}{z - z_0}\,dz \qquad \forall r \text{ s.t. } 0 < r \le R\,, \tag{1.20}$$

where, according to the usual convention (Remark 1.9), the integral path is the circumference with radius r and center at z_0, swept once with the positive (counter-clockwise) orientation.

This theorem entails that an holomorphic function in a disk which is continuous up to the boundary is uniquely prescribed in the whole disk by its value on the boundary.

Theorem 1.22. (Winding number formula).
Assume that $z_0 \in \mathbb{C}$ and γ is a piecewise C^1 closed curve that does not contain z_0. Then

$$\mathrm{Ind}(\gamma, z_0) = \frac{1}{2\pi i} \oint_\gamma \frac{dz}{z - z_0} . \tag{1.21}$$

Definition 1.23. *The point $z_0 \in \mathbb{C}$ is called isolated singularity of the analytic function f if there is a punctured disk: (say a set of the kind $B_R(z_0) \setminus \{z_0\}$ with $R > 0$) where f is analytic.*

An analytic function in a disk with center at z_0 can be represented by its Taylor series with center at z_0 (Definition 1.17). An analytic function in a punctured disk $B_R(z_0) \setminus \{z_0\}$ cannot be always represented by a Taylor series with center at z_0 : nevertheless it can be always represented by a series of both positive and negative powers of $(z - z_0)$, which is called Laurent series.

Theorem 1.24. *If f is holomorphic in $B_R(z_0) \setminus \{z_0\}$ then f admits the following expansion, which is called* Laurent expansion*:*

$$f(z) = \sum_{k=-\infty}^{+\infty} c_k \, (z - z_0)^k \qquad \forall z \in \mathbb{C} : \; 0 < |z - z_0| < R , \tag{1.22}$$

where the coefficients c_k are uniquely prescribed and they can be evaluated by the following formulas (r is any number s.t. $0 < r < R$):

$$c_k = \frac{1}{2\pi i} \oint_{\partial B_r(z_0)} \frac{f(w)}{(w - z_0)^{k+1}} \, dw \qquad \forall k \in \mathbb{Z}. \tag{1.23}$$

The series (1.22) *absolutely converges in the open set $\{ z \in \mathbb{C} : \; 0 < |z - z_0| < R \}$ and uniformly converges in all the compact subsets of the same punctured disk.*

Remark 1.25. Theorem 1.24 also includes the case when f is holomorphic in a disk $B_R(z_0)$: in such case the Laurent expansion (1.22) is a series of positive powers only (Taylor expansion of f) since all coefficients c_k evaluated by (1.23) vanish when k is negative; formulas (1.23) with $k \geq 0$ are called *Weierstrass coefficients formulas*; moreover any $f \in \mathcal{H}(B_R(z_0))$ fulfils the following formulas for derivatives: $\forall n \in \mathbb{N}$, $\forall z \in B_R(z_0)$ s.t. $|z - z_0| < r < R$

$$f^{(n)}(z) = \frac{n!}{2\pi i} \oint_{\partial B_r(z_0)} \frac{f(w)}{(w - z)^{n+1}} \, dw . \tag{1.24}$$

Theorem 1.24 entails that analytic functions admit only 3 kinds of isolated singularities, as it is clarified below.

Classification of isolated singularities

- **Removable singularity** or apparent singularity: an isolated singularity z_0 is removable if all c_k with $k < 0$ vanish in (1.22). In such case the extended definition of f with value $f(z_0) = c_0$ is analytic in $B_R(z_0)$. Hence formulas (1.23) provide also the coefficients of Taylor expansion of analytic functions in the disk $B_R(z_0)$.
- **Pole**: an isolated singularity z_0 is a pole if there are finitely many coefficients $c_k \neq 0$ with $k < 0$ in the expansion (1.22). In such case the order of the pole is the maximum natural number m s.t. $c_{-m} \neq 0$. A pole of order 1 is called *simple pole*
- **Essential singularity**: an isolated singularity z_0 is called essential singularity if there are infinitely many coefficients $c_k \neq 0$ with $k < 0$ in the expansion (1.22).

If z_0 is a pole of order $m \geq 1$ for f, then $\lim_{z \to z_0} |f(z)| = +\infty$; precisely $f(z) \asymp (z - z_0)^{-m}$ as $z \to z_0$.
Conversely, an analytic function strongly oscillates near an essential singularity, as precisely stated in the next statement.

Theorem 1.26. *(**Picard Theorem**, Charles Émile Picard 1856 − 1941) An analytic function assumes all complex values, except at most one, in any neighborhood of an essential singularity.*

Example 1.27. The function $f(z) = \exp\left(\dfrac{1}{z}\right) = \displaystyle\sum_{n=0}^{+\infty} \frac{1}{n!z^n}$ has an essential singularity at $z = 0$ and takes all complex values except 0 in any punctured disk with center at 0.
The function $g(z) = \sin\left(\dfrac{1}{z}\right) = \displaystyle\sum_{n=0}^{+\infty} \frac{(-1)^n}{(2n+1)!\, z^{2n+1}}$ has an essential singularity at $z = 0$ and takes every complex value in any punctured disk with center at 0 (see Exercise 1.9).

1.1.4 Residues

Definition 1.28. *If f is an analytic function with an isolated singularity at z_0, then the first coefficient of the singular part in Laurent expansion* (1.22) *centered at z_0 is called residue of f at z_0 and is denoted by* $\text{res}(f, z_0)$, *say*

$$\text{res}(f, z_0) \ := \ c_{-1} \ .$$

The evaluation of the residue does not always require the knowledge of the related Laurent expansion: often one can exploit some tricks and reduce the computations, as explained in the box below.

Practical computation of residue

- If $z_0 \in \mathbb{C}$ is a removable singularity of f then $\mathrm{res}(f, z_0) = 0$.
- If $z_0 \in \mathbb{C}$ is a simple pole, then

$$\mathrm{res}(f, z_0) = \lim_{z \to z_0} (z - z_0)\, f(z) . \tag{1.25}$$

- If $z_0 \in \mathbb{C}$ is a pole of order m for f, then

$$\mathrm{res}(f, z_0) = \frac{1}{(m-1)!} \lim_{z \to z_0} \frac{d^{m-1}}{dz^{m-1}} \Big((z - z_0)^m f(z) \Big) . \tag{1.26}$$

- Whenever f is analytic on $\{z \in \mathbb{C} : |z| > R\}$ for some $R > 0$, the following formula provides a conventional definition of the residue of f at infinity, in such a way that Theorem 1.31 below holds true:

$$\mathrm{res}(f, \infty) = \mathrm{res}\left(-\frac{1}{z^2} f\left(\frac{1}{z}\right), 0 \right) . \tag{1.27}$$

The following version of de L'Hôpital rule is a very useful tool to simplify the evaluation of limits in (1.25) and (1.26), when computing residues.

Theorem 1.29. *(**de l'Hôpital rule**, Guillaume François Antoine de l'Hôpital, 1661 – 1704). Assume that f and g are analytic functions in $B_R(z_0)$ s.t. $f(z_0) = g(z_0) = 0$ and $g'(z_0) \neq 0$. then*

$$\lim_{z \to z_0} \frac{f(z)}{g(z)} = \frac{f'(z_0)}{g'(z_0)} .$$

Theorem 1.30. (Residues Theorem). *Assume: $\Omega \subset \mathbb{C}$ is an open set, $S = \{z_1, z_2, \ldots, z_N\}$ is a finite set of points in Ω, $f : \Omega \setminus S \mapsto \mathbb{C}$ is an holomorphic function in $\Omega \setminus S$ and γ is a loop in $\Omega \setminus S$ which is Ω-homotopic to a constant. Then*

$$\oint_\gamma f(z)\, dz = 2\pi i \sum_{j=1}^{N} \mathrm{res}(f, z_j)\, \mathrm{Ind}(\gamma, z_j) . \tag{1.28}$$

In particular, if $f : \Omega \setminus S \mapsto \mathbb{C}$ is holomorphic, with $\Omega \subset \mathbb{C}$ open set, $S = \{z_1, z_2, \ldots, z_k\}$ finite set of points in Ω and A open set with piecewise C^1 boundary s.t. $\overline{A} \subset \Omega$, and $\partial A \cap S = \emptyset$, then

$$\int_{\partial A} f(z) dz = 2\pi i \sum_{z_j \in A} \mathrm{res}(f, z_j) .$$

Residues Theorem provides a simple way to compute path integral of analytic functions over loops; moreover in many cases it allows the evaluation of definite integrals on the real line even without knowing the antiderivative of the integrand. The method (when it works) consists in creating a loop by addition of a semi-circumference centered at 0 to the bounded interval $[-R,+R] \subset \mathbb{R}$, applying Residues Theorem and letting $R \to +\infty$ to achieve the principal value of the integral over $\mathbb{R}$ (see Definition 1.32 below). In order to accomplish this method several technical lemmas are available (see Section 1.1.5): they show how and when it is possible to prove that the additional integral on the half-circumference tends to 0 as $R \to +\infty$. Sometimes also the number of residues to be evaluated can be sensibly reduced, as shown by the next theorem.

Theorem 1.31. *Let f be an analytic function in $\mathbb{C}$ except at most a finite number of singularities $z_1, ..., z_N$, then*

$$\operatorname{res}(f,\infty) + \sum_{j=1}^{N} \operatorname{res}(f, z_j) = 0 .$$

Definition 1.32. ***(Principal value of an integral).*** *Given the function f, whose domain is $\mathbb{R}$ except a finite set of points x_j, $j = 1, \dots, J$, we define principal value (or Cauchy principal value) of the integral of f over $\mathbb{R}$, denoted by* p.v. $\int_{\mathbb{R}} f(x)\,dx$, *the value*

$$\text{p.v.} \int_{\mathbb{R}} f(x)\,dx := \lim_{\substack{R\to+\infty \\ \varepsilon_j \to 0_+}} \int_{[-R,+R] \setminus \bigcup_{j=1}^{J} [x_j - \varepsilon_j, x_j + \varepsilon_j]} f(x)\,dx ,$$

if the right-hand side limit exists and is finite.

Notice that the principal value may exists also when f is not even integrable in generalized sense; e.g., $1/x$ is not integrable in generalized sense in $\mathbb{R}$, nor in any neighborhood of 0, nevertheless p.v. $\int_{\mathbb{R}} 1/x\,dx = 0 =$ p.v. $\int_{[-1,1]} 1/x\,dx$ do exist.

1.1.5 Complex variable techniques for evaluating integrals

We make explicit some useful estimates of path integrals, valid when the path is an arc of circumference in $\mathbb{C}$. We introduce the notation:

$$C_R(\vartheta_1, \vartheta_2) = \{z \in \mathbb{C};\ z = Re^{i\vartheta},\ \vartheta \in [\vartheta_1, \vartheta_2]\} \quad \text{with } R > 0,\ \vartheta_1, \vartheta_2 \in [0, 2\pi] .$$

Lemma 1.33 *If f is integrable over $C_R(\vartheta_1, \vartheta_2)$, from (1.4) we get*

$$\left| \int_{C_R(\vartheta_1,\vartheta_2)} f(z)\,dz \right| \le (\vartheta_2 - \vartheta_1)\, R \sup_{|z|=R} |f(z)| \; . \tag{1.29}$$

In particular:

(i) If $\exists\, b > 1$ *s.t.* $|f(z)| \le \dfrac{C}{|z|^b}$ $\forall z : |z| \ge R_0$ *then*

$$\lim_{R\to\infty} \int_{C_R(\vartheta_1,\vartheta_2)} f(z)\,dz = 0 \; .$$

(ii) If $\exists\, a < 1$ *s.t.* $|f(z)| \le \dfrac{C}{|z|^a}$ $\forall z : |z| \le \varepsilon_0$ *then*

$$\lim_{\varepsilon\to 0} \int_{C_\varepsilon(\vartheta_1,\vartheta_2)} f(z)\,dz = 0 \; .$$

In some circumstances sharper estimates are needed: we state a general result which will be helpful in the computation of Fourier and Laplace transforms (Chapters 4 and 6).

Lemma 1.34 ***(Jordan Lemma).*** *Assume that* ϑ_1, ϑ_2 *belong to* $[0, 2\pi]$, $a > 0$, $R_0 > 0$, f *is continuous in the set* $\{z \in \mathbb{C} :\ |z| > R_0\}$ *and*

$$\lim_{R\to\infty} \sup_{z\in C_R(\vartheta_1,\vartheta_2)} |f(z)| = 0 \; .$$

Under the previous assumptions the following statements hold true.

(i) If $\vartheta_1 = 0$ *and* $\vartheta_2 = \pi$ *then* $\displaystyle\lim_{R\to\infty} \int_{C_R(\vartheta_1,\vartheta_2)} e^{iaz} f(z)\,dz = 0 \; .$

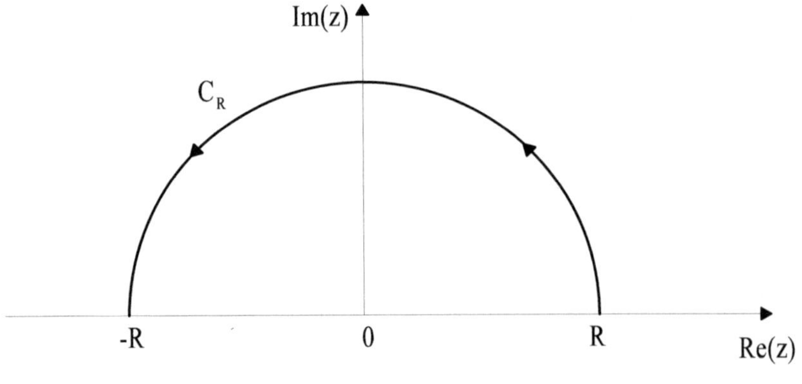

Figure 1.2 *- Loop fitting Jordan Lemma when the multiplier is* e^{iaz} $(\vartheta_1 = 0,\ \vartheta_2 = \pi)$.

(ii) If $\vartheta_1 = \pi$ *and* $\vartheta_2 = 2\pi$ *then* $\displaystyle\lim_{R\to\infty} \int_{C_R(\vartheta_1,\vartheta_2)} e^{-iaz} f(z)\, dz = 0$.

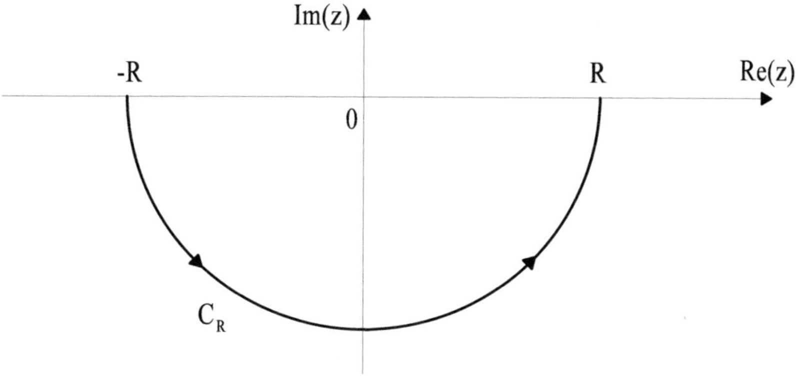

Figure 1.3 *- Loop fitting Jordan Lemma when the multiplier is* e^{-iaz} *(*$\vartheta_1 = \pi$, $\vartheta_2 = 2\pi$*).*

(iii) If $\vartheta_1 = \frac{\pi}{2}$ *and* $\vartheta_2 = \frac{3\pi}{2}$ *then* $\displaystyle\lim_{R\to\infty} \int_{C_R(\vartheta_1,\vartheta_2)} e^{az} f(z)\, dz = 0$.

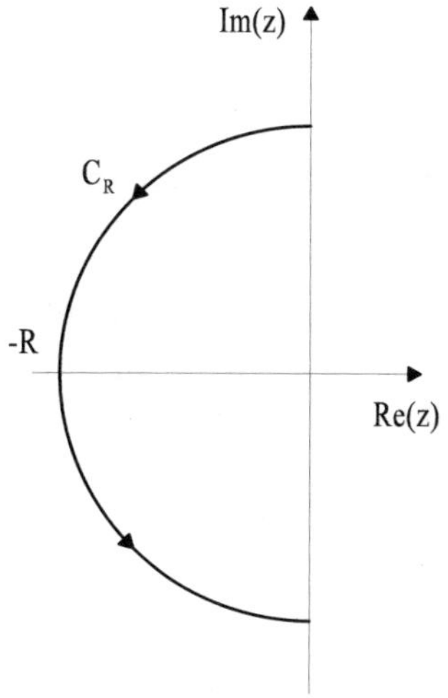

Figure 1.4 *- Loop fitting Jordan Lemma when the multiplier is* e^{az}
($\vartheta_1 = \pi/2$, $\vartheta_2 = 3\pi/2$*).*

(iv) If $\vartheta_1 = -\frac{\pi}{2}$ *and* $\vartheta_2 = \frac{\pi}{2}$ *then* $\lim_{R\to\infty} \int_{C_R(\vartheta_1,\vartheta_2)} e^{-az} f(z)\, dz = 0$.

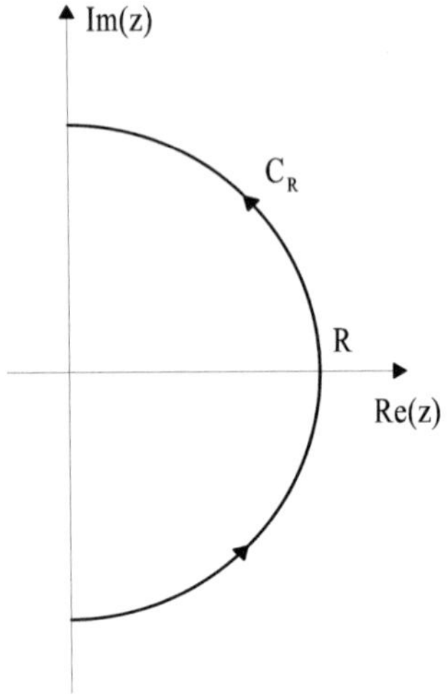

Figure 1.5 *- Loop fitting Jordan Lemma when the multiplier is* e^{-az} $(\vartheta_1 = -\pi/2,\ \vartheta_2 = \pi/2)$.

Lemma 1.35 *Let* $Q = Q(x, y)$ *be a rational function of two real variables with no singularity on the unitary disk* $\{(x, y) \in \mathbb{R}^2 :\ x^2 + y^2 = 1\}$. *Then*

$$\int_0^{2\pi} Q(\cos t, \sin t)\, dt \;=\; 2\pi \sum_{|z|<1} \operatorname{res}\left(\frac{1}{z}\, Q\left(\frac{1}{2}\left(z + \frac{1}{z}\right), \frac{1}{2i}\left(z - \frac{1}{z}\right)\right)\right) .$$

So far we examined only integrals along paths in $\mathbb{C}$ which avoid all singularities of the integrand. Now we state a result which applies to the evaluation of the principal value of a definite integral over $\mathbb{R}$ when simple poles do exist on the real axis.

Lemma 1.36 ***(Simple poles on the real axis)****.*

1) Assume $\varepsilon > 0$; f *is holomorphic in the half-plane* $\{z \in \mathbb{C} :\ \operatorname{Im} z > -\varepsilon\}$ *except a finite set of isolated singularities:* $\{x_1, \dots, x_N, z_1, \dots, z_J\}$; $x_n \in \mathbb{R}$, $n = 1, \dots, N$; $\operatorname{Im} z_j > 0$, $j = 1, \dots, J$; *all* x_n *are simple poles of* f *and*

$$\lim_{R\to+\infty} \int_{S_R} f(z)\, dz \;=\; 0 ,$$

where S_R *is the semi-circumference in the upper half-plane* $\{z \in \mathbb{C} : \operatorname{Im} z > 0\}$ *with radius* R*, centered at the origin and swept once in the counter-clockwise direction.*
Then

$$\text{p.v.} \int_{\mathbb{R}} f(x)\, dx \;=\; 2\pi i \sum_{j=1}^{J} \operatorname{res}(f, z_j) \,+\, \pi i \sum_{n=1}^{N} \operatorname{res}(f, x_n)\ .$$

2) Assume $\varepsilon > 0$; g *is holomorphic in the half-plane* $\{z \in \mathbb{C} : \operatorname{Im} z < \varepsilon\}$ *except a finite set of isolated singularities:* $\{x_1, \dots, x_N, z_1, \dots, z_J\}$*;* $x_n \in \mathbb{R}$, $n = 1, \dots, n$; $\operatorname{Im} z_j < 0$, $j = 1, \dots, J$*; all* x_n *are simple poles of* g *and*

$$\lim_{R \to +\infty} \int_{S_R} g(z)\, dz \;=\; 0\ ,$$

where Σ_R *denotes the half-circumference in the lower half-plane* $\{\operatorname{Im} z < 0\}$ *with radius* R*, center at the origin and swept once in the counter-clockwise direction.*
Then

$$\text{p.v.} \int_{\mathbb{R}} g(x)\, dx \;=\; -\,2\pi i \sum_{j=1}^{J} \operatorname{res}(g, z_j) \,-\, \pi i \sum_{n=1}^{N} \operatorname{res}(g, x_n)\ .$$

1.1.6 Multivalued functions of complex variable

Notice that the map $z \mapsto e^z$ is periodic with period $2\pi i$:

$$e^{z + 2k\pi i} \;=\; e^z \qquad \forall z \in \mathbb{C}\ \forall k \in \mathbb{Z}\,.$$

hence, for any $w \neq 0$ the equation $e^z = w$ has an infinite set of solutions given by $\{z_k = \ln|w| + i(\arg(w) + 2k\pi) : k \in \mathbb{Z}\}$, where ln denotes the (real-valued) natural logarithm with positive real argument and arg denotes the principal value of the argument of a complex number.
For this reason the complex logarithm, with domain $\mathbb{C} \setminus \{0\}$, is a multi-valued function, that is to say it takes more than one value for any single choice of the argument.
The square root is a multi-valued function too, precisely it is two-valued:

$$\sqrt{z} \;=\; \pm\sqrt{\varrho}\, e^{i\vartheta/2}\,, \qquad \text{if } z \;=\; \varrho\, e^{i\vartheta}\,.$$

Whereas the complex logarithm has infinitely many values:

$$\ln z \;=\; \ln \varrho + i\vartheta \,+\, 2k\pi i\,, \qquad \text{if } z \;=\; \varrho\, e^{i\vartheta}\,.$$

Unlike the real case, by varying the argument of a complex multi-valued function "a continuous motion from one value to another one is possible" without passing through a singularity. For instance, by following a loop around the origin (which is a *branching point* of the square root function) with winding number 2 and choosing one continuously varying single value of the square root of z, after the first round trip the value is changed into the other one (the opposite of the first one), after the second round trip the value turns back to the first one.

Remark 1.37. The above Section displays the properties of one-valued analytic functions of one complex variable. Nevertheless these properties can be exploited in the analysis of multi-valued functions which are local inverse functions of non injective single-valued analytic functions: e.g. complex logarithm (local inverse function of exponential), complex $n-$root (local inverse function of $n-$th power). This is possible if one single value is previously selected, in such a way that this value is a continuous function in a domain which is a subset of $\mathbb{C}$ with no path connecting different branching of the multi-valued function.
In practice, given a multi-valued function of this kind, first one performs cuts in the complex plane, all of them starting and ending at branching points of the given function, in order to get an open set with no loop with nontrivial winding number with respect to one of the branching points and null winding number with respect to the others (if any): then, one selects an analytic single-value of the function in this open set, by explicit definition through the polar representation with respect to the branching points of the complex variable and of the function. On this subject, see Exercises 1.31, 1.33, 1.34, 6.19, 6.20 and Remark 6.14 about the inverse Laplace transform of an analytic function whose extension to the left-plane exhibits multi-valued behavior.

1.2 Exercises

Exercise 1.1. Write the polar representation and the exponential representation of the following complex numbers:

1. $z = 1 + i\sqrt{3}$;
2. $z = 1 - i$;
3. $z = i - 1$;
4. $z = \sqrt{3} - i$.
5. $z = 1 + i \tan \dfrac{\pi}{4}$;
6. $z = \cos \dfrac{\pi}{3} - i \sin \dfrac{\pi}{3}$;
7. $z = i \cos \dfrac{\pi}{6} + \sin \dfrac{\pi}{6}$.

Solution -

1. $z = 2\left(\cos\left(\frac{\pi}{3}\right) + i \sin\left(\frac{\pi}{3}\right)\right) \; = \; 2 \, \exp(\pi i/3)$;
2. $z = \sqrt{2}\left(\cos\left(\frac{7\pi}{4}\right) + i \sin\left(\frac{7\pi}{4}\right)\right) = \sqrt{2}\left(\cos\left(-\frac{\pi}{4}\right) + i \sin\left(-\frac{\pi}{4}\right)\right) = \sqrt{2} \exp(-\pi i/4)$;
3. $z = \sqrt{2}\left(\cos\left(\frac{3}{4}\pi\right) + i \sin\left(\frac{3}{4}\pi\right)\right) \; = \; \sqrt{2} \, \exp(3\pi i/4)$;
4. $z = 2\left(\cos\left(-\frac{\pi}{6}\right) + i \sin\left(-\frac{\pi}{6}\right)\right) \; = \; 2 \, \exp(-\pi i/6)$.
5. $z = \sqrt{2}\left(\cos\left(\frac{\pi}{4}\right) + i \sin\left(\frac{\pi}{4}\right)\right) \; = \; \sqrt{2} \, \exp(\pi i/4)$;
6. $z = \left(\cos\left(-\frac{\pi}{3}\right) + i \sin\left(-\frac{\pi}{3}\right)\right) \; = \; \exp(-\pi i/3)$;
7. $z = \left(\cos\left(\frac{\pi}{3}\right) + i \sin\left(\frac{\pi}{3}\right)\right) \; = \; \exp(\pi i/3)$.

□

Exercise 1.2. Evaluate the complex derivative of the following functions of $z \in \mathbb{C}$:

$$z^5 - 5z \,, \qquad \sin(z^2) \,, \qquad \exp\big(\cos(z^6)\big) \,.$$

Solution - By applying elementary derivative formulae and the chain-rule to differentiate superposition of differentiable functions ($\frac{d}{dz} f(g(z)) \; = \; f'\big(g(z)\big)\, g'(z)$), we get:

$$\frac{d}{dz}\,(z^5 - 5z) \; = \; 5(z^4 - 1)\,, \qquad \frac{d}{dz}\,\sin(z^2) \; = \; 2z\,\cos(z^2)\,,$$

$$\frac{d}{dz}\,\exp\big(\cos(z^6)\big) \; = \; -\,6z^5\,\sin(z^6)\,\exp\big(\cos(z^6)\big)\,.$$

□

Exercise 1.3. Which ones among the following relationships hold true?
1) $|\cos z| \; \leq \; 1\,, \quad \forall z \, \in \, \mathbb{C}\,.$
2) $(\cos z)^2 + (\sin z)^2 \; = \; 1\,, \quad \forall z \, \in \, \mathbb{C}\,.$
3) $\sin(2z) \; = \; 2 \sin z \cos z\,, \quad \forall z \, \in \, \mathbb{C}\,.$

Solution - 1) is false, since $\cos(iy) \; = \; \cosh y\,,\ \forall y \in \mathbb{R}\,.$ 2) and 3) are true, due to the uniqueness of the analytic extension, since both functions $f(z) := (\cos z)^2 + (\sin z)^2 - 1$ and $g(z) := \sin(2z) - 2 \sin z \cos z$ vanish identically in $\mathbb{R}$. □

Exercise 1.4. Select holomorphic functions of complex variable in the list below, justify the answer in all cases and evaluate the complex derivative when the answer is positive.

$$|z|\,,\quad \operatorname{Re} z\,,\quad \operatorname{Im} z\,,\quad |(\cos z)^2+(\sin z)^2|\,,\quad \exp(z^3)\,,\quad \sin(e^z)\,.$$

Solution - $|z|$, $\operatorname{Re} z$ ed $\operatorname{Im} z$ are not holomorphic in z: actually they have no complex derivative at every point of the complex plane, since they are non constant and real-valued so they cannot fulfil the Cauchy-Riemann conditions (1.9).
Notwithstanding the superposition with the modulus, the function $|(\cos z)^2+(\sin z)^2|$ is holomorphic in $\mathbb{C}$: precisely it is identically equal to 1 (due to uniqueness of the analytic extension, since it is identically 1 in $\mathbb{R}$), hence it is everywhere differentiable with null derivative. Eventually, both $\exp(z^3)$ and $\sin(e^z)$ are holomorphic in $\mathbb{C}$ since they are superposition of entire analytic functions: $\frac{d}{dz}\exp(z^3) = 3z^2\exp(z^3)$, $\frac{d}{dz}\sin(e^z) = e^z\cos(e^z)$.
□

Exercise 1.5. Find the convergence disk and evaluate the sum of the *geometric series* $\sum_{n=0}^{+\infty} z^n$.
Solution - The coefficients are $c_n = 1$, $\forall n\in\mathbb{N}$: Hadamard criterion tells us that $R=1$ is the radius of convergence disk centered at $z=0$. By denoting $S_N(z)$ the partial sum we have $(1-z)S_N(z) = 1-z^{N+1}$ (if $z\in\mathbb{C}$), hence $S_N(z)\to 1/(1-z)$ as $N\to+\infty$, if $z\in\mathbb{C}:\ |z|<1$.

geometric series: $$\frac{1}{1-z} = \sum_{n=0}^{+\infty} z^n \quad \forall z\in\mathbb{C}:\ |z|<1\,. \tag{1.30}$$

□

Exercise 1.6. Deduce the power series expansion of $z\mapsto \arctan z$ from the sum of geometric series (Exercise 1.5)
Hint: expand $(1+z)^{-1}$ and $(1+z^2)^{-1}$, then integrate.

Exercise 1.7. Expand the function $f(z) = \dfrac{1}{z(z+1)}$ in integer powers series of z in the sets $A=\{z\in\mathbb{C}:\ 0<|z|<1\}$ and $B=\{z\in\mathbb{C}:\ 1<|z|\}$.
Solution - Recalling the sum of geometric series $\left(\sum_{n=0}^{+\infty} s^n = 1/(1-s),\ |s|<1\right)$:

$$f(z) = \frac{1}{z(z+1)} = \frac{1}{z}\sum_{n=0}^{+\infty}(-1)^n z^n = \sum_{n=0}^{+\infty}(-1)^n z^{n-1} = \sum_{m=-1}^{+\infty}(-1)^{m+1}z^m\,,\ 0<|z|<1\,,$$

$$f(z) = \frac{1}{z(z+1)} = \frac{1}{z^2}\frac{1}{1+\frac{1}{z}} = \frac{1}{z^2}\sum_{n=0}^{+\infty}(-1)^n z^{-n} = \sum_{n=0}^{+\infty}(-1)^n z^{-n-2}\,,\ 1<|z|\,.$$

□

Exercise 1.8. Solve the equation $\{z\in\mathbb{C}:\ \sin z = 2\}$.
Solution - Thanks to uniqueness of the analytic extension, trigonometric formulae for sum and subtraction of arcs (Table 10.1.1) hold true also when evaluated with complex variables; in particular $\sin(\alpha+\beta) = \sin\alpha\,\cos\beta + \sin\beta\,\cos\alpha$, $\forall\alpha,\beta\in\mathbb{C}$.
By setting $z=x+iy$ with $x,\,y\in\mathbb{R}$, we get

$$\sin z = \sin(x+iy) = \sin(x)\cos(iy)+\cos(x)\sin(iy) = \sin x\,\cosh y + i\,\cos x\,\sinh y\,.$$

Then:

$$\cos x \sinh y = 0 \quad \text{and} \quad \sin x \cosh y = 2 ,$$

hence $\{x = \frac{\pi}{2} + k\pi,\ k \in \mathbb{Z};\ \cosh y = (-1)^k\, 2\}$, and since the hyperbolic cosine has always positive value on the real line, we get $\{x = \pi/2 + 2k\pi;\ k \in \mathbb{Z},\ \cosh y = 2\}$, explicitly:

$$\left\{ x = \frac{\pi}{2} + 2\,k\pi,\ k \in \mathbb{Z};\ y = \pm \operatorname{arcosh} 2 = \ln(2 \pm \sqrt{3}) \right\} \text{ here ln denotes the real logarithm.}$$

Notice that the choice $y = 0$ (entailing null real part in the left-hand side of the equation) has been discarded, since such choice would have made impossible the equality of imaginary parts ($y = 0 \Rightarrow \cosh y = 1 \Rightarrow \sin x = 2$ with no solution if $x \in \mathbb{R}$). Moreover, in order to explicitly show the values of arcosh (inverse function of cosh), since $2 + \sqrt{3} = (2 - \sqrt{3})^{-1}$, one can write:

$$\cosh y = 2 \Leftrightarrow \frac{e^y + e^{-y}}{2} = 2 \Leftrightarrow e^{2y} - 4e^y + 1 = 0 \Leftrightarrow e^y = 2 \pm \sqrt{3}.$$

□

Exercise 1.9. Show that the equation $\{\, z \in \mathbb{C} :\ \sin z = \alpha \,\}$ has solutions for any $\alpha \in \mathbb{C}$.

Solution - The following chain of logical equivalences holds true:

$$\sin z = \alpha \quad \Leftrightarrow \quad \frac{1}{2i}\left(e^{iz} - e^{-iz}\right) = \alpha \quad \Leftrightarrow \quad e^{2iz} - 2i\alpha e^{iz} - 1 = 0 \quad \Leftrightarrow \quad e^{iz} = \alpha i \pm \sqrt{1 - \alpha^2}\,.$$

Notice that both numbers $\alpha i \pm \sqrt{1 - \alpha^2}$ are never null, whatever the value of $\alpha \in \mathbb{C}$.
Therefore we have infinitely many solutions, $\forall \alpha \in \mathbb{C}:\ z = -i\, \ln\left(i\alpha \pm \sqrt{1 - \alpha^2}\right)$.
The set of solutions is unbounded for any α. □

Remark - A visualization technique for the graph of complex *sine* function might be very helpful; but there is an obstruction: visualizing the graph of a complex-valued function f depending on a complex variable z requires 4 real dimension.
A classical trick to overcome this difficulty, consists in drawing several level lines of both $\operatorname{Re} f$ and $\operatorname{Im} f$ in the z plane: we recall that, due to the Cauchy-Riemann equations (1.9), the two families of lines cross each other with orthogonal tangents at any point z where $f'(z)$ exists and is different from 0. The resulting picture tells many qualitative things but it is somehow too flat and unsatisfactory as a visualization.
Another method consists in considering the graph of $|f|$ (the modulus of f is a real-valued function of two real variables: the real and imaginary parts of z), and then projecting on the surface some of the level lines of the argument of f. The method is cumbersome and often the resulting *analytic landscape* conceals many important details, due to perspective and occlusions.
A much more satisfactory approach is the *enhanced phase diagram*, which is allowed by modern computer graphics and consists in exploiting the *phase* of a complex number (phase(z) $:= z/|z|$, $\forall z \in \mathbb{C} \setminus \{0\}$). The phase removes all the indeterminacy of the multi-valued argument of a complex number and the ambiguity due to the different convention for the choice of its principal value or main branch. Moreover the phase belongs the unitary circle in $\mathbb{C}$ and we can associate continuously varying colors with points in $\partial B_1(0)$: e.g. red with 1, then according to visible light spectrum through yellow and green following a counterclockwise rotation, blue with -1 and eventually back red with $e^{2\pi i}$. The 2-*d* diagram obtained in this way (colored with color of phase($f(z)$) at the point z) may be further enhanced, by suitable addition of level lines for both modulus and phase of f (see [26] for a more detailed description and analysis of the technique). Figure 1.6 shows the resulting enhanced phase diagram for the *sine* function in a bounded portion of the complex plane.

Figure 1.6 - Enhanced phase diagram of $\sin(z)$, $z \in \mathbb{C}$. Notice the 2π periodicity.

Exercise 1.10. Classify the singularity of the following functions, at z_0:
(a) $f(z) = \dfrac{1}{(z-1)^3}$, $z_0 = 1$; (b) $f(z) = \dfrac{1-\cos z}{z^2}$, $z_0 = 0$; (c) $f(z) = e^{1/z}$, $z_0 = 0$; (d) $f(z) = \cot\dfrac{\pi}{z}$, $z_0 = \dfrac{1}{k}$ $(k \in \mathbb{Z})$; (e) $f(z) = \cot\dfrac{\pi}{z}$, $z_0 = 0$; (f) $f(z) = \dfrac{z^3}{z+1}$, $z_0 = \infty$.

Solution - (a) third order pole; (b) removable singularity; (c) essential singularity; (d) simple (or first order) pole; (e) z_0 is not an isolated singularity; (f) second order pole. □

Exercise 1.11. Evaluate the residue at $z = 0$ of the following functions:
(a) $f(z) = \dfrac{1}{\sin z}$ (b) $f(z) = \dfrac{1}{z^2}$ (c) $f(z) = z\, e^{1/z}$.

Solution - (a) 1, (b) 0, (c) $\frac{1}{2}$. □

Exercise 1.12. (i) Write the algebraic form of the following complex numbers:
$z_1 = e^i$, $z_2 = \cos(\pi + 3i)$, $z_3 = \sin(i\pi)$, $z_4 = e^{2+i\frac{\pi}{2}}$, $z_5 = \cos(\frac{3}{2}\pi - i)$, $z_6 = \sinh(1+i)$, $z_7 = \cosh(1-i)$.

(ii) Find all zeroes in $\mathbb{C}$ of the following functions:
(a) $z \mapsto \sin z$ (b) $z \mapsto \cos z$ (c) $z \mapsto \tan z$ (d) $z \mapsto \sinh z$ (e) $z \mapsto \cosh z$ (f) $z \mapsto e^z$.

(iii) Solve the following equations (unknown $z \in \mathbb{C}$):
(a) $e^z = i$ (b) $\sin z = 3$ (c) $\cos^2 z = -1$ (d) $e^z = |z|$ (e) $e^z = e^{\bar{z}} + 2i$ (f) $\sin z = \cos z$ (g) $(1+i)^z = \pi$.

Solution - (i) $z_1 = \cos(1) + i\sin(1)$, $z_2 = \cos(\pi)\cos(3i) - \sin(\pi)\sin(3i) = -\cosh(3)$, $z_3 = i\sinh(\pi)$, $z_4 = i\,e^2$, $z_5 = \cos(3\pi/2)\cos(i) + \sin(3\pi/2)\sin(i) = -i\sinh(1)$, $z_6 = \sinh(1)\cosh(i) - \cosh(1)\sinh(i) = \sinh(1)\cos(1) - i\cosh(1)\sin(1)$, $z_7 = \cosh(1)\cosh(i) - \sinh(1)\sin(i) = \cosh(1)\cos(1) - i\sinh(1)\sin(1)$.

(ii) (a) $k\pi,\ k \in \mathbb{Z}$, (b) $\frac{\pi}{2} + k\pi,\ k \in \mathbb{Z}$, (c) $k\pi,\ k \in \mathbb{Z}$, (d) $ik\pi,\ k \in \mathbb{Z}$,

(e) $i\frac{\pi}{2} + k\pi i,\ k \in \mathbb{Z}$, (f) $\emptyset$: the equation has no solution.

(iii) (a) $(\frac{\pi}{2} + 2k\pi)i,\ k \in \mathbb{Z}$, (b) $-i\ln(3 \pm 2\sqrt{2}) + \frac{\pi}{2} + 2k\pi,\ k \in \mathbb{Z}$,

(c) $-i\ln(\sqrt{2} \pm 1) + \frac{\pi}{2} + k\pi,\ k \in \mathbb{Z}$,

(d) $z = x + iy$ with $y = 2k\pi$ and x solution of $e^x = \sqrt{x^2 + 4k^2\pi^2}$, $k \in \mathbb{Z}$

(e) $z = x + iy$ with $x, y \in \mathbb{R}$ fulfilling $e^x \sin y = 1$

(f) $z = \frac{\pi}{4} + k\pi,\ k \in \mathbb{Z}$ (g) $\dfrac{\ln\pi + 2k\pi i}{\frac{1}{2}\ln 2 + i\frac{\pi}{4} + 2h\pi i}$, $h, k \in \mathbb{Z}$. □

Exercise 1.13. Prove the following claims:

1) if $z_1, z_2 \in \mathbb{C} \setminus \{0\}$ then $\{\ln(z_1 z_2)\} = \{\ln(z_1)\} + \{\ln(z_2)\}$,

2) $\{\ln(z_1) + \ln(z_1)\} \neq 2\,\{\ln(z_1)\}$.

Solution - Both claims refer to relationships between sets (not between numbers) since ln is a multi-valued function.

1) If w_j, $j = 1, 2$, are complex numbers fulfilling $e^{w_j} = z_j$, then exponential function properties entail $e^{w_1 + w_2} = e^{w_1} e^{w_2}$. Conversely, if $e^w = z = z_1 z_2$ and $e^{w_j} = z_j$, then $w = w_1 + w_2$. Hence 1) is proven.

2) The following relationship between sets holds true: $\{\ln(z_1) + \ln(z_1)\} = \{2\ln|z_1| + 2i\arg z_1 + 2k\pi i\ :\ k \in \mathbb{Z}\} \neq \{2\ln|z_1| + 2i\arg z_1 + 4k\pi i\ :\ k \in \mathbb{Z}\} = \{2\ln(z_1)\}$. Notice the analogy with the fact that one can get any natural number as the sum of two natural numbers ($\mathbb{N} + \mathbb{N} = \mathbb{N}$ where $\mathbb{N}$ denotes nonnegative integer numbers $= 0, 1, 2, \dots$), whereas doubling natural numbers generates only even numbers $2\mathbb{N} \neq \mathbb{N}$.
However, if we select exactly one determination of $\ln z$ in a simply connected open set $\Omega \subset \mathbb{C}$ which does not contain 0, then $\ln(z_1) + \ln(z_1) = 2\ln(z_1)\ \forall z_1 \in \Omega$. □

Exercise 1.14. Prove the following statements of **Mean-value Theorem**: the first one refers to harmonic functions of 2 real variables and the second one to holomorphic functions.

If u is an harmonic function of 2 real variables (i.e. $u_{xx} + u_{yy} = 0$) in Ω and $\overline{B_R((x_0, y_0))} \subset \Omega$, then

$$u(x_0, y_0) = \frac{1}{|B_R(x_0, y_0)|} \iint_{B_R(x_0, y_0)} u(x, y)\, dx\, dy\,.$$

If f is holomorphic in Ω and $\overline{B_R(z_0)} \subset \Omega$, then

$$f(z_0) = \frac{1}{2\pi} \int_0^{2\pi} f(z_0 + re^{i\vartheta})\, d\vartheta = \frac{1}{\pi R^2} \iint_{B_R(z_0)} f(x + iy)\, dx\, dy\,.$$

Solution - By exploiting the Cauchy formula (Theorem 1.21) together with the change of variable $z = z_0 + re^{i\vartheta}$, we get, $\forall r \in [0, R]$:

$$\begin{aligned} f(z_0) &= \frac{1}{2\pi i} \int_{\partial B_R(z_0)} \frac{f(z)}{z - z_0} = \\ &= \frac{1}{2\pi i} \int_0^{2\pi} \frac{f(z_0 + r\, e^{i\vartheta})\, i\, r\, e^{i\vartheta}}{re^{i\vartheta}}\, d\vartheta = \\ &= \frac{1}{2\pi} \int_0^{2\pi} f(z_0 + r\, e^{i\vartheta})\, d\vartheta\ . \end{aligned}$$

By multiplying both sides times r and integrating over $[0, R]$:

$$f(z_0) = \frac{1}{\pi R^2} \iint_{B_R(z_0)} f(x + iy)\, dx\, dy\ .$$

This proves the mean-value formula for holomorphic functions. If u is harmonic in B_R then there exists a holomorphic function f s.t. $\mathrm{Re}\, f = u$, hence the mean-value formula holds true for f and (by taking the real part of both sides) for u too. □

Exercise 1.15. 1) Prove that, if $\Omega \subset \mathbb{C}$ is a non empty connected open set and $f \in \mathcal{H}(\Omega)$, then the following equivalence holds true:

$$f'(z) = 0 \quad \forall z \in \Omega \qquad \Longleftrightarrow \qquad f \text{ is constant in } \Omega.$$

2) Prove that, if f is holomorphic in the open set $\Omega \subset \mathbb{C}$, then $u = u(x, y) = \mathrm{Re}\, f$ and $v = v(x, y) = \mathrm{Im}\, f$ are harmonic functions, say they fulfil $u_{xx} + u_{yy} = 0$ and $u_{xx} + u_{yy} = 0$ in Ω.

3) Starting from Cauchy-Riemann equations (1.9) deduce that $u = \mathrm{Re}\, f$ and $v = \mathrm{Im}\, f$, with f holomorphic, fulfil the following identities in polar coordinates:

$$u_\vartheta = -\varrho v_\varrho\,, \qquad v_\vartheta = \varrho u_\varrho\ .$$

Notice that these formulae hold true for $\varrho > 0$, since polar coordinates are singular at the origin.

Solution - 1) If f is constant then $f' \equiv 0$ trivially. Now we show that $f' \equiv 0$ in a connected open set entails f is constant. Due to (1.9) and (1.10) the gradients ∇u of $u = u(x, y) = \mathrm{Re}\, f$ and ∇v of $v = v(x, y) = \mathrm{Im}\, f$ vanish identically, then u and v are constant functions. Hence f is a constant function too.
Alternatively, we can exploit this property of open sets in $\mathbb{R}^n$: they are connected iff they are path-connected. Hence given $z_0 \in \Omega$, for every $z_1 \in \Omega$ we can find a continuous path $\gamma : [0, 1] \mapsto \Omega$ s.t. $\gamma(0) = z_0$, $\gamma(1) = z_1$. Then $f(z_1) - f(z_0) = \int_\gamma f'(z)\, dz = 0$.

2) Theorem 1.18 entails that an holomorphic function f in Ω is $C^\infty(\Omega)$ too, in particular u and v are $C^\infty(\Omega)$. By differentiating the first equation in (1.9) with respect to x and the second one with respect to y, we get $u_{xx} = v_{xy} = -u_{yy}$. By differentiating the first one in (1.9) with respect to y and the second one with respect to x, we get $v_{yy} = u_{xy} = -v_{xx}$.

3) By denoting a partial derivative with its related subscript:

$$\begin{aligned} u_\vartheta &= u_x\, x_\vartheta + u_y\, y_\vartheta = -\varrho\, u_x \sin\vartheta + \varrho\, u_y \cos\vartheta = -\varrho\, v_y \sin\vartheta - \varrho\, v_x \cos\vartheta = -\varrho\, v_\varrho\,, \\ v_\vartheta &= v_x\, x_\vartheta + v_y\, y_\vartheta = -\varrho\, v_x \sin\vartheta + \varrho\, v_y \cos\vartheta = \varrho\, u_y \sin\vartheta + \varrho\, u_x \cos\vartheta = \varrho\, u_\varrho\ . \end{aligned}$$

□

Exercise 1.16. Prove the following statement.

If $\Omega \subset \mathbb{C}$ is an open set and $f : \Omega \mapsto \mathbb{C}$ is continuous, then f is holomorphic in Ω if and only if the following formula holds true for every $\overline{B_R(z_0)} \subset \Omega$ with $R > 0$

$$f(z) = \frac{1}{2\pi i} \oint_{\partial B_R(z_0)} \frac{f(w)}{w - z}\, dw \qquad \forall z \in B_R(z_0). \tag{1.31}$$

Solution - If f is holomorphic then (1.31) holds true due to Theorem 1.21 (Cauchy formula). Conversely, if (1.31) holds true, then f is differentiable, since the right-hand side of (1.31) is differentiable with respect to z: differentiation with respect to a parameter of a definite integral on a bounded closed interval of a continuous function that is also C^1 with respect to the parameter. □

Exercise 1.17. Prove the following statement.

If $\Omega \subset \mathbb{C}$ is an open set, $u_n : \Omega \mapsto \mathbb{C}$ are holomorphic functions in Ω $\forall n \in \mathbb{N}$ and the sequence u_n uniformly converges in Ω to a function u, then also u is holomorphic in Ω.

Solution - Formula (1.31) holds true for every u_n:

$$u_n(z) = \frac{1}{2\pi i} \oint_{\partial B_R(z_0)} \frac{u_n(w)}{w - z}\, dw \qquad \forall z \in B_R(z_0). \tag{1.32}$$

The uniform limit u of u_n is continuous. By taking the limit as $n \to +\infty$ in (1.32), we get

$$u(z) = \frac{1}{2\pi i} \oint_{\partial B_R(z_0)} \frac{u(w)}{w - z}\, dw \qquad \forall z \in B_R(z_0). \tag{1.33}$$

Then u is holomorphic, due to Exercise 1.16. □

Exercise 1.18. Evaluate the following residues:

1) $\operatorname{res}(1/z, 0)$, 2) $\operatorname{res}(1/z^2, 0)$, 3) $\operatorname{res}(1/(\sin z), 0)$, 4) $\operatorname{res}(1/(\sin z)^2, 0)$,

5) $\operatorname{res}\left(\dfrac{1}{(\sin z)^3}, 0\right)$, 6) $\operatorname{res}\left(\dfrac{1}{z} - \dfrac{1}{\cosh z} + \exp\dfrac{1}{z^2}, 0\right)$.

Solution - 1) The residue is 1, due to Definition 1.28. 2) The residue is 0, since the function is even. 3) It is a simple pole, so thanks to (1.25) we get: $\operatorname{res}(1/(\sin z), 0) = \lim_{z\to 0} z/\sin z = 1$. 4) The residue is 0, since the function is even.

5) Point 0 is a triple pole; due to (1.26) with $m = 3$: $\operatorname{res}\left(\frac{1}{(\sin z)^3}, 0\right) = \frac{1}{2}\frac{d^2}{dz^2}\frac{z^3}{(\sin z)^3}$ and the conclusion could be achieved via de l'Hôpital rule. Alternatively, by Taylor expansion: $\sin z = z - z^3/6 + o(z^4)$, $(\sin z)^{-3} = \left(z - z^3/6 + o(z^4)\right)^{-3} = z^{-3}\left(1 + z^2/6 + o(z^3)\right)^3 = z^{-3}\left(1 + 3z^2/6 + o(z^3)\right)$ (in the previous computations, the same symbol o may denote different functions which are only estimated by their infinitesimal order) then $\operatorname{res}(1/(\sin z)^3) = 1/2$.

6) Residue linearly depends on the function: the residue of an algebraic sum equals the sum of the residues; the second and third terms in the sum are even functions (their residue vanishes); then $z = 0$ is an essential singularity with residue equal to 1. □

Exercise 1.19. Evaluate $\displaystyle\oint_\gamma \frac{z}{1+z^2}\,dz\,.$

Here: γ is the boundary (swept once with counter-clockwise orientation) of the square with center at 0, sides of length 5 and parallel to axis.

Solution - The integrand $f(z) := z/(1+z^2)$ has two simple poles $z_{1,2} = \pm i$.
The curve γ has winding number 1 with respect to both poles:

$$\operatorname{res}(f,i) = \lim_{z\to i} \frac{z}{z+i} = \frac{1}{2}\,, \qquad \operatorname{res}(f,-i) = \lim_{z\to -i} \frac{z}{z-i} = \frac{1}{2}\,.$$

Residues Theorem (Theorem 1.30) entails

$$\oint_\gamma f(z)\,dz = 2\pi i\Big(\operatorname{res}(f,i)+\operatorname{res}(f,-i)\Big) = 2\pi i\,.$$

Alternatively, one could exploit the fact that f has a finite number of isolated singularities, hence Theorem 1.31 gives $\operatorname{res}(f,\infty) + \sum_j \operatorname{res}(f,z_j) = 0$, where z_j are all singularities of f, except the point at ∞. Then

$$\begin{aligned}\oint_\gamma f(z)\,dz &= -2\pi i\operatorname{res}(f,\infty) = -2\pi i \operatorname{res}\Big(-\frac{1}{z^2}\,f\,(1/z\,,0)\Big) = \\ &= -2\pi i \operatorname{res}\Big(-\frac{1}{z(z^2+1)}\,,0\Big) = -2\pi i\lim_{z\to 0}\Big(-\frac{1}{z^2+1}\Big) = 2\pi i\,.\end{aligned}$$

□

Exercise 1.20. Set

$$f(z) = \frac{z^2\exp\dfrac{1}{z}}{1+z^4}\,.$$

1) Evaluate the residue of f at $z=0$.

2) Evaluate $\displaystyle\oint_\gamma f(z)\,dz$, where $\gamma(t) = 4e^{-it}$, $t\in[0,2\pi]$.

Solution - 1) By Taylor series expansion of exponential function and the geometric series:

$$\begin{aligned}\exp\frac{1}{z} &= \sum_{n=0}^{+\infty}\frac{1}{n!\,z^n} && |z|>0\,,\\ \frac{1}{z^4+1} &= \sum_{m=0}^{+\infty}(-1)^m\,z^{4m} && |z|<1\,.\end{aligned}$$

Since both series above converge absolutely and uniformly in any annulus of the kind $0<R_1\le|z|\le R_2<1$, performing the Cauchy product of the two series is allowed; hence the Laurent series of f centered at 0 is:

$$\sum_{n,\,m=0}^{+\infty}\frac{(-1)^m}{n!}\,z^{4m-n+2}\,,\qquad 0<|z|<1\,.$$

Then the origin is an essential singularity of f. We look for the coefficient of z^{-1}: by setting $4m-n+2=-1$ we deduce the related link between m and n, say $n=4m+3$, hence:

$$\operatorname{res}(f,0) = \sum_{m=0}^{+\infty}\frac{(-1)^m}{(4m+3)!}$$

2) The loop turns around five isolated singularities: four are simple poles (the 4 fourth roots z_j of -1) and the essential singularity at 0. The path γ sweeps once the circumference with radius 4 centered at 0, in the clockwise (negative) direction: then $\mathrm{Ind}(\gamma, 0) = -1$, $\mathrm{Ind}(\gamma, z_j) = -1\,,\; j = 1, 2, 3, 4$.

So it is more convenient to evaluate the residue at infinity and apply Theorem 1.31:

$$\mathrm{res}(f, \infty) \;=\; \mathrm{res}\left(-\frac{1}{z^2}\, f(1/z)\,,\, 0\right) \;=\; \mathrm{res}\left(\frac{-e^z}{z^4+1}\,,\, 0\right) \;=\; 0\,,$$

$$\oint_\gamma f(z)\,dz \;=\; -\,2\pi i\;\mathrm{res}(f, 0) \;-\; 2\pi i \sum_{j=1}^{4} \mathrm{res}(f, z_j) \;=\; +\,2\pi i\;\mathrm{res}(f, \infty) \;=\; 0\,.$$

□

Exercise 1.21. Evaluate $\displaystyle\int_C \frac{\exp(1/z)}{1-z}\,dz$ where C is the path in Figure 1.7 swept only once.

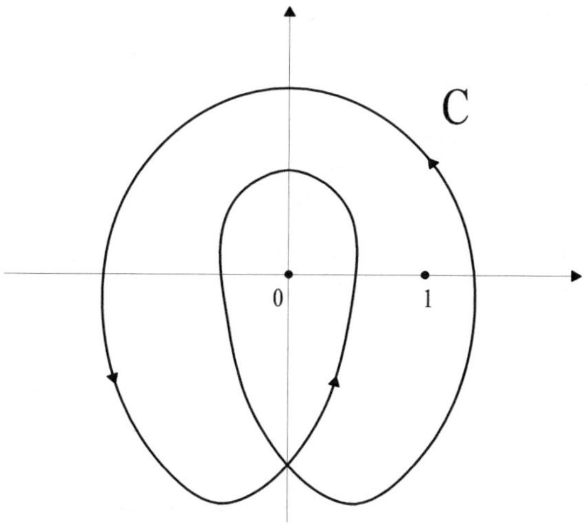

Figure 1.7 - Path C of Exercise 1.21.

Solution - The integrand f has a simple pole at $z = 1$ and an essential singularity at $z = 0$. Loop C has winding number 1 with respect to the point $z = 1$ and winding number 2 with respect to $z = 0$.

$$f(z) \;=\; \sum_{n=0}^{+\infty} \frac{1}{n!\,z^n} \sum_{m=0}^{+\infty} z^m \;=\; \sum_{n,m=0}^{+\infty} \frac{1}{n!}\, z^{m-n} \qquad 0 < |z| < 1\,,$$

$$\mathrm{res}(f, 0) \;=\; \sum_{m=0}^{+\infty} \frac{1}{(m+1)!} \;=\; e - 1\,, \qquad \mathrm{res}(f, 1) \;=\; -e\,.$$

Residues Theorem (Theorem 1.30) entails

$$\int_C f(z)\,dz \;=\; 2\,\pi\,i\,\big(\,\mathrm{res}(f, 1) \,+\, 2\;\mathrm{res}(f, 0)\,\big) \;=\; 2\,\pi\,i\,(e-2)\,.$$

□

Exercise 1.22. Evaluate

$$\int_0^{+\infty} \frac{\sin(x)}{x}\, dx, \qquad \int_{\mathbb{R}} \frac{\sin(x)}{x}\, dx\ ,$$

where both integrals are understood in the generalized sense.

Solution - The function $f(x) = \sin x/x$ (extended with value 1 at the origin) is a function $C^\infty(\mathbb{R})$: actually it is the restriction to $\mathbb{R}$ of the entire analytic function $\sin z/z$ (the singularity at 0 is removable). The function f is integrable in any bounded interval contained in $\mathbb{R}$. Nevertheless f is not Lebesgue integrable (Chapter 2) since $\int_{\mathbb{R}} |f(x)|dx > \frac{\pi}{2}\sum_{n=1}^{+\infty} \frac{1}{(n+1/2)\pi} = +\infty$: here the integral of $|f|$ is estimated from below by the total area of triangles with base $[n\pi, (n+1)\pi]$ and height equal to the value of f at $x = (n+1/2)\pi$ (for an alternative preliminary discussion of integrability see also Exercise 2.5).
Nevertheless both integrals proposed by the exercise exist and are finite in the Riemann generalized sense: this can be proven by applying Leibniz criterion (simple convergence of numerical series whose general term has alternating sign and modulus decreasing to 0) to the series of integrals of f over the intervals $[n\pi, (n+1)\pi]$. The generalized integral of f in $\mathbb{R}$ exists; hence also the Cauchy principal value of the integral (see Definition 1.32) of f in $\mathbb{R}$ does exist and they have the same value. We evaluate such value by using complex variable techniques. Principal value in $\mathbb{R}^+$ is exactly one half of the principal value in $\mathbb{R}$ since f is even.
The function $\sin z/z$ has an essential singularity at infinity: in order to apply Jordan Lemma (Lemma 1.34) and Lemma 1.36, we decompose f by Euler formula in a sum which splits the exponentials: $\sin x = \left(e^{ix} - e^{-ix}\right)/(2i)$.

$$\begin{aligned}
\text{p.v.} \int_{\mathbb{R}} \frac{\sin x}{x}\, dx &= \frac{1}{2i}\ \text{p.v.} \int_{\mathbb{R}} \frac{e^{ix}}{x}\, dx - \frac{1}{2i}\ \text{p.v.} \int_{\mathbb{R}} \frac{e^{-ix}}{x}\, dx =^{(*)} \\
&= \pi i\, \frac{1}{2i}\ \text{res}\left(\frac{e^{iz}}{z}, 0\right) - (-\pi i)\, \frac{1}{2i}\ \text{res}\left(\frac{e^{-iz}}{z}, 0\right)\ 0 \\
&= \frac{\pi}{2} + \frac{\pi}{2} = \pi\ .
\end{aligned}$$

The equality marked by $(*)$ exploits the following tools: variant (i) of Lemma 1.34 and case (1) in Lemma 1.36 to deal with the first term; variant (ii) of Lemma 1.34 and case (2) of Lemma 1.36 to deal with the second one. By summarizing, we proved:

$$\text{p.v.} \int_0^{+\infty} \frac{\sin(x)}{x}\, dx = \frac{\pi}{2}\ , \qquad \text{p.v.} \int_{\mathbb{R}} \frac{\sin(x)}{x}\, dx = \pi\ .$$

□

Exercise 1.23. Prove the following identities.

$$\int_0^{+\infty} \sin x^2\, dx = \int_0^{+\infty} \cos x^2\, dx = \frac{1}{2}\sqrt{\frac{\pi}{2}}\ . \tag{1.34}$$

The above integrals are called Fresnel integrals and they play an important role in diffraction phenomena in Optics.

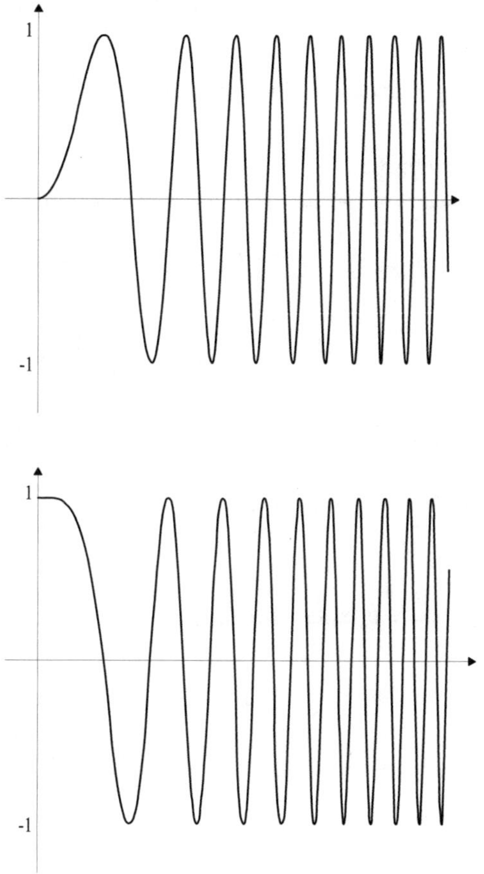

Figure 1.8 - Graphs of $\sin x^2$ and $\cos x^2$ (Exercise 1.23).

Solution - Consider the entire analytic function $f(z) = e^{iz^2}$ and the loop γ_R made by joining the following 3 paths $\gamma_{1,R}$, $\gamma_{2,R}$ and $\gamma_{3,R}$ along the subsequent sides of a triangle:

$$\gamma_{1,R}(t) = t\,, \quad \gamma_{2,R}(t) = R + it\,, \quad \gamma_{3,R}(t) = (R-t)(1+i)\,,\ t \in [0,R]\,.$$

$$\lim_{R\to+\infty} \int_{\gamma_{1,R}} e^{iz^2} = \lim_{R\to+\infty} \int_0^R e^{ix^2}\,dx = \int_0^{+\infty} \cos x^2\,dx + i \int_0^{+\infty} \sin x^2\,dx\,,$$

$$\int_{\gamma_{2,R}} e^{iz^2}\,dz = i \int_0^R e^{i(R^2-t^2)}\, e^{-2Rt}\,dt\,,$$

hence

$$\left|\int_{\gamma_{2,R}} e^{iz^2}\,dz\right| \le \left|\int_0^R e^{-2Rt}\,dt\right| \le \frac{1}{2R} \quad \overset{R\to+\infty}{\longrightarrow} \quad 0\,,$$

$$\int_{\gamma_{3,R}} e^{iz^2} = -(1+i) \int_0^R e^{i(2i)t^2}\,dt = -(1+i)\int_0^R e^{-2t^2}\,dt\,.$$

Then, due to the identity (2.11) (proven in the next Chapter):

$$\lim_{R\to+\infty} \int_{\gamma_{3,R}} e^{iz^2} = -(1+i)\int_0^{+\infty} e^{-2t^2}\,dt = -(1+i)\,\frac{1}{2}\sqrt{\frac{\pi}{2}}\,.$$

Theorem 1.20 provides $\int_{\gamma_R} f(z)\,dz = 0$. By letting $R \to \infty$ we achieve the equalities to be proven. □

Exercise 1.24. Prove Lemma 1.35.

Solution - Consider the parametrization $\gamma(t) = e^{it}$, $t \in [0, 2\pi]$. Then $\gamma'(t) = ie^{it}$ and

$$\begin{aligned}\int_0^{2\pi} Q(\cos t, \sin t)\,dt &= \int_0^{2\pi} Q\left(\frac{1}{2}\left(e^{it}+e^{-it}\right), \frac{1}{2i}\left(e^{it}-e^{-it}\right)\right)dt = \\ &= \int_\gamma \frac{1}{i\,z} Q\left(\frac{1}{2}\left(z+\frac{1}{z}\right), \frac{1}{2\,i}\left(z-\frac{1}{z}\right)\right)dz = \\ &= 2\,\pi\,i \sum_{|z|<1} \operatorname{res}\left(\frac{1}{i\,z} Q\left(\frac{1}{2}\left(z+\frac{1}{z}\right), \frac{1}{2i}\left(z-\frac{1}{z}\right)\right)\right) = \\ &= 2\,\pi \sum_{|z|<1} \operatorname{res}\left(\frac{1}{z} Q\left(\frac{1}{2}\left(z+\frac{1}{z}\right), \frac{1}{2i}\left(z-\frac{1}{z}\right)\right)\right).\end{aligned}$$

□

Exercise 1.25. Evaluate the following integrals:

(a) $\displaystyle\int_0^{2\pi} \frac{d\vartheta}{a+b\sin\vartheta}$ $(a > b > 0)$, (b) $\displaystyle\int_{-\infty}^{+\infty} \frac{dx}{x^6+64}$,

(c) $\displaystyle\int_{-\infty}^{+\infty} \frac{\cos(mx)}{x^2+1}dx$ $(m \in \mathbb{R})$, (d) $\displaystyle\int_{-\infty}^{+\infty} \frac{3x^2-6x+1}{x^4+1}\,dx$.

Solution - (a) $\dfrac{2\pi}{\sqrt{a^2-b^2}}$, (b) $\dfrac{\pi}{48}$, (c) πe^{-m}, (d) $2\sqrt{2}\,\pi$. □

Exercise 1.26. 1) Prove that the following generalized integral exists, for any $z \in \mathbb{C}$ s.t. $\operatorname{Re} z > 0$,

$$\Gamma(z) = \int_0^{+\infty} e^{-t}\,t^{z-1}\,dt\,, \tag{1.35}$$

where: $t^{z-1} = \exp\big((z-1)\ln t\big)$, $\ln t \in \mathbb{R}$. Hence the function $\Gamma : \{z \in \mathbb{C} :\ \operatorname{Re} z > 0\} \mapsto \mathbb{C}$ is well defined. Γ is called *Euler Gamma function.*

2) Prove that Γ is an analytic function in the half-plane $\{z \in \mathbb{C} :\ \operatorname{Re} z > 0\}$.

3) Prove that Γ fulfils

$$\Gamma(z+1) = z\,\Gamma(z) \qquad \forall z \in \mathbb{C} :\ \operatorname{Re} z > 0\,, \tag{1.36}$$

$$\Gamma(n+1) = n! \qquad \forall n \in \mathbb{N}\,. \tag{1.37}$$

4) Prove that: Γ has a unique analytic extension to the complex plane pruned of negative integers and 0, moreover all discarded points are simple poles of the function Γ with residues

$$\operatorname{res}\big(\Gamma(z), z=-k\big) = \frac{(-1)^k}{k!} \qquad \forall k \in \mathbb{N}$$

and (1.36) holds true in the whole domain where Gamma has been extended.

Solution - 1) The integrals defining Γ in the right half-plane exist for any z with $\operatorname{Re} z > 0$, due to the inequality $|e^{-t}\,t^{z-1}| \le e^{-t}\,t^{\operatorname{Re} z-1}$, where the integrand is dominated by an integrable function in $(0, +\infty)$, for any $z \in \mathbb{C}$ s.t. $\operatorname{Re} z > 0$.

2) We examine separately two functions F and G which split Γ into a sum:

$$F(z) = \int_0^1 e^{-t}t^{z-1}\,dt\,, \qquad G(z) = \int_1^\infty e^{-t}t^{z-1}\,dt\,.$$

- G is entire analytic. This is true since, for $T > 1$ the map $z \to G_T(z) = \int_1^T e^{-t}t^{z-1}\,dt$ is entire analytic by the theorem about derivation of a definite integral. Let $R > 0$, then R is the maximum of abscissae for points in $\overline{B_R(0)}$ and $\forall \varepsilon > 0$ $\exists T_\varepsilon > 1$ such that, for every $T_2 > T_1 > T_\varepsilon$ and every $z \in B_R(0)$, we have

$$|G_{T_1}(z) - G_{T_2}(z)| = \left|\int_{T_1}^{T_2} e^{-t}\,t^{z-1}\,dt\right| \le \int_{T_1}^{T_2} e^{-t}\,t^{R-1}\,dt \le \varepsilon\,.$$

  By Cauchy criterion we obtain the uniform convergence of G_T to G if $z \in \overline{B_R(0)}$ and $T \to +\infty$. Thanks to Exercise 1.17 also G is analytic in $B_R(0)$ and G is entire, by arbitrariness of R.

- F is analytic in the half-plane $\mathrm{Re}\,z > 0$.
  This is true since $\forall \tau \in (0,1)$, the map $z \to F_\tau(z) = \int_\tau^1 e^{-t}t^{z-1}\,dt$ is entire analytic, in particular it is analytic in the open disk $B_R(R+1/R)$, with center $R+1/R$ and radius R, contained in the right half-plane. Moreover $\forall \varepsilon > 0$ $\exists \tau_\varepsilon \in (0,1)$ such that, for any choice $0 < \tau_1 < \tau_2 < \tau_\varepsilon$, we have

$$|F_{\tau_1}(z) - F_{\tau_2}(z)| = \left|\int_{\tau_1}^{\tau_2} e^{-t}\,t^{z-1}\,dt\right| \le \int_{\tau_1}^{\tau_2} e^{-t}\,t^{(1/R)-1}\,dt < R\,{\tau_2}^{1/R} \le \varepsilon\,.$$

  Then F_τ uniformly converges to F in $\overline{B_R(R+1/R)}$, therefore F is analytic in the open disk. Since these disks cover the whole open right half-plane, F is analytic in such half-plane.

3) Integrating by parts, if $\mathrm{Re}\,z > 0$ we get:

$$\Gamma(z+1) = \int_0^{+\infty} e^{-t}t^z\,dt = -e^{-t}\,t^z\,\Big|_{t=0}^{t=+\infty} + z\int_0^{+\infty} e^{-t}\,t^{z-1}\,dt = z\,\Gamma(z)\,.$$

By substituting $z = n \in \mathbb{N}$ in the previous formula we get

$$\Gamma(n+1) = n\Gamma(n) = n(n-1)\Gamma(n-1) = \cdots = n(n-1)(n-2)\ldots 3\;2\;1\;\Gamma(1) = n!\,.$$

4) The analytic extension (which is unique due to Theorem 1.19) is built iteratively, starting in the vertical strip $\{z \in \mathbb{C} :\ -1 < \mathrm{Im}\,z \le 0\}$. We define:

$$F_1(z) = \begin{cases} \Gamma(z) & \text{if } \mathrm{Re}\,z > 0\,, \\[2ex] \dfrac{\Gamma(z+1)}{z} & \text{if } -1 < \mathrm{Re}\,z \le 0 \text{ and } z \ne 0\,, \end{cases}$$

since (1.36) holds true in $\{\mathrm{Re}\,z > 0\}$, we get that F_1 is analytic in $\{\mathrm{Re}\,z > -1\} \setminus \{0\}$ and has a simple pole at $z = 0$ (Γ is holomorphic in $B_1(1)$). From $\Gamma(1) = 1$ we get $\mathrm{res}(\Gamma, 0) = 1$.
By definition F_1 fulfils: $F_1(z+1) = z\,F_1(z)$ $\forall z \in \mathbb{C} :\ \mathrm{Re}\,z > -1$ and $z \ne 0$.
Then we can define F_2 by analytic extension of F_1 in the strip $-2 < \mathrm{Re}\,z \le -1$ except for the point $z = -1$:

$$F_2(z) = \frac{F_1(z+1)}{z} \qquad \text{if} \qquad -2 < \mathrm{Re}\,z \le -1\,.$$

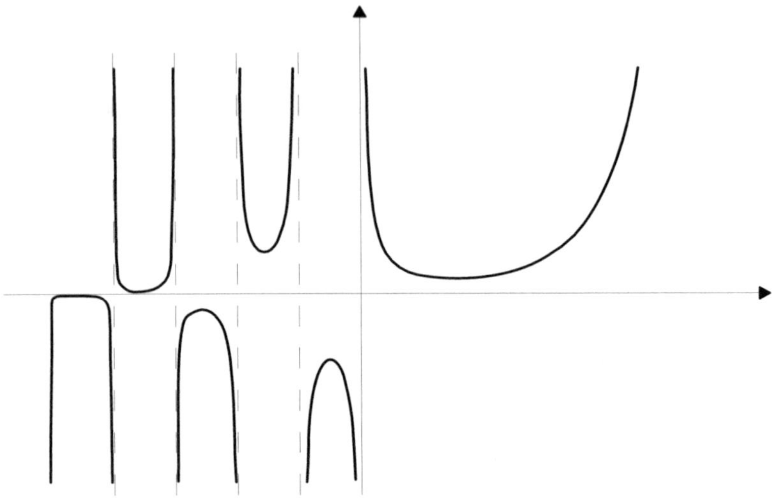

Figure 1.9 - Graph of the Gamma function (only with real variable): $\{(x, \Gamma(x)) :\ x \in \mathbb{R}\}$.

In this way $z = -1$ is a simple pole too and its residue is $\text{res}(\Gamma, -1) = \text{res}(\Gamma, 0)/(-1)$. The procedure can be repeated on every vertical strip between to negative integers. Eventually $\text{res}(\Gamma, -k) = \text{res}(\Gamma, -(k-1))/(-k)$. □

Exercise 1.27. Find nontrivial solutions of the following differential equations of second order in the unknown $y = y(x)$, with $x \in \mathbb{R}$, according to these choices of the parameter: $n = 0, 1, 2, \ldots$:

$$y'' - 2xy' + 2ny = 0\,. \tag{1.38}$$

The previous equation is named *Hermite equation* (Charles Hermite, 1822–1901).

Solution - An ordinary differential equation in normal form with analytic coefficients has analytic solutions: in this case the solutions form a linear space of dimension 2 (equal to the equation's order). After setting $y(x) = \sum_{k=0}^{+\infty} A_k x^k$, we look for a recursive formula that ties coefficients of the power series representing the analytic solution.
By term-wise derivation and substitution in the equation:

$$\sum_{k=2}^{+\infty} k(k-1) A_k x^{k-2} - 2 \sum_{k=1}^{+\infty} k A_k x^k + 2n \sum_{k=0}^{+\infty} A_k x^k = 0\ .$$

By changing the index ($m = k - 2$) in the first series, we get:

$$\sum_{m=0}^{+\infty} (m+1)(m+2) A_{m+2} x^m - 2 \sum_{m=1}^{+\infty} m A_m x^m + 2n \sum_{m=0}^{+\infty} A_m x^m = 0$$

$$\sum_{m=0}^{+\infty} \left[(m+1)(m+2) A_{m+2} + 2(n-m) A_m\right] x^m = 0.$$

Since all the coefficients must vanish, we get the recursive relationship:

$$A_{m+2} = \frac{2(m-n) A_m}{(m+1)(m+2)} \qquad m \in \mathbb{N}\,.$$

If n is even, by choosing $A_0 = 1$, $A_1 = 0$, we find $A_{2m+1} = 0\ \forall m$, $A_{2m} = 0\ \forall m > n/2$: the related solution is a polynomial of degree n whose nontrivial coefficients are $A_{2m} =$

$(-1)^m 2^m (n-2m+2)!!/(2m)!$ if $0 \le m \le n/2$. If n is odd, by choosing $A_0 = 0$, $A_1 = 1$, we find again a polynomial solution of degree n.
The function $H_n(x) := (-1)^n \, e^{x^2/2} \dfrac{d^n}{dx^n} e^{-x^2/2}$ is called *Hermite polynomial* of order n. Hermite polynomials H_n are solutions of (1.38), they fulfil: $H_n(-x) = (-1)^n H_n(x)$ and

$$\int_{\mathbb{R}} H_n(x)\, H_m(x)\, e^{-x^2/2}\, dx \;=\; 0 \;, \qquad \forall n \neq m \;.$$

The last relationship is referred to by saying that all H_n are mutually orthogonal on the real line with respect to the weight function $e^{-x^2/2}$. □

Exercise 1.28. Find the solutions of the following differential equations of second order in the unknown $y = y(x)$, with $x \in \mathbb{R}$, according to these choices of the parameter: $n = 0, 1, 2, \ldots$:

$$x^2 y'' + x y' + \left(x^2 - n^2\right) y \;=\; 0 \;. \tag{1.39}$$

the previous equation is named *Bessel equation* (Friedrich Wilhelm Bessel, 1784–1846).

Solution - As in Exercise 1.27 we deal with a linear ordinary differential equation with analytic coefficients (polinomyals), so we can look for an iterative formula linking the coefficients of a power series expansion of the analytic solutions.
The novelty (and the difficulty) here consists in the degenerate coefficient of y'', vanishing at $x = 0$: due to this fact we cannot write in normal form the associated Cauchy problem with initial value at $x = 0$. By setting $y\,(x) = x^h \sum_{k=0}^{+\infty} A_k x^k$, $h \ge 0$ and $A_0 \neq 0$, term-wise evaluating the derivatives and substituting them in the equation, we deduce that the following series must vanish identically:

$$\sum_{k=0}^{+\infty} (k+h)(k+h-1) A_k x^{k+h} + \sum_{k=0}^{+\infty} (k+h)\, A_k x^{k+h} + \sum_{k=0}^{+\infty} A_k x^{k+h+2} - n^2 \sum_{k=0}^{+\infty} A_k x^{k+h} \;.$$

By changing the index $m = k+2$ in the third series and collecting terms, we get

$$\sum_{m=0}^{+\infty} \left[(m+h)^2 - n^2\right] A_m x^{m+h} + \sum_{m=2}^{+\infty} A_{m-2} x^{m+h} = 0.$$

Then, by matching the coefficients of terms with the same power of x:

$$\begin{cases} \left(h^2 - n^2\right) A_0 = 0 \\ ((h+1)^2 - n^2) A_1 = 0 \\ ((m+h)^2 - n^2) A_m + A_{m-2} = 0 \qquad\qquad \forall m \ge 2 \,. \end{cases}$$

Since $A_0 \neq 0$, the first equation entails $h = n$. Then the second one entails $A_1 = 0$. So all A_k with k odd vanish, while for any even k, A_k can be deduced from A_0 via the iterative relationship

$$A_{2k} = -\frac{A_{2k-2}}{4k\,(k+n)} \qquad\quad k \ge 1.$$

Explicitly the solutions we have found can be expressed (up to the irrelevant numerical factor $n!A_0$) by the convergent power series

$$J_n(z) \;:=\; \sum_{k=0}^{\infty} \frac{(-1)^k}{k!(k+n)!} \left(\frac{z}{2}\right)^{2k+n} \;.$$

The functions J_n are named *Bessel functions of first kind and order n*. All J_n have an analytic extension which is entire: indeed by Hadamard criterion (1.12) their convergence radius is $+\infty$. Moreover (by term-wise integration of the series) we obtain that the Bessel function of first kind and order 0 has the following integral representation:

$$J_0(z) \;=\; \sum_{k=0}^{\infty} \frac{(-1)^k}{(k!)^2} \left(\frac{z}{2}\right)^{2k} \;=\; \frac{1}{\pi} \int_{-1}^{1} \frac{\cos(tz)}{\sqrt{1-t^2}}\, dt \,, \qquad z \in \mathbb{C} \,.$$

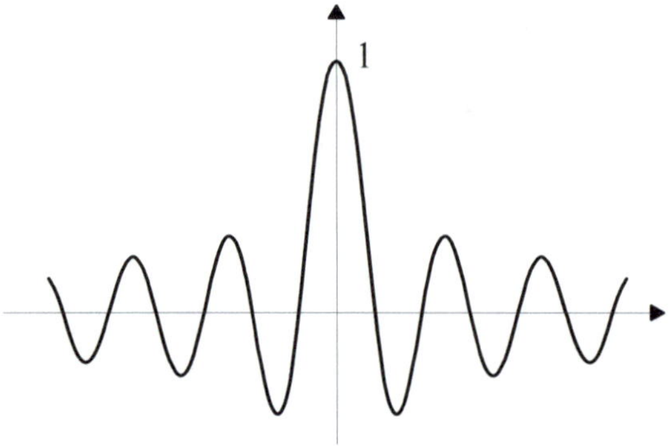

Figure 1.10 - Graph of the Bessel function J_0 : $\{(x, J_0(x)) : x \in \mathbb{R}\}$.

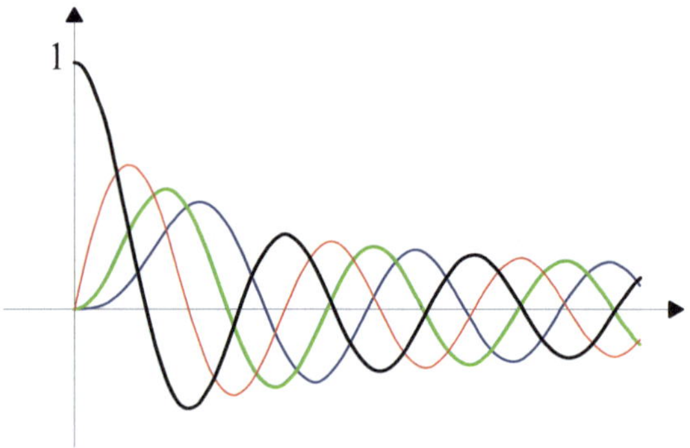

Figure1.11 - Graphs of the Bessel functions J_0, J_1, J_2, J_3 : $\{(x, J_n(x)) : x \in \mathbb{R}^+\}$.

In addition to the solution J_n and to the linear space spanned by J_n, there are other solutions of (1.39), linearly independent from J_n and defined on $\{x \in \mathbb{R} : x \neq 0\}$. These solutions are called *Bessel functions of second kind and order n* or *Neumann functions* (John von Neumann, born János Neumann, 1903-1957). Bessel functions of second kind, denoted by Y_n, are given by $Y_n(x) = J_n(x) \ln x + x^{-n} \sum_{k=0}^{+\infty} b_k x^k$ with suitable choices of coefficients b_k (here the series is convergent $\forall x \in (0, +\infty)$); more precisely, the following explicit expansion can be obtained by long computations.

$$Y_n(x) = J_n(x)\ln(x) - \frac{1}{2}\left(\frac{x}{2}\right)^{-n} \sum_{k=0}^{n-1} \frac{(n-k-1)!}{k!} \left(\frac{x}{2}\right)^{2k} - \frac{1}{2}\left(\frac{x}{2}\right)^{k} \sum_{k=0}^{+\infty} (-1)^k \frac{h_k + h_{k+n}}{k!(k+n)!} \left(\frac{x}{2}\right)^{2k},$$

where $h_0 = 0$, $h_k = 1 + 1/2 + \cdots + 1/k$, $k > 1$, and the series converges for any $x \in \mathbb{R}$. By summarizing, the general solution in $(0, +\infty)$ of the equation (1.39) is: $A\,J_n(x) + B\,Y_n(x)$.

Exercise 1.29. Find the entire analytic solutions of the following differential equations of second order in the unknown $y = y(x)$ with $x \in \mathbb{R}$, according to these choices of the parameter: $n = 0, 1, 2, \ldots$:

$$x^2y'' + xy' - \left(x^2 + n^2\right) y = 0 \,. \tag{1.40}$$

The above equation is named *modified Bessel equation.*

Solution - As in the case of the equation (1.39), the coefficients are analytic (polynomials) and the coefficient of y'' degenerates at $x = 0$. By the same procedure applied in Exercise 1.28 we get entire analytic solutions I_n (*modified Bessel functions of first kind and order* n):

$$I_n(z) \;=\; \sum_{k=0}^{+\infty} \frac{(z/2)^{2k+n}}{k!\,(k+n)!} \,.$$

Equation (1.40) admits also other solutions, which are linearly independent from I_n and singular at 0. These solutions are denoted by K_n and named *modified Bessel functions of second kind and order* n or *MacDonald functions* (Hector Munro MacDonald, 1865 − 1935). The general solution in $(0, +\infty)$ of the equation (1.40) is: $A\, I_n(x) \,+\, B\, K_n(x)$.
We emphasize that the Bessel functions J_n, Y_n, I_n and K_n are called also cylindrical functions, since they appear in the solution of differential problems in cylinder-shaped domains (see Exercises 4.48, 4.51). □

Exercise 1.30. 1) Classify the singularity at $z = 0$ of the function

$$f(z) \;=\; \frac{1}{1 \,-\, \cosh(z\sqrt{z})} \,.$$

2) Evaluate the residue of f at $z = 0$.

Solution - 1) Due to $\cosh z = \cosh(-z)\,,\ \forall z \in \mathbb{C}$, the value of $\cosh(z\sqrt{z})$ is independent of the selected choice of the square root value, hence $\cosh(z\sqrt{z})$ is a single-valued function, moreover it is holomorphic in $\mathbb{C}\backslash\{0\}$ and non constant. Then $z = 0$ is not a branching point of f, but an isolated singularity at most. By exploiting the Taylor expansion of hyperbolic cosine (Table 10.2) we find:

$$\cosh(z\sqrt{z}) \;=\; \sum_{n=0}^{+\infty} \frac{(z\sqrt{z})^{2n}}{(2n)!} \;=\; \sum_{n=0}^{+\infty} \frac{z^{3n}}{(2n)!} \,,$$

$$1 \,-\, \cosh(z\sqrt{z}) \;=\; -\sum_{n=1}^{+\infty} \frac{z^{3n}}{(2n)!} \,,$$

$$\lim_{z\to 0}\; z^3 f(z) \;=\; \lim_{z\to 0}\; \frac{z^3}{-\displaystyle\sum_{n=1}^{+\infty} \frac{z^{3n}}{(2n)!}} \;=\; -2 \,.$$

Actually the origin is an isolated singularity of f: it is a third order pole.

2) $f(z) = -(z)^{-3}\big(1 + z^3/24 + o(z^5)\big)^{-1} = -(z)^{-3}\big(1 - z^3/24 + o(z^5)\big) \;\Rightarrow\; \operatorname{res}(f, 0) = 0.$ □

Exercise 1.31. Evaluate the following integral, by exploiting complex variable tools:

$$I \;=\; \int_0^{+\infty} \frac{\sqrt{x}}{x^2 + 4}\, dx \,.$$

Solution - The extension to complex domains of the arithmetic square root function is a multi-valued function. Let us consider the curve in Figure 1.12, where the two circumferences have radii ε and R, $0 < \varepsilon < 2 < R$, while AB and CD are two copies of the segment $[\varepsilon, R]$ respectively the upper and lower edge of a cut of $\mathbb{C}$ along positive real number semi-axis.

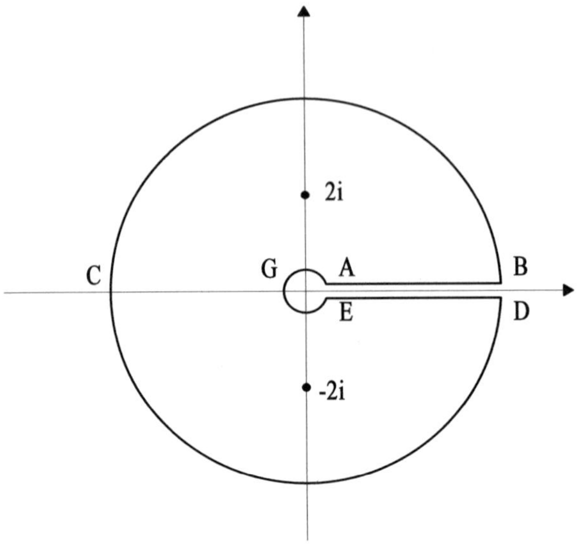

Figure 1.12 - Integration path of Exercise 1.31, circumventing the branching point.

In the so cut complex plane, we define the following single-valued analytic choice of the multi-valued function $\sqrt{z}$, where $\varrho^{1/2}$ denotes the arithmetic root, coherently with the usual convention:

$$\sqrt{z} = \varrho^{1/2} e^{i\vartheta/2}, \qquad z = \varrho e^{i\vartheta}, \qquad \varrho > 0, \qquad 0 \leq \vartheta \leq 2\pi$$

and, by referring to the previous choice for the value of $\sqrt{z}$, we define also

$$F(z) = \frac{\sqrt{z}}{z^2+4}.$$

To simplify notations, we set: $\lim\limits_{\varepsilon} = \lim\limits_{\varepsilon \to 0}$, $\lim\limits_{R} = \lim\limits_{R \to +\infty}$, $\lim\limits_{\varepsilon, R} = \lim\limits_{\varepsilon \to 0,\, r \to +\infty}$. Then

$$\lim_{\varepsilon, R} \int_{AB} F(z)\, dz = \int_0^{\infty} \frac{\sqrt{\varrho}}{\varrho^2+4}\, d\varrho = I,$$

$$\lim_{\varepsilon, R} \int_{DE} F(z)\, dz = \int_{\infty}^{0} \frac{\sqrt{\varrho}\exp(\pi i)}{\varrho^2+4}\, d\varrho = I.$$

Moreover, if $\varepsilon < 1/2$ and $R > 3$, then there is a constant C s.t.

$$\left| \int_{BCD} F(z)\, dz \right| \leq C R \frac{\sqrt{R}}{R^2}, \qquad \left| \int_{EGA} F(z)\, dz \right| \leq C \varepsilon \sqrt{\varepsilon}.$$

Then $$\lim_{R} \int_{BCD} F(z)\, dz = 0, \qquad \lim_{\varepsilon} \int_{EGA} F(z)\, dz = 0.$$

By summing and applying Residues Theorem:

$$I = \frac{1}{2}\lim_{\varepsilon,R}\int_{ABCDEGA} F(z)\,dz = \pi i\,[\,\mathrm{res}(F,2i) + \mathrm{res}(F,-2i)\,] =$$

$$= \pi i \left[\lim_{z\to 2i} (z-2i)F(z) + \lim_{z\to -2i} (z+2i)F(z) \right] =$$

$$= \pi i \left[\frac{\sqrt{2}\exp(i\,\pi/4)}{4i} + \frac{\sqrt{2}\exp(i\,3\pi/4)}{-4i} \right] = \frac{\pi}{2}\,.$$

□

Exercise 1.32. Set $f(z) = \dfrac{1}{z}\exp\left(-z^2 + \dfrac{1}{z^2}\right)$, then evaluate $\displaystyle\int_{\partial B_1(0)} f(z)\,dz$.

Solution - The function f has an essential singularity at 0. By taking into account the Taylor expansion for J_0, Bessel function of first kind and order 0 (Table 10.2),

$$J_0(x) = \sum_{n=0}^{+\infty} \frac{(-1)^n}{(n!)^2}\left(\frac{x}{2}\right)^2 ,$$

we get:

$$f(z) = \frac{1}{z}\exp(-z^2)\exp\left(\frac{1}{z^2}\right) = \frac{1}{z}\sum_{n=0}^{+\infty}\frac{(-1)^n z^{2n}}{n!}\sum_{m=0}^{+\infty}\frac{1}{z^{2m}\,m!}$$

$$= \sum_{n,m=0}^{+\infty}\frac{(-1)^n}{n!\,m!}\,z^{2n-2m-1} ,$$

$$\mathrm{res}(f,0) = \sum_{n=0}^{+\infty}\frac{(-1)^n}{(n!)^2} = J_0(2)\,.$$

Then Residues Theorem (Theorem 1.30) entails

$$\int_{\partial B_1(0)} f(z)\,dz = 2\pi i\ \mathrm{res}(f,0) = 2\pi i\,J_0(2)\,.$$

□

Exercise 1.33. 1) By exploiting complex analysis tools, evaluate

$$A = \int_0^{+\infty}\frac{dx}{\sqrt{x}\,(x^2+1)}\,.$$

2) By taking into account the answer to item 1) evaluate

$$B = \int_0^{+\infty}\frac{\ln x\ dx}{\sqrt{x}\,(x^2+1)}\,.$$

Solution - The generalized integral A exists with finite value since the integrand behaves like $1/\sqrt{x}$ around 0 (0 is the only finite singularity on the real line) and like $1/x^{5/2}$ at infinity.

The generalized integral B exists with finite value since the integrand behaves like $(\ln x)/\sqrt{x}$ around 0 (0 is the only finite singularity on the real line) and like $(\ln x)/x^{5/2}$ at infinity.

1) $f(z) = \left(\sqrt{z}(z^2+1)\right)^{-1}$ is a multi-valued function: we fix the choice

$$\left\{\, z = \varrho\exp(i\vartheta)\,,\quad \varrho > 0\,,\quad 0 \le \vartheta \le 2\pi\,,\quad \sqrt{z} = \varrho^{1/2}\exp(i\vartheta/2)\,\right\}.$$

We refer to the closed oriented curve $C_{R,\varepsilon}$ in Figure 1.13, $0 < \varepsilon < R < +\infty$.

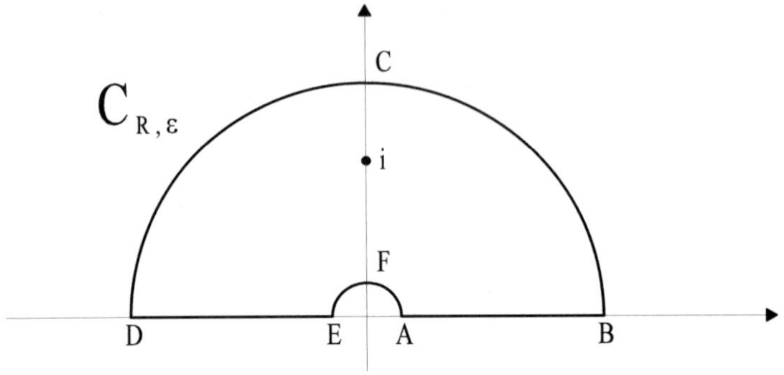

Figure 1.13 - Integration path of Exercise 1.33.

$\int_{BCD} f(z)\,dz \overset{R\to+\infty}{\longrightarrow} 0$ by (i) of Lemma 1.33: we have $|f(z)| \leq \dfrac{C}{R^{5/2}}$ $\forall z \in BCD$, $\forall R > 2$.

$\int_{EFA} f(z)\,dz \overset{\varepsilon\to 0}{\longrightarrow} 0$ due to the estimate $\left|\int_{EFA} f(z)\,dz\right| \leq \dfrac{C}{\sqrt{\varepsilon}}\,\varepsilon$, $\forall \varepsilon < \dfrac{1}{2}$.

$$\int_{AB} f(z)\,dz = \int_{\varepsilon}^{R} \frac{d\varrho}{\sqrt{\varrho}\,(\varrho^2+1)} \underset{R\to+\infty}{\overset{\varepsilon\to 0}{\longrightarrow}} \int_0^{+\infty} \frac{d\varrho}{\sqrt{\varrho}\,(\varrho^2+1)}\,.$$

On DE we have: $z = -\varrho = \varrho \exp(i\pi)$, $\sqrt{z} = i\,\varrho^{1/2}$, $dz = -\,d\varrho$ and

$$\int_{DE} f(z)\,dz = \int_{R}^{\varepsilon} \frac{-\,d\varrho}{i\sqrt{\varrho}\,(\varrho^2+1)} \underset{R\to+\infty}{\overset{\varepsilon\to 0}{\longrightarrow}} \int_0^{+\infty} \frac{d\varrho}{i\sqrt{\varrho}\,(\varrho^2+1)}\,.$$

In the lower half-plane there is only one singularity: the simple pole $z = +i$, with residue $\text{res}(f,i) = \dfrac{\sqrt{2}}{2i-2}$. Residues Theorem (Theorem 1.30) together with the previous estimates provide $A = \frac{\pi}{\sqrt{2}}$.

2) Let us consider again the loop $C_{R,\varepsilon}$. let f be defined as in item 1). To deal with the multi-valued function $g(z) = f(z)\ln(z)$ we select the value such that the complex logarithm has argument in $[0, 2\pi]$. Then

$$\int_{BCD} g(z)\,dz \overset{R\to+\infty}{\longrightarrow} 0\,, \qquad \int_{EFA} g(z)\,dz \overset{\varepsilon\to 0}{\longrightarrow} 0\,, \qquad \text{res}(g,i) = \frac{\pi}{2\sqrt{2}\,(1+i)}\,.$$

$$\lim_{\varepsilon,R} \left(\int_{AB} g(z)\,dz + \int_{DE} g(z)\,dz\right) = \int_0^{+\infty} \frac{\ln\varrho\,d\varrho}{\sqrt{\varrho}\,(\varrho^2+1)} + \int_0^{+\infty} \frac{(\ln\varrho + i\pi)\,d\varrho}{i\sqrt{\varrho}\,(\varrho^2+1)} =$$
$$= (1-i)\int_0^{+\infty} g(x)\,dx \;+\; \pi\,A\,.$$

By the Cauchy Theorem together with previous estimates we get $B = -\dfrac{\pi^2}{2\sqrt{2}}$.

□

Exercise 1.34. By exploiting complex analysis tools, evaluate the integral

$$I \;=\; \text{p.v.} \int_0^{+\infty} \frac{x^{1/6}}{(x+i)(x^2-1)}\, dx \;.$$

Solution - The function $x \mapsto \varphi(x) := \frac{x^{1/6}}{(x+i)(x^2-1)}$ (where $x \in \mathbb{R},\ x > 0$ and $x^{1/6}$ denotes the arithmetic sixth root of x) is neither Riemann integrable nor Lebesgue integrable, because of the singularity at $x = 1$ which behaves like $\psi(x)/(x-1)$ with smooth ψ and $\psi(1) \neq 0$. Nevertheless there exists the principal value (Definition 1.32) of the integral of φ over $(0, +\infty)$.

To deal with the multi-valued function $z \mapsto \varphi(z) := \dfrac{z^{1/6}}{(z+i)(z^2-1)}$, $z \in \mathbb{C} \setminus \{0, \pm 1, -i\}$, we select the following value:

$$\left\{ z = \varrho\, e^{i\vartheta} \quad \varrho > 0\,, \quad 0 \le \vartheta \le 2\pi\,, \quad \varphi(z) \;=\; \frac{\varrho^{1/6}\, e^{i\vartheta/6}}{(\varrho \exp(i\vartheta) + i)(\varrho^2 \exp(2\, i\, \vartheta) - 1)} \right\} .$$

Though this is a single-valued analytic function, we have to examine four simple poles: $z = \pm 1,\ z = -i,\ z = \exp(2\pi i)$. In fact in addition to the poles -1 and $-i$, the point $z = 1$ produces two different singularities of the multi-valued function φ on the two edges of the cut.

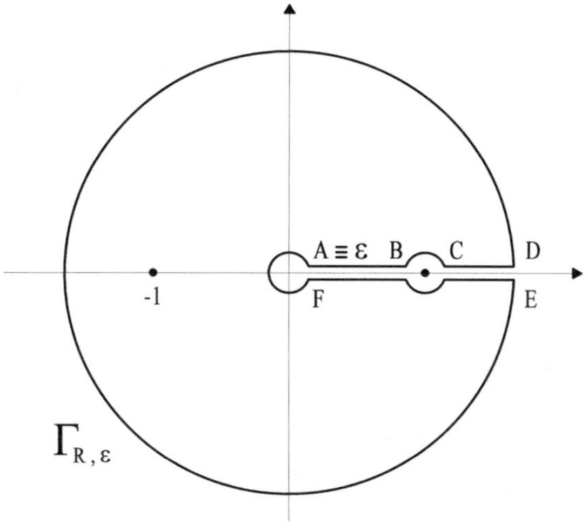

Figure 1.14 - Integration path of Exercise 1.34. The path sweeps both edges of the cut and circumvents the branching point and all other singularities.

By referring to the path $\Gamma_{R,\varepsilon}$ in Figure 1.14 ($0 < \varepsilon < R < +\infty$) we get:

$$|\varphi(z)| \le C\, R^{-17/6} \;\; \forall z \in DE\,, \qquad |\varphi(z)| \le C\, \varepsilon^{1/6} \;\; \forall z \in FA\,,$$

$$I \;=\; \lim_{\varepsilon \to 0,\, R \to +\infty} \int_{AB \cup CD} \varphi(z)\, dz \;.$$

By applying the Residues Theorem and Lemma 1.36 (to deal with simple poles on the real axis) and by carefully taking into account of the orientation when turning around 1 and $e^{2\pi i}$ with small half-circumferences, we get:

$$\big(1 - \exp(i\pi/3)\big)\, I \;-\; \pi\, i \Big(\operatorname{res}(\varphi, \exp(2\pi i)) + \operatorname{res}(\varphi, 1) \Big) \;=\; 2\,\pi\, i \Big(\operatorname{res}(\varphi, -1) + \operatorname{res}(\varphi, -i) \Big)\,,$$

$$\mathrm{res}(\varphi,1) = \frac{1}{(1+i)2} = \frac{1-i}{4}\,; \quad \mathrm{res}(\varphi,\exp(2\pi i)) = \frac{e^{i\pi/3}}{2(1+i)} = \frac{1}{8}(\sqrt{3}+1+i(\sqrt{3}-1))\,;$$

$$\mathrm{res}(\varphi,-i) = \frac{e^{i\pi/4}}{-2} = -\frac{\sqrt{2}}{4} - i\frac{\sqrt{2}}{4}\,; \quad \mathrm{res}(\varphi,-1) = \frac{e^{i\pi/6}}{-2(-1-i)} = \frac{\sqrt{3}-1}{8} + i\,\frac{\sqrt{3}+1}{8}\,.$$

$$\begin{aligned} I &= \frac{2\pi i}{1-\sqrt{3}}\left(-\frac{\sqrt{2}}{2} - i\frac{\sqrt{2}}{2} + \frac{\sqrt{3}-1}{4} + i + \frac{\sqrt{3}+1}{4} + \frac{1}{4} - \frac{i}{4} + \frac{1}{8}(3+1+i(3-1))\right) = \\ &= \frac{\pi i}{2}\,\frac{1+i\sqrt{3}}{8}\left((3\sqrt{3}+1-4\sqrt{2}) + i(3\sqrt{3}-1-4\sqrt{2})\right) = \\ &= \frac{\pi}{4}\left((-2-\sqrt{3}+\sqrt{2}+\sqrt{6}) + i(-2+\sqrt{3}-\sqrt{2}+\sqrt{6})\right). \end{aligned}$$

□

Exercise 1.35. Evaluate the integral

$$I = \int_1^{+\infty} \frac{x^2-2}{x^2\,(x^3+1)}\,dx\,.$$

Solution - The integral exists since the integrand $g(z) = \frac{z^2-2}{z^2(z^3+1)}$ has no singularity on $[1,+\infty]$ and at infinity it is asymptotic to $1/x^3$. The integrand $g(z)$ exhibits no symmetry property, therefore we introduce the multivalued function

$$f(z) = g(z)\,\ln(z-1)$$

that has a branching point at $z=1$ and is such that any consecutive pair of single-valued branches differ by $2\pi i$ times the integrand g.
After cutting the complex plane along the half-line $[1,+\infty]$ and having chosen this single-valued branch of the multivalued function $\ln(z-1)$ as follows

$$\{\; z-1 = \varrho\,\exp(i\vartheta)\,, \quad \varrho>0\,, \quad 0\le\vartheta\le 2\pi\,, \quad \ln(z-1) = \ln\varrho + i\vartheta\;\}\,,$$

we integrate the related branch of f along the path $\gamma_{R,\varepsilon} = ABCDEA$ drawn in Figure 1.15.
The analytic function f has 3 simple poles, $z_1 = 1/2+i\sqrt{3}/2$, $z_2=-1$, $z_3 = 1/2-i\sqrt{3}/2$, and a double pole at $z=0$. Since

$$\int_{AE} f(z)\,dz \overset{\varepsilon\to 0}{\longrightarrow} 0\,, \qquad \int_{BCD} f(z)\,dz \overset{R\to+\infty}{\longrightarrow} 0\,,$$

taking into account

$$\begin{aligned} &\text{in } AB: \quad dz = dx\,, \qquad f(z) = \frac{x^2-2}{x^2(x^3+1)}\,\ln(x-1)\,, \\ &\text{in } DE: \quad dz = -dx\,, \qquad f(z) = \frac{x^2-2}{x^2(x^3+1)}\left(\ln(x-1)+2\pi i\right), \end{aligned}$$

by Residues Theorem we obtain

$$2\pi i\sum_{j=1}^{4}\mathrm{res}(f,z_j) = \int_{\gamma_{R,\varepsilon}} f(z)\,dz \underset{\varepsilon\to 0}{\overset{R\to+\infty}{\longrightarrow}} \lim_{R\to+\infty}\left(\int_{AB} f(z)\,dz + \int_{DE} f(z)\,dz\right) = -2\pi i\,I$$

hence $I = -\sum_{j=1}^{4}\mathrm{res}(f,z_j)$. We evaluate the residues:

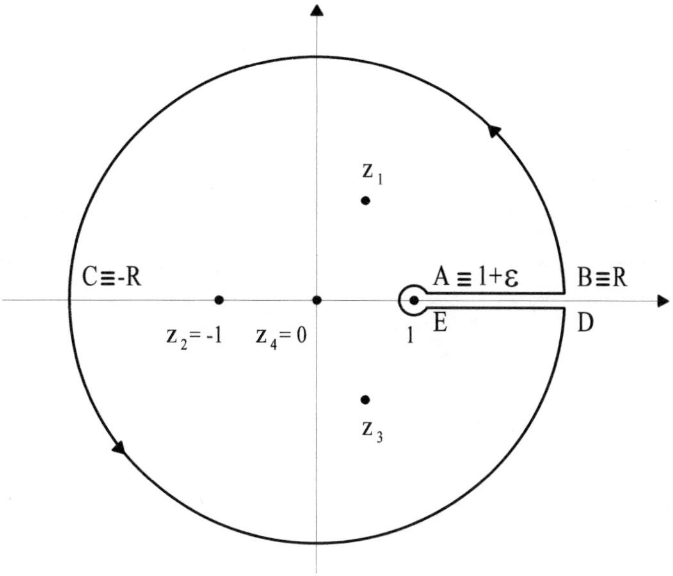

Figure 1.15 - Integration path of Exercise 1.35. The path sweeps both edges of the cut and circumvents the branching point.

- At $z_4 = 0$ (double pole), we get: $z_4 - 1 = -1 = e^{\pi i}$, $\ln(z_4 - 1) = \pi i$, hence

$$\operatorname{res}(f, z_4) = \left[\frac{d}{dz}\left(\frac{z^2-2}{z^3+1}\ln(z-1)\right)\right]_{z=0} =$$
$$= \left[\frac{2z(z^3+1) - 3z^2(z^3+1)}{(z^3+1)^2}\ln(z-1) + \frac{z^2-2}{(z^3+1)(z-1)}\right] = 2\ ;$$

- At $z_1 = 1/2 + i\sqrt{3}/2$ (simple pole), we get: $z_1 - 1 = -1/2 + i\sqrt{3}/2 = e^{2\pi i/3}$, $\ln(z_1 - 1) = 2\pi i/3$, $z_1^2 = -1/2 + i\sqrt{3}/2$, $z_1^4 = -1/2 - i\sqrt{3}/2$, hence

$$\operatorname{res}(f, z_1) \overset{\text{Hôpital}}{=} \frac{z_1^2 - 2}{z_1^2\, 3z_1^2}\frac{2}{3}\pi i = \frac{-5 + i\sqrt{3}}{-1 - i\sqrt{3}}\frac{2}{9}\pi i = \frac{\pi i}{9} + \pi\,\frac{\sqrt{3}}{3}\ ;$$

- At $z_2 = -1$ (simple pole), we get: $z_2 - 1 = -2 = e^{\ln 2}e^{\pi i}$, $\ln(z_2 - 1) = \ln(-2) = \ln 2 + \pi i$, hence

$$\operatorname{res}(f, z_2) \overset{\text{Hôpital}}{=} \frac{-1}{3(-1)^2}(\ln 2 + \pi i) = -\frac{\ln 2}{3} - \frac{\pi i}{3}\ ;$$

- At $z_3 = 1/2 - i\sqrt{3}/2 = e^{5\pi i/3}$ (simple pole), we get: $z_3 - 1 = -1/2 - i\sqrt{3}/2 = e^{4\pi i/3}$, $\ln(z_3 - 1) = 4\pi i/3$, $z_3^2 = e^{4\pi i/3}$, $z_3^4 = e^{2\pi i/3} = -1/2 + i\sqrt{3}/2$, hence

$$\operatorname{res}(f, z_3) \overset{\text{Hôpital}}{=} \frac{z_3^2 - 2}{z_3^2\, 3z_3^2}\ln(z_3 - 1) = \frac{-5 - i\sqrt{3}}{-1 + i\sqrt{3}}\frac{4}{9}\pi i = \frac{2}{9}\pi i - \frac{2}{3}\sqrt{3}\pi\ .$$

By summarizing the integral value is $I = -\sum_{j=1}^{4}\operatorname{res}(f, z_j) = -2 + \frac{1}{3}\ln 2 + \frac{\sqrt{3}}{3}\pi$. □

Exercise 1.36. If f is an holomorphic function in a nonempty open domain Ω and does not vanish identically, then it is possible to define the

$$\textit{logarithmic derivative} \text{ di } f : \qquad \frac{f'(z)}{f(z)} .$$

Prove the following statements, which show how the logarithmic derivative provides an efficient tool for counting zeroes and poles of analytic functions.

1) (**Argument principle**) Assume f is an holomorphic function not identically vanishing in an open set $\Omega \subset \mathbb{C}$ except possibly isolated singularities which are all poles. Then for every nonempty open set U whose boundary is piecewise C^1 with $\overline{U} \subset \Omega$ and every function f which has neither zeroes nor poles on ∂U, we have

$$Z - P = \frac{1}{2\pi i} \int_{\partial U} \frac{f'(z)}{f(z)} \, dz , \tag{1.41}$$

where $Z = \sum_j m_j$ is the number of zeros in U of f, all counted with their multiplicity m_j, and $P = \sum_j p_j$ is the number of poles in U of f, all counted with their orders p_j.

The number $\displaystyle\frac{1}{2\pi i} \int_{\partial U} \frac{f'(z)}{f(z)} \, dz$ is called **logarithmic index** of f.

2) (**Rouché Theorem**, Eugène Rouché, 1832 – 1910)
Let f and g be two holomorphic functions in an open set $\Omega \subset \mathbb{C}$, U a nonempty connected open set with piecewise C^1 boundary s.t. $\overline{U} \subset \Omega$ and $f(z) \neq 0 \ \forall z \in \partial U$ and $|f(z)| > |g(z)| \ \forall z \in \partial U$. Then the two equations

$$\{ z \in U : \ f(z) = 0 \} \qquad \text{and} \qquad \{ z \in U : \ f(z) + g(z) = 0 \}$$

have the same number of solutions.

Solution - 1) Due to the Residues Theorem, all contributions to the logarithmic index come from singularities of the logarithmic derivative f'/f, say from zeroes and poles of f. Then knowing the behavior of f'/f close to these points is enough for the computations.
If z_0 is a zero of f with multiplicity $m > 0$, then in a suitable neighborhood of z_0 there exists a function φ holomorphic s.t. $f(z) = (z - z_0)^m \varphi(z)$ and $\varphi(z_0) \neq 0$, hence $f'(z) = k(z - z_0)^{m-1} \varphi(z) + (z - z_0)^m \varphi'(z)$ and

$$\frac{f'(z)}{f(z)} = \frac{m}{z - z_0} + \frac{\varphi'(z)}{\varphi(z)} .$$

Analogously, if z_0 is a pole of order $p > 0$ of f, then $f(z) = (z - z_0)^{-p} \varphi(z)$ with φ holomorphic and $\varphi(z_0) \neq 0$, hence

$$\frac{f'(z)}{f(z)} = \frac{-p}{z - z_0} + \frac{\varphi'(z)}{\varphi(z)} .$$

2) We apply formula (1.41) to the holomorphic function $f(z) + s\,g(z)$, $s \in [0,1]$, which is poles free in Ω, by noticing that $f(z) + sg(z) \neq 0 \ \forall z \in \partial U$ and $s \in [0,1]$:

$$Z(s) = \frac{1}{2\pi i} \int_{\partial U} \frac{f'(z) + sg'(z)}{f(z) + sg(z)} \, dz = \frac{1}{2\pi i} \int_0^1 \frac{f'(\gamma(t)) + sg'(\gamma(t))}{f(\gamma(t)) + sg(\gamma(t))} \gamma'(t) \, dt ,$$

where $\gamma : [0,1] \mapsto \Omega$ is a parametrization of ∂U coherent with the orientation of the boundary (see Remark 1.9).
Since $s \mapsto Z(s)$ is an integer-valued continuous function in an interval, necessarily it must be constant. Then: the number of zeroes of f in U is $Z(0) = Z(1)$ which is also the number of zeroes of $f + g$ in U. □

Exercise 1.37. Prove that all roots of the polynomial $z^7 - 5z^3 + 13$ belong to the annulus $\{ z \in \mathbb{C} :\ 1 < |z| < 2 \}$.

Solution - We apply the Rouché Theorem (item 2 of Exercise 1.36).
i) By choosing $f(z) \equiv 13$ and $g(z) = z^7 - 5z^3$, since

$$|g(z)| = |z^7 - 5z^3| \leq |z|^7 + 5|z|^3 = 6 < 13 \equiv |f(z)| \qquad \text{if } |z| = 1\,,$$

we deduce that $f + g$ and f have the same number of roots in $B_1(0)$, say 0.
ii) By choosing $f(z) = z^7$ and $g(z) = 13 - 5z^3$, since

$$|g(z)| = |13 - 5z^3| \leq 13 + 5|z|^3 = 53 < 2^7 = |f(z)| \qquad \text{if } |z| = 2\,,$$

we deduce that $f + g$ and f have the same number of roots in $B_2(0)$, counted with their multiplicity, say 7. □

Chapter 2
L^p spaces

If we put on a tiled floor a certain number of coins of different values, we have two possible ways to compute the total amount of money on the floor. The first way consists in computing the total value of the coins on each tile and, subsequently, to add all the values of each tile. The second way consists in creating groups of coins all having the same value and then to add all the values of each group. In order to compute integrals we need to go from a discrete setting to a continuous setting and from a finite number to an infinite number of possible values. In such case, the first way to proceed is the main idea for the computation of integrals following Riemann's theory (Bernhard Riemann, 1826 –1866) whereas the second way is the main idea for the computation of integrals following Lebesgue's theory (Henri Lebesgue, 1875 –1941). If one is not interested in fine phenomena one could follow the simpler way suggested by Riemann. The advantages of the Lebesgue's theory are both its natural applications in Functional Analysis and the simplicity of certain methods to compute the limit of integrals of a sequence of functions, see Theorem 2.5. In this Chapter we will not define Lebesgue's integral (for which we refer to [2, 5, 14, 25]) but we will merely state its main properties and applications to L^p spaces, widely used in this monograph.

2.1 Definition and properties

In agreement with the Lebesgue theory of measure and integration, we say that a given property is satisfied *almost everywhere* (from now on, a.e.) on a measurable set $\Omega \subseteq \mathbb{R}^n$ if it holds for every $\mathbf{x} \in \Omega$ except at most a set of measure zero (typical examples are a single point or a countable number of points). In this respect, in the sequel by *function* we mean an *equivalence class of functions* which are equal a.e. We denote by $|\Omega|$ the Lebesgue measure of a measurable set Ω.

Definition 2.1. *Let $p \in \mathbb{R}$ with $1 \leq p < \infty$ and assume that $\Omega \subseteq \mathbb{R}^n$ is measurable. We define*

$$L^p(\Omega) = \left\{ f : \Omega \to \mathbb{R};\ f \,\text{Lebesgue measurable}\,,\ \int_\Omega |f(\mathbf{x})|^p \, d\mathbf{x} < +\infty \right\}.$$

Moreover, we endow $L^p(\Omega)$ with the norm

$$\|f\|_{L^p(\Omega)} = \|f\|_p := \left(\int_\Omega |f(\mathbf{x})|^p \, d\mathbf{x} \right)^{\frac{1}{p}} \tag{2.1}$$

for every $f \in L^p(\Omega)$.

Only when it will be necessary to specify the set of integration, we will use the extended notation $\|\cdot\|_{L^p(\Omega)}$ instead of $\|\cdot\|_{L^p}$; in many cases we will simplify further the notation and use $\|\cdot\|_p$. Note that the exponent p needs not to be an integer number.

For any nonnegative measurable function $f : \Omega \to \mathbb{R}$, we introduce the following definition.

$$\operatorname*{ess\,sup}_{\mathbf{x}\in\Omega} f(\mathbf{x}) := \inf\{\alpha \geq 0;\ |f^{-1}(\alpha, \infty)| = 0\} \in [0, \infty] \tag{2.2}$$

and we call such value *the essential supremum* of f over Ω. Put

$$L^\infty(\Omega) = \left\{ f : \Omega \to \mathbb{R};\ f \,\text{Lebesgue measurable}\,,\ \operatorname*{ess\,sup}_{\mathbf{x}\in\Omega} |f(\mathbf{x})| < +\infty \right\}.$$

On the space $L^\infty(\Omega)$ we define the norm

$$\|f\|_{L^\infty(\Omega)} = \|f\|_\infty := \operatorname*{ess\,sup}_{\mathbf{x}\in\Omega} |f(\mathbf{x})|$$

for every $f \in L^\infty(\Omega)$. The functions in the space L^∞ are bounded up to a set of measure zero, namely if $f \in L^\infty(\Omega)$ then there exists a constant $C > 0$ such that $|f(\mathbf{x})| \leq C$ for almost every $\mathbf{x} \in \Omega$. More precisely,

$$|f(\mathbf{x})| \leq \|f\|_{L^\infty} \qquad \text{a.e. } \mathbf{x} \in \Omega.$$

Definition 2.2. (Conjugate exponent)
Let $1 \leq p \leq \infty$; we say that p' is the conjugate exponent of p if

$$\frac{1}{p} + \frac{1}{p'} = 1$$

with the convention that $\frac{1}{\infty} = 0$.

Note that, necessarily, $1 \leq p' \leq \infty$. For instance: $p = \infty \Rightarrow p' = 1$, $p = 1 \Rightarrow p' = \infty$, $p = 3 \Rightarrow p' = \frac{3}{2}$, $p = 2 \Rightarrow p' = 2$.

The conjugate exponent plays a crucial role in the following inequality.

Theorem 2.3. (Hölder inequality)
Let $\Omega \subseteq \mathbb{R}^n$ be measurable e let $p, p' \in [1, \infty]$ be conjugate. If $f \in L^p(\Omega)$ and $g \in L^{p'}(\Omega)$, then $fg \in L^1(\Omega)$ and

$$\int_\Omega |fg| = \|fg\|_{L^1} \leq \|f\|_{L^p} \cdot \|g\|_{L^{p'}}. \tag{2.3}$$

A first important consequence of Hölder inequality (2.3) is the inclusion between L^p spaces when $\Omega \subset \mathbb{R}^n$ is a set of finite measure. More precisely,

$$\begin{aligned} &\text{if } |\Omega| < \infty \text{ and } 1 \leq p < q \leq \infty \quad \text{then} \quad L^q(\Omega) \subset L^p(\Omega) \quad \text{and} \\ &\|u\|_{L^p(\Omega)} \leq |\Omega|^{\frac{1}{p} - \frac{1}{q}} \|u\|_{L^q(\Omega)} \qquad \forall u \in L^q(\Omega) \end{aligned} \tag{2.4}$$

with the convention that $1/q = 0$ whenever $q = \infty$.
Note that the assumption $|\Omega| < \infty$ is crucial in (2.4): indeed if, for instance, $u \equiv 1$ in $\mathbb{R}$, we have $u \in L^\infty(\mathbb{R})$ but $u \notin L^p(\mathbb{R})$ for every $1 \leq p < \infty$.

The $L^p(\Omega)$ spaces are vector spaces and $\|\cdot\|_p$ is a norm for every $p \in [1, \infty]$ (provided that functions a.e. equal are identified). This enables us to define the convergence in the norm topology of $L^p(\Omega)$. We say that a sequence $\{f_m\}$ converges to f in $L^p(\Omega)$ ($f_m \to f$ in $L^p(\Omega)$) if

$$\lim_{m \to \infty} \|f_m - f\|_p = 0.$$

In some cases, this convergence is also called *strong convergence* in $L^p(\Omega)$.

Theorem 2.4. *If* $1 \le p \le \infty$ *and if* $f_m \to f$ *in* $L^p(\Omega)$, *then there exists a subsequence* $\{f_{m_k}\}$ *such that* $f_{m_k} \to f$ *a.e. in* Ω, *namely*

$$f_{m_k}(\mathbf{x}) \to f(\mathbf{x}) \quad \text{for a.e.} \quad \mathbf{x} \in \Omega\,.$$

Moreover, if $p = \infty$, *then the whole sequence converges.*

Except for the case $p = \infty$, it is not possible to obtain pointwise convergence a.e. without extracting a subsequence, see Exercise 2.28.
Let us now state the most important result concerning sequences in L^p spaces, namely a rule under which two limit processes commute: the Lebesgue integration and a.e. convergence:

Theorem 2.5. (Lebesgue dominated convergence Theorem). *Let* $\{f_k\}$ *be a sequence of measurable functions defined on a measurable set* $\Omega \subset \mathbb{R}^n$ *with values in* $\mathbb{R}$. *If:*
- f_k *converges pointwise a.e. to* f *in* Ω,
- *there exists a dominating function* $g \in L^1(\Omega)$ *such that* $|f_k(\mathbf{x})| \le g(\mathbf{x})$ *for a.e.* $\mathbf{x} \in \Omega$ *and for every* k, *then*

$$f \in L^1(\Omega) \qquad \text{and} \qquad \lim_{k\to\infty} \int_\Omega f_k = \int_\Omega f.$$

The dominated convergence Theorem is somehow the inverse of Theorem 2.4: pointwise convergence implies convergence in L^1 provided it is dominated. For a counterexample when the dominating function is missing, see Exercise 2.7.
Let $\Omega \subset \mathbb{R}^n$ be an open set and let $f \in C^0(\Omega)$. We call *support* of f and we denote it by

$$\operatorname{spt} f = \overline{\{\mathbf{x} \in \Omega; f(\mathbf{x}) \neq 0\}} \cap \Omega \tag{2.5}$$

the closure (relatively to Ω) of the set where $f \neq 0$. Note that the intersection with Ω is necessary since the closure of the set could contain points of $\partial\Omega$ which are not in the open set Ω.
Next, we define the set of *test functions* in Ω:

$$\mathcal{D}(\Omega) = \{f \in C^\infty(\Omega); \quad \operatorname{spt} f \text{ compact}\}.$$

Roughly speaking, $\mathcal{D}(\Omega)$ contains the smooth functions in Ω which vanish before reaching the boundary $\partial\Omega$ and before reaching infinity if Ω is unbounded.

An explicit example of function in $\mathcal{D}(\mathbb{R}^n)$ is the following

$$u(\mathbf{x}) = \begin{cases} e^{1/(|\mathbf{x}|^2-1)} & \text{if } |\mathbf{x}| < 1 \\ 0 & \text{if } |\mathbf{x}| \geq 1 \,. \end{cases}$$

The functions in $\mathcal{D}(\Omega)$ are "rare" and have nice properties. Moreover, they well approximate functions in L^p. By this, we mean that if $p \in [1, \infty)$ then the space $\mathcal{D}(\Omega)$ is dense in $L^p(\Omega)$ and hence

$$\forall f \in L^p(\Omega),\ \forall \varepsilon > 0\ \ \exists f_\varepsilon \in \mathcal{D}(\Omega),\ \int_\Omega |f - f_\varepsilon|^p < \varepsilon.$$

It is sometimes useful to emphasize the local properties of functions in L^p. We denote by $L^p_{loc}(\Omega)$ the space

$$L^p_{loc}(\Omega) = \{f \in L^p(K),\ \forall K \text{ compact} \subset \Omega\}.$$

Functions in $L^p_{loc}(\Omega)$ (where the subscript "*loc*" stands for local) are those functions which have a good behavior inside of Ω, but for which we have no information close to $\partial\Omega$. For instance, the function $f(x) = \frac{1}{x}$ is not integrable at 0 and at $+\infty$, whence it does not belong to $L^1(0, \infty)$; nevertheless, it belongs to the space $L^1_{loc}(0, \infty)$. We also say that a sequence $\{f_m\}$ converges in $L^p_{loc}(\Omega)$ to $f \in L^p_{loc}(\Omega)$ if $f_m \to f$ in $L^p(K)$ for every compact $K \subset \Omega$.
It is well-known that a *primitive* of a given function of one real variable has more regularity than the function itself. For instance, if $f \in C^k$ then any primitive $\int f$ belongs to C^{k+1}, thanks to the fundamental Theorem of integral calculus. It is therefore natural to wonder which are the properties of a primitive of an integrable function: in order to give an answer, we give the following Definition and Theorem.

Definition 2.6. (Absolutely continuous functions).
Let $I \subset \mathbb{R}$ be an interval. A function $F: I \to \mathbb{R}$ is said to be absolutely continuous in I (denoted by $F \in AC(I)$) if there exists a function f in $L^1(I)$ s.t. for every $x, x_0 \in I$ we have

$$F(x) = F(x_0) + \int_{x_0}^{x} f(t)\, dt \,. \tag{2.6}$$

A function $F: I \to \mathbb{R}$ is said locally absolutely continuous in I (denoted by $F \in AC_{loc}(I)$) if there exists a function f in $L^1_{loc}(I)$ s.t. (2.6) holds for every $x, x_0 \in I$.

Theorem 2.7. *Let $I \subset \mathbb{R}$ be an interval and let $F \in AC_{loc}(I)$. Then*

- *F is continuous in I;*
- *F is differentiable in classical sense at a.e. $x \in I$ (that is, a.e. in I the limit of the difference quotient of F exists and is finite);*
- *The following equality holds:*

$$F' = f \qquad \text{a.e. in } I\,, \tag{2.7}$$

  where F' denotes the function which coincides with the derivative of F in the points of I where the derivative exists and vanishes in the complementary set (which has measure zero!);
- *$F' = f$ is also the weak derivative of F in I (see Definition 3.18).*

The previous Theorem is an extension of the fundamental Theorem of integral calculus and implies the following strict embeddings.

$$C^1(I) \underset{\neq}{\subset} AC_{loc}(I) \underset{\neq}{\subset} C^0(I)\,. \tag{2.8}$$

Example 2.8. The function $F(x) = \max(1 - |x|, 0)$, $x \in \mathbb{R}$, does not belong to $C^1(\mathbb{R})$ but it belongs to $AC(\mathbb{R}) \subset C^0(\mathbb{R})$ and we have $F'(x) = -\operatorname{sign}(x)\,\chi_{(-1,+1)}$.

Let us now define complex-valued L^p spaces: these spaces are the natural framework for the integral transform defined in Chapters 4 and 6.
Given a function f defined on a measurable set $\Omega \subseteq \mathbb{R}^n$ with values in $\mathbb{C}$ we denote, respectively, by $\mathrm{Re}f$ and $\mathrm{Im}f$ the real part and the imaginary part of f:

$$f = \mathrm{Re}f + i\,\mathrm{Im}f.$$

We also denote by $|\cdot|$ the usual modulus in $\mathbb{C}$. The space $L^p(\Omega, \mathbb{C})$ is the space of functions $f : \Omega \to \mathbb{C}$ such that $\mathrm{Re}f$, $\mathrm{Im}f \in L^p(\Omega)$, with the norm

$$\|f\|_p := \left(\int_\Omega |f(\mathbf{x})|^p d\mathbf{x} \right)^{1/p} \qquad \text{if } 1 \le p < \infty\,,$$

$$\|f\|_\infty := \operatorname*{ess\,sup}_{\mathbf{x}\in\Omega} |f(\mathbf{x})| \qquad \text{if } p = \infty\,.$$

The integral of given a function $f \in L^1(\Omega, \mathbb{C})$ is defined by

$$\int_\Omega f(\mathbf{x}) d\mathbf{x} := \int_\Omega \mathrm{Re}f(\mathbf{x}) d\mathbf{x} + i \int_\Omega \mathrm{Im}f(\mathbf{x}) d\mathbf{x}.$$

Note that (as in the real case) a measurable function $f : \Omega \to \mathbb{C}$ belongs to $L^1(\Omega, \mathbb{C})$ if and only if $|f| \in L^1(\Omega)$.

Remark 2.9. Given a measurable set $\Omega \subset \mathbb{R}^n$ and a measurable function $f : \Omega \mapsto \mathbb{C}$. we have that

$$f \text{ is integrable in } \Omega \qquad \Longleftrightarrow \qquad |f| \text{ is integrable in } \Omega\ .$$

In particular, if a function belongs to $L^1(\Omega)$ then its modulus is integrable in Ω.
Hence, if a function is Riemann integrable in a bounded set, then it is also Lebesgue integrable in the same set.
However, the existence of a generalized Riemann integral does not guarantee integrability in the Lebesgue sense (see Exercise 2.5).

We conclude the list of properties of the Lebesgue integral with a statement about the reduction of multiple integrals and with the possibility of changing the order of integrations.

Theorem 2.10. *(**Fubini-Tonelli Theorem**). Let $f \in L^1(\mathbb{R}^{n+m})$.
Then, if we denote by $(\mathbf{x}, \mathbf{y}) \in \mathbb{R}^{n+m}$ a generic point (with $\mathbf{x} \in \mathbb{R}^n$, $\mathbf{y} \in \mathbb{R}^m$), we have:*

(i) for a.e. $\mathbf{x} \in \mathbb{R}^n$ the integral $\displaystyle\int_{\mathbb{R}^m} f(\mathbf{x}, \mathbf{y})\, d\mathbf{y}$ exists and is finite

and the function $\mathbf{x} \mapsto \displaystyle\int_{\mathbb{R}^m} f(\mathbf{x}, \mathbf{y})\, d\mathbf{y}$ is integrable with respect to $\mathbf{x}$;

(ii) for a.e. $\mathbf{y} \in \mathbb{R}^m$ the integral $\displaystyle\int_{\mathbb{R}^n} f(\mathbf{x}, \mathbf{y})\, d\mathbf{x}$ exists and is finite

and the function $\mathbf{y} \mapsto \displaystyle\int_{\mathbb{R}^n} f(\mathbf{x}, \mathbf{y})\, d\mathbf{x}$ is integrable with respect to $\mathbf{y}$;

(iii) the following reduction formulas hold

$$\int_{\mathbb{R}^{n+m}} f(\mathbf{x}, \mathbf{y})\, d\mathbf{x}\, d\mathbf{y} = \int_{\mathbb{R}^n} \left(\int_{\mathbb{R}^m} f(\mathbf{x}, \mathbf{y})\, d\mathbf{y} \right) d\mathbf{x} = \int_{\mathbb{R}^m} \left(\int_{\mathbb{R}^n} f(\mathbf{x}, \mathbf{y})\, d\mathbf{x} \right) d\mathbf{y}\, .$$

2.2 Exercises

Let us first recall some integrability criteria which will be repeatedly used in the subsequent exercises.
We say that $f(x) \sim g(x)$ (f is asymptotic to g) for $x \to x_0 \in [-\infty, +\infty]$ if

$$\lim_{x \to x_0} \frac{f(x)}{g(x)} = 1 .$$

We say that $f(x) \asymp g(x)$ (f has the same order as g) for $x \to x_0 \in [-\infty, +\infty]$ if

$$\exists C \in (-\infty, 0) \cup (0, +\infty) \quad \text{such that} \quad \lim_{x \to x_0} \frac{f(x)}{g(x)} = C .$$

Obviously, being asymptotic is a more precise information than just having the same order: $f(x) \sim g(x) \;\Rightarrow\; f(x) \asymp g(x)$. These symbols are very useful to study the convergence of integrals because they allow to reduce "almost" all the integrands to powers of the variable. We know that

$$\int_0^1 \frac{dx}{x^\alpha} < \infty \iff \alpha < 1; \qquad \int_1^{+\infty} \frac{dx}{x^\alpha} < \infty \iff \alpha > 1;$$

$$\int_0^{1/2} \frac{dx}{x|\ln x|^\alpha} < \infty \iff \alpha > 1; \qquad \int_2^{+\infty} \frac{dx}{x(\ln x)^\alpha} < \infty \iff \alpha > 1 .$$

If $f, g \in C^0((a,b])$, $f, g \geq 0$ in $(a,b]$ and $f(x) \asymp g(x)$ as $x \to a$ then

$$\int_a^b f(x)dx < +\infty \iff \int_a^b g(x)dx < +\infty ;$$

for instance, if $f \in C^0((0,1])$ and $f(x) \asymp x^{-\alpha}$ as $x \to 0$, then $\int_0^1 f(x)dx$ converges iff $\alpha < 1$.
Since the integrability of a function depends on its behavior in a neighborhood of a certain point x_0 (possibly infinity), we denote by $\mathcal{U}(x_0)$ a general neighborhood of x_0 which contains no further singularities of the considered function. In particular, this notation allows to write quite simply the well-known comparison principle: if $f, g \in C^0((a,b])$ and $0 \leq f(x) \leq g(x)$ in $\mathcal{U}(a_+)$, then

$$\int_a^b g(x)dx < +\infty \implies \int_a^b f(x)dx < +\infty .$$

These criteria also extend to multiple integrals:
let $B_1 = B_1(\mathbf{0}) = \{\mathbf{x} \in \mathbb{R}^n ;\ |\mathbf{x}| < 1\}$, then

$$\int_{B_1} \frac{d\mathbf{x}}{|\mathbf{x}|^\alpha} < \infty \iff \alpha < n;$$

$$\int_{\mathbb{R}^n \setminus B_1} \frac{d\mathbf{x}}{|\mathbf{x}|^\alpha} < \infty \iff \alpha > n.$$

Finally, let us recall that if f is a radially symmetric function, $f = f(|\mathbf{x}|)$, then for every $R > 0$ we have

$$\int_{B_R} f(\mathbf{x})\, d\mathbf{x} = \sigma_{n-1} \int_0^R r^{n-1} f(r)\, dr \tag{2.9}$$

where σ_{n-1} denotes the surface measure of the unit sphere ∂B_1, see (2.13). When $n = 2$ we find the well-known formula for polar coordinates:

$$\int_{B_1} f(\mathbf{x})\, d\mathbf{x} = 2\pi \int_0^1 r f(r)\, dr.$$

We conclude with a remark on the use of Theorem 2.5. In the worked solutions of the exercises below we suggest a possible dominating function but, if some estimates do not seem straightforward, the reader may find different ones: it is enough to find just one (of course, when it exists!).

2.2.1 Exercises with solutions

Exercise 2.1. Let $f : (0, +\infty) \to (0, +\infty)$ be defined by:

$$f(x) = \begin{cases} x^{-\alpha} & \text{if } x \in (0,1) \\ x^{-\beta} & \text{if } x \in [1, +\infty). \end{cases}$$

Determine the values of $\alpha > 0$ and $\beta > 0$ for which f is Lebesgue integrable.

Solution - For any $\alpha < 1$ and $\beta > 1$. □

Exercise 2.2. Determine the values of $\alpha > 0$ for which the function

$$f(x) = \sum_{k=1}^{+\infty} \frac{1}{k^\alpha} \chi_{[k-1,k]}(x),$$

is Lebesgue integrable, where $\chi_{[k-1,k]}(x)$ is the characteristic function of the interval $[k-1, k]$.

Solution - Any $\alpha > 1$. □

Exercise 2.3. Find $f_1 \in L^1(0, +\infty)$ such that $f_1 \notin L^2(0, +\infty)$ and $f_1 \notin L^\infty(0, +\infty)$.
Find $f_2 \in L^2(0, +\infty)$ such that $f_2 \notin L^1(0, +\infty)$ and $f_2 \notin L^\infty(0, +\infty)$.
Find $f_3 \in L^\infty(0, +\infty)$ such that $f_3 \notin L^2(0, +\infty)$ and $f_3 \notin L^1(0, +\infty)$.

Solution - One can take

$$f_1(x) = \frac{1}{(x+1)\sqrt{x}}, \qquad f_2(x) = \frac{\ln x}{x+1}, \qquad f_3(x) = 1.$$

Of course, different examples are available! □

Exercise 2.4. Let

$$f(x) = \frac{\ln(1+x)}{x^{3/2}} \qquad x \in (0, +\infty).$$

Find the values of $p \in [1, +\infty)$ such that $f \in L^p(0, +\infty)$.

Solution - Two behaviors need to be analyzed.

$$x \to 0_+ : \quad |f(x)|^p \asymp \left(\frac{x}{x^{3/2}}\right)^p = \frac{1}{x^{p/2}} \in L^1(\mathcal{U}(0_+)) \iff 1 \le p < 2$$

$$x \to +\infty : \quad |f(x)|^p \asymp \frac{[\ln(1+x)]^p}{x^{3p/2}} \in L^1(\mathcal{U}(+\infty)) \qquad \iff p \ge 1$$

hence $f \in L^p(0, +\infty)$ iff $1 \le p < 2$. □

Exercise 2.5. Consider the function

$$f(x) = \frac{\sin x}{x} \qquad x \in (0, +\infty).$$

Prove that f is Riemann integrable in generalized sense but $f \notin L^1(0, +\infty)$.

Solution - The function f is Riemann integrable in generalized sense if the limit

$$\int_0^{+\infty} \frac{\sin x}{x}\, dx = \lim_{k \to +\infty} \int_0^k \frac{\sin x}{x}\, dx$$

exists and is finite. For $k > 1$ we decompose the integral and we integrate by parts:

$$\int_0^k \frac{\sin x}{x}\, dx = \int_0^1 \frac{\sin x}{x}\, dx - \int_1^k \frac{\cos x}{x^2}\, dx - \left[\frac{\cos x}{x}\right]_1^k .$$

The first summand is finite since $\sin x/x$ may be extended to a continuous function on $[0, 1]$. When $k \to +\infty$, we have

$$\lim_{k \to +\infty} \int_1^k \frac{\cos x}{x^2}\, dx = \int_1^{+\infty} \frac{\cos x}{x^2}\, dx$$

and this limit is finite due to the comparison principle since $|\cos x/x^2| \le |1/x^2|$. Finally,

$$\lim_{k \to +\infty} \left[\frac{\cos x}{x}\right]_1^k = -\cos 1.$$

It follows that f is Riemann integrable in generalized sense; for its exact value, see Exercise 1.22.

On any interval $[m\pi, (m+1)\pi]$, with $m \in \mathbb{N}$, we have

$$|f(x)| = \left|\frac{\sin x}{x}\right| \ge \frac{1}{2x} \qquad \forall x \in \left[m\pi + \frac{\pi}{6}, (m+1)\pi - \frac{\pi}{6}\right].$$

Hence,

$$\begin{aligned}\int_0^{+\infty} |f(x)|\, dx &= \sum_{m=0}^{+\infty} \int_{m\pi}^{(m+1)\pi} |f(x)|\, dx \ge \sum_{m=0}^{+\infty} \int_{m\pi+\pi/6}^{(m+1)\pi-\pi/6} \frac{dx}{2x} \\ &= \sum_{m=0}^{+\infty} \ln\left(1 + \frac{4}{6m+1}\right) = +\infty\end{aligned}$$

since $\ln(1+\frac{4}{6m+1}) \sim \frac{2}{3m}$. Therefore, $f \notin L^1(0,+\infty)$. □

Exercise 2.6. For any $k \in \mathbb{N}$, consider the function defined on $[0,1]$ by

$$f_k(x) = \sqrt[k]{x}.$$

Find the pointwise limit f of the sequence $\{f_k\}$, prove that the convergence $f_k \to f$ is not uniform and compute

$$\lim_{k\to+\infty} \int_0^1 f_k(x)\,dx \ .$$

Solution - By letting $k \to +\infty$ we find

$$f_k(x) = x^{1/k} \to f(x) = \begin{cases} 0 & \text{if } x = 0 \\ 1 & \text{if } x \in (0,1]. \end{cases}$$

Since $f \notin C^0([0,1])$ the convergence cannot be uniform. Consequently, we cannot apply the theory of the Riemann integral to compute the limit. Nevertheless, $|f_k(x)| \le 1 \in L^1(0,1)$ and hence we may apply the Lebesgue Theorem (Theorem 2.5) and find

$$\lim_{k\to+\infty} \int_0^1 f_k(x)\,dx = \int_0^1 f(x)\,dx = 1 \ .$$

□

Exercise 2.7. For any $k \in \mathbb{N}$, consider the function defined on $[0,1]$ by

$$f_k(x) = \begin{cases} k & \text{if } x \in [0,\frac{1}{k}], \\ 0 & \text{if } x \in (\frac{1}{k},1]. \end{cases}$$

Find the pointwise limit f of the sequence $\{f_k\}$, and prove that

$$\lim_{k\to+\infty} \int_0^1 f_k(x)\,dx \neq \int_0^1 f(x)\,dx \ .$$

Solution - By letting $k \to +\infty$ we find

$$f_k(x) \to f(x) = \begin{cases} +\infty & \text{if } x = 0, \\ 0 & \text{if } x \in (0,1]. \end{cases}$$

Therefore, $f_k \to f = 0$ a.e. in $[0,1]$ (everywhere except for $x = 0$). Moreover, we have

$$\int_0^1 f_k(x)\,dx = 1 \qquad \forall k \ge 1.$$

Hence,

$$\lim_{k\to+\infty} \int_0^1 f_k(x)\,dx = 1 \neq 0 = \int_0^1 f(x)\,dx \ .$$

The statement of Theorem 2.5 does not hold; this does not contradict the Theorem because there exists no dominating function $g \in L^1(0,1)$ such that $|f_k(x)| \le g(x)$ for a.e. $x \in [0,1]$.
□

Exercise 2.8. For any $b \in \mathbb{R}$ and $k \in \mathbb{N}$, consider the function defined on $(0,+\infty)$ by

$$f_k(x) = \frac{x^b}{1+(-1)^k\, e^{kx}}.$$

1) Determine for which values of b we have

$$(i)\ f_k \in L^1(0,+\infty) \qquad (ii)\ f_k \in L^2(0,+\infty) \qquad (iii)\ f_k \in L^\infty(0,+\infty)$$

for every $k \in \mathbb{N}$.
2) Find the pointwise limit f of $\{f_k\}$ and determine for which values of b the following step is correct

$$\lim_{k\to\infty} \int_0^\infty f_k(x)\, dx = \int_0^\infty f(x)\, dx.$$

Solution - 1) In $\mathcal{U}(+\infty)$ there are no integrability problems (nor boundedness), since f_k is infinitesimal of exponential order. In $\mathcal{U}(0_+)$:

$$|f_k(x)| \asymp x^b \text{ if } k \text{ is even }, \qquad |f_k(x)| \asymp x^{b-1} \text{ if } k \text{ is odd.}$$

Therefore:

$$\begin{aligned} &f_k \in L^1(0,+\infty) \quad \forall k \in \mathbb{N} \Leftrightarrow b > 0 \\ &f_k \in L^2(0,+\infty) \quad \forall k \in \mathbb{N} \Leftrightarrow b > \frac{1}{2} \\ &f_k \in L^\infty(0,+\infty) \ \forall k \in \mathbb{N} \Leftrightarrow b \geq 1. \end{aligned}$$

2) $f(x) = 0$ on $(0,+\infty)$. For the validity of the limit computation it is necessary that $f_k \in L^1$ for every k, and hence that $b > 0$. Such condition is also sufficient: indeed

$$|f_k(x)| \leq f_2(x) \ \ \forall x \ \text{ if } k \text{ even }, \qquad |f_k(x)| \leq |f_1(x)| \ \ \forall x \ \text{ if } k \text{ odd.}$$

The function $\varphi(x) = \max\{|f_1(x)|, f_2(x)\}$ is a dominant because $\varphi \in L^1(0,+\infty)$ and $|f_k(x)| \leq \varphi(x)$ for every x: we may then apply Theorem 2.5. □

Exercise 2.9. Consider the sequence of functions define on $(0,+\infty)$ by

$$f_k(x) = \begin{cases} x^{-1/k} & \text{if } 0 < x \leq 1\,, \\ x^{-k} & \text{if } x > 1\,. \end{cases}$$

Determine for which $k \geq 1$ we have $f_k \in L^1(0,+\infty)$ and compute

$$\lim_{k\to\infty} \int_0^\infty f_k(x)\, dx\,.$$

Solution - One can prove that $f_k \in L^1(0,+\infty)$ for every $k \geq 2$. Moreover, $f_k(x) \leq f_2(x)$ for every $x > 0$ and for every $k \geq 2$. We have

$$f_k(x) \to f(x) = \begin{cases} 1 & \text{if } 0 < x \leq 1\,, \\ 0 & \text{if } x > 1\,. \end{cases}$$

For $k \geq 2$ the assumptions of Theorem 2.5 are fulfilled and hence

$$\lim_{k\to\infty} \int_0^\infty f_k(x)\, dx = \int_0^\infty f(x)\, dx = 1.$$

□

Exercise 2.10. Consider the sequence of functions $\{f_k\}$ defined a.e. on $(0,+\infty)$ by

$$f_k(x) = \frac{(\arctan(kx))^{1/3}\, \ln(1+x)}{|1-x|^{1/k}\, x^{3/2}}\,.$$

1) Find the values of $k \in \mathbb{N} \setminus \{0\}$ for which $f_k \in L^1(0,+\infty)$.
2) Find the pointwise limit f and find the values of $p \geq 1$ for which $f \in L^p(0,+\infty)$.

3) Establish if the following step is allowed

$$\lim_{k\to\infty}\int_0^\infty f_k(x)\,dx = \int_0^\infty f(x)\,dx.$$

Solution - 1) The asymptotic behavior of f_k in the three "delicate points" is the following:

$$x\to 0_+: \quad f_k(x) \asymp \frac{x^{1/3}\,x}{x^{3/2}} = x^{-1/6} \in L^1(\mathcal{U}(0_+)) \quad \forall k\geq 1$$

$$x\to 1: \quad f_k(x) \asymp \frac{1}{|1-x|^{1/k}} \in L^1(\mathcal{U}(1)) \qquad \text{sse } k\geq 2$$

$$x\to: +\infty \quad f_k(x) \asymp \frac{\ln(1+x)}{x^{3/2+1/k}} \in L^1(\mathcal{U}(+\infty)) \qquad \forall k\geq 1$$

hence $f_k\in L^1(0,+\infty)$ iff $k\geq 2$.
2) The pointwise limit is

$$f(x) = \left(\frac{\pi}{2}\right)^{1/3}\frac{\ln(1+x)}{x^{3/2}} \qquad \forall x\neq 1$$

and its asymptotic behavior at 0 and ∞ is

$$x\to 0_+: \quad |f(x)|^p \asymp \left(\frac{x}{x^{3/2}}\right)^p = \frac{1}{x^{p/2}} \in L^1(\mathcal{U}(0_+)) \text{ iff } 1\leq p<2$$

$$x\to +\infty: \quad |f(x)|^p \asymp \frac{[\ln(1+x)]^p}{x^{3p/2}} \in L^1(\mathcal{U}(+\infty)) \qquad \forall p\geq 1$$

hence $f\in L^p(0,+\infty)$ iff $1\leq p<2$.
3) Let

$$\varphi(x) = \begin{cases} \left(\dfrac{\pi}{2}\right)^{1/3}\dfrac{\ln(1+x)}{x^{3/2}} & \text{if } x\geq 2 \\[2ex] \left(\dfrac{\pi}{2}\right)^{1/3}\dfrac{\ln(1+x)}{|1-x|^{1/2}\,x^{3/2}} & \text{if } 0<x<2,\ x\neq 1 \end{cases}$$

then $\varphi\in L^1(0,+\infty)$ and $|f_k(x)|\leq\varphi(x)$ a.e. and $\forall k\geq 2$.
We may then apply Theorem 2.5: the step is correct. □

Exercise 2.11. Consider the sequence of functions $\{f_k\}$ defined on $(0,+\infty)$ by

$$f_k(x) = \frac{\ln(e^{kx}-k\,\sin x)}{kx(1+x^2)}.$$

1) Determine for which values of $p\in[1,\infty]$ one has $f_1\in L^p(0,+\infty)$.
2) Find the pointwise limit f of f_k.
3) Compute

$$\lim_{k\to\infty}\int_0^\infty f_k(x)\,dx.$$

Solution - 1)

$$f_1(x) = \frac{\ln(e^x-\sin x)}{x(1+x^2)} = \frac{\ln(e^x(1-e^{-x}\sin x))}{x(1+x^2)} = \frac{x+\ln(1-e^{-x}\,\sin x)}{x(1+x^2)}$$

and, recalling that $e^x>x>\sin x$ if $x>0$, we obtain

$$x \to 0_+ \Rightarrow f_1(x) \to 0\,, \qquad x \to +\infty \Rightarrow f_1(x) \le \frac{c}{x^2}.$$

Hence $f_1 \in L^p(0,+\infty)$ for every $p \in [1,\infty]$.
2) We have

$$f_k(x) = \frac{kx + \ln(1 - k\,e^{-kx}\,\sin x)}{kx(1+x^2)} = \frac{1}{1+x^2} + \frac{\ln(1 - k\,e^{-kx}\,\sin x)}{kx(1+x^2)}.$$

But since

$$|\ln(1 - k\,e^{-kx}\,\sin x)| \le (2\ln 2)\,k\,e^{-kx}\,|\sin x| \tag{2.10}$$

we infer that

$$\left|\frac{\ln(1 - k\,e^{-kx}\,\sin x)}{kx(1+x^2)}\right| \le \frac{(2\ln 2)k\,e^{-kx}\,|\sin x|}{kx(1+x^2)} = \frac{(2\ln 2)e^{-kx}\,|\sin x|}{x(1+x^2)} \to 0$$

and, in turn, that $f_k(x) \to \dfrac{1}{1+x^2}$ in $(0,+\infty)$.
The estimate (2.10) follows from the inequality $k\,e^{-kx}\,\sin x < e^{-1}\ \forall x \in [0,\pi]\ \forall k \ge 1$, which is absolutely nontrivial but it is left as an exercise to the reader.
3) By using the estimate (2.10) and taking into account that $\dfrac{|\sin x|}{x} \le 1$, we infer that

$$\left|\frac{\ln(1 - k\,e^{-kx}\,\sin x)}{kx(1+x^2)}\right| \le \frac{(2\ln 2)e^{-kx}}{1+x^2} \le \frac{2\ln 2}{1+x^2}$$

and, subsequently,

$$|f_k(x)| \le \frac{1+2\ln 2}{1+x^2} \in L^1(0,+\infty).$$

Then by applying Theorem 2.5 we obtain

$$\lim_{k\to\infty}\int_0^\infty f_k(x)\,dx = \int_0^\infty \frac{dx}{1+x^2} = \frac{\pi}{2}.$$

□

Exercise 2.12. a) Compute $\displaystyle\int_{\mathbb{R}} e^{-x^2}\,dx\,.$ b) Compute $\displaystyle\int_{\mathbb{R}^n} e^{-|\mathbf{x}|^2}\,d\mathbf{x}\,.$

c) Compute the area σ_{n-1} of the n-dimensional unit sphere S^{n-1}, where $S^{n-1} := \partial B_1(\mathbf{0})$.

d) Compute the volume ω_n of the n-dimensional unit ball $B_1(\mathbf{0}) \subset \mathbb{R}^n$.

Solution - a) Put $I = \displaystyle\int_{\mathbb{R}} e^{-x^2}\,dx$ then by renaming the integration variable, we get

$$I^2 = \left(\int_{\mathbb{R}} e^{-x^2}\,dx\right)^2 = \int_{\mathbb{R}} e^{-x^2}\,dx \cdot \int_{\mathbb{R}} e^{-y^2}\,dy = \int_{\mathbb{R}^2} e^{-|\mathbf{x}|^2}\,d\mathbf{x} = \int_0^{2\pi}\int_0^\infty e^{-\varrho^2}\,\varrho\,d\varrho\,d\vartheta = \pi\,,$$

and hence

$$\int_{\mathbb{R}} e^{-x^2}\,dx = \sqrt{\pi}\,. \tag{2.11}$$

b) From item a) and the identity $\displaystyle\int_{\mathbb{R}^n} e^{-|\mathbf{x}|^2}\, d\mathbf{x} = \prod_{j=1}^{N} \int_{\mathbb{R}} e^{-x_j^2}\, dx_j$ (consequence of Fubini's Theorem) we obtain

$$\int_{\mathbb{R}^n} e^{-|\mathbf{x}|^2}\, d\mathbf{x} = \pi^{n/2}. \tag{2.12}$$

c) Let us use both cartesian coordinates $\mathbf{x} = (x_1, \dots, x_n)$ and polar coordinates $\varrho, \boldsymbol{\vartheta}$. Here $\varrho = |\mathbf{x}|$ denotes the modulus of $\mathbf{x}$, whereas $\boldsymbol{\vartheta} = \mathbf{x}/\varrho$ denotes the unit radial vector (which is n-dimensional but can be described by $n-1$ angular coordinates) and $d\boldsymbol{\vartheta}$ denotes the surface element on $\partial B_1(\mathbf{0})$. For instance: if $n = 2$ then $d\boldsymbol{\vartheta} = d\vartheta$, $\vartheta \in [0, 2\pi]$, if $n = 3$ then $d\boldsymbol{\vartheta} = \sin\varphi\, d\vartheta\, d\varphi$, $\vartheta \in [0, 2\pi]$, $\varphi \in [0, \pi]$, see Section 10.6.2.
Therefore, in any space dimension, the surface element over $\partial B_\varrho(\mathbf{0})$ is given by $d\sigma = \varrho^{n-1} d\boldsymbol{\vartheta}$, while the n-dimensional volume element $d\mathbf{x}$ satisfies $d\mathbf{x} = \varrho^{n-1}\, d\boldsymbol{\vartheta}\, d\varrho$.
With these notations let us integrate in polar coordinates the identity (2.12) and, denoting with Γ the Euler Gamma function (see Exercise 1.26),

$$\pi^{n/2} = \int_{\mathbb{R}^n} e^{-|\mathbf{x}|^2}\, d\mathbf{x} = \int_0^\infty \int_{\partial B_1(\mathbf{0})} e^{-\varrho^2} \varrho^{n-1}\, d\boldsymbol{\vartheta}\, d\varrho =$$

$$= \sigma_{n-1} \int_0^\infty e^{-\varrho^2} \varrho^{n-1}\, d\varrho = \boxed{\varrho^2 = t,\ 2\varrho d\varrho = dt}$$

$$= \frac{1}{2}\sigma_{n-1} \int_0^\infty e^{-t}\, t^{n/2-1}\, dt = \frac{1}{2}\Gamma\left(\frac{n}{2}\right)\sigma_{n-1}.$$

Area of the unit sphere S^{n-1} in $\mathbb{R}^n$: $$\sigma_{n-1} = \frac{2\,\pi^{n/2}}{\Gamma(n/2)} \tag{2.13}$$

d) If we apply the divergence Theorem in $\mathbb{R}^n$ (Table 10.5, Formula 3) with the choice $\mathbf{u}(\mathbf{x}) = \mathbf{x}$, we obtain $\operatorname{div}\mathbf{u} = \operatorname{div}\mathbf{x} = n$ and taking into account that on $\partial B_1(\mathbf{0})$ one has $|\mathbf{x}| = 1$ and that the unit normal vector outgoing from $B_1(\mathbf{0})$ is $\boldsymbol{\nu} = \mathbf{x}/|\mathbf{x}|$, we also get

$$\omega_n = \int_{B_1(\mathbf{0})} 1\, d\mathbf{x} = \frac{1}{n}\int_{\partial B_1(\mathbf{0})} \mathbf{x} \cdot \frac{\mathbf{x}}{|\mathbf{x}|}\, d\boldsymbol{\vartheta} = \frac{1}{n}\int_{\partial B_1(\mathbf{0})} d\boldsymbol{\vartheta} = \frac{\sigma_{n-1}}{n}.$$

Hence, by invoking (2.13),

Volume of the unit disc $B_1(\mathbf{0}) \subset \mathbb{R}^n$: $$\omega_n = \frac{\sigma_{n-1}}{n} = \frac{2\,\pi^{n/2}}{n\,\Gamma(n/2)} \tag{2.14}$$

Remark - In the simplest cases ($n = 1, 2, 3$) we find the well-known values (since $\Gamma(1/2) = \sqrt{\pi}$, $\Gamma(1) = 1$, $\Gamma(3/2) = (1/2)\,\Gamma(1/2) = \sqrt{\pi}/2$):

- if $n = 1$, then $\sigma_0 = 2$ (count the number of points on the boundary of the interval $(-1, 1)$) and $\omega_1 = 2$ (length of $(-1, 1)$;

- if $n = 2$, then $\sigma_1 = 2\pi$ (length of the unit circle) and $\omega_2 = \pi$ (area of the unit disc);
- if $n = 3$, then $\sigma_2 = 4\pi$ (surface of the 3-dimensional unit sphere) and $\omega_3 = (4/3)\pi$ (volume of the unit ball).

$\square$

Exercise 2.13. Let B the unit ball in $\mathbb{R}^n$ ($n \geq 2$) and consider the functions

$$f_1(\mathbf{x}) = \frac{1 - \cos|\mathbf{x}|}{|\mathbf{x}|^{5/2}}, \qquad f_2(\mathbf{x}) = \frac{1}{e^{|\mathbf{x}|} - 1}, \qquad f_3(\mathbf{x}) = \frac{1}{\sin|\mathbf{x}|^2} + \ln|\mathbf{x}|,$$

defined a.e. in B. Find for which $p \in [1, +\infty]$ one has $f_i \in L^p(B)$ per $i = 1, 2, 3$.

Solution - In radial coordinates:

$$f_1(|\mathbf{x}|) = f_1(r) = \frac{1 - \cos r}{r^{5/2}} \sim \frac{1}{2\sqrt{r}} \quad \text{as } r \to 0.$$

Therefore, by (2.9),

$$f_1 \in L^p(B_1) \Leftrightarrow \int_0^1 \frac{r^{n-1}}{r^{p/2}}\, dr < +\infty \Leftrightarrow n - 1 - \frac{p}{2} > -1 \Leftrightarrow p < 2n.$$

In radial coordinates:

$$f_2(|\mathbf{x}|) = f_2(r) = \frac{1}{e^r - 1}.$$

Therefore, by (2.9),

$$f_2 \in L^p(B) \Leftrightarrow \int_0^1 \frac{r^{n-1}}{(e^r - 1)^p}\, dr < \infty \Leftrightarrow \int_0^1 \frac{r^{n-1}}{r^p}\, dr < \infty \Leftrightarrow n-1-p > -1 \Leftrightarrow p < n.$$

In radial coordinates:

$$f_3(|\mathbf{x}|) = f_3(r) = \frac{1}{\sin r^2} + \ln r.$$

Hence, $f_3(r) \asymp \frac{1}{r^2}$ as $r \to 0$ and, by (2.9),

$$f_3 \in L^p(B) \Leftrightarrow \int_0^1 r^{n-1}|f_3(r)|^p\, dr < \infty \Leftrightarrow \int_0^1 \frac{r^{n-1}}{r^{2p}}\, dr < \infty \Leftrightarrow p < \frac{n}{2}.$$

$\square$

Exercise 2.14. Consider the functions defined a.e. in $\mathbb{R}^n$

$$f_1(\mathbf{x}) = \frac{\ln(1 + |\mathbf{x}|)}{|\mathbf{x}|^2}, \qquad f_2(\mathbf{x}) = \frac{e^{-1/|\mathbf{x}|} - 1}{|\mathbf{x}|^3},$$

$$f_3(\mathbf{x}) = \frac{\ln(1 + |\mathbf{x}|^2)}{|\mathbf{x}|^3}, \qquad f_4(\mathbf{x}) = \frac{\arctan|\mathbf{x}|}{|\mathbf{x}|^4(1 + |\mathbf{x}|^2)}.$$

For any $i = 1, ..., 4$, determine the values of $p \in [1, +\infty]$ for which $f_i \in L^p(\mathbb{R}^n)$.

Solution - By (2.9):

$$\int_{\mathbb{R}^n} |f_1(\mathbf{x})|^p\, d\mathbf{x} = \sigma_{n-1} \int_0^\infty r^{n-1} \frac{[\ln(1 + r)]^p}{r^{2p}}\, dr = \sigma_{n-1} \int_0^\infty \varphi(r)\, dr;$$

$$r \to 0_+ : \quad \varphi(r) \sim \frac{r^{n-1} r^p}{r^{2p}} = r^{n-p-1} \in L^1(\mathcal{U}(0_+)) \quad \Leftrightarrow \quad n - p > 0$$

$$r \to +\infty : \quad \varphi(r) \sim \frac{r^{n-1} (\ln r)^p}{r^{2p}} \in L^1(\mathcal{U}(+\infty)) \quad \Leftrightarrow \quad n - 2p < 0.$$

Hence, $f_1 \in L^p(\mathbb{R}^n) \Leftrightarrow \frac{n}{2} < p < n$.
By (2.9):

$$\int_{\mathbb{R}^n} |f_2(\mathbf{x})|^p \, d\mathbf{x} = \sigma_{n-1} \int_0^\infty r^{n-1} \frac{|e^{-1/r} - 1|^p}{r^{3p}} \, dr = \sigma_{n-1} \int_0^\infty \psi(r) \, dr \, ;$$

$$r \to 0_+ : \quad \psi(r) \sim r^{n-1-3p} \in L^1(\mathcal{U}(0_+)) \quad \Leftrightarrow \quad n - 3p > 0$$

$$r \to +\infty : \quad \psi(r) \sim r^{n-1-4p} \in L^1(\mathcal{U}(+\infty)) \quad \Leftrightarrow \quad n - 4p < 0.$$

Hence, $f_2 \in L^p(\mathbb{R}^n) \Leftrightarrow \frac{n}{4} < p < \frac{n}{3}$.
By (2.9):

$$f_3 \in L^p(\mathbb{R}^n) \Leftrightarrow \int_0^\infty r^{n-1} \frac{[\ln(1 + r^2)]^p}{r^{3p}} \, dr < +\infty \, ;$$

$$r \to 0_+ : \quad r^{n-1} f_3(r) \sim \frac{r^{n-1} r^{2p}}{r^{3p}} = r^{n-p-1} \in L^1(\mathcal{U}(0_+)) \quad \Leftrightarrow \quad p < n$$

$$r \to +\infty : \quad r^{n-1} f_3(r) \asymp r^{n-1-3p} (\ln r)^p \in L^1(\mathcal{U}(+\infty)) \quad \Leftrightarrow \quad p > \frac{n}{3}.$$

Hence, $f_3 \in L^p(\mathbb{R}^n) \Leftrightarrow \frac{n}{3} < p < n$.
By (2.9):

$$f_4 \in L^p(\mathbb{R}^n) \Leftrightarrow \int_0^\infty r^{n-1} \frac{[\arctan r]^p}{r^{4p}(1 + r^2)^p} \, dr < +\infty \, ;$$

$$r \to 0_+ : \quad r^{n-1} f_4(r) \sim \frac{r^{n-1} r^p}{r^{4p}} = r^{n-3p-1} \in L^1(\mathcal{U}(0_+)) \quad \Leftrightarrow \quad p < \frac{n}{3}$$

$$r \to +\infty : \quad r^{n-1} f_4(r) \asymp \frac{r^{n-1}}{r^{6p}} = r^{n-1-6p} \in L^1(\mathcal{U}(+\infty)) \quad \Leftrightarrow \quad p > \frac{n}{6}.$$

Hence, $f_4 \in L^p(\mathbb{R}^n) \Leftrightarrow \frac{n}{6} < p < \frac{n}{3}$. □

2.2.2 Exercises without detailed solution

Exercise 2.15. Determine the values of $p \in [1, +\infty]$ for which $f \in L^p(0, 1)$ when

$$(a) \; f(x) = \frac{\sin x}{x^{3/2}} \quad (b) \; f(x) = \ln x \quad (c) \; f(x) = \frac{e^{2x} - 1}{x} \quad (d) \; f(x) = \frac{\sin x}{\sqrt[4]{x(1 - x)}}.$$

Solution - (a) $1 \leq p < 2$, (b) $1 \leq p < +\infty$, (c) $1 \leq p \leq +\infty$, (d) $1 \leq p < 4$. □

Exercise 2.16. Determine the values of $p \in [1, +\infty]$ for which $f \in L^p(1, +\infty)$ when

$$(a) \; f(x) = x^3 e^{-x^2} \quad (b) \; f(x) = \sqrt{x} \sin \frac{1}{x} \quad (c) \; f(x) = \frac{\arctan x}{\sqrt[3]{x + 1}} \quad (d) \; f(x) = \frac{1}{\ln(x + 1)}.$$

Solution - (a) $1 \leq p \leq +\infty$, (b) $2 < p \leq +\infty$, (c) $3 < p \leq +\infty$, (d) $p = +\infty$. □

Exercise 2.17. Let $I = (0, 1/2)$, $J = (2, +\infty)$ and

$$g(x) = \frac{1}{\sqrt{x}\,\ln x}, \qquad x \in I \cup J.$$

For which values of $p \geq 1$ we have $g \in L^p(I)$ and $g \in L^p(J)$?

Solution - $g \in L^p(I)$ for every $1 \leq p \leq 2$; $g \in L^p(J)$ for every $2 \leq p \leq +\infty$. □

Exercise 2.18. Let

$$f_k(x) = x^k\, e^{-kx} \qquad x \in (0, +\infty), \quad k \geq 1.$$

Prove that $f_k(x) \to 0$ a.e., in $L^1(0, +\infty)$ and in $L^\infty(0, +\infty)$.

Exercise 2.19. On the interval $(0, +\infty)$ consider the functions:

$$f_k(x) = k^2\, x^2\, e^{-kx} \quad k \geq 1 \quad \text{and} \quad f(x) \equiv 0.$$

Determine for which $p \geq 1$ one has $f_k \in L^p(0, +\infty)$ and if $f_k \to f$ in $L^1(0, +\infty)$ or in $L^\infty(0, +\infty)$, as $k \to +\infty$.

Solution - For any $p \geq 1$; $f_k \to f$ in $L^1(0, +\infty)$ but not in $L^\infty(0, +\infty)$. □

Exercise 2.20. Find a sequence $f_k : \mathbb{R} \to \mathbb{R}$ such that $f_k \to 0$ in $L^\infty(\mathbb{R})$ but not in $L^2(\mathbb{R})$.
Find a sequence $g_k : \mathbb{R} \to \mathbb{R}$ such that $g_k \to 0$ in $L^1(\mathbb{R})$ but not in $L^2(\mathbb{R})$.
Find a sequence $h_k : \mathbb{R} \to \mathbb{R}$ such that $h_k \to 0$ in $L^2(\mathbb{R})$ but not in $L^1(\mathbb{R})$.

Solution - For instance: $f_k(x) = \dfrac{1}{\sqrt{k}}\chi_{[0,k]}(x)$, $g_k(x) = \sqrt{k}\,\chi_{[0,1/k]}(x)$, $h_k(x) = \dfrac{1}{k}\,\chi_{[0,k]}(x)$.
□

Exercise 2.21. For all $k \geq 1$, let

$$f_k(x) = \frac{\sqrt{k}}{1 + (kx)^2} \qquad x \in \mathbb{R}.$$

Compute the pointwise limit f of f_k and determine if $f_k \to f$ in $L^1(\mathbb{R})$ or in $L^2(\mathbb{R})$.

Solution - We have

$$f(x) = \begin{cases} +\infty & \text{if } x = 0, \\ 0 & \text{if } x \neq 0. \end{cases}$$

Moreover, $f_k \to 0$ in L^1 but not in L^2. □

Exercise 2.22. Consider the sequence of functions ($k \geq 1$)

$$f_k(x) = \frac{k}{\sqrt{x}}\left(e^{\frac{1}{k(x+1)}} - 1\right), \qquad x \in (0, +\infty).$$

1) Prove that $f_k \in L^1(0, +\infty)$ for every $k \geq 1$.
2) Find the pointwise limit f of f_k.
3) Prove that Theorem 2.5 may be applied and compute

$$I = \lim_{k \to \infty} \int_0^\infty f_k(x)\, dx.$$

Solution - Hint: prove and use the inequality $e^t - 1 \leq et$, $\forall t \in [0, 1]$. By letting $k \to \infty$ one finds

$$f(x) = \frac{1}{\sqrt{x}(x+1)}, \qquad I = \pi.$$

□

Exercise 2.23. Consider the sequence of functions ($k \geq 1$)

$$f_k(x) = \frac{k \sin \frac{e^x}{k}}{1 + ke^{-kx} + e^{2x}}, \qquad x \in [0, +\infty).$$

1) Find the pointwise limit f of f_k and determine for which $p \geq 1$ one has $f \in L^p(0, +\infty)$.
2) Prove that Theorem 2.5 may be applied and compute

$$I = \lim_{k \to \infty} \int_0^{\infty} f_k(x)\, dx.$$

Solution - $f(x) = \dfrac{e^x}{1 + e^{2x}}$ if $x \neq 0$, $f(0) = 0$, $I = \frac{\pi}{4}$. □

Exercise 2.24. Consider the sequence of functions ($k \geq 1$)

$$f_k(x) = \frac{e^{-k/x}}{x[\ln(1 + kx)]^2}, \qquad x \in (0, +\infty).$$

1) Find the pointwise limit f of f_k.
2) Find the values of $p \geq 1$ for which $f_k \in L^p(0, +\infty)$.
3) Prove that Theorem 2.5 may be applied and compute

$$\lim_{k \to \infty} \int_0^{\infty} f_k(x)\, dx.$$

Solution - $f(x) = 0\ \forall x > 0$, $f_k \in L^p(0, +\infty)\ \forall p \in [1, +\infty]$. □

Exercise 2.25. (i) Prove that if $f : \Omega \to \mathbb{R}$ is measurable and nonnegative we have

$$\operatorname*{ess\,sup}_{\mathbf{x} \in \Omega} f(\mathbf{x}) \leq \sup_{\mathbf{x} \in \Omega} f(\mathbf{x}); \tag{2.15}$$

if, moreover $f(\mathbf{y}) \leq \operatorname*{ess\,sup}_{\mathbf{x} \in \Omega} f(\mathbf{x})$ for every $\mathbf{y} \in \Omega$, then equality holds in (2.15).
(ii) Prove that if $f = \widetilde{f}$ a.e. in Ω then

$$\operatorname*{ess\,sup}_{\mathbf{x} \in \Omega} f = \operatorname*{ess\,sup}_{\mathbf{x} \in \Omega} \widetilde{f}.$$

(iii) Deduce from items (i)-(ii) that if f e g are measurable nonnegative functions defined on Ω then

$$\operatorname*{ess\,sup}_{\mathbf{x} \in \Omega} [f(\mathbf{x}) + g(\mathbf{x})] \leq \operatorname*{ess\,sup}_{\mathbf{x} \in \Omega} f(\mathbf{x}) + \operatorname*{ess\,sup}_{\mathbf{x} \in \Omega} g(\mathbf{x}).$$

(iv) Let $f : \mathbb{R} \to \mathbb{R}$ be defined by

$$f(x) = \begin{cases} x & \text{if } x \in \mathbb{N} \\ 0 & \text{if } x \notin \mathbb{N}\,. \end{cases}$$

Compute $\|f\|_\infty$ and $\sup_{x \in \mathbb{R}} f(x)$.

Solution - (iv) $\|f\|_\infty = 0$, $\sup_{x \in \mathbb{R}} f(x) = +\infty$. □

Exercise 2.26. Let $1 \leq p < q \leq +\infty$ and assume that $\Omega \subseteq \mathbb{R}^n$ is measurable. Prove that $L^q_{loc}(\Omega) \subset L^p_{loc}(\Omega)$. Hint: use (2.4).

Exercise 2.27. Find examples of functions such that:
1) $f \in L^1(\mathbb{R})$ and $f \notin L^p(\mathbb{R})\ \forall p > 1$.

2) $g \in L^\infty(\mathbb{R})$ and $g \notin L^p(\mathbb{R})$ $\forall p \in [1, +\infty)$.
3) For a given $q \in (1, +\infty)$, $w \in L^q(\mathbb{R})$ and $w \notin L^p(\mathbb{R})$ $\forall p \neq q$.

Solution - 1) $u(x) = \dfrac{\chi_{(-1/2,1/2)}(x)}{|x|\,(\ln|x|)^2}$.

2) $$g(x) = \begin{cases} x/e & \text{if } |x| < e\,, \\ \operatorname{sign}(x)/\ln|x| & \text{if } |x| > e\,. \end{cases}$$

3)
$$w(x) = |u(x)|^{1/q} + \frac{\chi_{(1,+\infty)}(x)}{|x|^{1/q}\,(\ln|x|)^{2/q}} \quad \text{where } u \text{ is defined in item 1)}\,.$$

□

Exercise 2.28. For any $k \in \mathbb{N}$ and for every $h \in \mathbb{N}$ such that $h = 0, ..., 2^k - 1$, put

$$f_{2^k+h}(x) = \chi_{[h2^{-k},(h+1)2^{-k}]}(x)\,.$$

Note that as k and h vary as above, the indices $m = 2^k + h$ fully cover $\mathbb{N} \setminus \{0\}$: $2^0 + 0 = 1$, $2^1 + 0 = 2$, $2^1 + 1 = 3$, $2^2 + 0 = 4$, $2^2 + 1 = 5$, $2^2 + 2 = 6$, $2^2 + 3 = 7$, $2^3 + 0 = 8$,...
(i) Prove that $f_m \to 0$ in $L^1(0, 1)$.
(ii) Prove that f_m does not converge pointwise a.e. to 0 in $(0, 1)$.
(iii) Find a subsequence of $\{f_m\}$ which converges pointwise a.e. to 0 in $(0, 1)$.

Solution - (iii) One can take $\{f_{2^k}\}$, with $h = 0$ for every k. □

Exercise 2.29. Consider a sequence $f_k : [a, b] \mapsto \mathbb{R}$ of Riemann-integrable functions on the interval $[a, b] \subset \mathbb{R}$ such that f_k converges uniformly to f in $[a, b]$. Prove that:
1) f is Riemann-integrable,
2) the limit $\lim_{k\to\infty} \int_a^b f_k(x)\,dx = \int_a^b f(x)\,dx$ exists and is finite.

Solution - 1) The uniform convergence implies that for every $\varepsilon > 0$ there exists $k \in \mathbb{N}$ s.t. $|f_k(x) - f(x)| < \varepsilon$ for every $x \in [a, b]$. The Riemann-integrability of f_k ensures that there exists a finite partition of the interval $[a, b]$ such that $S(f_k) - s(f_k) < \varepsilon$, where we denoted by $S(f_k)$ and $s(f_k)$ the upper sums and the lower sums of f_k associated to the considered partition. We infer:

$$\begin{aligned} S(f) - s(f) &\leq |S(f) - S(f_k)| + |S(f_k) - s(f_k)| + |s(f_k) - s(f)| \leq \\ &\leq \varepsilon\,(b-a) + \varepsilon + \varepsilon\,(b-a) = \varepsilon\,\big(1 + 2(b-a)\big)\,. \end{aligned}$$

The statement follows by arbitrariness of ε.

2) $$\left|\int_a^b f_k(x)\,dx - \int_a^b f(x)\,dx\right| = \left|\int_a^b \big(f_k(x) - f(x)\big)\,dx\right| \leq \int_a^b |f_k(x) - f(x)|dx$$

so that the statement follows from uniform convergence since the last term may be estimated with $(b-a)\,\sup_x |f_k(x) - f(x)|$.

Remark - For a partition I_j of a given interval $[a, b]$ we call upper and lower sums of a function w the quantities $S(w) = \sum_j \left(\sup_{I_j} w\right) |I_j|$ and $s(w) = \sum_j \left(\inf_{I_j} w\right) |I_j|$. □

Exercise 2.30. Prove the next statement.

Let f_k be a given a sequence of Riemann-integrable functions $f_k : [a, b] \mapsto \mathbb{R}$.
If
- f_k converges pointwise to f for every $x \in [a, b]$,
- there exists a Riemann-integrable dominating function g such that $|f_k(x)| \leq g(x)$ for every $x \in [a, b]$ and for every k,
- f is Riemann-integrable,

then

$$\lim_{k\to\infty} \int_a^b f_k(x)\,dx = \int_a^b f\,dx\,.$$

Solution - Riemann-integrable functions are also Lebesgue-integrable on the same interval, hence Theorem 2.5 ensures the convergence of the integrals (computed with the Lebesgue theory) which, in this case, coincide with the Riemann integral.

Remark - We have shown that, if we assume that the pointwise limit is Riemann-integrable, we obtain the analogue of Theorem 2.5 for the Riemann integral. However, we point out that the (everywhere) pointwise convergence and the existence of a Riemann-integrable dominating function does not ensure the Riemann-integrability of the limit. On the other hand, the a.e. pointwise convergence and the existence of a Lebesgue-integrable dominating function do ensure the Lebesgue-integrability of the limit and the possibility to switch the order of the limit and integral sign.
The statement, the proof and this Remark also hold in case of functions f, f_k of more variables defined on a (measurable) domain of finite measure. □

Chapter 3
Distributions

Distributions or generalized functions are mathematical objects which differ from the elementary functions in the sense that they cannot be computed pointwise for every value of the independent variable; nevertheless they keep many typical features of functions. The main issue is that one can evaluate the action of any distribution on compactly supported smooth functions (test functions): such evaluation, called *duality pairing* (*pairing* for short; *dualità* in Italian and *crochet* in French), provides an operative and quantitative description of natural phenomena and of devices for instantaneous measurements (snapshots of physical quantities). The pairing of a distribution with a test function has formal properties analogous to the integration of the product of an integrable function times a smooth function. Moreover distributions have all derivatives in a suitable weak sense (distributional derivatives); this property allows to work with distributions as solutions or as given source terms of differential equations: for instance we can study Maxwell's system of equations for electromagnetic field when concentrated charges, dipoles and/or currents in thin conducting structures (wires or metal foils) are present.
Historically, distributions showed up first as a natural description of physical phenomena notably in electrical circuits analysis, subsequently they were given an effective and rigorous theoretical framework by the work of Laurent Schwartz (1915–2002) in the Forties of the XXth Century .

3.1 Definitions and tools

We use the following notation: $\mathbb{R}^n$ is the n-dimensional Euclidean space; any point $\mathbf{x} \in \mathbb{R}^n$ is represented by its coordinates $\mathbf{x} = (x_1, x_2, \cdots, x_n)$ with respect to the canonical basis $\{\mathbf{e}_1, \mathbf{e}_2, \cdots, \mathbf{e}_n\}$; the Euclidean norm of $\mathbf{x} \in \mathbb{R}^n$ is

$$|\mathbf{x}| = (x_1^2 + x_2^2 + \cdots + x_n^2)^{1/2} \qquad \forall \mathbf{x} \in \mathbb{R}^n.$$

$B_R(\mathbf{x})$ denotes the open disk (or ball) with center at $\mathbf{x}$ and radius R:

$$B_R(\mathbf{x}) = \{ \mathbf{y} \in \mathbb{R}^n : |\mathbf{y} - \mathbf{x}| < R \} .$$

In the following Ω denotes a generic nonempty open subset of $\mathbb{R}^n$; the case $\Omega = \mathbb{R}^n$ is allowed. Derivation with respect to the j th variable is denoted by

$$D_j = \frac{\partial}{\partial x_j} \qquad j = 1, \cdots, n .$$

$\mathbb{N}$ denotes the set of nonnegative integers; $\mathbb{N}^n$ denotes the set of multi-indices $\boldsymbol{\alpha} = (\alpha_1, \alpha_2, \cdots, \alpha_n)$ where α_j is a nonnegative integer, $\forall\, j = 1, \ldots, n$.
If $\boldsymbol{\alpha} \in \mathbb{N}^n$ we set

$$\operatorname{ord}(\boldsymbol{\alpha}) = \alpha_1 + \alpha_2 + \cdots + \alpha_n , \qquad \boldsymbol{\alpha}! = \alpha_1! \alpha_2! \cdots \alpha_n! \quad ,$$

$$x^{\boldsymbol{\alpha}} = x_1^{\alpha_1} x_2^{\alpha_2} \cdots x_n^{\alpha_n} , \qquad D_x^{\boldsymbol{\alpha}} = D_1^{\alpha_1} D_2^{\alpha_2} \cdots D_n^{\alpha_n} ,$$

If $\boldsymbol{\beta} \leq \boldsymbol{\alpha}$ (say $\boldsymbol{\alpha}, \boldsymbol{\beta} \in \mathbb{N}^n$ and $\beta_j \leq \alpha_j$ for $j = 1, \cdots, n$) we also set

$$\binom{\boldsymbol{\alpha}}{\boldsymbol{\beta}} = \binom{\alpha_1}{\beta_1} \binom{\alpha_2}{\beta_2} \cdots \binom{\alpha_N}{\beta_N} .$$

The nonnegative integer $\operatorname{ord}(\boldsymbol{\alpha})$ is called *order of the multi-index* $\boldsymbol{\alpha}$, and it is also denoted by $|\boldsymbol{\alpha}|$. Beware avoiding confusion with Euclidean norm: the meaning of $|\cdot|$ is clear by the context, since whenever it refers to a multi-index then it is the order and not the norm, elsewhere $|\cdot|$ is the Euclidean norm.

Example 3.1. If $n = 3$, $\mathbf{x} = (x_1, x_2, x_3)$, $\boldsymbol{\alpha} = \{3, 0, 1\}$, $\boldsymbol{\beta} = \{0, 0, 0\}$ and $v \in C^4(\mathbb{R}^3)$, then

$$|\boldsymbol{\alpha}| = 4,\ \mathbf{x}^{\boldsymbol{\alpha}} = x_1^3 x_3,\ D^{\boldsymbol{\alpha}} v = \frac{\partial^{|\boldsymbol{\alpha}|} v}{\partial \mathbf{x}^{\boldsymbol{\alpha}}} = \frac{\partial^3}{\partial x_1^3} \frac{\partial}{\partial x_3} v,\ |\boldsymbol{\beta}| = 0,\ \mathbf{x}^{\boldsymbol{\beta}} = 1,\ D^{\boldsymbol{\beta}} v = v.$$

3.1.1 Test functions and distributions

Definition 3.2. *The support of a continuous function* v, *whose domain is the open set* $\Omega \subset \mathbb{R}^n$, *is the set:*

$$\operatorname{spt} v := \overline{\{ \mathbf{x} \in \Omega : v(\mathbf{x}) \neq \mathbf{0} \}} \cap \Omega .$$

Definition 3.3. $C^k(\Omega)$ *denotes the space of complex-valued functions that are defined in* Ω *and have continuous derivatives up to the order* k *in* Ω, *where* k *is a nonnegative integer.*
$C^\infty(\Omega)$ *denotes the space of complex-valued functions that are defined in* Ω *and have continuous derivatives of any order in* Ω.
Also the symbol $\mathcal{E}(\Omega)$ *is used to denote the space* $C^\infty(\Omega)$.

Example 3.4. $|x| \in C^0(\mathbb{R}) \setminus C^1(\mathbb{R})$. $(x^+)^2 \in C^1(\mathbb{R}) \setminus C^2(\mathbb{R})$, if $x^+ = \max(x, 0)$. $x^2 + y^2 \in \mathcal{E}(\mathbb{R}^2) = C^\infty(\mathbb{R}^2)$, where $(x, y) = \mathbf{x} \in \mathbb{R}^2$. $|\mathbf{x}| \in C^0(\mathbb{R}^n) \setminus C^1(\mathbb{R}^n)$.

Definition 3.5. ***Space of test functions:*** $\mathcal{D}(\Omega)$.
$C^\infty(\Omega)$ functions whose support is a compact subset of Ω are called test functions.
The vector space of test functions is denoted by $\mathcal{D}(\Omega)$ (or by $C_0^\infty(\Omega)$).

Convergence in the space $\mathcal{D}(\Omega)$ of test functions.
The sequence φ_j is convergent in $\mathcal{D}(\Omega)$ to $\varphi \in \mathcal{D}(\Omega)$
(notation: $\varphi_j \to \varphi$ in $\mathcal{D}(\Omega)$) if:
1) there exists a compact set $K \subset \Omega$: $\operatorname{spt} \varphi_j \subset K \ \forall j$,
2) $D^{\boldsymbol{\alpha}} \varphi_j$ uniformly converge to $D^{\boldsymbol{\alpha}} \varphi$ if $j \to \infty$, $\forall$ multi-index $\boldsymbol{\alpha}$.

Space of distributions: $\mathcal{D}'(\Omega)$.
A *distribution* is an operator u which associates a (real or complex) number $\langle u, \varphi \rangle$ to any test function φ; the value $\langle u, \varphi \rangle$ is called *pairing* (or *duality*) of u and φ, and must fulfil the two following properties:
1) $\langle u, a\varphi + b\psi \rangle = a \langle u, \varphi \rangle + b \langle u, \psi \rangle \qquad \forall a, b \in \mathbb{C}, \ \forall \varphi, \psi \in \mathcal{D}(\Omega)$,
2) $\varphi_j \longrightarrow \varphi$ in $\mathcal{D}(\Omega) \implies \langle u, \varphi_j \rangle \longrightarrow \langle u, \varphi \rangle$.
The above definition can be shortly rephrased as follows:
"a distribution is a linear continuous functional on $\mathcal{D}(\Omega)$".

The set of distributions forms a linear space, denoted by $\mathcal{D}'(\Omega)$, with the following operations of sum and product times a scalar:

$$\begin{aligned} \langle u + v, \varphi \rangle &= \langle u, \varphi \rangle + \langle v, \varphi \rangle, && \forall u, v \in \mathcal{D}'(\Omega), \ \forall \varphi \in \mathcal{D}(\Omega), \\ \langle a\,u, \varphi \rangle &= a \langle u, \varphi \rangle && \forall a \in \mathbb{C}, \ \forall u \in \mathcal{D}'(\Omega), \ \forall \varphi \in \mathcal{D}(\Omega). \end{aligned}$$

Convergence in the space of distributions
A sequence of distributions u_j is convergent in the space of distributions to $u \in \mathcal{D}'(\Omega)$ (notation: $u_j \to u$ in $\mathcal{D}'(\Omega)$) if:

$$\langle u_j, \varphi \rangle \longrightarrow \langle u, \varphi \rangle \qquad \forall \varphi \in \mathcal{D}(\Omega).$$

- Pointwise values of a distribution are meaningless.
- A distribution is known when a rule is given for evaluating its duality with any test function.
- Two distributions u, v coincide if and only if $\langle u, \varphi \rangle = \langle v, \varphi \rangle \ \forall \varphi \in \mathcal{D}(\Omega)$.
- No elementary function is a test function (except the trivial example of the identically vanishing constant function): it follows from uniqueness of analytic extension (Theorem 1.19).

Example 3.6. An example of test function depending on n variables and with radial symmetry is

$$\eta : \mathbb{R}^n \to \mathbb{R} \ \text{ s.t. } \ \eta(\mathbf{x}) = \exp\left(\frac{1}{|\mathbf{x}|^2 - 1}\right) \ \text{if } |\mathbf{x}| < 1, \ \eta(\mathbf{x}) = 0 \text{ if } |\mathbf{x}| \geq 1.$$

The restriction of this function η in a subset $\Omega \subset \mathbb{R}^n$ may be or may not be a test function in Ω, according to whether $\overline{B_1(\mathbf{0})}$ is disjoint or intersects the boundary: $\eta \in \mathcal{D}(\mathbb{R}^n)$ but $\eta \notin \mathcal{D}(B_1(\mathbf{0}))$.

Example 3.7. The main example of distribution in $\mathcal{D}'(\mathbb{R}^n)$ is the *Dirac delta* (P.A.M. Dirac 1902 – 1984), denoted by δ (or by $\delta_{\mathbf{0}}$) and defined as follows:

$$\langle \delta, \varphi \rangle = \varphi(\mathbf{0}) \qquad \forall \varphi \in \mathcal{D}(\mathbb{R}^n).$$

To describe a unitary concentration at a point $\widetilde{\mathbf{x}} \in \mathbb{R}^n$, we set:

$$\delta_{\widetilde{\mathbf{x}}}(\mathbf{x}) := \delta(\mathbf{x} - \widetilde{\mathbf{x}}) \in \mathcal{D}'(\mathbb{R}^n), \qquad \langle \delta_{\widetilde{\mathbf{x}}}, \varphi \rangle = \varphi(\widetilde{\mathbf{x}}) \quad \forall \varphi \in \mathcal{D}(\mathbb{R}^n).$$

Example 3.8. Any locally integrable function $f \in L^1_{loc}(\Omega)$ identifies uniquely a distribution $u_f \in \mathcal{D}'(\Omega)$, via:

$$\langle u_f, \varphi \rangle = \int_\Omega f(\mathbf{x})\varphi(\mathbf{x})\, d\mathbf{x} \qquad \forall \varphi \in \mathcal{D}(\Omega). \tag{3.1}$$

In order to avoid heavy notation, we shortly denote by f also the distribution u_f associated to the function f in (3.1). Identification between f and u_f is expressed precisely by the embedding

$$L^1_{loc}(\Omega) \subset \mathcal{D}'(\Omega). \tag{3.2}$$

We emphasize that not all distributions can be represented by an L^1_{loc} function via the identification (3.1): for instance no L^1_{loc} function can represent the Dirac delta. Anyway the above example clarifies how duality between a distribution and a test function generalizes the integral of a locally integrable function multiplied by a compactly supported smooth function.
For this reason in the following we denote the duality $\langle u, \varphi \rangle$ (between the distribution u and the test function φ) also by the notation

$$\int_\Omega u\,\varphi \qquad \forall u \in \mathcal{D}'(\Omega), \ \forall \varphi \in \mathcal{D}(\Omega);$$

referring to this last notation one must be aware that the symbol $\int$ does not represents an integral (except in particular cases: when u is a distribution associated to an L^1_{loc} function). This representation is very useful. For instance the *change of variables formula* in multiple integrals extends to the duality between distributions and test functions, in the case of bijective affine linear change of coordinates only:

$$\int_{\Omega} u(\mathbb{A}\mathbf{x}+\mathbf{b})\,\varphi(\mathbf{x})\,d\mathbf{x} = \int_{\mathbb{A}(\Omega)+\mathbf{b}} u(\mathbf{y})\,\varphi(\mathbb{A}^{-1}(\mathbf{y}-\mathbf{b}))\,|det\,\mathbb{A}|^{-1}\,d\mathbf{y} \\ \forall \text{ invertible matrix } \mathbb{A},\ \forall\mathbf{b}\in\mathbb{R}^n,\ \forall u\in\mathcal{D}'(\Omega),\ \forall\varphi\in\mathcal{D}(\Omega). \tag{3.3}$$

We notice that expressions like $u(\mathbf{x})$ and $u(\mathbb{A}\mathbf{x}+\mathbf{b})$ are meaningless whenever $\mathbf{x}$ is fixed, since the pointwise value of a distribution is not well defined (not even almost everywhere in Ω). The identity (3.3) has to be understood as the definition of the distribution $u(\mathbb{A}\mathbf{x}+\mathbf{b})$ whenever $u(\mathbf{x})$ is known.
The independent variable of a distribution plays the role of the "integration variable" or "dummy variable", since a distribution is not defined through pointwise values, while it is defined by its duality with all test functions in $\mathcal{D}(\Omega)$. So, the knowledge of a distribution u amounts to the capability of evaluating $\langle u,\varphi\rangle$, $\forall\varphi\in\mathcal{D}(\Omega)$.

Example 3.9. The Dirac δ is a distribution $(-n)$ homogeneous in $\mathbb{R}^n$.
In fact, from (3.3) with the choices $\mathbb{A}=a\,\mathbb{I}_n$, $a\neq 0$ and $\mathbf{b}=\mathbf{0}$ we get

$$\int_{\mathbb{R}^n}\delta(a\,\mathbf{x})\,\varphi(\mathbf{x})\,d\mathbf{x} \;=\; |a|^{-n}\int_{\mathbb{R}^n}\delta(\mathbf{y})\,\varphi(\mathbf{y}/a)\,d\mathbf{y} \;=\; |a|^{-n}\,\varphi(\mathbf{0}) \;=$$

$$= |a|^{-n}\int_{\mathbb{R}^n}\delta(\mathbf{x})\,\varphi(\mathbf{x})\,d\mathbf{x} \qquad \forall\varphi\in\mathcal{D}(\Omega),$$

say:

$$\delta(a\,\mathbf{x}) \;=\; |a|^{-n}\,\delta(\mathbf{x}).$$

Test functions are rare but they are "sufficiently many" for approximating any distribution, as stated precisely in the next Theorem.

Theorem 3.10. *$\mathcal{D}(\Omega)$ is dense in $\mathcal{D}'(\Omega)$, say*

$$\forall u\in\mathcal{D}'(\Omega)\ \exists\ u_n\in\mathcal{D}(\Omega)\ \textit{s.t.}\ u_n\longrightarrow u\ \textit{in}\ \mathcal{D}'(\Omega).$$

We state a necessary and sufficient condition in order that a linear functional on $\mathcal{D}(\Omega)$ is a distribution.

Theorem 3.11. *A linear functional u defined on $\mathcal{D}(\Omega)$ with values in $\mathbb{C}$ is a distribution if and only if for every compact set $K\subset\Omega$ there exist a constant $C=C(K)$ and nonnegative integer $m=m(K)$ s.t.*

$$\left|\int_{\Omega} u\,\varphi\right| \;\le\; C\sum_{|\boldsymbol{\alpha}|\le m}\ \sup_x\,|\,D^{\boldsymbol{\alpha}}\varphi(x)\,| \qquad \forall\varphi\in\mathcal{D}(\Omega):\ \operatorname{spt}\varphi\subset K.$$

We say that the distribution has finite order if the integer m can be chosen independently from K (in this case the least integer m with this property is the order of the distribution), otherwise the distribution has infinite order.
So Theorem 3.11 can be expressed by saying that any distribution locally has finite order: in any case evaluating a duality involves only a finite number of derivatives, according to the support of the test function selected for the evaluation. Zero order distributions are named *measures*.

Example 3.12. The Dirac delta is a measure, in fact:

$$|\langle \delta, \varphi \rangle| = |\varphi(\mathbf{0})| \leq \|\varphi\|_{L^\infty} .$$

Any distribution which is associated to a function $f \in L^1_{loc}$ is a measure since:

$$\left| \int f \varphi \right| \leq \|f\|_{L^1(\operatorname{spt} \varphi)} \|\varphi\|_{L^\infty} .$$

Definition 3.13. *It is possible to define the product of a distribution in Ω times a function $C^\infty(\Omega)$: if $\psi \in C^\infty(\Omega)$ and $u \in \mathcal{D}'(\Omega)$ then*

$(\psi u) \in \mathcal{D}'(\Omega)$ *is the distribution s.t.* $\langle \psi u , \varphi \rangle = \langle u , \psi \varphi \rangle \quad \forall \varphi \in \mathcal{D}(\Omega)$.

Definition 3.14. *The support of a distribution $u \in \mathcal{D}'(\Omega)$, denoted by* $\operatorname{spt} u$, *is the smallest subset $C \subset \Omega$ relatively closed in Ω such that u vanishes identically in the complement, say*

$$\int_\Omega u \varphi = 0 \qquad \forall \varphi : \operatorname{spt} \varphi \cap C = \emptyset .$$

Recall that C is relatively closed in $\Omega \subset \mathbb{R}^n$ if and only if there exists a closed set $D \subset \mathbb{R}^n$ such that $C = D \cap \Omega$.

Though the previous definition may appear unnecessarily complicated with respect to Definition 3.2, actually Definition 3.14 is necessary for a correct notion of support even for $L^1(\Omega)$ functions which are not continuous: for instance one can recall the Dirichlet function which belongs to the equivalence class of the function that vanishes identically (they are equal a.e. in $\mathbb{R}$).

Definition 3.15. ***Compactly supported distributions:*** $\mathcal{E}'(\Omega)$.
The symbol $\mathcal{E}'(\Omega)$ denotes the space of distributions whose support is compact, say it is bounded and has positive distance from the boundary of the open set Ω.

- Any compactly supported distribution has finite order.
- The Dirac δ has compact support: $\delta \in \mathcal{E}'(\mathbb{R}^n)$, $\operatorname{spt} \delta = \{\mathbf{0}\}$.
- A distribution has compact support if and only if it can be extended to a linear continuous functional on $\mathcal{E}$; in such case duality acts between $\mathcal{E}'(\Omega)$ and $\mathcal{E}(\Omega)$; moreover, for any $v \in \mathcal{E}'(\Omega)$ we have:

1) $\langle v, a\varphi + b\psi \rangle = a \langle v, \varphi \rangle + b \langle v, \psi \rangle \qquad \forall a, b \in \mathbb{C},\ \forall \varphi, \psi \in \mathcal{E}(\Omega)$,
2) $\varphi_j \longrightarrow \varphi$ in $\mathcal{E}(\Omega) \implies \langle v, \varphi_j \rangle \longrightarrow \langle v, \varphi \rangle$.

In 2) and in the sequel: the convergence $\varphi_j \longrightarrow \varphi$ in $\mathcal{E}(\Omega)$ stands for $D^{\boldsymbol{\alpha}} \varphi_j \to D^{\boldsymbol{\alpha}} \varphi$ uniformly in any compact subset of Ω , $\forall$ multi-index $\boldsymbol{\alpha}$.

Remark that a sequence of C^∞ functions may be convergent in $\mathcal{E}(\Omega)$ and not uniformly convergent in Ω : e.g., $\Omega = \mathbb{R}$, $\varphi_j(x) = x/j$, $\varphi(x) \equiv 0$.
Here we quote the definition of a space of smooth functions in $\mathbb{R}^n$ fast decaying at infinity and the one of its dual space: a proper subspace of $\mathcal{D}'(\mathbb{R}^n)$. They both play an important role in the applications of the Fourier transform.

Definition 3.16. *(**Schwartz space** $\mathcal{S}(\mathbb{R}^n)$ **,** or **space of fast decaying smooth functions**).*
A function ψ of n real variables belongs to $\mathcal{S}(\mathbb{R}^n)$ iff $\psi \in C^\infty$ and

$$P(\mathbf{x})\, D^\alpha \psi \;\in\; L^2(\mathbb{R}^n) \qquad \forall \text{ polinomial } P,\ \forall \text{ multi-index } \alpha. \tag{3.4}$$

A sequence ψ_m of functions in $\mathcal{S}(\mathbb{R}^n)$ converges in $\mathcal{S}(\mathbb{R}^n)$ to the function $\psi \in \mathcal{S}(\mathbb{R}^n)$ iff

$$|\mathbf{x}|^k D^\alpha \psi_m \text{ converges to } |\mathbf{x}|^k D^\alpha \psi \text{ in } L^2(\mathbb{R}^n) \qquad \forall k \in \mathbb{N}\ \forall \alpha \in \mathbb{N}^n .$$

Definition 3.17. *(**Tempered distributions:** space $\mathcal{S}'(\mathbb{R}^n)$)*
*The symbol $\mathcal{S}'(\mathbb{R}^n)$ denotes the space of **tempered distributions** , which is the linear space of all distributions whose action as linear functional can be extended in order to act on $\mathcal{S}(\mathbb{R}^n)$ and continuously with respect to the convergence in the space $\mathcal{S}(\mathbb{R}^n)$.*
A sequence u_m of distributions in $\mathcal{S}'(\mathbb{R}^n)$ converges to the tempered distribution $u \in \mathcal{S}'(\mathbb{R}^n)$ in $\mathcal{S}'(\mathbb{R}^n)$ iff

$$\int_{\mathbb{R}^n} u_m\, v \;\to\; \int_{\mathbb{R}^n} u\, v \qquad \forall v \in \mathcal{S}(\mathbb{R}^n) .$$

The following proper embeddings hold true:

$$\mathcal{D}(\mathbb{R}^n) \subset \mathcal{S}(\mathbb{R}^n) \subset \mathcal{E}(\mathbb{R}^n) , \tag{3.5}$$

$$\mathcal{E}'(\mathbb{R}^n) \subset \mathcal{S}'(\mathbb{R}^n) \subset \mathcal{D}'(\mathbb{R}^n) . \tag{3.6}$$

Moreover, the following convergence relationships hold true:

$$\psi_m \to \psi \text{ in } \mathcal{D}(\mathbb{R}^n) \;\Rightarrow\; \psi_m \to \psi \text{ in } \mathcal{S}(\mathbb{R}^n) \;\Rightarrow\; \psi_m \to \psi \text{ in } \mathcal{E}(\mathbb{R}^n) , \tag{3.7}$$

$$u_m \to u \text{ in } \mathcal{E}'(\mathbb{R}^n) \;\Rightarrow\; u_m \to u \text{ in } \mathcal{S}'(\mathbb{R}^n) \;\Rightarrow\; u_m \to u \text{ in } \mathcal{D}'(\mathbb{R}^n) . \tag{3.8}$$

3.1.2 Distributional derivatives

The derivatives of a distribution are defined in such a way that they always exist, are distributions themselves and coincide with the classical derivatives (defined as limits of difference quotients), whenever the latter exist.

Definition 3.18. *If $u \in \mathcal{D}'(\mathbb{R})$ then the distributional derivative of u is the distribution $u' \in \mathcal{D}'(\mathbb{R})$ uniquely defined by*

$$\int_{\mathbb{R}} u'\,\varphi = -\int_{\mathbb{R}} u\,\varphi' \qquad \forall \varphi \in \mathcal{D}(\mathbb{R})\,. \tag{3.9}$$

The distribution u' is also denoted by the symbol $\dfrac{du}{dx}$.

Definition 3.19. *If $\Omega \subset \mathbb{R}^n$, $n \geq 1$ and $u \in \mathcal{D}'(\Omega)$, then the distributional derivative of u with respect to the variable x_j (j th component of $\mathbf{x}$) is the distribution $\dfrac{\partial u}{\partial x_j} \in \mathcal{D}'(\Omega)$ uniquely defined by*

$$\int_{\Omega} \frac{\partial u}{\partial x_j}\,\varphi \;=\; -\int_{\Omega} u\,\frac{\partial \varphi}{\partial x_j} \qquad \forall \varphi \in \mathcal{D}(\Omega)\,. \tag{3.10}$$

The distributional derivative $D^{\boldsymbol{\alpha}} u = \dfrac{\partial^{|\boldsymbol{\alpha}|} u}{\partial \mathbf{x}^{\boldsymbol{\alpha}}} \in \mathcal{D}'(\Omega)$ is defined for any multi-index $\boldsymbol{\alpha}$, by:

$$\int_{\Omega} D^{\boldsymbol{\alpha}} u\,\varphi \;=\; (-1)^{|\boldsymbol{\alpha}|} \int_{\Omega} u\; D^{\boldsymbol{\alpha}} \varphi \qquad \forall \varphi \in \mathcal{D}(\Omega)\,. \tag{3.11}$$

• If a classical derivative exists and is continuous at every point in Ω then it coincides with the corresponding distributional derivative.
• If $f \in C^k$ and $|\boldsymbol{\alpha}| \leq k$, then, by referring to the Example 3.8, we have

$$D^{\boldsymbol{\alpha}}\, u_f \;=\; u_{D^{\boldsymbol{\alpha}} f}\;.$$

• Every distribution has derivatives of any order.
• The distributional derivative is continuous with respect to the convergence in $\mathcal{D}'(\Omega)$:

$$u_k \to u \;\text{ in } \mathcal{D}'(\Omega) \quad \Longrightarrow \quad D^{\boldsymbol{\alpha}} u_k \to D^{\boldsymbol{\alpha}} u \;\text{ in } \mathcal{D}'(\Omega), \quad \forall \boldsymbol{\alpha} \in \mathbb{N}^n.$$

• The *Leibniz formula* about derivatives of a product holds true also in $\mathcal{D}'(\Omega)$:

$$D^{\boldsymbol{\alpha}}(vu) = \sum_{\boldsymbol{\beta} \leq \boldsymbol{\alpha}} \binom{\boldsymbol{\alpha}}{\boldsymbol{\beta}} (D^{\boldsymbol{\beta}} v)(D^{\boldsymbol{\alpha}-\boldsymbol{\beta}} u) \;\; \forall\, v \in \mathcal{E}(\Omega),\; u \in \mathcal{D}'(\Omega),\; \boldsymbol{\alpha} \in \mathbb{N}^n \tag{3.12}$$

this is a consequence of the following computations: $\forall \varphi \in \mathcal{D}(\Omega), \forall j$,

$$\int_\Omega \frac{\partial}{\partial x_j}(vu)\,\varphi = -\int_\Omega (vu)\,\frac{\partial \varphi}{\partial x_j} = -\int_\Omega u\left(v\frac{\partial \varphi}{\partial x_j}\right) =$$
$$= -\int_\Omega u\,\frac{\partial}{\partial x_j}(v\varphi) + \int_\Omega u\left(\frac{\partial v}{\partial x_j}\varphi\right) =$$
$$= \int_\Omega \left(v\frac{\partial u}{\partial x_j}\right)\varphi + \int_\Omega \left(\frac{\partial v}{\partial x_j}u\right)\varphi .$$

• The *Schwarz Theorem* (about equality of mixed derivatives) holds true (without any additional assumption!):

$$\frac{\partial}{\partial x_j}\frac{\partial}{\partial x_k}u = \frac{\partial}{\partial x_k}\frac{\partial}{\partial x_j}u \qquad \forall u \in \mathcal{D}'(\Omega),\ \forall j,k = 1,\dots,n. \tag{3.13}$$

Example 3.20. The Heaviside function H is defined as follows:

$$H(t) = \begin{cases} 1 & \text{if } t \ge 0, \\ 0 & \text{if } t < 0. \end{cases}$$

The distributional derivative of the Heaviside function is the Dirac delta:

$$H' = \delta .$$

Indeed, $\forall \varphi \in \mathcal{D}(\mathbb{R})$ the following equalities hold true:

$$\int_{\mathbb{R}} H'\,\varphi = -\int_{\mathbb{R}} H(t)\,\varphi'(t)\,dt = -\int_0^{+\infty}\varphi'(t)\,dt = \varphi(0) = \int_{\mathbb{R}}\delta\,\varphi .$$

By computing the first and second distributional derivative of H we find a precise formalism to deal with some important concepts of Physics.

Example 3.21. Derivatives of the Dirac δ :

$$\int_{\mathbb{R}} \delta'\,\varphi = -\int_{\mathbb{R}}\delta\,\varphi' = -\varphi'(0),$$

$$\int_{\mathbb{R}} \delta^{(k)}\,\varphi = (-1)^{(k)}\int_{\mathbb{R}}\delta\,\varphi^{(k)} = (-1)^{(k)}\,\varphi^{(k)}(0) \quad \forall k \in \mathbb{N}.$$

The distributions δ and $(-\delta')$ have relevant interpretations in Electrostatic: respectively *concentrated unitary electrical charge* and *unitary moment electric dipole*. The electric dipole describes the limit of a system of two electric charges of intensity $\frac{1}{2\varepsilon}$ and $-\frac{1}{2\varepsilon}$ and located at points of abscissa ε and $-\varepsilon$: notice that this system has unitary moment (independently of ε), moreover, the two-charges system converges in the space of distributions, as $\varepsilon \to 0_+$, to a limit distribution which provides a rigorous definition of an electric dipole:

$$\frac{1}{2\varepsilon}\left(\delta_\varepsilon(x) - \delta_{-\varepsilon}(x)\right) = \frac{1}{2\varepsilon}\left(\delta(x-\varepsilon) - \delta(x+\varepsilon)\right) \to -\delta'(x).$$

The above relationship is justified by:

$$\langle \frac{1}{2\varepsilon}\left(\delta_\varepsilon(x) - \delta_{-\varepsilon}(x)\right),\, \varphi\rangle = \frac{1}{2\varepsilon}\left(\varphi(\varepsilon) - \varphi(-\varepsilon)\right) \to \varphi'(0) = -\langle \delta', \varphi\rangle .$$

Actually the three-dimensional physical case, of a dipole which is oriented along the direction $\mathbf{e}_j$, is described by the following distributional limit:

$$\frac{1}{2\varepsilon}\left(\delta(\mathbf{x}-\varepsilon\mathbf{e}_j)-\delta(\mathbf{x}+\varepsilon\mathbf{e}_j)\right) \to -\frac{\partial}{\partial x_j}\,\delta(\mathbf{x}).$$

Theorem 3.22. ***(Primitive of a distribution)*** *For any given distribution $f \in \mathcal{D}'(\mathbb{R})$, the differential equation*

$$u \in \mathcal{D}'(\mathbb{R}): \qquad u' = f \qquad \text{in } \mathcal{D}'(\mathbb{R})$$

has infinitely many solutions. Any two solutions differ by a constant.

Theorem 3.23. ***(Potential of a conservative field)*** *Assume that n distributions $f_j \in \mathcal{D}'(\mathbb{R})$, $j = 1, \dots, n$, are given.*
Then there exists a potential u of the field $\mathbf{f} := (f_1, \dots, f_n)$ in $\mathbb{R}^n$, say

$$\exists u \in \mathcal{D}'(\mathbb{R}^n) \ \ s.t.\ \nabla_{\mathbf{x}} u = \mathbf{f} \qquad in\ \mathcal{D}'(\mathbb{R}^n),$$

if and only if the following conditions hold true

$$\frac{\partial f_i}{\partial x_j} = \frac{\partial f_j}{\partial x_i} \qquad \forall i, j = 1, \dots, n. \tag{3.14}$$

If (3.14) is verified, then there are infinitely many potentials and they differ by a constant.

Division problems in $\mathcal{D}'(\mathbb{R})$. Assume $n = 1$ and $m \in \mathbb{N}$.
Then a distribution $u \in \mathcal{D}'(\mathbb{R})$ fulfils

$$x^m\, u = 0 \quad \text{in } \mathcal{D}'(\mathbb{R}) \tag{3.15}$$

if and only if, for suitable constants c_h,

$$u = \sum_{h=0}^{m-1} c_h\, \delta^{(h)} \qquad \text{(finite sum)}. \tag{3.16}$$

Division problem in $\mathcal{D}'(\mathbb{R}^n)$.
Every distribution whose support is contained in a single point $\widetilde{\mathbf{x}} \in \Omega \subset \mathbb{R}^n$ is a (finite) linear combination of the Dirac delta $\delta_{\widetilde{\mathbf{x}}} = \delta(\mathbf{x}-\widetilde{\mathbf{x}})$ concentrated at the same point and of its partial derivatives:

$$\sum_{\boldsymbol{\alpha}} c_{\boldsymbol{\alpha}}\, D^{\boldsymbol{\alpha}}\, \delta_{\widetilde{\mathbf{x}}} \qquad \text{(finite sum)}. \tag{3.17}$$

The previous two statements allow to deal with several division problems that appear when looking for solutions of differential equations with the Fourier transform method (Chapter 4).

3.1.3 Convolution

First we consider the convolution product of two integrable functions:

$$u * v\,(\mathbf{x}) := \int_{\mathbb{R}^n} u(\mathbf{x}-\mathbf{y})v(\mathbf{y})\,d\mathbf{y} = \int_{\mathbb{R}^n} v(\mathbf{x}-\mathbf{y})u(\mathbf{y})\,d\mathbf{y}\,.$$

If $u, v \in L^1(\mathbb{R}^n)$ then $u * v \in L^1(\mathbb{R}^n)$.
If $u \in L^1_{loc}(\mathbb{R}^n)$ and $v \in L^1(\mathbb{R}^n)$ has compact support, then $u * v \in L^1_{loc}(\mathbb{R}^n)$.

Theorem 3.24. *Assume* $u \in L^1_{loc}(\mathbb{R}^n)$, $v \in L^1(\mathbb{R}^n)$ *and* $\mathrm{spt}(v)$ *is compact. If* $v \in C^k(\mathbb{R}^n)$ *then*

$$u * v \in C^k(\mathbb{R}^n), \qquad D^{\boldsymbol{\alpha}}(u * v) = u * D^{\boldsymbol{\alpha}} v \quad \forall \boldsymbol{\alpha}\ s.t.\ \mathrm{ord}(\boldsymbol{\alpha}) \leq k.$$

If in addition $v \in \mathcal{E}(\mathbb{R}^n)$ *then* $u * v \in \mathcal{E}(\mathbb{R}^n)$.
We emphasize that the assumptions on u *and* v *can be exchanged.*

Definition 3.25. ***(Convolution of a distribution times a function)*** *It is defined whenever at least one of the factors is compactly supported: assume* $\varphi \in \mathcal{D}(\mathbb{R}^n)$ *and* $u \in \mathcal{D}'(\mathbb{R}^n)$, *then the function*

$$\mathbf{y} \mapsto \varphi(\mathbf{x}-\mathbf{y}) \quad \text{belongs to } \mathcal{D}(\mathbb{R}^n)\,,$$

and the function $u * \varphi$ *is defined at every point* $\mathbf{x} \in \mathbb{R}^n$ *by*

$$u * \varphi\ (\mathbf{x}) = \int_{\mathbb{R}^n} u(\mathbf{y})\,\varphi(\mathbf{x}-\mathbf{y})\,d\mathbf{y}\,,$$

this function is called convolution of u *and* φ.
This definition can be extended also to the cases:

- $\varphi \in \mathcal{E}(\mathbb{R}^n)$ *and* $u \in \mathcal{E}'(\mathbb{R}^n)$,
- $\varphi \in \mathcal{S}(\mathbb{R}^n)$ *and* $u \in \mathcal{S}'(\mathbb{R}^n)$.

Example 3.26. The following identity holds true: $\delta * \varphi = \varphi\,, \forall\, \varphi \in \mathcal{D}(\mathbb{R}^n)$.
In fact, $\delta(\mathbf{y}) = \delta(-\mathbf{y})$, entails $\forall \varphi \in \mathcal{D}(\mathbb{R}^n)$:

$$(\delta * \varphi)(\mathbf{x}) = \int_{\mathbb{R}^n} \delta(\mathbf{x}-\mathbf{y})\,\varphi(\mathbf{y})\,d\mathbf{y} = \int_{\mathbb{R}^n} \delta(\mathbf{y}-\mathbf{x})\,\varphi(\mathbf{y})\,d\mathbf{y} = \varphi(\mathbf{x})\,.$$

Theorem 3.27. *Assume* $u \in \mathcal{D}'(\mathbb{R}^n)$, $v \in \mathcal{E}(\mathbb{R}^n)$ *and at least one is compactly supported (i.e.* $u \in \mathcal{E}'(\mathbb{R}^n)$ *or* $\varphi \in \mathcal{D}(\mathbb{R}^n)$*), then* $u * v \in \mathcal{E}(\mathbb{R}^n)$ *and*

$$D^{\boldsymbol{\alpha}}(u * v) = (D^{\boldsymbol{\alpha}} u) * v = u * (D^{\boldsymbol{\alpha}} v) \qquad \forall \alpha \in \mathbb{N}^n.$$

The convolution product of two distributions is based on the identity recalled by the next Lemma, which clarifies the action of duality with the convolution product of two functions.

Lemma 3.28 *$\forall u \in \mathcal{D}'(\mathbb{R}^n)$, $\forall \varphi, \psi \in \mathcal{E}(\mathbb{R}^n)$ such that at least two among them are compactly supported, we have*

$$\int_{\mathbb{R}^n} (u * \varphi)\, \psi \;=\; \int_{\mathbb{R}^n} u\, (\check{\varphi} * \psi) \qquad \textit{where } \check{\varphi}(\mathbf{x}) := \varphi(-\mathbf{x})\,.$$

Definition 3.29. ***(Convolution product of distributions)***
Given $u,\, v \in \mathcal{D}'(\mathbb{R}^n)$ such that at least one is compactly supported, we define the convolution $u * v$ by

$$\int_{\mathbb{R}^n} (u*v)\, \psi = \int_{\mathbb{R}^n} u\, (\check{v}*\psi) \qquad \forall \psi \in \mathcal{D}(\mathbb{R}^n) \qquad \textit{where } \check{v}(\mathbf{x}) := v(-\mathbf{x})\,.$$

In this way $u * v$ is a distribution in $\mathcal{D}'(\mathbb{R}^n)$ which fulfils

$$\operatorname{spt}(u * v) \subset \overline{\operatorname{spt}(u) + \operatorname{spt}(v)}\,, \tag{3.18}$$

$$D^{\boldsymbol{\alpha}}(u * v) = (D^{\boldsymbol{\alpha}}u) * v = u * (D^{\boldsymbol{\alpha}}v) \qquad \forall \boldsymbol{\alpha}\,. \tag{3.19}$$

We emphasize that, if both u and v have compact support then $v * \psi$ has compact support too.

A remarkable identity is the following one:

$$u * \delta = u = \delta * u \quad \forall u \in \mathcal{D}'(\mathbb{R}^n)\,. \tag{3.20}$$

The convolution is continuous with respect to balanced distributional convergences, for instance:

$$\begin{array}{lll}
u_k \to u \;\; \mathcal{D}'(\mathbb{R}^n),\;\; v_k \to v \;\; \mathcal{E}'(\mathbb{R}^n)\,, & \Rightarrow & u_k * v_k \to u * v \;\; \mathcal{D}'(\mathbb{R}^n)\,; \\
u_k \to u \;\; \mathcal{E}'(\mathbb{R}^n),\;\; v_k \to v \;\; \mathcal{E}'(\mathbb{R}^n)\,, & \Rightarrow & u_k * v_k \to u * v \;\; \mathcal{E}'(\mathbb{R}^n)\,; \\
u_k \to u \;\; \mathcal{D}'(\mathbb{R}^n),\;\; v_k \to v \;\; \mathcal{D}(\mathbb{R}^n)\,, & \Rightarrow & u_k * v_k \to u * v \;\; \mathcal{E}(\mathbb{R}^n)\,; \\
u_k \to u \;\; \mathcal{E}'(\mathbb{R}^n),\;\; v_k \to v \;\; \mathcal{D}(\mathbb{R}^n)\,, & \Rightarrow & u_k * v_k \to u * v \;\; \mathcal{D}(\mathbb{R}^n)\,.
\end{array}$$

The convolution of distributions is well posed when at least one factor is compactly supported. Nevertheless, when dealing with distributions of one single variable, this assumption can be weakened by substituting it with the request that both factors have support contained in $[0, +\infty]$ (see the definitions in (6.19) e (6.37)): in such case the ensuing convolution has support contained in $[0, +\infty]$ too and its derivatives fulfil (3.19).

3.1.4 Fundamental solution of a differential operator

Let L be a linear differential operator with constant coefficients:

$$L = \sum_{|\boldsymbol{\alpha}|\leq m} C_{\boldsymbol{\alpha}} D^{\boldsymbol{\alpha}} \qquad \text{with } m \in \mathbb{N}\,,\ C_{\boldsymbol{\alpha}} \in \mathbb{C}\,.$$

A distribution $U \in \mathcal{D}'(\mathbb{R}^n)$ which fulfils

$$LU \;=\; \delta \qquad \text{in } \mathcal{D}'(\mathbb{R}^n) \tag{3.21}$$

is called *fundamental solution* of the operator L in $\mathbb{R}^n$.

Given $f \in \mathcal{E}'(\mathbb{R}^n)$, we study the differential equation in the unknown u :

$$u \in \mathcal{D}'(\mathbb{R}^n): \qquad L\,u \;=\; f \;\text{ in } \mathcal{D}'(\mathbb{R}^n). \tag{3.22}$$

If U is fundamental solution of the operator L in $\mathbb{R}^n$, then a solution of the problem (3.22) can be obtained by

$$u \;=\; U * f\,. \tag{3.23}$$

In fact

$$L(U * f) \;=\; (LU) * f \;=\; \delta * f \;=\; f\,.$$

Some relevant examples of linear differential operators with constant coefficients are:

the **Laplacian**:	$\displaystyle\sum_{i=1}^{n} \frac{\partial^2}{\partial x_i^2}$	(symbol $\Delta_{\mathbf{x}}$) ,
the **heat operator**:	$\partial_t \,-\, \Delta_{\mathbf{x}}$,	
the **wave operator** or **d'Alembertian**:	$\partial_{tt} \,-\, c^2\,\Delta_{\mathbf{x}}$	(symbol $\Box_c$) .

The constant $c > 0$ in the wave operator denotes the speed of propagation.

Integral transforms (Chapters 4 and 6) are efficient tools to find the fundamental solution of linear operators. We list some relevant results below; for their proofs we refer to Exercises 3.20, 3.30, 3.31, 4.30, 4.49, 4.54, 4.57, 6.31 and 6.32.

The fundamental solution of $L = I - d^2/dt^2$ in $\mathcal{D}'(\mathbb{R})$ is: $U = \dfrac{1}{2}\, e^{-|t|}$.

The *fundamental solution of the Laplacian* $\Delta_{\mathbf{x}}$ *in* $\mathbb{R}^n$ *is:*

$$U(\mathbf{x}) = \frac{1}{2}\,|\mathbf{x}| \qquad n = 1 \qquad (3.24)$$

$$U(\mathbf{x}) = \frac{1}{2\pi}\,\ln|\mathbf{x}| \qquad n = 2 \qquad (3.25)$$

$$U(\mathbf{x}) = -\frac{1}{(n-2)\,\sigma_{n-1}}\,\frac{1}{|\mathbf{x}|^{n-2}} \qquad n \geq 3 \qquad (3.26)$$

where σ_{n-1} is the $(n-1)$ dimensional measure of $S^{n-1} = \partial B_1(\mathbf{0})$:

$$\sigma_{n-1} = |\,\partial B_1\,|\,, \qquad \omega_n = |\,B_1\,|\,, \qquad \sigma_{n-1} = n\,\omega_n\,,$$

for the computation of values ω_n and σ_{n-1} see (2.13) e (2.14).

In particular the *fundamental solution of the Laplacian* in the 3D space is

$$U(\mathbf{x}) = -\frac{1}{4\pi}\,\frac{1}{|\mathbf{x}|} \qquad n = 3\,. \qquad (3.27)$$

The function U defined by (3.27) represents: in Mechanics, the Newtonian gravitational potential associated to a material point of unitary density set at the origin; in Electrostatics, the Coulomb potential associated to a unitary electric charge concentrated at the origin.

The *fundamental solution of the heat operator*, in $\mathbb{R}^n_{\mathbf{x}} \times \mathbb{R}_t$, is

$$U(\mathbf{x},t) = (4\pi t)^{-\frac{n}{2}}\,\exp\left(-\frac{|\mathbf{x}|^2}{4t}\right)\,H(t)$$

The *fundamental solution of the wave operator* $\Box_c$, in $\mathbb{R}^n_{\mathbf{x}} \times \mathbb{R}_t$, is

$$U(\mathbf{x},t) = \frac{1}{2c}\,H(ct - |\mathbf{x}|) \qquad n = 1 \qquad (3.28)$$

$$U(\mathbf{x},t) = \frac{1}{2\pi c\,\sqrt{c^2\,t^2 - |\mathbf{x}|^2}}\,H(ct - |\mathbf{x}|) \qquad n = 2 \qquad (3.29)$$

$$U(\mathbf{x},t) = \frac{1}{4\pi\,c^2\,t}\,\delta_{\{|\mathbf{x}| = c\,t\}} \qquad n = 3 \qquad (3.30)$$

Explicitly, in the case of three space variables, the fundamental solution of the wave operator $\Box_c$ in $\mathbb{R}^3_{\mathbf{x}} \times \mathbb{R}_t$ verifies, $\forall \varphi \in \mathcal{D}(\mathbb{R}^3_{\mathbf{x}} \times \mathbb{R}_t)$:

$$\int_{\mathbb{R}^3_{\mathbf{x}} \times \mathbb{R}_t} U\varphi \, d\mathbf{x}dt \;=\; \begin{cases} 0 & \text{if } t \leq 0, \\ t \displaystyle⨍_{\partial B_{ct}(\mathbf{0})} \varphi(\mathbf{x},t)\; d\sigma(\mathbf{x}) & \text{if } t > 0\,, \end{cases}$$

where $⨍_{\partial B_{ct}(\mathbf{0})} \varphi \, d\sigma \;=\; \int_{\partial B_{ct}(\mathbf{0})} \varphi \, d\sigma / (4\pi c^2 t^2)\,.$

Remark 3.30. (Fundamental solution concentrated at a generic point y in $\mathbb{R}^n$)
Often the study of a problem similar to (3.21) is important, say:

$$L\,K \;=\; \delta_{\mathbf{y}} \qquad \text{in } \mathcal{D}'(\mathbb{R}^n) \tag{3.31}$$

where $\mathbf{y} \in \mathbb{R}^n$ is fixed and $\delta_{\mathbf{y}}(\mathbf{x}) = \delta(\mathbf{x} - \mathbf{y})\,.$
In some context also the distribution K is called *fundamental solution of the operator* L, more precisely *fundamental solution focussed at* $\mathbf{y}$ or *fundamental solution with pole at* y. It is important to avoid confusion between K and the fundamental solution U concentrated at the origin of $\mathbb{R}^n$:

$$K(\mathbf{x}, \mathbf{y}) = U(\mathbf{x} - \mathbf{y}) \qquad\qquad \forall \mathbf{x},\, \mathbf{y} \in \mathbb{R}^n \tag{3.32}$$

The focussed fundamental solution K is effective in the solution of linear differential equations of the kind (3.22). In fact, if U is a function L^1_{loc} (hence the convolution (3.23) is expressed by an integral) then K coincides with the kernel appearing in the convolution integral in (3.23).
Explicitly, the solution of the problem

$$u \in \mathcal{D}'(\mathbb{R}^n): \quad Lu = f$$

is given by

$$u(\mathbf{x}) \;=\; U{*}f\,(\mathbf{x}) \;=\; \int_{\mathbb{R}^n} U(\mathbf{x} - \mathbf{y})\, f(\mathbf{y})\, d\mathbf{y} \;=\; \int_{\mathbb{R}^n} K(\mathbf{x}, \mathbf{y})\, f(\mathbf{y})\, d\mathbf{y}\,. \tag{3.33}$$

In Chapter 4 (which deals with the Fourier transform) we will make use of the representation U of the fundamental solution, defined in (3.21). Also in Chapters 7 and 8 (concerning differential equations which govern transport and diffusion phenomena) we will make use of the fundamental solution, but exploiting the representation K, defined in (3.31).

3.2 Solved exercises

Notation - In all the exercises and solutions of this section, the following equivalent notations

$$\langle u, \varphi \rangle \qquad \text{and} \qquad \int_{\Omega} u(\mathbf{x})\,\varphi(\mathbf{x})\,d\mathbf{x}$$

are indifferently used to denote the pairing of $u \in \mathcal{D}'(\Omega)$ and $\varphi \in \mathcal{D}(\Omega)$ (or the pairing of $u \in \mathcal{E}'(\Omega)$ and $\varphi \in \mathcal{E}(\Omega)$), according to the context and the convenience in the computations. The exercises are collected in two subsections of this chapter: Subsection 3.2.1 refers to distributions of one single variable, Subsection 3.2.2 refers to distributions of several variables. Additional exercises concerning distributions, which are solved by using Fourier and Laplace transforms, are proposed in different chapters: see Subsections 4.2.2, 4.2.3 and 6.2.3.

3.2.1 Exercises about distributions of one variable

Exercise 3.1. Prove that $\delta \in \mathcal{D}'(\mathbb{R}^n)$ is an even distribution, say δ fulfils the identity

$$\check{\delta} = \delta \qquad \text{in } \mathcal{D}'(\mathbb{R}^n)\,.$$

Solution - By definition $\check{\delta}(\mathbf{x}) = \delta(-\mathbf{x})$. We use the change of variables formula (3.3) with $\mathbb{A} = -\mathbb{I}_n$. We get $|\det \mathbb{A}| = 1$ and

$$\int_{\mathbb{R}^n} \check{\delta}\,\varphi = \int_{\mathbb{R}^n} \delta\,\check{\varphi} = \varphi(-\mathbf{x})|_{\mathbf{x}=\mathbf{0}} = \varphi(\mathbf{0}) = \int_{\mathbb{R}^n} \delta\,\varphi \qquad \forall \varphi \in \mathcal{D}(\mathbb{R}^n)\,.$$

□

Exercise 3.2. Prove the following identities in $\mathcal{D}'(\mathbb{R})$,

1) $x\,\delta(x) \;=\; 0$,
2) $x\,\delta'(x) \;=\; -\,\delta(x)$,
3) $x\,\delta^{(k+1)}(x) \;=\; -\,(k+1)\,\delta^{(k)}(x)$,

Solution - For any $\varphi \in \mathcal{D}(\mathbb{R})$ we have
$\langle x\delta, \varphi\rangle = \langle \delta, x\varphi\rangle = 0\,\varphi(0) = 0$; $\langle x\delta', \varphi\rangle = \langle \delta', x\varphi\rangle = -\langle \delta, (x\varphi)'\rangle = -\langle \delta, \varphi + x\varphi'\rangle = -\varphi(0)$; hence 1) and 2) follow.
Alternatively, 2) can be deduced from 1) by differentiating $x\delta$ and exploiting the Leibniz formula (3.12): $0 = (d/dx)0 = (x\delta)' = \delta + x\delta'$.
Identity 3) can be verified by applying the induction principle, in fact: if $k = 0$ then 3) becomes 2); by assuming that 3) holds true for a fixed value $k \in \mathbb{N}$ and differentiating we get 3) with k substituted by $k+1$. □

Exercise 3.3. If $\delta(x)$ denotes the Dirac delta in one variable, evaluate

1) $\langle \delta(x-1), \varphi(x)\rangle$,
2) $\langle \delta(x+\pi), \varphi(x)\rangle$,

3) $\langle \delta(5x), \varphi(x) \rangle$,
4) $\delta(5x)$
5) $\langle \delta(-5x), \varphi(x) \rangle$,
6) $\delta(-5x)$
7) $\langle \delta(3x-4), \varphi(x) \rangle$,
8) $\delta(3x-4)$
9) $\langle \delta'(x+3), \varphi(x) \rangle$,
10) $\langle \cos(x)\delta'(x+3), \arctan(x) \rangle$.
11) $\langle\, 5x^2\delta''(x), \varphi(x) \rangle$.
12) $5x^2\delta''(x)$.

Solution - By exploiting the change of variables formula (3.3) and taking into account that δ is even, has compact support and is -1 homogeneous (in one variable), we obtain the following answers.

1) $\langle \delta(x-1), \varphi(x) \rangle \;=\; \langle \delta(x), \varphi(x+1) \rangle \;=\; \varphi(1)$.
2) $\langle \delta(x+\pi), \varphi(x) \rangle \;=\; \varphi(-\pi)$.
3) $\langle \delta(5x), \varphi(x) \rangle \;=\; \dfrac{1}{5}\varphi(0)$.
4) $\delta(5x) \;=\; \dfrac{1}{5}\delta(x)$.
5) $\langle \delta(-5x), \varphi(x) \rangle \;=\; \dfrac{1}{5}\varphi(0)$.
6) $\delta(-5x) \;=\; \dfrac{1}{5}\delta(x)$.
7) $\langle \delta(3x-4), \varphi(x) \rangle \;=\; \frac{1}{3}\langle \delta(y), \varphi\big((y+4)/3\big) \rangle \;=\; \frac{1}{3}\varphi(4/3)$
8) $\delta(3x-4) \;=\; \dfrac{1}{3}\delta(x-4/3)$.
9) $\langle \delta'(x+3), \varphi(x) \rangle \;=\; -\varphi'(-3)$
10) $\langle \cos(x)\delta'(x+3), \arctan(x) \rangle = -\big(\cos(x)\arctan(x)\big)'|_{x=-3} = \sin(3)\arctan(3) - \cos(3)/10$.
11) $\langle\, 5x^2\delta''(x), \varphi(x) \rangle \;=\; 5\langle\, \delta''(x), x^2\varphi(x) \rangle \;=\; 5\langle\, \delta(x), (x^2\varphi(x))'' \rangle \;=$
$\;=\; 5\langle\, \delta(x), 2\varphi(x) + 4x\varphi'(x) + x^2\varphi''(x)) \rangle \;=\; 10\,\varphi(0)$.
12) $5x^2\delta''(x) \;=\; 10\,\delta(x)$. □

Exercise 3.4. The distribution p.v. $\dfrac{1}{x} \in \mathcal{D}'(\mathbb{R})$ (*principal value of* $\dfrac{1}{x}$) is defined as follows (in analogy with Definition 1.32 of principal value of an integral):

$$\left\langle \text{p.v.}\ \frac{1}{x}, \varphi \right\rangle \;:=\; \lim_{\varepsilon \to 0_+} \int_{|x|>\varepsilon} \frac{\varphi(x)}{x}\, dx \;=\; \text{p.v.} \int_{\mathbb{R}} \frac{\varphi(x)}{x}\, dx \qquad \forall \varphi \in \mathcal{D}(\mathbb{R}) .$$

1) Prove that actually p.v.$(1/x)$ is a distribution and that p.v.$(1/x)$ has order 1.
2) Prove that $\dfrac{d}{dx}\ \ln|x| \;=\; \text{p.v.}\ \dfrac{1}{x}$,

Solution - 1) p.v. $\frac{1}{x}$ is obviously a linear functional. In order to show that it is a distribution and has order 1 we exploit the characterization provided by Theorem 3.11. For any fixed $\varphi \in \mathcal{D}(\mathbb{R})$, there exists $a \in \mathbb{R}$ such that $\operatorname{spt}\varphi \;\subset\; [-a, +a]$. By applying the mean value Theorem, we get

$$\left|\left\langle \text{p.v.}\,\frac{1}{x},\varphi\right\rangle\right| = \left|\text{p.v.}\int_{\mathbb{R}}\frac{\varphi(x)}{x}\,dx\right| = \left|\text{p.v.}\int_{-a}^{a}\frac{\varphi(x)-\varphi(0)+\varphi(0)}{x}\,dx\right| \le$$

$$\le \left|\varphi(0)\ \text{p.v.}\int_{-a}^{a}\frac{dx}{x}\,dx\,\right| + \left|\text{p.v.}\int_{-a}^{a}\frac{\varphi(x)-\varphi(0)}{x}\,dx\right| \le 2\,a\,\sup_{x}|\varphi'(x)|\,.$$

2) Since $\ln|x| \in L^1_{loc}(\mathbb{R}) \subset \mathcal{D}'(\mathbb{R})$, we get for any $\varphi \in \mathcal{D}(\mathbb{R})$

$$\langle d/dx\ \ln|x|\,,\varphi\rangle = -\langle \ln|x|,\varphi'\rangle = -\int_{\mathbb{R}}\ln|x|\,\varphi'(x)\,dx = \tag{3.34}$$

$$= -\lim_{\varepsilon\to 0_+,\ R\to+\infty}\left(\int_{-R}^{-\varepsilon}\ln|x|\,\varphi'(x)\,dx + \int_{\varepsilon}^{R}\ln|x|\,\varphi'(x)\,dx\right) =$$

$$= \text{p.v.}\int_{\mathbb{R}}\frac{1}{x}\,\varphi(x)\,dx = \langle \text{p.v.}\,\frac{1}{x}\,,\,\varphi\rangle\,.$$

An integration by parts has been performed in the last step (by taking into account that $\operatorname{spt}\varphi$ has compact support and $\lim\,\big(\varphi(\varepsilon)-\varphi(-\varepsilon)\big)\,\ln\varepsilon = 0$ as $\varepsilon\to 0$). □

Exercise 3.5. The distribution f.p. $\frac{1}{x^2} \in \mathcal{D}'(\mathbb{R})$ (*finite part of* $1/x^2$) is defined as the distribution that acts by duality on any $\varphi \in \mathcal{D}(\mathbb{R})$ as follows:

$$\left\langle \text{f.p.}\,\frac{1}{x^2},\varphi\right\rangle := \lim_{\varepsilon\to 0_+}\left(\int_{|x|>\varepsilon}\frac{\varphi(x)}{x^2}\,dx - \frac{2}{\varepsilon}\,\varphi(0)\right) = \lim_{\varepsilon\to 0_+}\left(\int_{|x|>\varepsilon}\frac{\varphi(x)}{x^2}\,dx - \frac{2}{\varepsilon}\,\langle\delta,\varphi\rangle\right).$$

Prove the identity

$$\frac{d}{dx}\ \text{p.v.}\ \frac{1}{x} = -\ \text{f.p.}\ \frac{1}{x^2}\,.$$

Solution - Since $\varphi \in \mathcal{D}(\mathbb{R})$, we have:

$$(*)\quad \int_{-\infty}^{-\varepsilon}\frac{\varphi'(x)}{x}\,dx = \int_{-\infty}^{-\varepsilon}\frac{\varphi(x)}{x^2}\,dx + \frac{\varphi(-\varepsilon)}{-\varepsilon}\,,\qquad \int_{+\varepsilon}^{+\infty}\frac{\varphi'(x)}{x}\,dx = \int_{+\varepsilon}^{+\infty}\frac{\varphi(x)}{x^2}\,dx - \frac{\varphi(\varepsilon)}{\varepsilon}\,,$$

$$\varphi(\varepsilon) = \varphi(0) + \varepsilon\varphi'(0) + o(\varepsilon)\,,\qquad \varphi(-\varepsilon) = \varphi(0) - \varepsilon\varphi'(0) + o(\varepsilon)\,,$$

$$(**)\qquad \frac{1}{\varepsilon}\,(\varphi(\varepsilon)+\varphi(-\varepsilon)) = \frac{2}{\varepsilon}\,\varphi(0) + o(1)\,.$$

By computing the pairing with $\varphi \in \mathcal{D}'(\Omega)$, we get:

$$-\left\langle \frac{d}{dx}\left(\text{p.v.}\,\frac{1}{x}\right),\varphi\right\rangle = \left\langle \text{p.v.}\,\frac{1}{x}\,,\,\varphi'\right\rangle = \lim_{\varepsilon\to 0_+}\int_{|x|>\varepsilon}\frac{\varphi'(x)}{x}\,dx =$$

$$= \lim_{\varepsilon\to 0_+}\left(\int_{-\infty}^{-\varepsilon}\frac{\varphi'(x)}{x}\,dx + \int_{\varepsilon}^{+\infty}\frac{\varphi'(x)}{x}\,dx\right) \overset{\boxed{(*)}}{=}$$

$$= \lim_{\varepsilon\to 0_+}\left(\int_{|x|>\varepsilon}\frac{\varphi(x)}{x^2}\,dx - \frac{\varphi(\varepsilon)+\varphi(-\varepsilon)}{\varepsilon}\right) \overset{\boxed{(**)}}{=}$$

$$= \lim_{\varepsilon\to 0_+}\left(\int_{|x|>\varepsilon}\frac{\varphi(x)}{x^2}\,dx - \frac{2}{\varepsilon}\,\varphi(0)\right) = \left\langle \text{f.p.}\,\frac{1}{x^2},\varphi\right\rangle.$$

□

Exercise 3.6. Prove that

1) $x\left(\text{p.v.}\ \dfrac{1}{x}\right) = 1 \qquad \text{in } \mathcal{D}'(\mathbb{R})\,,$

2) $x^2\left(\text{f.p.}\ \dfrac{1}{x^2}\right) = 1 \quad \text{in } \mathcal{D}'(\mathbb{R})\,.$

Solution - For any $\varphi \in \mathcal{D}(\mathbb{R})$ we have

1) $\langle x\ (\text{p.v. } 1/x)\,, \varphi(x)\rangle = \langle (\text{p.v. } 1/x)\,,\ x\,\varphi\rangle = \lim_{\varepsilon\to 0_+}\int_{|x|>\varepsilon}\frac{1}{x}(x\varphi(x))dx = \int_{\mathbb{R}}\varphi(x)dx =$
$= \langle 1, \varphi\rangle.$

2) $\langle x^2\left(\text{f.p. } 1/x^2\right)\,,\ \varphi\rangle = \langle\left(\text{f.p. } 1/x^2\right)\,,\ x^2\varphi\rangle = \langle -d/dx\,(\text{p.v.}\,1/x)\,,\ x^2\varphi\rangle =$

$$= \langle (\text{p.v.}\,1/x)\,,\ d/dx(x^2\varphi)\rangle = \langle (\text{p.v.}\,1/x)\,,\ (2x\varphi + x^2\varphi')\rangle =$$

$$= \lim_{\varepsilon\to 0_+}\int_{|x|>\varepsilon}(2\varphi + x\varphi')\,dx = \int_{\mathbb{R}}(2\varphi + x\varphi')\,dx \overset{\boxed{\text{integrating by parts}}}{=}$$

$$= \int_{\mathbb{R}}(2\varphi - \varphi)\,dx = \langle 1, \varphi\rangle\,. \qquad \square$$

Exercise 3.7. Solve the following division problem in $\mathcal{D}'(\mathbb{R})$, with unknown u.

$$u \in \mathcal{D}'(\mathbb{R}) : \qquad 3x^2(1-x)^4\,u(x) = 0 \text{ in } \mathcal{D}'(\mathbb{R})\,.$$

Solution - The product is a well-defined distribution since the multiplier $3x^2(1-x)^4$ is a polynomial, hence it belongs to $\mathcal{E}(\mathbb{R})$. Moreover, the multiplier has two isolated zeroes, both of finite and integer order: $x = 0$ has order 2, $x = 1$ has order 4. Due to (3.15)-(3.16) we get

$$u(x) = A\,\delta(x) + B\,\delta'(x) + C\,\delta(x-1) + D\,\delta'(x-1) + E\,\delta''(x-1) + F\,\delta'''(x-1)\,,$$

where A, B, C, D, E and F are arbitrary constants. $\square$

Exercise 3.8. Solve the following division problem in $\mathcal{D}'(\mathbb{R})$, with unknown u.

$$u \in \mathcal{D}'(\mathbb{R}) : \qquad (x-\pi)^3\,u(x) = 0 \text{ in } \mathcal{D}'(\mathbb{R})\,.$$

Solution - The same reasoning of the previous exercise applies to this case. The solution is

$$C_0\delta(x-\pi) + C_1\delta'(x-\pi) + C_2\delta''(x-\pi)\,. \qquad \square$$

Exercise 3.9. Given $\varphi \in \mathcal{D}(\mathbb{R})$, evaluate

1) $\displaystyle\int_{\mathbb{R}}\delta\,\varphi\ ,$

2) $\delta * \varphi\ ,$

3) $\delta'(x-1) * \varphi(x)\,,$

4) $\displaystyle\int_{\mathbb{R}}\delta'(x-1)\,\varphi(x)\,dx\,.$

Solution - 1) $\displaystyle\int_{\mathbb{R}}\delta\varphi = \varphi(0)\,.$

2) $(\delta * \varphi)(x) = \varphi(x)\ .$

3) $\delta'(x-1) * \varphi(x) = \varphi'(x-1)\,.$

4) $\displaystyle\int_{\mathbb{R}}\delta'(x-1)\,\varphi(x)\,dx = -\varphi'(1)\,.$

$\square$

Exercise 3.10. Given $v \in \mathcal{D}'(\mathbb{R})$, evaluate
1) $\delta * v$,
2) $\delta' * v$,
3) $\delta^{(k)} * v , \quad k \in \mathbb{N}$,
4) $\delta'(x-1) * v(x)$,
5) $H * H$,
6) $H * H * H$,
7) $\dfrac{d^3}{dx^3}\left(H * (x^+)^2\right)$, where $x^+ = \max(x, 0)$.

Solution - All convolution products are well posed since the first factor has compact support in items 1), 2), 3), while both factors have support contained in $[0, +\infty)$ in items 4), 5) and 6). We use Definition 3.29 and the property (3.19).
1) $\delta * v = v$.
2) $\delta' * v = v'$.
3) $\delta^{(k)} * v = v^{(k)} , \quad k \in \mathbb{N}$,
4) $\delta'(x-1) * v(x) = v'(x-1)$.
5) $(H * H)(x) = \displaystyle\int_0^{x^+} 1\, dy = x^+$.
6) $H * H * H = (H * H) * H = x^+ * H = \displaystyle\int_0^{x^+} y\, dy = \frac{1}{2}(x^+)^2$.
7) $\dfrac{d^3}{dx^3}\left(H * (x^+)^2\right) = H * \left(\dfrac{d^2}{dx^2}(2\,x^+)\right) = 2\,H * \left(\dfrac{d}{dx} H\right) = 2\,H * \delta = 2H$.

□

Exercise 3.11. Distributional derivative of a jump discontinuity ($n = 1$)
Prove the following statement.

Given $a \in \mathbb{R}$, $f \in C^1(-\infty, a)$, $g \in C^1(a, +\infty)$ with $f' \in L^1(-\infty, a)$, $g' \in L^1(a, +\infty)$, define a function u as $u(x) = f(x)$ if $x < a$, $u(x) = g(x)$ if $x > a$.
Then u is well defined a.e. in $\mathbb{R}$ and its distributional derivative u' is the sum of a locally integrable function, equal to f' in $(-\infty, a)$ and to g' in $(a, +\infty)$, plus a singular part: the Dirac delta concentrated at $x = a$ times the jump size of u, say $\left(u(a_+) - u(a_-)\right)\delta(x - a)$.

Solution - We write $u(a_-) = f(a) := f(a_-)$, $u(a_+) = g(a) := g(a_+)$ for short; we notice that these values are finite since f' and g' are integrable. Then, $\forall \varphi \in \mathcal{D}(\mathbb{R})$, we evaluate:

$$
\begin{aligned}
\langle u', \varphi \rangle &= -\langle u, \varphi' \rangle = -\int_{-\infty}^{+\infty} u(x)\varphi'(x)\, dx = \\
&= -\int_{-\infty}^{a} f(x)\varphi'(x)\, dx - \int_{a}^{+\infty} g(x)\varphi'(x)\, dx = \quad \boxed{\text{integrating by parts}} \\
&= \int_{-\infty}^{a} f'(x)\varphi(x)\, dx - f(a)\varphi(a) + \int_{a}^{+\infty} g'(x)\varphi(x)\, dx + g(a)\varphi(a) = \\
&= \int_{-\infty}^{a} f'(x)\varphi(x)\, dx + \int_{a}^{+\infty} g'(x)\varphi(x)\, dx + \left(g(a) - f(a)\right)\varphi(a) = \\
&= \int_{-\infty}^{a} f'(x)\varphi(x)\, dx + \int_{a}^{+\infty} g'(x)\varphi(x)\, dx + \left(u(a_+) - u(a_-)\right)\langle \delta_a, \varphi \rangle .
\end{aligned}
$$

Remark - The continuity assumption on f' and g' can be weakened: it is sufficient to assume that the distributional derivatives f', g' belong respectively to $L^1_{loc}(-\infty, a)$, $L^1_{loc}(a, +\infty)$ and that $f(a_-)$ and $g(a_+)$ exist with finite value.
In any case the function u has classical derivative at a.e. $x \in \mathbb{R}$, but in general the L^1_{loc} function defined a.e. in $\mathbb{R}$ as its classical derivative **is not** the distributional derivative of u: it is the distributional derivative too only if $u(a_+) = u(a_-)$.
For instance: $\text{sign}(x)$ has classical derivative equal to 0 a.e. but $\text{sign}'(x) = 2\delta(x)$ in $\mathcal{D}'(\mathbb{R})$; on the contrary the absolutely continuous function $\exp(-|x|)$ has classical derivative with value $-\text{sign}(x)\exp(-|x|)$ a.e. and its distributional derivative coincides with such function, say $(d/dx))\exp(-|x|) = -\text{sign}(x)\exp(-|x|)$. □

Exercise 3.12. Evaluate the distributional derivative of the following distributions in $\mathcal{D}'(\mathbb{R})$.

1) $v(x) \;=\; \dfrac{1}{2}\,\text{sign}(x)\, e^{-|x|}\,,$

2) $w(x) \;=\; \text{sign}(\sin(x))\,,$

3) $u(x) \;=\; 3\ln|x| \,+ \text{sign}(2-4x)\,.$

Solution - We exploit the solutions of Exercises 3.4 and 3.11:

1) v is a function C^1 in $\mathbb{R}^+$ and in $\mathbb{R}^-$, v and v' have finite (but different) limits at 0_- and at 0_+, v exhibits a jump at 0 of size $1: v(0_+) - v(0_-) = 1$. So:

$$v'(x) \;=\; -\frac{1}{2}\, e^{-|x|} + \delta(x)\,.$$

2) $w(x) = 1$ if $2k\pi < x < (2k+1)\pi$, $w(x) = -1$ if $(2k+1)\pi < x < (2k+2)\pi$, $\forall k \in \mathbb{Z}$. Then w is piece-wise constant with alternating positive and negative jumps, all of size 2. Hence:

$$w'(x) \;=\; 2\left(\sum_{k=-\infty}^{+\infty} \delta(x - 2k\pi) - \sum_{k=-\infty}^{+\infty} \delta(x-(2k+1)\pi)\right).$$

3) u is a locally integrable function (it has a Lebesgue integrable singularity at 0) and a jump of size -2 in $x = 1/2$. Hence

$$u'(x) \;=\; 3\,\text{p.v.}\,\frac{1}{x} \;-\; 2\,\delta(x - 1/2)\,.$$

□

Exercise 3.13. Set $v_k(x) = k/(1 + k^2x^2)$ and prove that $v_k \to \pi\delta$ in $\mathcal{S}'(\mathbb{R})$ (hence, it converges in $\mathcal{D}'(\mathbb{R})$ too).

Solution - It is sufficient to prove that, as $k \to \infty$,

$$\int_{\mathbb{R}} \frac{k\,\varphi(x)}{1 + k^2\,x^2}\,dx = \int_{\mathbb{R}} \frac{\varphi(t/k)}{1+t^2}\,dt \to \pi\,\varphi(0) \qquad \forall \varphi \in \mathcal{S}(\mathbb{R}).$$

By an elementary approach, we show that this convergence actually takes place for any bounded φ which is continuous at 0, through the following estimate of the difference:

$$\left|\int_{\mathbb{R}} \frac{\varphi(t/k)}{1+t^2}\,dt \;-\; \pi\,\varphi(0)\right| \le \left|\int_{|t|\le T} \frac{\varphi(t/k) - \varphi(0)}{1+t^2}\,dt\right| + \left|\int_{|t|>T} \frac{\varphi(t/k) - \varphi(0)}{1+t^2}\,dt\right| = I_1 + I_2$$

$(1+t^2)^{-1} \in L^1(\mathbb{R})$, $\int_{\mathbb{R}} 1/(1+t^2) = \pi$, then $\forall \varepsilon > 0$, $\exists T = T(\varepsilon)\,:\, \int_{|t|>T} 1/(1+t^2)dt \le \varepsilon$.
Moreover, $\forall\, \varepsilon > 0$, $T < +\infty$, $\exists k\,:\, |\varphi(t/k) - \varphi(0)| \le \varepsilon\;\forall t:\; |t| \le T$.
Then $I_1 + I_2 \le \pi\,\varepsilon \,+\, 2\,\varepsilon\,\|\varphi\|_\infty$ and the proof follows by the arbitrariness of ε.
Alternatively, in order to show the convergence of the integrals, it is sufficient to notice that we can apply the dominated convergence theorem (Theorem 2.5) since the following inequality holds true a.e.: $|\varphi(t/k)/(1+t^2)| \le \|\varphi\|_\infty/(1+t^2) \in L^1(\mathbb{R})$. □

Exercise 3.14. Evaluate the limit in $\mathcal{D}'(\mathbb{R})$ of $v_k(x) = ke^{-k|x|}$ as $k \to +\infty$, and explain all the computations.

Solution - $v_k(x) \to 2\,\delta(x)$ in $\mathcal{D}(\mathbb{R})$. In fact the change of variables $kx = y$, $dx = dy/k$, provides

$$\langle v_k, \varphi\rangle = \int_{\mathbb{R}} ke^{-k|x|}\varphi(x)\,dx = \int_{\mathbb{R}} e^{-|y|}\varphi(y/k)\,dy \to \varphi(0)\int_{\mathbb{R}} e^{-|y|}\,dy = 2\varphi(0) = \langle 2\delta, \varphi\rangle$$

for any $\varphi \in \mathcal{D}(\mathbb{R})$, hence the answer follows. The interchange of the limit and the integration is justified by Theorem 2.5 (dominated convergence): for any k the integrands are dominated by the integrable function $\|\varphi\|_\infty e^{-|y|}$. □

Exercise 3.15. Consider the sequences

$$1)\; f_k(x) \;=\; \frac{1}{\sqrt{k}}\,\chi_{[0,k]}(x)\,, \qquad 2)\; g_k(x) \;=\; \sin(kx)\,\chi_{[0,7]}(x)\,.$$

For each sequence, establish whether it is convergent or not in $L^1(\mathbb{R})$, $L^2(\mathbb{R})$, $L^\infty(\mathbb{R})$, $\mathcal{D}'(\mathbb{R})$, $\mathcal{S}'(\mathbb{R})$, by justifying the answer.

Solution - 1) $|f_k(x)| \le 1/\sqrt{k}$, $\forall x \in \mathbb{R}$, then f_k converges in L^∞ to $f \equiv 0$, hence it converges also in L^1_{loc} and in $\mathcal{D}'(\mathbb{R})$, $\mathcal{S}'(\mathbb{R})$.
Nevertheless f_k converges neither in $L^1(\mathbb{R})$ nor in $L^2(\mathbb{R})$; in fact, in case of convergence, the possible limit should coincide with the distributional limit ($f \equiv 0$), while $\|f_k - 0\|_1 = \|f_k\|_1 = \sqrt{k}$, $\|f_k - 0\|_2 = \|f_k\|_2 = 1$ for $k = 1, 2, 3, \ldots$.
2) g_k converges in $\mathcal{S}'(\mathbb{R})$ and hence in $\mathcal{D}'(\mathbb{R})$. In fact, integrating by parts we get

$$\int_{\mathbb{R}} \sin(kx)\,\varphi(x)\,dx = \frac{1}{k}\int_0^7 \cos(kx)\,\varphi'(x)\,dx - \frac{1}{k}\left[\cos(kx)\,\varphi(x)\right]_0^7 \quad \overset{k\to+\infty}{\longrightarrow} \quad 0 \quad \forall\varphi \in \mathcal{D}(\mathbb{R}).$$

Finally, g_k does not converge in any L^p norm, since the possible limit should be $g \equiv 0$, due to the convergence in $\mathcal{S}'(\mathbb{R})$, but this leads to a contradiction: $\|g_k - g\|_\infty = \|g_k\|_\infty = 1\;\forall k$ if $p = \infty$ and, if $p \in [1, \infty)$, (by denoting $⨍_E w = (\int_E w)/|E|$)

$$\begin{aligned}\|g_k - g\|_p \;=\; \|g_k\|_p \;=\; &\int_0^7 |\sin(kx)|^p\,dx = \frac{1}{k}\int_0^{7k} |\sin(y)|^p\,dy = \\ =\; 7&⨍_0^{7k} |\sin(y)|^p\,dy \quad \overset{k\to+\infty}{\longrightarrow} \quad 7⨍_0^{2\pi} |\sin(y)|^p\,dy \;>\; 0\,.\end{aligned}$$

The convergence in the last step is trivial if 7 is substituted by 2π. Actually, the limit does not change if, in place of $[0, 2\pi]$, we consider the interval $[0, 7]$ or any other interval $[0, a]$ whose right-end is a real number $a > 0$, since

$$(*) \qquad a⨍_0^{ka} |\sin(y)|^p\,dy \;=\; \frac{2h\pi}{k}⨍_0^{2h\pi} |\sin(y)|^p\,dy + \frac{1}{k}\int_{2h\pi}^{ka} |\sin(y)|^p\,dy$$

(where $h = h(k)$ is the unique integer s.t. $2h\pi \le ka < 2(h+1)\pi$)
while $2h\pi/k \to 1$, $⨍_0^{2h\pi} |\sin(y)|^p dy \;=\; ⨍_0^{2\pi} |\sin(y)|^p dy\;\forall h$ and the last summand in $(*)$ is an infinitesimal, as $k \to \infty$. □

Exercise 3.16. Study the convergence in the space of distributions (of one real variable) of the series

$$\sum_{k=0}^{+\infty} \delta(x - 1/k)\,.$$

Solution - The series does not converge in $\mathcal{D}'(\mathbb{R})$, therefore it represents no distribution in $\mathbb{R}$: e.g. the duality of its partial sums with a function $\varphi \in \mathcal{D}(\mathbb{R})$ s.t. $\varphi(0) = 1$ leads to a divergent numerical series.
However, the given series is convergent in $\mathcal{D}'(0, +\infty)$, hence it defines a distribution in $\mathcal{D}'(0, +\infty)$: in fact the duality with any test function in $\mathcal{D}(0, +\infty)$ corresponds to a finite sum. □

Exercise 3.17. Study the convergence in $\mathcal{D}'(\mathbb{R})$ of the series

$$\sum_{k=0}^{+\infty} \delta^{(k)}(x-k).$$

Solution - The given series converges in the space of distributions in $\mathcal{D}'(\mathbb{R})$, therefore it represents a distribution in $\mathbb{R}$: in fact the pairing with any test function in $\mathcal{D}(\mathbb{R})$ amounts to a finite sum. Notice that the series defines a distribution of infinite order; there is no contradiction with the theorem about the structure of distributions (Theorem 3.11) since the order is finite in any compact subset of $\mathbb{R}$. □

Exercise 3.18. Set

$$v = \sum_{k=0}^{+\infty} c_k\, \delta(x-k), \qquad x \in \mathbb{R}.$$

1) Prove that the series converges in $\mathcal{D}'(\mathbb{R})$ and hence it defines a distribution $v \in \mathcal{D}'(\mathbb{R})$ for any choice of the coefficients c_k.
2) Prove that, if $c_k = e^k$, then the series converges in $\mathcal{D}'(\mathbb{R})$ but it does not converge in $\mathcal{S}'(\mathbb{R})$, thus $v \notin \mathcal{S}'(\mathbb{R})$.

Solution - 1) This item follows by Theorem 3.11, since the pairing with any test function in $\mathcal{D}(\mathbb{R})$ amounts to a finite sum, whose modulus is estimated by the L^∞ norm of finitely many derivatives with uniformly bounded order in any compact set.
2) By choosing $c_k = e^k$ and $\psi(x) = \exp(-\sqrt{1+x^2}) \in \mathcal{S}(\mathbb{R})$, we get

$$\int_{\mathbb{R}} \left(\sum_{k=0}^{K} c_k\, \delta(x-k) \right) \psi = \sum_{k=0}^{K} e^k\, \psi(k) \sim \sum_{k=0}^{K} 1 \qquad \text{which is divergent as } K \to +\infty.$$

□

Exercise 3.19. Express the the following differential equations, whose unknowns u, v belong to $\mathcal{D}'(\mathbb{R})$,

$$3\,u'' - 5\,u' + 2\,u = f\,, \qquad v'''(x) = 2\,v(x-7) + g(x)$$

in the form $F * u = f$ and $G * v = g$, where F and G are suitable distributions in $\mathcal{E}'(\mathbb{R})$.
Solution - By (3.19): $F(x) = 3\,\delta''(x) - 5\,\delta'(x) + 2\,\delta(x)\,,\ G(x) = \delta'''(x) - 2\,\delta(x-7)\,.$ □

Exercise 3.20. Verify that $U(x) = \frac{1}{2}\, e^{-|x|}$ solves $-U'' + U = \delta$ in $\mathcal{S}'(\mathbb{R})$, say U is a fundamental solution of the operator $I - d^2/dt^2$ in $\mathbb{R}$.
Solution - $U \subset L^\infty(\mathbb{R})$, then U is a tempered distribution. We compute the derivatives:

$$U'(x) = -\frac{1}{2}\,\text{sign}(x)\, e^{-|x|}\,, \quad U'(0_+) - U'(0_-) = 1\,, \quad U''(x) = \frac{1}{2}\, e^{-|x|} + \delta(x)\,.$$

By substitution we get: $-U'' + U = -\frac{1}{2}\, e^{-|x|} + \left(\frac{1}{2}\, e^{-|x|} + \delta(x)\right) = \delta(x)\,.$ □

Exercise 3.21. Verify that $W(x) = \frac{1}{2}\sin(|x|)$ solves $W'' + W = \delta$ in $\mathcal{S}'(\mathbb{R})$, say W is a fundamental solution of the operator $I + d^2/dt^2$ in $\mathbb{R}$.

Solution - $W \subset L^\infty(\mathbb{R})$, then W is a tempered distribution.
By using the Leibniz formula (3.12) and

$$\sin(|x|) = \sin(x)\,\mathrm{sign}(x)\,, \quad \mathrm{sign}' = 2\delta\,, \quad \mathrm{sign}'' = 2\delta'\,, \quad \sin(x)\,\delta'(x) = -\delta(x)\,,$$

we obtain

$$\begin{aligned} W''(x) &= 1/2\left(\,\sin''(x)\,\mathrm{sign}(x) + 2\,\sin'(x)\,\mathrm{sign}'(x) + \sin(x)\,\mathrm{sign}''(x)\right) = \\ &= 1/2\left(-\,\sin(x)\,\mathrm{sign}(x) + 4\,\cos(x)\,\delta(x) + 2\,\sin(x)\,\delta'(x)\right) = \\ &= -\,1/2\,\sin(|x|) + 2\,\delta(x) + \sin(x)\,\delta'(x) = -\,1/2\,\sin(|x|) + 2\,\delta(x) - \delta(x) = \\ &= -W(x) + \delta(x)\,. \end{aligned}$$

□

Exercise 3.22. (**Incremental quotients**) Let u be a function or a distribution of one real variable; for any $h \in \mathbb{R}$, with $h \neq 0$, we define

$$w_h(x) = \frac{u(x+h) - u(x)}{h}\,.$$

Prove that

1) if $u \in C^1(\mathbb{R})$, then $w_h(x) \to u'(x)$ pointwise for any x, as $h \to 0$;
2) if $u \in C^1(\mathbb{R})$ and $\mathrm{spt}\,u$ is compact, then $w_h(x) \to u'(x)$ uniformly in $\mathbb{R}$, as $h \to 0$;
3) if $u \in \mathcal{D}(\mathbb{R})$, then $w_h(x) \to u'(x)$ in $\mathcal{D}(\mathbb{R})$, as $h \to 0$;
4) if $u \in \mathcal{D}'(\mathbb{R})$, then $w_h(x) \to u'(x)$ in $\mathcal{D}'(\mathbb{R})$, as $h \to 0$.

Solution - 1) This item is a consequence of the definition of classical derivative.

2) Due to the Lagrange Theorem, $w_h(x) = u'(x+th)$ with $t = t(u,x,h) \in (0,1)$, moreover u' is uniformly continuous. Then $w_h(x) \to u'(x)$ uniformly in $\mathbb{R}$.

3) The task is to fulfil Definition 3.5 about convergence in the space of test functions. The first condition in the Definition is fulfilled: the w_h with $|h| < 1$ all have support contained in the compact set $\{x \in \mathbb{R} :\ \exists y \in \mathrm{spt}\,u\ s.t.\ |x-y| \leq 1\}$.
Moreover the derivatives $u^{(k)}$ of any order k are uniformly continuous (since they are continuous and compactly supported), then the same argument applied to u' in item 2) of the present exercise can be applied to all derivatives:

$$w_h^{(k)} = u^{(k+1)}(x + t_k h) \quad \text{for suitable } t_k = t_k(u,x,h) \in (0,1)\,,$$

then also the condition 2) in the definition is fulfilled.

4) For any fixed $\varphi \in \mathcal{D}(\Omega)$, "by shifting the difference quotient to the test function" we get (in the sequel all the integrations are performed in $\mathbb{R}$):

$$\begin{aligned} \int w_h\,\varphi &= \int \frac{u(x+h) - u(x)}{h}\,\varphi(x)\,dx = \\ &= \frac{1}{h}\int u(x+h)\,\varphi(x)\,dx - \frac{1}{h}\int u(x)\,\varphi(x)\,dx = \\ &= \int u(x)\,\frac{1}{h}\,(\varphi(x-h) - \varphi(x))\,dx\,. \end{aligned}$$

By applying item 3) of the present exercise to φ, the last integral converges to

$$-\int u\,\varphi'\,dx = \int u'\,\varphi\,dx\,.$$

□

3.2.2 Exercises about distributions of several variables

Exercise 3.23. Denoting by $\delta(\mathbf{x})$ the Dirac delta in n variables, write the explicit formula of duality with a test function, for the following distributions in $\mathcal{D}'(\mathbb{R}^n)$:

1) $5\,\delta(3\mathbf{x})$.

2) $\delta(3\mathbf{x}+\mathbf{b})$ where $n=2$, $\mathbf{b}=(1,2)$

3) $\delta(\mathbb{A}\mathbf{x}+\mathbf{b})$ where $n=2$, $\mathbb{A}=\begin{bmatrix} 0 & -1 \\ 1 & 0 \end{bmatrix}$, $\mathbf{b}=(1,2)$

Solution - 1) For every $\varphi\in\mathcal{D}'(\mathbb{R}^n)$ we have

$$\int_{\mathbb{R}^n} 5\,\delta(3\mathbf{x})\varphi(\mathbf{x})\,d\mathbf{x} \;=\; 5\int_{\mathbb{R}^n}\delta(\mathbf{y})\varphi(\mathbf{y}/3)\,3^{-n}d\mathbf{y} \;=\; \frac{5}{3^n}\varphi(0) \;=\; \frac{5}{3^n}\int_{\mathbb{R}^n}\delta(\mathbf{x})\varphi(\mathbf{x})\,d\mathbf{x}\,,$$

then $5\,\delta(3\mathbf{x}) \;=\; (5/3^n)\,\delta(\mathbf{x})$.

2) For every $\varphi\in\mathcal{D}'(\mathbb{R}^n)$ we have

$$\begin{aligned}\int_{\mathbb{R}^n}\delta(3\mathbf{x}+\mathbf{b})\varphi(\mathbf{x})\,d\mathbf{x} \;&=\; \int_{\mathbb{R}^n}\delta(\mathbf{y})\varphi\big((\mathbf{y}-\mathbf{b})/3\big)\,3^{-n}d\mathbf{y} \;=3^{-n}\varphi(-\mathbf{b}/3) \;= \\ &=\; 3^{-n}\int_{\mathbb{R}^n}\delta(\mathbf{x}+\,\mathbf{b}/3)\,\varphi(\mathbf{x})\,d\mathbf{x}\,,\end{aligned}$$

then, by setting $\mathbf{x}=(x,y)$, we obtain $\delta(3\mathbf{x}+\mathbf{b}) \;=\; 3^{-n}\,\delta(\mathbf{x}+\,\mathbf{b}/3) \;=\; \delta(x+1/3,y+2/3)$.

3) For every $\varphi\in\mathcal{D}'(\mathbb{R}^n)$, by setting $\mathbf{x}=(x,y)$, we obtain

$$\begin{aligned}\int_{\mathbb{R}^n}\delta(\mathbb{A}\mathbf{x}{+}\mathbf{b})\varphi(\mathbf{x})\,d\mathbf{x} \;&=\; \int_{\mathbb{R}^n}\delta(\mathbf{y})\varphi\big(\mathbb{A}^{-1}(\mathbf{y}{-}\mathbf{b})\big)\,|\det\mathbb{A}|^{-1}d\mathbf{y} = \varphi\big(\mathbb{A}^{-1}(-\mathbf{b})\big) = \varphi(2,-1) = \\ &= \int_{\mathbb{R}^n}\delta(x-2,y+1)\varphi(x,y)\,dxdy = \int_{\mathbb{R}^n}\delta(\mathbf{x}-\mathbb{A}^{-1}\mathbf{b})\varphi(\mathbf{x})d\mathbf{x},\end{aligned}$$

then: $\delta(\mathbb{A}\mathbf{x}+\mathbf{b}) \;=\; \delta(\mathbf{x}-\mathbb{A}^{-1}\mathbf{b}) \;=\; \delta(x-2,y+1)$.

□

Exercise 3.24. (The integral mean on a disk converge the to Dirac δ in $\mathcal{D}'(\mathbb{R}^n)$, as the radius of the disk tends to 0)
Show that $\left|B_{1/k}(\mathbf{0})\right|^{-1}\chi_{B_{1/k}(\mathbf{0})}(\mathbf{x})$ converges to $\delta(\mathbf{x})$ in the space of distributions as k tends to $+\infty$.
Solution - We set: $u_k(\mathbf{x}) \;=\; \left|B_{1/k}(\mathbf{0})\right|^{-1}\chi_{B_{1/k}(\mathbf{0})}(\mathbf{x})$ and

$$⨍_E v(\mathbf{x})\,d\mathbf{x} \;=\; \frac{1}{|E|}\int_E v(\mathbf{x})\,d\mathbf{x} \qquad \forall E\subset\mathbb{R}^n \text{ measurable: } |E|<+\infty,\ \forall v\in L^1(E).$$

Then for any $\varphi\in\mathcal{D}(\mathbb{R}^n)$ we get

$$\begin{aligned}\int_{\mathbb{R}^n}u_k\,\varphi \;&=\; \left|B_{1/k}(\mathbf{0})\right|^{-1}\int_{B_{1/k}(\mathbf{0})}\varphi = ⨍_{B_{1/k}(\mathbf{0})}\varphi(\mathbf{x})\,d\mathbf{x} = \\ &= ⨍_{B_{1/k}(\mathbf{0})}\varphi(\mathbf{0})\,d\mathbf{x} \;+\; ⨍_{B_{1/k}(\mathbf{0})}\big(\varphi(\mathbf{x})-\varphi(\mathbf{0})\big)\,d\mathbf{x}\end{aligned}$$

the first integral in the last line is equal to $\varphi(\mathbf{0}) \;=\; \int\delta\varphi$ and the second integral is an infinitesimal as $k\to 0$, due to the continuity of φ at 0. □

Exercise 3.25. (Approximating the Dirac δ in $\mathcal{S}'(\mathbb{R}^n)$ by Gaussian functions)
Set $v_k(\mathbf{x}) = \pi^{-n/2}\, k^n \exp(-k^2|\mathbf{x}|^2)$, then, by explaining the computations, prove that, $v_k \to \delta$ in $\mathcal{S}'(\mathbb{R}^n)$ as $k \to +\infty$, and hence $v_k \to \delta$ in $\mathcal{D}'(\mathbb{R}^n)$ too.

Solution - Notice that v_k belongs to $\mathcal{S}(\mathbb{R}^n) \subset L^1(\mathbb{R}^n)$ for any k. By performing the change of variables $k\mathbf{x} = \mathbf{y}$, $d\mathbf{x} = k^{-n} d\mathbf{y}$, we obtain, $\forall \psi \in \mathcal{S}$,

$$\int_{\mathbb{R}^n} v_k\, \psi = \int_{\mathbb{R}^n} \pi^{-n/2} k^n \exp(-k^2|\mathbf{x}|^2)\, \psi(\mathbf{x})\, d\mathbf{x} = \int_{\mathbb{R}^n} \pi^{-n/2} \exp(-|\mathbf{y}|^2) \psi\left(\frac{\mathbf{y}}{k}\right) d\mathbf{y}.$$

The last integral fulfils the assumptions of the dominated convergence Theorem (the modulus of all the integrands is estimated by $\|\psi\|_\infty \pi^{-n/2} \exp(-|\mathbf{y}|^2) \in L^1(\mathbb{R}^n)$ and the integrands converge pointwise everywhere to $\pi^{-n/2} \exp(-|\mathbf{y}|^2)\, \psi(\mathbf{0})$: then, due to Theorem 2.5 and (2.11), such integral tends to

$$\pi^{-n/2} \int_{\mathbb{R}^n} \exp(-|\mathbf{y}|^2)\, \psi(\mathbf{0})\, d\mathbf{y} \;=\; \psi(\mathbf{0}) \;=\; \int_{\mathbb{R}^n} \delta\, \psi\,.$$

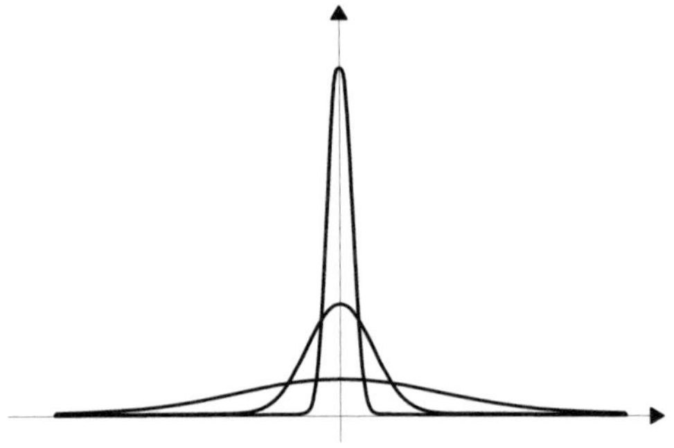

Figure 3.1 - Exercise 3.25: graphs of $v_k(\mathbf{x})$, for $n = 1$ and $k = 1,\ 1/10,\ 1/100$. □

Exercise 3.26. (Approximating the constant function 1 in $\mathcal{S}'(\mathbb{R}^n)$ by Gaussian functions)
Set $w_k(\mathbf{x}) = \exp(-|\mathbf{x}|^2/k^2)$, then, by explaining the computations, prove that, $w_k \to \delta$ in $\mathcal{S}'(\mathbb{R}^n)$ as $k \to +\infty$, hence $w_k \to \delta$ in $\mathcal{D}'(\mathbb{R}^n)$ too.

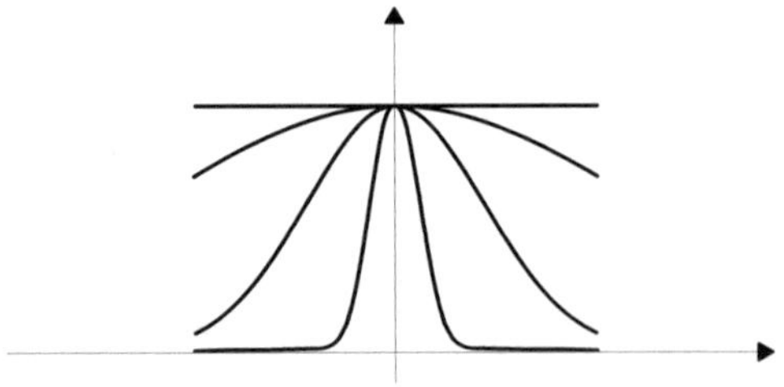

Figure 3.2 - Exercise 3.26: graphs of $w_k(\mathbf{x})$ as $n = 1$ and $k = 1,\ 1/10,\ 1/100$.

Solution - We notice that the function w_k belongs to $\mathcal{S}(\mathbb{R}^n) \subset L^1_{loc}(\mathbb{R}^n)$ for any k and they converge to 1 pointwise and uniformly on compact subsets. We get, $\forall \psi \in \mathcal{S}$,

$$\int_{\mathbb{R}^n} w_k\,\psi = \int_{\mathbb{R}^n} \exp(-|\mathbf{x}|^2/k^2)\,\psi(\mathbf{x})\,d\mathbf{x} \;\to\; \int_{\mathbb{R}^n} \psi(\mathbf{y})d\mathbf{y}\,,$$

in fact the assumptions of Theorem 2.5 (dominated convergence) are fulfilled: everywhere pointwise convergent integrand to ψ; every integrand is dominated by $|\psi| \in L^1(\mathbb{R}^n)$. □

Exercise 3.27. Derivatives of functions with jump discontinuity ($n > 1$)
Prove the following statement.

Assume $n \geq 2$, $\Omega \subset \mathbb{R}^n$ is an open set, Γ is a simple smooth ($n-1$ dimensional) surface which splits Ω in two connected components, denoted by Ω_+ and Ω_-, u is a function defined in $\Omega \setminus \Gamma$ which belongs to $C^1(\Omega_+) \cap C^1(\Omega_-)$ and u admits both extensions in $C^1(\overline{\Omega_+})$ and in $C^1(\overline{\Omega_-})$.
Then u is defined a.e. in Ω and belongs to $L^1_{loc}(\Omega) \subset \mathcal{D}'(\Omega)$, moreover, for any $j = 1, \ldots, n$, its distributional derivative $D_{x_j}u$ is given by the sum of a locally integrable function, which coincides with the *j–th* classical derivative of u in $\Omega_+ \cup \Omega_-$ plus a singular term: a Dirac delta concentrated on the surface Γ times a continuous function equal to the width of the jump of u and weighted by the component along $\mathbf{e}_j$ of the normal vector to Γ. More precisely:

$$D_{x_j}u = D_{x_j}u_+ + D_{x_j}u_- + [u]\,\nu_j\,\delta_\Gamma\,, \tag{3.35}$$

where $D_{x_j}u_+$, $D_{x_j}u_-$ denote the derivatives of u in Ω_+ and Ω_-, $[u] = u_+ - u_-$ denotes the jump of u on Γ, $\nu_j = \nu \cdot \mathbf{e}_j$ the component in the direction of $\mathbf{e}_j$ of the unit normal vector to Γ with orientation pointing toward Ω_+ and $[u]\nu_j\delta_\Gamma \in \mathcal{D}'(\Omega)$ is the distribution s.t.

$$\langle [u]\nu_j\delta_\Gamma, \varphi\rangle = \int_\Gamma [u](\mathbf{x})\,\nu_j(\mathbf{x})\,\varphi(\mathbf{x})\,d\sigma(\mathbf{x}) \qquad \forall\varphi \in \mathcal{D}(\Omega)\,.$$

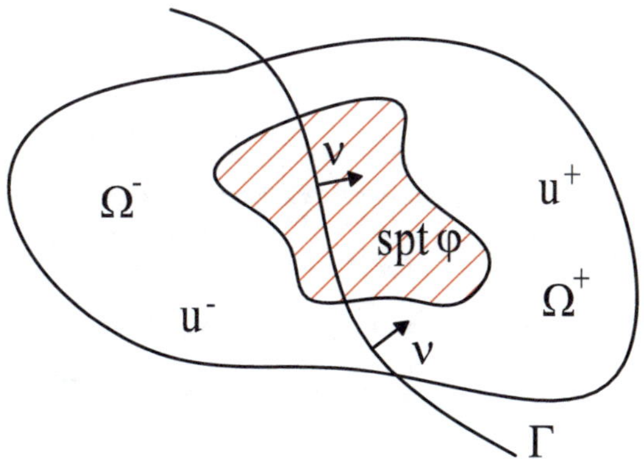

Figure 3.3 - Derivatives of functions with jump discontinuity (Exercise 3.27).

Solution - We choose $\varphi \in \mathcal{D}'(\Omega)$, then we apply the definition of distributional derivative and integrate by parts in Ω_+ and in Ω_-:

$$\begin{aligned}\langle D_{x_j} u , \varphi \rangle &= - \int_{\Omega} u\,(D_{x_j}\varphi)\,d\mathbf{x} = \\ &= - \int_{\Omega_+} u_+\,(D_{x_j}\varphi)\,d\mathbf{x} - \int_{\Omega_-} u_-\,(D_{x_j}\varphi)\,d\mathbf{x} = \end{aligned}$$

$$= \int_{\Omega_+} (D_{x_j}u_+)\,\varphi\;d\mathbf{x} + \int_{\Gamma} u_+\,\varphi\,\nu_j\,d\sigma(\mathbf{x}) + \int_{\Omega_-} (D_{x_j}u_-)\,\varphi\,d\mathbf{x} - \int_{\Gamma} u_-\,\varphi\,\nu_j\,d\sigma(\mathbf{x})\,.$$

□

Exercise 3.28. Prove the formula (3.17) which is stated at the end of Section 3.1 and recalled below (for reader's convenience).
Division problem in $\mathcal{D}'(\mathbb{R}^n)$: every distribution whose support is a single point $\widetilde{\mathbf{x}} \in \Omega \subset \mathbb{R}^n$ is a finite sum of derivatives of Dirac delta $\delta_{\widetilde{\mathbf{x}}} = \delta(\mathbf{x}-\widetilde{\mathbf{x}})$ concentrated at the same point, that is it has a representation of the following kind:

$$\sum_{\boldsymbol{\alpha}} c_{\boldsymbol{\alpha}}\, D^{\boldsymbol{\alpha}}\,\delta_{\widetilde{\mathbf{x}}}\,.$$

Solution - We know that $\operatorname{spt} u \subset \{\widetilde{\mathbf{x}}\}$, that is $u \in \mathcal{E}'(\mathbb{R}^n)$. Then u has finite order m, due to Theorem 3.11. Without loss of generality we may assume that $\widetilde{\mathbf{x}} = \mathbf{0}$.
Every $v \in \mathcal{E}(\mathbb{R}^n)$ can be represented by its Taylor polynomial of degree m together with a remainder:

$$v(\mathbf{x}) = \sum_{|\boldsymbol{\alpha}|\le m} \frac{D^{\boldsymbol{\alpha}}v(\mathbf{0})}{\boldsymbol{\alpha}!}\,\mathbf{x}^{\boldsymbol{\alpha}} + r(\mathbf{x})\,.$$

So $\qquad \langle u, v\rangle = \langle u, r\rangle + \sum_{|\boldsymbol{\alpha}|\le m} c_{\boldsymbol{\alpha}} D^{\boldsymbol{\alpha}} v(\mathbf{0}) \qquad$ where $c_{\boldsymbol{\alpha}} \stackrel{\text{def}}{=} \dfrac{D^{\boldsymbol{\alpha}}v(\mathbf{0})}{\boldsymbol{\alpha}!}\,\langle u, \mathbf{x}^{\boldsymbol{\alpha}}\rangle\,.$

In order to prove (3.17), it is sufficient to show that $\langle u, r\rangle = 0$. We choose $v \in \mathcal{E}(\mathbb{R}^n)$ vanishing in neighborhood of $\mathbf{0}$ and $v \equiv 1$ if $|\mathbf{x}| \ge 1$. Then $\langle u, v(k\mathbf{x})r(\mathbf{x})\rangle = 0$ for any $k \in \mathbb{N}$.
The difference $v(k\mathbf{x})r(\mathbf{x}) - r(\mathbf{x})$ has support in $\{|\mathbf{x}| \le 1/k\}$. Then, to prove that $\langle u, r\rangle = 0$, it is enough to show that $\langle u, \big(v(k\mathbf{x})r(\mathbf{x}) - r(\mathbf{x})\big)\rangle$ tends to 0 as $k \to \infty$.
This last convergence holds true if $\|D^{\boldsymbol{\alpha}}\big(v(k\mathbf{x})r(\mathbf{x}) - r(\mathbf{x})\big)\|_\infty \to 0$, for any $|\boldsymbol{\alpha}| \le m$. Moreover

$$D^{\boldsymbol{\alpha}}\big(v(k\mathbf{x})r(\mathbf{x}) - r(\mathbf{x})\big) = \big(v(k\mathbf{x}) - 1\big) D^{\boldsymbol{\alpha}} r(\mathbf{x}) + \sum_{\boldsymbol{\beta}+\boldsymbol{\gamma}=\boldsymbol{\alpha},\,|\boldsymbol{\beta}|>0} \binom{\boldsymbol{\alpha}}{\boldsymbol{\beta}} k^{|\boldsymbol{\beta}|}\, \partial^{\boldsymbol{\beta}} v(k\mathbf{x})\, \partial^{\boldsymbol{\gamma}} r(\mathbf{x}). \tag{3.36}$$

Since all derivatives of r up to the order m vanish at $\mathbf{0}$, we know that the first summand in in the right-hand side of (3.36) is $O(k^{-m-1+|\boldsymbol{\alpha}|})$. The other summands in the right-hand side are $O(k^{|\boldsymbol{\beta}|})O(k^{-m-1-|\boldsymbol{\gamma}|}) = O(k^{-1})$, since $|\boldsymbol{\beta}| + |\boldsymbol{\gamma}| = |\boldsymbol{\alpha}|$. □

Exercise 3.29. Set $\delta = \delta_{\mathbf{0}}(\mathbf{x})$, $\mathbf{x} = (x, y) \in \mathbb{R}^2$. Prove that

$$(x^2 + y^2)\,(\partial_{xxx}\,\delta - 3\partial_{xyy}\,\delta) = 0 \qquad \text{in } \mathcal{D}'(\mathbb{R}^2)\,.$$

Solution - By exploiting the Leibniz formula (3.12) in the computation of the derivatives of products, for any $v \in \mathcal{D}(\mathbb{R}^2)$, $w \in \mathcal{E}(\mathbb{R}^2)$, we obtain

$$\partial_{xxx}(vw) = v_{xxx} + 3v_{xx}w_x + 3v_x w_{xx} + v w_{xxx}$$

$$\partial_{xyy}(vw) = (v_x w + v w_x)_{yy} = v_{xyy}w + 2v_{xy}w_y + v_x w_{yy} + v_{yy}w_x + 2v_y w_{xy} + v w_{xyy}\,.$$

From Definition 3.19 of distributional derivative we deduce, for any $v \in \mathcal{D}(\mathbb{R}^2)$,

$$\begin{aligned}\langle (x^2+y^2)(\partial_{xxx}\delta - 3\,\partial_{xyy}\delta), v\rangle &= \langle \partial_{xxx}\delta - 3\,\partial_{xyy}\delta\,,\,(x^2+y^2)v\,\rangle = \\ &= \langle\, \delta\,,\,(-\partial_{xxx}+3\,\partial_{xyy})\,\big((x^2+y^2)v\big)\,\rangle = \\ &= -\partial_{xxx}\,\big((x^2+y^2)v\big)\big|_{(0,0)} + 3\,\partial_{xyy}\big((x^2+y^2)v\big)\big|_{(0,0)} \\ &= -3\times 2\times \partial_x v(\mathbf{0}) + 3\times 2\times \partial_x v(\mathbf{0}) = 0\,.\end{aligned}$$

Alternative method of solution - The left-hand side of the identity to be proved has compact support (derivatives and products times a polynomial of a distribution with compact support), then the left-hand side belongs to $\mathcal{S}'$ and is $\mathcal{F}$ transformable (see Chapter 4 about Fourier transform). Set $\boldsymbol{\xi} = (\xi,\eta)$

$$\mathcal{F}\left\{(x^2+y^2)\,(\partial_{xxx}\delta - 3\partial_{xyy}\delta)\right\} = -\Delta_{\boldsymbol{\xi}}\Big((i)^3\,(\xi^3-3\xi\eta^2)\Big) = 0$$

since $(\xi^3 - 3\xi\eta^2) = \mathrm{Re}\,\big((\xi+i\eta)^3\big)$ is an harmonic function of $(\xi,\eta)\in\mathbb{R}^2$. □

Exercise 3.30. Show that

$$\Delta\left(\frac{-1}{4\pi\,|\mathbf{x}|}\right) = \delta(\mathbf{x}) \qquad \mathcal{D}'(\mathbb{R}^3)\,.$$

Solution - We simplify the computations by studying the given function times a constant:

Figure 3.4 - $L^1(\mathbb{R}^3)$ approximation of the fundamental solution of the Laplacian operator (multiplied times (-4π)), built by truncation (Exercise 3.30).

$u(\mathbf{x}) := 1/|\mathbf{x}|$ is a radial function and $u\in L^1_{loc}(\mathbb{R}^3)\subset\mathcal{D}'(\mathbb{R}^3)$ since, by integrating in polar coordinates (see the Table in Section 10.6.2), the Jacobian $\varrho^2\sin\varphi$ cancels with the singularity of u at $\mathbf{0}$). Moreover $u\in C^\infty(\mathbb{R}^3\setminus\{\mathbf{0}\})$, then by using the expression of Laplacian operator in polar coordinates (Table 10.6.4)

$$\Delta = \frac{\partial^2}{\partial\varrho^2} + \frac{2}{\varrho}\frac{\partial}{\partial\varrho} \qquad \text{if } n=3\,,$$

we find

$$\Delta u \equiv 0 \qquad \text{in } \mathbb{R}^3\setminus\{\mathbf{0}\}\,.$$

We are left to compute the possible singular part of Δu with support contained at $\mathbf{0}$, which in any case must have a representation of the kind (3.17).
We consider the following approximations of u :

$$u_k(\mathbf{x}) := \min(u(\mathbf{x}), k) = \min(|\mathbf{x}|^{-1}, k).$$

Then

$$u_k \overset{L^1_{loc}(\mathbb{R}^3)}{\longrightarrow} u \implies u_k \overset{\mathcal{D}'(\mathbb{R}^3)}{\longrightarrow} u \implies \Delta u_k \overset{\mathcal{D}'(\mathbb{R}^3)}{\longrightarrow} \Delta u.$$

Moreover $\Delta u_k = 0$ in $\mathbb{R}^3 \setminus \{\partial B_{1/k}(\mathbf{0})\}$ and Δu_k has a singular part (radially symmetric) concentrated on the boundary $\partial B_{1/k}(\mathbf{0})$ of the small disk with radius $1/k$, a contribution due to the jump of the derivative of u_k in the direction of the exterior normal to $B_{1/k}(\mathbf{0})$:

$$\begin{aligned}\Delta u_k &= \left[\frac{\partial u_k}{\partial \varrho}\right]_{\varrho=1/k} \delta_{\partial B_{1/k}(\mathbf{0})} = \left[\frac{\partial u_k}{\partial x_1}\right]_{\mathbf{x}=(1/k,0,0)} \delta_{\partial B_{1/k}(\mathbf{0})} = \\ &= -\left.\frac{1}{|\mathbf{x}|^2}\right|_{\varrho=1/k} \delta_{\partial B_{1/k}(\mathbf{0})} = -k^2\, \delta_{\partial B_{1/k}(\mathbf{0})}\end{aligned}$$

where $[w]_\mathbf{x} = w^+ - w^-$ denotes the jump in the radial direction of the function w at $\mathbf{x}$. Since, $\forall \psi \in \mathcal{E}(\mathbb{R}^3)$,

$$-\int_{\mathbb{R}^3} k^2 \delta_{\partial B_{1/k}(\mathbf{0})}\, \psi(\mathbf{x})\, d\mathbf{x} = -4\pi \fint_{\partial B_{1/k}(\mathbf{0})} \psi(\mathbf{x})\, d\mathbf{x} \overset{k\to+\infty}{\longrightarrow} -4\pi\, \psi(\mathbf{0}) = -4\pi \int_{\mathbb{R}^3} \delta\, \psi,$$

we obtain $\Delta u_k \to -4\pi\, \delta$, then $\Delta u = -4\pi\, \delta$ and $\Delta\big(-1/(4\pi|\mathbf{x}|)\big) = \delta$. □

Exercise 3.31. Given $J > 0$, prove that the vector field

$$\mathbf{B} = \frac{J}{2\pi}\left(\frac{-y}{x^2+y^2}, \frac{+x}{x^2+y^2}, 0\right) \tag{3.37}$$

fulfils the following system of differential equations in the space of distributions in $\mathbb{R}^3$

$$\mathbf{curl\, B} = J\, \mathbf{e}_3\, \delta_{\{x=y=0\}}, \tag{3.38}$$

where $\mathbf{x} = (x, y, z)$, $\mathbf{e}_3$ is the unit vector with the same orientation of the z axis and

$$\delta_{\{x=y=0\}}(\mathbf{x}) = \delta(x) \otimes \delta(y) \otimes 1(z)$$

denotes the Dirac delta uniformly concentrated on the z axis.

Solution - Obviously $\mathbf{curl\, B}$ can have only one nontrivial component: the third one. By using the **curl** representation in Table 10.4, we compute such component in $\mathbb{R}^3 \setminus \{x = y = 0\}$, say in the region where $\mathbf{B}$ is a C^∞ vector field:

$$\mathbf{curl\, B} = \mathbf{0} \qquad \text{in } \mathbb{R}^3 \setminus \{x = y = 0\}.$$

Alternatively, to show that $\mathbf{B}$ is an irrotational vector field in $\mathbb{R}^3 \setminus \{x = y = 0\}$, one can simply remark that $\mathbf{B}$ is locally a gradient in $\mathbb{R}^3 \setminus \{(0,0,z) : z \in \mathbb{R}\}$. In fact:

by setting $\vartheta = \arctan(y/x)$, one finds $\quad d\vartheta = \dfrac{-y\, dx + x\, dy}{x^2+y^2} \quad$ in $\mathbb{R}^2 \setminus \{(0,0)\}$.

To evaluate the possible singular part of the distribution $\mathbf{curl\, B}$ with support on the z axis, we consider the approximation B_k built by modifying $\mathbf{B}$ in the infinite cylinder $C_k := \{\mathbf{x} = (x, y, z) : x^2 + y^2 < 1/k^2\}$:

$$\mathbf{B}_k(\mathbf{x}) := \begin{cases} \mathbf{B}(\mathbf{x}) & \text{if } x^2+y^2 > 1/k^2, \\ \dfrac{J}{2\pi} k^2\, (-y, x, 0) & \text{if } x^2+y^2 \le 1/k^2. \end{cases}$$

Notice that both $\mathbf{B}_k$ and $\mathbf{B}$ belong to $L^1_{loc}(\mathbb{R}^3)$, as one can easily verify by using cylindrical coordinates (Table 10.6.3), since the Jacobian cancels with the singularity on the z axis. Then

$$\mathbf{B}_k \overset{L^1_{loc}(\mathbb{R}^3)}{\longrightarrow} \mathbf{B} \implies \mathbf{B}_k \overset{\mathcal{D}'(\mathbb{R}^3)}{\longrightarrow} u \implies \mathbf{curl}\,\mathbf{B}_k \overset{\mathcal{D}'(\mathbb{R}^3)}{\longrightarrow} \mathbf{curl}\,\mathbf{B}\,.$$

Since

$$\mathbf{B}_k \in C^0(\mathbb{R}^3) \bigcap C^\infty(\mathbb{R}^3 \setminus C_k) \bigcap C^\infty(\overline{C_k})\,,$$

by taking into account Exercise 3.27 we deduce that also $\mathbf{curl}\,\mathbf{B}_k$ is a function $L^1_{loc}(\mathbb{R}^3)$ without singular part, precisely, due to

$$\mathbf{curl}\,\big(-y,x,0\big) = \det\begin{bmatrix} \mathbf{e}_1 & \mathbf{e}_2 & \mathbf{e}_3 \\ \frac{\partial}{\partial x} & \frac{\partial}{\partial y} & \frac{\partial}{\partial z} \\ -y & x & 0 \end{bmatrix} = \Big(\frac{\partial}{\partial x}x - \frac{\partial}{\partial y}(-y)\Big)\,\mathbf{e}_3 = 2\,\mathbf{e}_3\,,$$

we obtain

$$\mathbf{curl}\,\mathbf{B}_k = \begin{cases} \mathbf{curl}\,\mathbf{B} = \mathbf{0} & \text{in } \mathbb{R}^3 \setminus C_k\,, \\ \\ \dfrac{J}{\pi}\,k^2\,\mathbf{e}_3 & \text{in } C_k\,. \end{cases}$$

Eventually, we evaluate the distributional limit of $\mathbf{curl}\,\mathbf{B}_k$: it is enough to consider the only nontrivial component of $\mathbf{curl}\,\mathbf{B}_k$, say the third one. For every $\varphi \in \mathcal{D}'(\mathbb{R}^3)$ we find

$$\begin{aligned}
\int_{\mathbb{R}^3} (\mathbf{e}_3 \cdot \mathbf{curl}\,\mathbf{B}_k)\,\varphi\,d\mathbf{x} &= \int_{C_k} (\mathbf{e}_3 \cdot \mathbf{curl}\,\mathbf{B}_k)\,\varphi\,d\mathbf{x} \\
&= \int_{\mathbb{R}}\Big(\int_{x^2+y^2<k^{-2}} (\mathbf{e}_3 \cdot \mathbf{curl}\,\mathbf{B}_k)\,\varphi\,dx\,dy\Big)\,dz = \\
&= \int_{\mathbb{R}}\Big(\frac{J}{\pi}k^2 \int_{x^2+y^2<k^{-2}} \varphi\,dx\,dy\Big)\,dz = \\
&= J\int_{\mathbb{R}}\Big(\fint_{x^2+y^2<k^{-2}} \varphi\,dx\,dy\Big)\,dz \overset{k\to+\infty}{\longrightarrow} J\int_{\mathbb{R}}\varphi(0,0,z)\,dz = \\
&= J\,\langle\,\delta_{\{x=y=0\}}\,,\varphi\,\rangle\,.
\end{aligned}$$

Remark 1 - Notice that the approximation $\mathbf{B}_k$ introduced in the solution of Exercise 3.31 quantitatively expresses the magnetic field associated to the steady D.C. current of unitary intensity in the direction $\mathbf{e}_3$ along a straight infinite wire with circular cross section of radius $1/k$ and set on the z axis. In fact $\mathbf{B}_k$ solves

$$\mathbf{curl}\,\mathbf{B}_k = \frac{J}{\pi}\,k^2\,\mathbf{e}_3\,\chi_{C_k} \qquad \text{in } \mathcal{D}'(\mathbb{R}^3)\,.$$

Remark 2 - The last two exercises provide a quantitative picture of two relevant examples in Electromagnetism, respectively:

- Exercise 3.30: potential v and electric field $\mathbf{E}$ associated to a point charge located at the origin and of the same intensity of the unitary charge:

$$v(\mathbf{x}) = -\frac{1}{4\,\pi\,|\mathbf{x}|}\,, \qquad \mathbf{E} = \frac{1}{4\pi}\frac{\mathbf{x}}{|\mathbf{x}|^3} = \nabla v(\mathbf{x})\,, \qquad \operatorname{div}\mathbf{E}(\mathbf{x}) = \delta(\mathbf{x}) \qquad \text{in } \mathcal{D}'(\mathbb{R}^3)\,;$$

- Exercise 3.31: magnetic field $\mathbf{B}$ given by (3.38) produced by the steady D.C. current of unitary intensity flowing in an infinite straight thin wire.

Both examples describe steady fields, nevertheless in the second one the electric charges move. Recall the *Maxwell system of equations*

$$-\mu_0\,\epsilon_0\,\frac{\partial \mathbf{E}}{\partial t} + \mathbf{curl}\,\mathbf{B} = \mu_0\,\mathbf{j} \qquad \textbf{(Ampere's circuital law)}, \tag{3.39}$$

$$\operatorname{div}\mathbf{E} = \varrho/\epsilon_0 \qquad \textbf{(Gauss's law)}, \tag{3.40}$$

$$+\frac{\partial \mathbf{B}}{\partial t} + \mathbf{curl}\,\mathbf{E} = \mathbf{0} \qquad \textbf{(Faraday's law)}, \tag{3.41}$$

$$\operatorname{div}\mathbf{B} = 0 \qquad \textbf{(Gauss's magnetic law)}, \tag{3.42}$$

where $\mathbf{j}$ denotes the electric current and ϱ the charge density, moreover we choose measure units such that the dielectric constant is normalized ($\epsilon_0 = 1$) and the magnetic permeability too ($\mu_0 = 1$).
In the first case (Exercise 3.30: $\varrho = \delta$, $\mathbf{j} = \mathbf{0}$, potential $v = -1/(4\pi|\mathbf{x}|)$ and $\partial E/\partial t = \partial B/\partial t = 0$) the Maxwell system reads

$$\mathbf{curl}\,\mathbf{B} = \mathbf{0}\,, \quad \operatorname{div}\mathbf{E} = \Delta v = \delta\,, \quad \mathbf{curl}\,\mathbf{E} = \mathbf{0}\,, \quad \operatorname{div}\mathbf{B} = 0\,, \quad \text{in } \mathcal{D}'(\mathbb{R}^3),$$

and the solutions are: $\mathbf{E} = \nabla \dfrac{-1}{4\pi|\mathbf{x}|} = \dfrac{1}{4\pi}\dfrac{\mathbf{x}}{|\mathbf{x}|^3}\,, \quad \mathbf{B} = \mathbf{0}\,.$

In the second case (Exercise 3.31 - Figure 3.5 : $\varrho = \delta_{\{x=y=0\}}$, $\mathbf{j} = J\,\mathbf{e}_3\,\delta_{\{x=y=0\}}$ and $\partial E/\partial t = \partial B/\partial t = 0$),

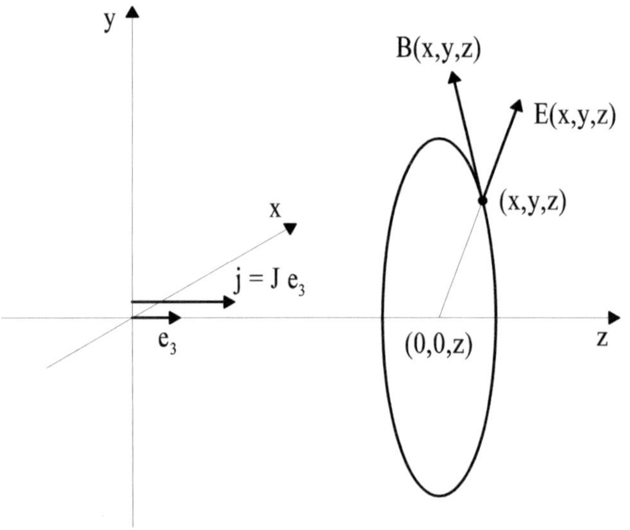

Figure 3.5 - Electric and magnetic field produced by the steady D.C. current $\mathbf{j} = J\mathbf{e}_3$ in the infinite straight wire set along the z axis. (Exercise 3.31).

the Maxwell system reduces to the following one

$$\mathbf{curl}\,\mathbf{B} = J\,\mathbf{e}_3\,\delta_{\{x=y=0\}}\,, \quad \operatorname{div}\mathbf{E} = J\delta_{\{x=y=0\}}\,, \quad \mathbf{curl}\,\mathbf{E} = \mathbf{0}\,, \quad \operatorname{div}\mathbf{B} = 0\,, \quad \text{in } \mathcal{D}'(\mathbb{R}^3).$$

This system describes the fields produced by the charges on the z axis, where the steady D.C. current of constant intensity J flows: the solution is given by the electric field $\mathbf{E}$

orthogonal to the z axis at every space point $\mathbf{x}$ off the z axis

$$\mathbf{E}(\mathbf{x}) = |\mathbf{E}(\mathbf{x})| \frac{\widetilde{\mathbf{x}}}{|\widetilde{\mathbf{x}}|}, \quad |\mathbf{E}(\mathbf{x})| = \frac{J}{2\,\pi\,r}, \quad \widetilde{\mathbf{x}} = (x, y, 0), \quad r = |\widetilde{\mathbf{x}}| = (x^2 + y^2)^{1/2}, \quad (3.43)$$

and by the magnetic field $\mathbf{B}$ which is singular on the z axis, has field lines which are circumferences turning around the z axis and intensity

$$|\mathbf{B}(\mathbf{x})| = \frac{J}{2\,\pi\,r} \qquad \textbf{(Biot-Savart law)}, \qquad (3.44)$$

where $r > 0$ is the distance of $\mathbf{x} = (x, y, z)$ from the z axis (Jean-Baptiste Biot, 1774-1862; Félix Savart, 1791-1841).

Formula (3.43) can be proved by noticing that $\mathbf{E}(\mathbf{x}) = |\mathbf{E}(\mathbf{x})|\,\widetilde{\mathbf{x}}/|\widetilde{\mathbf{x}}|$ due to symmetry and the evaluating and summing only the component of the Coulomb force which is perpendicular to the wire:

$$\begin{aligned}
|\mathbf{E}(\mathbf{x})| &= \frac{J}{4\pi} \int_{\mathbb{R}} \frac{r}{(r^2+z^2)^{3/2}}\, dz = \\
&= \frac{J}{2\pi} \int_0^{+\infty} \frac{r}{(r^2+z^2)^{3/2}}\, dz = \boxed{z = r\sinh t,\ dz = r\cosh t\, dt} \\
&= \frac{J}{2\pi r} \int_0^{+\infty} \frac{dt}{\big(\cosh t\big)^2} = \frac{J}{2\pi} \int_0^{+\infty} \frac{4}{(e^t + e^{-t})^2}\, dt = \boxed{e^t = s,\ dt = ds/s} \\
&= \frac{2J}{\pi r} \int_1^{+\infty} \frac{s}{(s^2+1)^2} = \frac{J}{2\pi r}\,.
\end{aligned}$$

Alternatively, the value $|\mathbf{E}(\mathbf{x})|$ can be obtained by integrating the equation (3.40), with the choice $\varepsilon_0 = 1$, over the cylinder $\{\mathbf{x} :\ x^2 + y^2 < r^2,\ z_1 < z < z_2\}$, applying formula 3 of Table 10.5 and taking into account that there is a nontrivial contribution to the flux of $\mathbf{E}$ only on the lateral boundary of the cylinder.

Formula (3.44) is a straightforward consequence of (3.37).
The reader is invited to verify that the magnetic field $\mathbf{B}$ given by (3.37) fulfils formula 7 in Table 10.5, with

$$\mathbf{u} = \mathbf{B}, \quad S = \{\mathbf{x} :\ x^2 + y^2 < r^2,\ z = \widetilde{z}\}, \quad \gamma = \partial S = \{\mathbf{x} :\ x^2 + y^2 = r^2,\ z = \widetilde{z}\}\,.$$

□

Chapter 4
Fourier Transform

In this chapter we examine the properties of the Fourier transform (Jean Baptiste Joseph Fourier, 1768-1830).
This integral transform is a classical tool; first it was introduced for the study of differential equations, later it proved effective in many different fields, notably in recent engineering applications to signal analysis.
Algebraic and functional properties of Fourier transform provide valid and simple methods for solving a wide class of problems. This fact justify the effort at broadening the classic Fourier transform domain of integrable functions to more general sets, as L^2 and the space of tempered distributions.

4.1 Fourier transform method

We consider only scalar-valued functions and distributions, with real or complex values. In ensuing sections we gradually consider various frameworks where the Fourier transform can be used. The theory is displayed without proofs: the more instructive ones are given later in Section 4.2. Further in-depth analysis can be found in [1], [4], [5], [7], [12], [17], [25].

4.1.1 Fourier transform in $L^1(\mathbb{R})$

Assume that $u \in L^1(\mathbb{R})$.
Then the Fourier transform of u is the function $\widehat{u} : \mathbb{R} \to \mathbb{C}$ defined by

$$\widehat{u}(\xi) = \int_{\mathbb{R}} e^{-i\xi x}\, u(x)\, dx \qquad \forall \xi \in \mathbb{R} . \tag{4.1}$$

The operator $\mathcal{F}$ that maps a function u to its Fourier transform is linear, that is $\mathcal{F} : u \to \widehat{u}$ fulfils:

$$\mathcal{F}\{\alpha u + \beta v\} = \alpha \mathcal{F}\{u\} + \beta \mathcal{F}\{v\} \qquad \forall \alpha, \beta \in \mathbb{R}, \forall u, v \in L^1(\mathbb{R}) . \quad (4.2)$$

Linearity (4.2) holds true also for every extensions of the transform definition, that will be given subsequently. Moreover

$$\{ u \in L^1(\mathbb{R}),\ u' \in L^1(\mathbb{R}) \} \quad \Rightarrow \quad \widehat{u'}(\xi) = i\xi \widehat{u}(\xi) , \quad (4.3)$$

$$\{ u \in L^1(\mathbb{R}),\ xu \in L^1(\mathbb{R}) \} \quad \Rightarrow \quad \widehat{u} \in C^1(\mathbb{R}), \text{ e } \widehat{xu}(\xi) = i \frac{d}{d\xi} \widehat{u}(\xi) . \quad (4.4)$$

The transform fulfils the following algebraic rules ($a, \xi \in \mathbb{R}$, $a \neq 0$):

$$u(x) \quad \to \quad \widehat{u}(\xi) = \mathcal{F}\{u\}(\xi) = \int_{\mathbb{R}} e^{-i\,\xi x} u(x) dx ,$$

$$u(ax) \quad \to \quad \frac{1}{|a|} \widehat{u} \left(\frac{\xi}{a} \right) \qquad \textit{scaling} , \quad (4.5)$$

$$u(x-a) \quad \to \quad e^{-ia\xi}\, \widehat{u}(\xi) \qquad \textit{shifting} , \quad (4.6)$$

$$u(x)\, e^{iax} \quad \to \quad \widehat{u}(\xi - a) \qquad \textit{modulation} , \quad (4.7)$$

$$u(x) \cos(\xi_0 x) \quad \to \quad \frac{1}{2} \left(\widehat{u}(\xi - \xi_0) + \widehat{u}(\xi + \xi_0) \right) \qquad \textit{modulation} \quad (4.8)$$

The transform preserves symmetries:

$$u \text{ even} \quad \Rightarrow \quad \widehat{u} \text{ even} , \quad (4.9)$$

$$u \text{ odd} \quad \Rightarrow \quad \widehat{u} \text{ odd} , \quad (4.10)$$

$$u \text{ real and even} \quad \Rightarrow \quad \widehat{u} \text{ real and even} , \quad (4.11)$$

$$u \text{ real and odd} \quad \Rightarrow \quad \widehat{u} \text{ imaginary and odd} . \quad (4.12)$$

We recall that a function u depending on one real variable is called even or odd if it fulfils, respectively, $u(-x) = u(x)$ or $u(-x) = -u(x)$, $\forall x \in \mathbb{R}$.
Fourier transform is injective, so we can define the inverse transform $\mathcal{F}^{-1}$ as follows, with domain in the linear space of Fourier transforms of L^1 functions:

$$u, \widehat{u} \in L^1(\mathbb{R}) \Rightarrow \quad u(x) = \frac{1}{2\pi} \int_{\mathbb{R}} e^{+i\,\xi x}\, \widehat{u}(\xi)\, d\xi \quad \forall x \in \mathbb{R} , \quad (4.13)$$

$$u \in L^1(\mathbb{R}) \Rightarrow \quad u(x) = \lim_{\varepsilon \to 0_+} \frac{1}{2\pi} \int_{\mathbb{R}} e^{+i\xi x}\, \widehat{u}(\xi)\, e^{-(\varepsilon^2 \xi^2)/4}\, d\xi , \quad (4.14)$$

the limit in (4.14) denotes the strong $L^1(\mathbb{R})$ convergence.

4.1.2 Fourier transform in $L^1(\mathbb{R}^n)$

Definition 4.1. *Assume that* $u \in L^1(\mathbb{R}^n)$.
Then the **Fourier transform** *of* u *is the function* $\widehat{u} : \mathbb{R}^n \to \mathbb{C}$ *defined by*

$$\widehat{u}(\boldsymbol{\xi}) = \int_{\mathbb{R}^n} e^{-i\boldsymbol{\xi}\cdot\mathbf{x}}\, u(\mathbf{x})\, d\mathbf{x} \qquad \forall \boldsymbol{\xi} \in \mathbb{R}^n \tag{4.15}$$

where $\boldsymbol{\xi} = (\xi_1, \ldots, \xi_n)$, $\mathbf{x} = (x_1, \ldots, x_n)$, $\boldsymbol{\xi}\cdot\mathbf{x} = \xi_1 x_1 + \cdots + \xi_n x_n$ *is the scalar product and* $d\mathbf{x} = dx_1 dx_2 \cdots dx_n$.

Theorem 4.2. *If* $u \in L^1(\mathbb{R}^n)$, $\mathbf{y} \in \mathbb{R}^n$ *and* A *is a real invertible* $n \times n$ *matrix, then* $\mathcal{F} : u(\mathbf{x}) \to \widehat{u}(\boldsymbol{\xi})$ *acts as follows:*

$$u(\mathbf{x} - \mathbf{y}) \quad \to \quad e^{-i\boldsymbol{\xi}\cdot\mathbf{y}}\widehat{u}(\boldsymbol{\xi}), \tag{4.16}$$

$$e^{i\mathbf{y}\cdot\mathbf{x}}u(\mathbf{x}) \quad \to \quad \widehat{u}(\boldsymbol{\xi} - \mathbf{y}), \tag{4.17}$$

$$u(A\mathbf{x}) \quad \to \quad |\det A|^{-1}\, \widehat{u}(A^{-T}\boldsymbol{\xi}) \tag{4.18}$$

where A^{-T} *is the transposed inverse matrix of* A,

$$\overline{u(\mathbf{x})} \quad \to \quad \overline{\widehat{u}(-\boldsymbol{\xi})}, \tag{4.19}$$

moreover, recalling that u *is* radially symmetric *if* $u(\mathbf{x}) = u(\mathbf{y})$ $\forall \mathbf{x}$ *and* $\mathbf{y}$ *in* $\mathbb{R}^n$ *such that* $|\mathbf{x}| = |\mathbf{y}|$, u *is* even *if* $u(-\mathbf{x}) = u(\mathbf{x})$ $\forall \mathbf{x} \in \mathbb{R}^n$ *and* u *is* odd *if* $u(-\mathbf{x}) = -u(\mathbf{x})$ $\forall \mathbf{x} \in \mathbb{R}^n$, *we obtain*

$$u \text{ radially symmetric} \quad \Rightarrow \quad \widehat{u} \text{ radially symmetric}, \tag{4.20}$$

$$u \text{ even} \quad \Rightarrow \quad \widehat{u} \text{ even}, \tag{4.21}$$

$$u \text{ odd} \quad \Rightarrow \quad \widehat{u} \text{ odd}; \tag{4.22}$$

$$u \text{ real and even} \quad \Rightarrow \quad \widehat{u} \text{ real and even}, \tag{4.23}$$

$$u \text{ real and odd} \quad \Rightarrow \quad i\widehat{u} \text{ real and odd}. \tag{4.24}$$

Theorem 4.3. ***(Riemann-Lebesgue Lemma)***
Assume that u *belongs to* $L^1(\mathbb{R}^n)$. *Then*

$$\widehat{u} \in C^0(\mathbb{R}^N) \cap L^\infty(\mathbb{R}^n), \tag{4.25}$$

$$\|\widehat{u}\|_{L^\infty(\mathbb{R}^n)} \leq \|u\|_{L^1(\mathbb{R}^n)}, \tag{4.26}$$

$$\lim_{|\boldsymbol{\xi}| \to +\infty} \widehat{u}(\boldsymbol{\xi}) = 0. \tag{4.27}$$

Theorem 4.4. *Assume* $u \in C^1(\mathbb{R}^n) \cap L^1(\mathbb{R}^n)$ *and* $\dfrac{\partial u}{\partial x_j} \in L^1(\mathbb{R}^n)$, *where* j *is an index in* $\{1, \cdots, n\}$. *Then*

$$\mathcal{F}\left\{\frac{\partial u}{\partial x_j}\right\} = i\,\xi_j\,\widehat{u}(\boldsymbol{\xi}) \,. \tag{4.28}$$

Remark 4.5. If $n = 1$ then the conditions of Theorem 4.4 can be weakened, by only assuming that $u \in L^1(\mathbb{R}) \cap AC(\mathbb{R})$ (see Definition 2.6).

Theorem 4.6. *Assume that* $u \in L^1(\mathbb{R}^n)$ *and* $x_j u(\mathbf{x}) \in L^1(\mathbb{R}^n)$, *where* j *is an index in* $\{1, \cdots, n\}$. *Then* $\widehat{u}$ *has a continuous derivative* $\dfrac{\partial \widehat{u}}{\partial \xi_j}$ *fulfilling*

$$\widehat{x_j u} = i\,\frac{\partial \widehat{u}}{\partial \xi_j} \tag{4.29}$$

If u *and* $|\mathbf{x}|u(\mathbf{x})$ *belong to* $L^1(\mathbb{R}^n)$ *then* $\widehat{u} \in C^1(\mathbb{R}^n)$.

Theorem 4.7. $u, v \in L^1(\mathbb{R}^n) \;\Rightarrow$

$$\int_{\mathbb{R}^n} \widehat{u}\, v = \int_{\mathbb{R}^n} u\, \widehat{v}\,. \tag{4.30}$$

Inverse Fourier transform ($\mathcal{F}^{-1}$):

$$u,\, \widehat{u} \in L^1(\mathbb{R}^n) \;\Rightarrow\; u(\mathbf{x}) = \frac{1}{(2\pi)^n}\int_{\mathbb{R}^n} e^{+i\,\boldsymbol{\xi}\cdot\mathbf{x}}\,\widehat{u}(\boldsymbol{\xi})\,d\boldsymbol{\xi} \quad \forall \mathbf{x} \in \mathbb{R}^n\,;$$

$$u \in L^1(\mathbb{R}^n) \;\Rightarrow\; u(\mathbf{x}) = \lim_{\varepsilon\to 0_+} \frac{1}{(2\pi)^n}\int_{\mathbb{R}^n} e^{+i\boldsymbol{\xi}\cdot\mathbf{x}}\,\widehat{u}(\boldsymbol{\xi})\; e^{-(\varepsilon^2|\boldsymbol{\xi}|^2)/4}\,d\boldsymbol{\xi}$$

where the limit denotes the strong convergence in $L^1(\mathbb{R}^n)$.

4.1.3 Fourier transform in $L^2(\mathbb{R}^n)$

Assume that $u \in L^2(\mathbb{R}^n)$.
Then the Fourier transform of u is the function $\widehat{u} : \mathbb{R}^n \to \mathbb{C}$ defined by

$$\widehat{u}(\boldsymbol{\xi}) = \lim_{R\to+\infty} \int_{B_R(\mathbf{0})} e^{-i\boldsymbol{\xi}\cdot\mathbf{x}}\, u(\mathbf{x})\, d\mathbf{x} \quad \forall \boldsymbol{\xi} \in \mathbb{R}^n \tag{4.31}$$

where where the limit denotes the strong convergence in $L^2(\mathbb{R}^n)$.

If $u \in L^1(\mathbb{R}^n) \cap L^2(\mathbb{R}^n)$, then the definition (4.31) coincides with (4.1).

The transform fulfils the properties listed below.

$$u \in L^2(\mathbb{R}^n) \Rightarrow \widehat{u} \in L^2(\mathbb{R}^n), \tag{4.32}$$

$$u \in L^p(\mathbb{R}^n) \Rightarrow \widehat{u} \in L^{p'}(\mathbb{R}^n),$$
$$\forall p, p' : p \in [1,2],\ p' = \frac{p}{p-1} \in [2, \infty]. \tag{4.33}$$

If $n = 1$, by evaluating the principal value of the integral (4.31) via complex variable methods we can find the point-wise value of $\widehat{u}(\xi)$ for a.e. $\xi \in \mathbb{R}$:

$$\widehat{u}(\xi) = v.p. \int_{\mathbb{R}} e^{-i\xi x}\, u(x)\, dx = \lim_{R\to+\infty} \int_{R}^{+R} e^{-i\xi x}\, u(x)\, dx \text{ a.e. } \xi \in \mathbb{R}. \tag{4.34}$$

Theorem 4.8. ***(Plancherel identity)*** *Every $u \in L^2(\mathbb{R}^n)$ fulfils:*

$$\int_{\mathbb{R}^n} |\,\widehat{u}(\boldsymbol{\xi})\,|^2\, d\boldsymbol{\xi} = (2\,\pi)^n \int_{\mathbb{R}^n} |\,u(\mathbf{x})\,|^2\, d\mathbf{x}\,, \tag{4.35}$$

that is

$$\|\,\widehat{u}\,\|_{L^2(\mathbb{R}^n)} = (2\,\pi)^{n/2}\, \|\,u\,\|_{L^2(\mathbb{R}^n)}\,. \tag{4.36}$$

A straightforward consequence of the Plancherel identity is the fact that

$$v_k \to v \text{ in } L^2(\mathbb{R}^n) \qquad \Leftrightarrow \qquad \widehat{v_k} \to \widehat{v} \text{ in } L^2(\mathbb{R}^n)\,.$$

4.1.4 Fourier transform of tempered distributions

Definition 4.9. ***(Schwartz space $\mathcal{S}(\mathbb{R}^n)$, or space of fast decaying smooth functions)***
A function v of n real variables belongs to $\mathcal{S}(\mathbb{R}^n)$ iff $v \in C^\infty$ and

$$P(\mathbf{x})\, D^{\boldsymbol{\alpha}} v \in L^2(\mathbb{R}^n) \qquad \forall \text{ polinomial } P,\ \forall \text{ multi-index } \boldsymbol{\alpha}. \tag{4.37}$$

This definition entails:

$$\mathcal{S}(\mathbb{R}^n) \subset L^1(\mathbb{R}^n) \cap \mathcal{F}\left(L^1(\mathbb{R}^n)\right), \tag{4.38}$$

$$u \in \mathcal{S}(\mathbb{R}^n) \;\Rightarrow\; \widehat{u} \in \mathcal{S}(\mathbb{R}^n).$$

A sequence of functions $v_m \in \mathcal{S}(\mathbb{R}^n)$ converges in $\mathcal{S}(\mathbb{R}^n)$ to $v \in \mathcal{S}(\mathbb{R}^n)$ iff
$|\mathbf{x}|^k D^{\boldsymbol{\alpha}} v_m$ converges to $|\mathbf{x}|^k D^{\boldsymbol{\alpha}} v$ in $L^2(\mathbb{R}^n)$ $\quad \forall k \in \mathbb{N}\ \forall \boldsymbol{\alpha} \in \mathbb{N}^n$.

So

$$u_m \to u \text{ in } \mathcal{S}(\mathbb{R}^n) \qquad \Rightarrow \qquad \widehat{u_m} \to \widehat{u} \text{ in } \mathcal{S}(\mathbb{R}^n). \tag{4.39}$$

$\mathcal{S}'(\mathbb{R}^n)$ denotes the space of **tempered distributions**, that is the distributions that can be extended as linear functionals acting continuously on $\mathcal{S}(\mathbb{R}^n)$ with respect to the convergence in $\mathcal{S}(\mathbb{R}^n)$ (see Definition 3.17).

A sequence u_m of distributions in $\mathcal{S}'(\mathbb{R}^n)$ converges in $\mathcal{S}'(\mathbb{R}^n)$ to $u \in \mathcal{S}'(\mathbb{R}^n)$ (notation $v_m \overset{\mathcal{S}'(\mathbb{R}^n)}{\to} v$) iff

$$\int_{\mathbb{R}^n} u_m\, v \;\to\; \int_{\mathbb{R}^n} u\, v \qquad \forall v \in \mathcal{S}(\mathbb{R}^n).$$

Hence

$$u_m \overset{\mathcal{S}'(\mathbb{R}^n)}{\to} u \;\Rightarrow\; |\mathbf{x}|^k D^{\boldsymbol{\alpha}} u_m \overset{\mathcal{S}'(\mathbb{R}^n)}{\to} |\mathbf{x}|^k D^{\boldsymbol{\alpha}} u,\ \forall k \in \mathbb{N}, \forall \boldsymbol{\alpha} \in \mathbb{N}^n. \tag{4.40}$$

Definition 4.10. *Assume that $u \in \mathcal{S}'(\mathbb{R}^n)$.*
Then the **Fourier transform** *of u is the tempered distribution $\widehat{u} \in \mathcal{S}'(\mathbb{R}^n)$ defined by*

$$\int_{\mathbb{R}^n} \widehat{u}\, v = \int_{\mathbb{R}^n} u\, \widehat{v} \qquad \forall v \in \mathcal{S}(\mathbb{R}^n). \tag{4.41}$$

Then

$$u_m \to u \text{ in } \mathcal{S}'(\mathbb{R}^n) \qquad \Rightarrow \qquad \widehat{u_m} \to \widehat{u} \text{ in } \mathcal{S}'(\mathbb{R}^n). \tag{4.42}$$

The definition of Fourier transform in $\mathcal{S}'(\mathbb{R}^n)$ extends the definition in L^1; for this reason it is denoted by same the symbol $\mathcal{F}$.

The Fourier transform is one-to-one in $\mathcal{S}'(\mathbb{R}^n)$ and fulfils

$$\mathcal{F}^{-1}u(\mathbf{x}) = (2\pi)^{-n} (\mathcal{F}u)(-\mathbf{x}) \qquad \forall u \in \mathcal{S}'(\mathbb{R}^n). \tag{4.43}$$

All algebraic properties of Fourier transform $\mathcal{F}$ stated by Theorem 4.2 for $u \in L^1(\mathbb{R}^n)$ hold true for any $u \in \mathcal{S}'(\mathbb{R}^n)$ too.
Also the usual relations between differentiation and polynomial multipliers, that are useful in solving differential equations, keep unchanged in $\mathcal{S}'(\mathbb{R}^n)$.

Theorem 4.11. *If* $u \in \mathcal{S}'(\mathbb{R}^n)$ *and* $j = 1, \ldots, n$, *then*

$$\widehat{\frac{\partial u}{\partial x_j}}(\xi) = i\,\xi_j\,\widehat{u}(\xi), \tag{4.44}$$

$$\widehat{x_j u} = i\,\frac{\partial \widehat{u}}{\partial \xi_j}. \tag{4.45}$$

Theorem 4.12. *If* $u \in \mathcal{S}'(\mathbb{R}^n)$ *then:*

$$u(\mathbf{x}) = (2\pi)^{-n}\,\widehat{\widehat{\widehat{u}}}(\mathbf{x}) \tag{4.46}$$

$$\widehat{\widehat{u}}(\mathbf{x}) = (2\pi)^n\,u(-\mathbf{x}) \tag{4.47}$$

$$\frac{1}{(2\pi)^{2n}}\,\widehat{\widehat{\widehat{\widehat{u}}}} = u \tag{4.48}$$

$$\|\,\widehat{u}\,\|_{L^2(\mathbb{R}^n)} = (2\pi)^{n/2}\,\|\,u\,\|_{L^2(\mathbb{R}^n)}. \tag{4.49}$$

The speed of decay at infinity for u is deeply connected with the regularity properties of its transform $\widehat{u}$.

The next Theorem states precisely the strong regularity characterizing the Fourier transform of compactly supported distributions (that is $u \in \mathcal{E}' \subset \mathcal{S}'$).

Theorem 4.13. ***(Paley-Wiener)*** *Assume that* $u \in \mathcal{S}'(\mathbb{R}^n)$. *Then*

$\operatorname{spt} u \subset \overline{B_R(\mathbf{0})}$ *iff* $\widehat{u}$ *fulfils the following two properties:*

- $\widehat{u}$ *is an analytic function in* $\mathbb{R}^n$ *that can be extended to an entire function in* $\mathbb{C}^n$ *(that is* $\widehat{u}$ *is an analytic function of* n *complex variables with no singularity),*
- *there exists* A, B *s.t.* $|\widehat{u}(\mathbf{z})| \le A\,(1+|\mathbf{z}|)^B \exp\left(R|\operatorname{Im}\mathbf{z}|\right) \quad \forall \mathbf{z} \in \mathbb{C}^n$, *where* $\operatorname{Im}\mathbf{z} = (\operatorname{Im} z_1, \operatorname{Im} z_2, \ldots, \operatorname{Im} z_n)$.

We emphasize that in order to evaluate the Fourier transform of a distribution u with compact support we can skip the implicit formula (4.41): indeed it is sufficient the evaluation of the pairing: $\int_{\mathbb{R}^n} u\, e^{-i\boldsymbol{\xi}\cdot\mathbf{x}}\, d\mathbf{x}$.
For instance $\delta \in \mathcal{S}'(\mathbb{R}^n)$ is a distribution with compact support and fulfils: $\widehat{\delta}(\boldsymbol{\xi}) = \int_{\mathbb{R}^n} \delta(\mathbf{x}) e^{-i\boldsymbol{\xi}\cdot\mathbf{x}} d\mathbf{x} = e^0 = 1$, hence

$$\widehat{\delta} = 1\ .$$

4.1.5 Convolution

We consider the convolution and its main properties, in both frameworks of functions and of tempered distributions.

Definition 4.14. *Given* $u, v \in L^1(\mathbb{R}^n)$, *we define another function* $u * v$, *called* **convolution of** u **and** v, *by*

$$u * v\ (\mathbf{x}) \ = \ \int_{\mathbb{R}^n} u(\mathbf{x}-\mathbf{y}) v(\mathbf{y})\, d\mathbf{y}$$

Convolution is bilinear, commutative e associative in $L^1(\mathbb{R}^n)$. Moreover

- $\left\{ u \in L^1(\mathbb{R}^n),\ v \in L^1_{loc}(\mathbb{R}^n), \right\} \quad \Rightarrow \quad u * v \in L^1_{loc}(\mathbb{R}^n)$.
- $\left\{ u \in L^1(\mathbb{R}^n),\ v \in L^p(\mathbb{R}^n),\ p \in [1,+\infty] \right\} \quad \Rightarrow$
$$\{ u * v \in L^p(\mathbb{R}^n), \quad \|u * v\|_{L^p} \le \|u\|_{L^1} \|v\|_{L^p} \}\ .$$
- $\left\{ u,\ v \in L^2(\mathbb{R}^n) \right\} \ \Rightarrow \ \{ u * v \in L^\infty(\mathbb{R}^n), \quad \|u * v\|_{L^\infty} \le \|u\|_{L^2} \|v\|_{L^2} \}$.
- **Young Theorem**
$$\left\{ u \in L^p(\mathbb{R}^n),\ v \in L^q(\mathbb{R}^n), \quad p,q,r \in [1,+\infty]\,, \quad \frac{1}{r} = \frac{1}{p} + \frac{1}{q} - 1 \right\} \quad \Rightarrow$$
$$\{ u * v \in L^r(\mathbb{R}^n), \quad \|u * v\|_{L^r} \le \|u\|_{L^p} \|v\|_{L^q} \}\ .$$

Definition 4.15. *Given* $u \in \mathcal{S}'(\mathbb{R}^n)$ *and* $v \in \mathcal{S}(\mathbb{R}^n)$ *we define another function* $u * v \in C^\infty \cap \mathcal{S}'(\mathbb{R}^n)$, *called* **convolution of** u **and** v, *by*

$$u * v\ (\mathbf{x}) \ = \ \int_{\mathbb{R}^n} u(\mathbf{x}-\mathbf{y})v(\mathbf{y})\, d\mathbf{y}\,.$$

Definition 4.16. *Given* $u,\, v \ \in \ \mathcal{S}'(\mathbb{R}^n)$, *such that at least one of the two has compact support, we define another tempered distribution* $u * v$, *called* **convolution of** u **and** v *by*

$$\int_{\mathbb{R}^n} (u * v)\ \psi \ = \ \int_{\mathbb{R}^n} u\,(\check{v} * \psi) \qquad \forall \psi \in \mathcal{S}(\mathbb{R}^n)\,.$$

Theorem 4.17. *If* $u, \,, v \ \in \mathcal{S}(\mathbb{R}^n)$, *then*

$$\widehat{u * v} \ = \ \widehat{u}\,\widehat{v} \tag{4.50}$$

$$\widehat{u\,v} \ = \ (2\pi)^{-n}\ \widehat{u} * \widehat{v}\ . \tag{4.51}$$

Formula (4.50) holds true in the following cases too, where $p' = p/(p-1)$:

i) $\quad u \in L^1(\mathbb{R}^n),\ v \in L^p(\mathbb{R}^n) : \quad p \in [1,2]\,;$

ii) $\quad u \in L^1(\mathbb{R}^n),\ v \in L^p(\mathbb{R}^n) \cap \mathcal{F}\Big(L^{p'}(\mathbb{R}^n)\Big) : \quad p \in [2,\infty]\,;$

iii) $\quad u \in L^p(\mathbb{R}^n) \cap \mathcal{F}\Big(L^{p'}(\mathbb{R}^n)\Big),\ v \in L^q(\mathbb{R}^n) \cap \mathcal{F}\Big(L^{q'}(\mathbb{R}^n)\Big),\ \exists r :$

$$\frac{1}{r} = \frac{1}{p} + \frac{1}{q} - 1\,, \qquad p,\, q \in [1,\infty]\,, \qquad r \in [1,2]\,;$$

iv) $\quad u \in \mathcal{S}'(\mathbb{R}^n),\ v \in \mathcal{S}(\mathbb{R}^n)\,.$

Formula (4.51) holds true in the following cases too:

i) $\quad u,\, v \in L^2(\mathbb{R}^n)\,;$

ii) $\quad u \in L^1(\mathbb{R}^n),\ \ v \in L^\infty(\mathbb{R}^n) \cap \mathcal{F}\Big(L^1(\mathbb{R}^n)\Big)\,;$

iii) $\quad u \in L^p(\mathbb{R}^n) \cap \mathcal{F}\Big(L^{p'}(\mathbb{R}^n)\Big),\ v \in L^{p'}(\mathbb{R}^n) \cap \mathcal{F}\big(L^p(\mathbb{R}^n)\big) : \quad p \in [1,\infty]\,;$

iv) $\quad u \in \mathcal{S}'(\mathbb{R}^n),\ v \in \mathcal{S}(\mathbb{R}^n)\,.$

Remark 4.18. Beyond the properties described in Riemann-Lebesgue Lemma (Theorem 4.3), the set of Fourier transform of $L^1(\mathbb{R}^n)$ functions cannot be simply characterized.
We introduce the symbol $\mathcal{V}$ to shortly denote the linear space which is the image of $L^1(\mathbb{R}^n)$ via Fourier transform:

$$\mathcal{V} \ := \ \{ v \in C^0(\mathbb{R}^n) \cap L^\infty(\mathbb{R}^n) :\ v = \widehat{w},\ \text{with } w \in L^1(\mathbb{R}^n) \}\,. \tag{4.52}$$

The function g, defined by

$$g(\xi) \;=\; \begin{cases} \xi/e & \text{if } |\xi| < e\,, \\ \operatorname{sign}(\xi)/\ln|\xi| & \text{if } |\xi| > e\,, \end{cases} \tag{4.53}$$

is an example of bounded and continuous function of one variable which is infinitesimal at infinity, but it is not the Fourier transform of any $L^1(\mathbb{R})$ function (see Exercise 4.23 for the details).
On the other hand, the Bessel function J_0 provides an example of function in $\mathcal{V}$ that belongs neither to $L^1(\mathbb{R})$ nor to $L^2(\mathbb{R})$. In fact

$$J_0(\xi) = \frac{1}{\pi}\mathcal{F}\left\{\frac{\chi_{(-1,1)}(x)}{(1-x^2)^{1/2}}\right\} \qquad \text{with} \qquad \frac{\chi_{(-1,1)}(x)}{(1-x^2)^{1/2}} \in L^1(\mathbb{R})\setminus L^2(\mathbb{R})$$

(see Remarks 3 and 4 in Exercise 4.21).
Nevertheless, referring to the Fourier transform of tempered distributions, these relations holds true:

$$\mathcal{V} \subset L^\infty(\mathbb{R}^n) \subset \mathcal{S}'(\mathbb{R}^n)\,, \qquad \mathcal{F}(\,\mathcal{V}\,) \;=\; L^1(\mathbb{R}^n)\,. \tag{4.54}$$

4.1.6 Solving differential equations with Fourier transform

We consider the differential equation

$$\{\,\text{given } f \in \mathcal{S}'(\mathbb{R}), \quad \text{find } u \in \mathcal{S}'(\mathbb{R}) : \quad Lu = f \text{ in } \mathbb{R}\,\} \tag{4.55}$$

where $L = \sum_{k=0}^{m} c_k\, d^k/dx^k$ is a linear differential operator of order m with constant coefficients, that is the c_k's are given constants.
Explicitly the differential equation in (4.55) reads

$$\sum_{k=0}^{m} c_k\, u^{(k)} = f\,.$$

We denote by P the *symbol of the differential operator* L, that is the polynomial $P(\xi) = \sum_{k=0}^{m} c_k\, i^k\, \xi^k$ of degree m depending on the variable ξ.
Transforming both sides of the differential equation we obtain

$$P(\xi)\,\widehat{u}(\xi) \;=\; \widehat{f}(\xi)\,.$$

- If P has no real zeroes (that $P(\xi) \neq 0\ \forall\xi \in \mathbb{R}$),
  then the equation (4.55) has a unique solution $u = \mathcal{F}^{-1}\left(\widehat{f}(\xi)/P(\xi)\right)$.
- If P has real zeroes,
  then the solution of (4.55) is not unique: all solutions can be found by solving a division problem in in the distributional framework (see (3.15),(3.16) and Exercises 4.32, 4.33.

- A fundamental solution of (4.55) is a solution U of the problem

$$\text{find } U \in \mathcal{S}'(\mathbb{R}) : \quad LU = \delta \text{ in } \mathbb{R}. \tag{4.56}$$

- If $\operatorname{spt} f$ is compact, then the knowledge of the fundamental solution U allows the evaluation of the solution u of (4.55) by the following formula

$$u = U * f. \tag{4.57}$$

In fact, since to compute derivatives of a convolution product it is enough to evaluate them on one single factor, we find

$$Lu = L(U * f) = (LU) * f = \delta * f = f.$$

4.1.7 Alternative definitions

Several variations for Definition 4.15 of Fourier transform are present in specialist literature: in these variations the multiplier $e^{-i\boldsymbol{\xi}\cdot\mathbf{x}}$ that appears in the integral may be substituted by different ones, like

(a) $e^{i\boldsymbol{\xi}\cdot\mathbf{x}}$,

(b) $e^{\pm 2\pi i\boldsymbol{\xi}\cdot\mathbf{x}}$,

(c) $\dfrac{1}{(2\pi)^{n/2}}\, e^{\pm i\boldsymbol{\xi}\cdot\mathbf{x}}$.

For this reason it is expedient being careful when using formulas and transform tables, by double-checking which definitions they refer to.
All different definitions amount only to slight changes in formulas, like substitution of some multiplicative constants or variables dilation.
For instance, the variation (a) keeps unchanged the Plancherel identity given by (4.36); whereas, under variation (b) or (c) the constant 1 replaces $(2\pi)^{n/2}$ in the Plancherel identity.
So the choices (b) and (c) make $\mathcal{F}$ an isometry in $L^2(\mathbb{R}^n)$, the same for $\mathcal{F}^{-1}$.
On the other hand all qualitative properties of $\mathcal{F}$ are left unchanged by any variant of the definition: for instance (4.32) and Theorem 4.3 hold too, with no change under variations (a),(b) and (c).

Actually, recalling that the transform $\mathcal{F}$ is exploited as a wide-range problem-solving tool, we emphasize that **the solutions of the problems are not affected by the selected definition**, while the transform of the solutions is modified according to the choice of the multiplier.
We advise that software Mathematica 9. © (Wolfram) provides the option "FourierParameters" to select the transform parameters according to different conventions used in Mathematics, Physics, Signal Theory and Systems Engineering ([27]).

4.2 Exercises

4.2.1 Exercises on $\mathcal{F}$ transform in $L^1(\mathbb{R})$ and $L^2(\mathbb{R})$

Exercise 4.1. Given $u(x) = \dfrac{1}{1+x^2}$, $x \in \mathbb{R}$. Evaluate $\mathcal{F}\{u\}$.

Solution - We consider the analytic function f depending on the complex variable z : $f(z) = \dfrac{1}{1+z^2}$, analytic extension of u. The singularities of f are the simple poles $\pm i$.

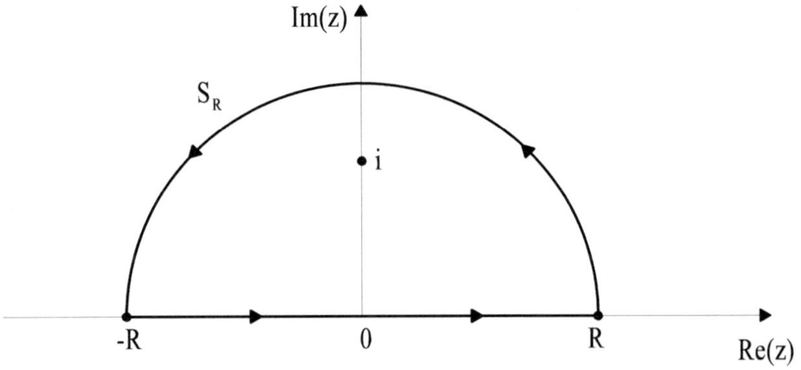

Figure 4.1 - Loop Σ_R of Exercise 4.1 .

let Σ_R be the loop composed by the segment $[-R, +R]$ contained in the real axis and by the half-circumference S_R with radius $R > 1$ centered at the origin and contained in the upper half-plane: Σ_R is oriented in the positive sense (counterclockwise) and swept once. Since $|f(z)| \sim 1/|z|^2$ as $|z| \to \infty$ we can apply the Residues Theorem and Jordan Lemma:

$$\int_{-\infty}^{+\infty} \frac{\exp(-i\xi x)}{1+x^2}\, dx = 2\pi i \,\mathrm{Res}\left(\frac{\exp(-i\xi z)}{1+z^2}, z = i\right) = \pi\, e^{\xi} \qquad \forall \xi < 0$$

Since f is an even function, $\widehat{f}$ is even too, and its values at $\xi > 0$ can be deduced from the values at $\xi < 0$, by $\widehat{f}(-\xi) = \widehat{f}(\xi)$.
Moreover $f \in L^1$ implies that $\widehat{f} \in C^0$, then $\widehat{f}(0)$ can be deduced from values $\widehat{f}(\xi)$ with $\xi \neq 0$, by continuity. Summarizing

$$\widehat{f}(\xi) = \pi \exp(-|\xi|) \qquad \forall \xi \in \mathbb{R}.$$

□

Exercise 4.2. Evaluate the Fourier transform of the Gaussian function e^{-x^2}, $x \in \mathbb{R}$.

Solution - The function of real variable e^{-x^2} admits a unique entire analytic extension $f(z) = e^{-z^2}$, $z \in \mathbb{C}$ this extension is not infinitesimal at infinity: indeed, the point at infinity of the complex plane is an essential singularity of f. So in this case neither it is possible to apply the Jordan Lemma nor there is hope for other kind of estimates entailing that the integral on half-circumferences centered at the origin is an infinitesimal. Therefore we use an "ad hoc" strategy, by exploiting the following identity

$$\int_{-\infty}^{+\infty} \exp\left[-(x+ib)^2\right] dx = \int_{-\infty}^{+\infty} \exp(-x^2)\, dx \overset{(2.11)}{=} \sqrt{\pi} \quad \forall b \in \mathbb{R}. \tag{4.58}$$

Hence

$$\begin{aligned} \mathcal{F}\{\exp(-x^2)\}(\xi) &= \int_{-\infty}^{+\infty} \exp(-i\xi x) \exp(-x^2)\, dx = \\ &= \exp(-\xi^2/4) \int_{-\infty}^{+\infty} \exp\left[-(x+i\xi/2)^2\right] dx = \\ &= \sqrt{\pi}\exp(-\xi^2/4) . \end{aligned}$$

We are left only to prove (4.58). To this aim we consider the entire function $z \to \exp(-z^2)$ (analytic extension of $\exp(-x^2)$, $x \in \mathbb{R}$) and the loop Γ_R in Figure 4.2.

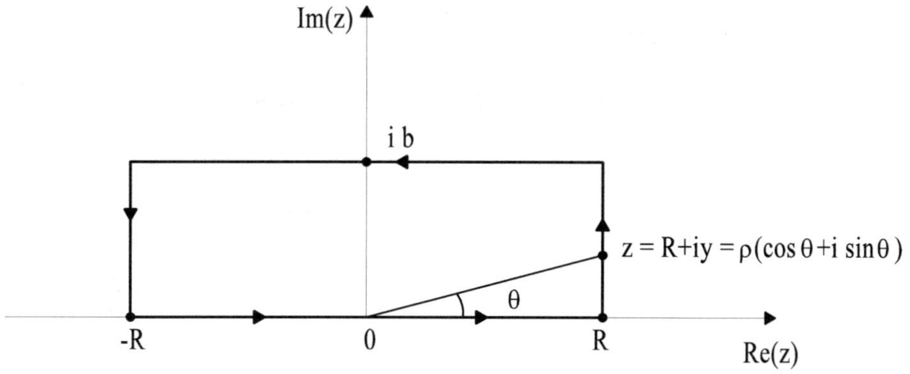

Figure 4.2 - Loop Γ_R of Exercise 4.2 .

Due to the Cauchy Theorem

$$\oint_{\Gamma_R} \exp(-z^2)\, dz = 0 \quad \forall R > 0.$$

Moreover, setting $A = -R$, $B = R$, $C = R + ib$, $D = -R + ib$, we have

$$\lim_{R \to +\infty} \int_B^C \exp(-z^2)\, dz = 0$$

since, if $R > 2b$, we obtain, for any $z \in BC$, $z = |z|(\cos\theta + i\sin\theta)$ e

$$\begin{aligned} |\exp(-z^2)| &= \exp(-|z|^2 \cos 2\theta) \le \exp(-R^2 \cos 2\theta) = \\ &= \exp\left(-R^2(\cos^2\theta - \sin^2\theta)\right) = \exp\left(-R^2 \cos(2\vartheta)\,.\right) \le \\ &\le \exp(-R^2/4)\,. \end{aligned}$$

Analogously we can prove $$\lim_{R\to+\infty}\int_D^A \exp(-z^2)\,dz = 0\,.$$

Then
$$\int_A^B \exp(-z^2)\,dz = -\int_C^D \exp(-z^2)\,dz + o(1/R)$$
and, passing to the limit as $R\to+\infty$, we get (4.58).
So we have found the transform of the Gaussian function for $n=1$:

$$\mathcal{F}\{\exp(-x^2)\}(\xi) = \sqrt{\pi}\,\exp(-\xi^2/4)\,.$$

Another method for solving the exercise: to achieve the transform, we can consider the first order homogeneous linear differential equation

$$u'(x) = -2xu(x) \qquad x\in\mathbb{R}$$

whose solution is $\exp(-x^2)$ (notice that both $x\exp(-x^2)$ and $\dfrac{d}{dx}\exp(-x^2)$ are integrable functions), then transform this equation and find the differential equation:

$$i\xi\widehat{u}(\xi) = -2i\,\widehat{u}'(\xi)\,.$$

The general integral is of the kind $C\exp(-\xi^2/4)$, where we have to choose

$$C = \widehat{u}(0) = \int_{\mathbb{R}} \exp(-t^2)\,dt = \sqrt{\pi}\,.$$

□

Exercise 4.3. Set $f(x) = 5x\cos(7x)\,(9x^2+4)^{-2}$ $x\in\mathbb{R}$ and evaluate $\widehat{f}(\xi)$.
Solution - By setting
$$\psi(x) = (9x^2+4)^{-1}$$
we find
$$\psi'(x) = -\,18\,x\,(9x^2+4)^{-2}\,,$$
hence
$$f(x) = -\frac{5}{18}\psi'(x)\cos 7x = -\frac{5}{36}\psi'(x)\left[e^{7ix}+e^{-7ix}\right].$$
Formulas $\mathcal{F}\{u(x)\,e^{iax}\} = \widehat{u}(\xi-a)$ and $\widehat{u'} = +i\xi\widehat{u}$ imply

$$(*) \qquad \widehat{f}(\xi) = -\frac{5}{36}\,i\left[(\xi+7)\widehat{\psi}(\xi+7)+(\xi-7)\widehat{\psi}(\xi-7)\right].$$

To evaluate $\widehat{\psi}$, we notice that it is an even and continuous function, since ψ is even and belongs to L^1. We apply complex variable techniques: the Jordan Lemma (Lemma 1.34 (i)) and Residues Theorem (Theorem 1.30). We select $\xi<0$ in the half-plane $\operatorname{Im} z>0$, then

$$\widehat{\psi}(\xi) = 2\pi i\,\operatorname{Res}\left(e^{-i\xi z}(9z^2+4)^{-1},\frac{2}{3}i\right) = 2\pi i\frac{e^{2/3\xi}}{12i} = \frac{\pi}{6}e^{2/3\xi} \qquad \xi<0$$

$$\widehat{\psi}(\xi) = \frac{\pi}{6}e^{-2/3|\xi|} \qquad \forall\xi\in\mathbb{R}\,.$$

From $(*)$ we deduce

$$\widehat{f}(\xi) = -\frac{5\pi i}{216}\left[(\xi+7)\,e^{-2|\xi+7|/3} + (\xi-7)\,e^{-2|\xi-7|/3}\right] \qquad \forall\xi\in\mathbb{R}\,.$$

□

Exercise 4.4. With the same notation of Exercise 4.3, set

$$F(x) = \int_x^{+\infty} f(t)\, dt \qquad x \in \mathbb{R}.$$

Show that F is well defined and Fourier transformable.
Perform a qualitative *a priori* study of $\widehat{F}$.
Then evaluate $\widehat{F}$.

Solution - f has no singularity on the real axis and fulfils $|f(x)| \leq \dfrac{C}{|x|^3}$ for $|x| \gg 0$.
Then f belongs to $L^1(\mathbb{R})$ and F is well defined (because it is an antiderivative of f with sign changed).Since f is odd, F is even. It suffices to study F for ≥ 0.

$$|F(x)| \leq \int_x^{\infty} |f(t)|\, dt \leq \int_x^{\infty} \frac{5t}{(9t^2+4)^2}\, dt = \frac{5}{18}\, \frac{1}{9x^2+4}\,.$$

The properties $(9x^2+4)^{-1} \in L^1 \cap L^2$ and $x(9x^2+4)^{-1} \in L^2$ imply

$$F \in L^1(0,\infty) \cap L^2(0,\infty) \qquad \text{e} \qquad xF \in L^2(0,\infty)\,.$$

We observe that F even, $F \in L^1(\mathbb{R}) \cap L^2(\mathbb{R})$ and $xF \in L^2(\mathbb{R})$. Hence $\widehat{F}$ is well defined, continuous, infinitesimal at infinity and its squared modulus is integrable.
Moreover $\widehat{F} \in AC_{loc}(\mathbb{R})$ (see Definition 2.6) and $d\widehat{F}/d\xi \in L^2(\mathbb{R})$.
Since F is real and even, we deduce that $\widehat{F}$ is real and even too.
Since $f \in C^\infty$ and $F' = -f$, we deduce that $F \in C^\infty(\mathbb{R})$.
Moreover it is easy to verify that $f^{(n)} \in L^1 \cap L^2$ for any n; hence $F^{(n)} \in L^1 \cap L^2$ for any n, $\xi^n \widehat{F} \in L^2$ and is an infinitesimal at infinity, for any n. Summarizing $\widehat{F}$ is an infinitesimal faster than any negative real power of ξ at infinity.
To compute explicitly $\widehat{F}$ we start from $F' = -f$: we get $+i\xi\widehat{F}(\xi) = -\widehat{f}(\xi)$ and

$$\widehat{F}(\xi) = \frac{5\pi}{216\xi}\left[(\xi+7)e^{-2|\xi+7|/3} + (\xi-7)e^{-2|\xi-7|/3}\right] \qquad \xi \neq 0$$

eventually $\widehat{F}(0) = \int_{\mathbb{R}} f(x)dx = 0$ since f is odd e integrable in $\mathbb{R}$. □

Exercise 4.5. Evaluate the Fourier transform of the function

$$f(x) = \frac{\big(\sin(3x)\big)^2}{x^2+2x+2}\,, \qquad x \in \mathbb{R}.$$

Solution - After setting

$$g(x) = (x^2+2x+2)^{-1},$$

we obtain

$$f(x) = \left(\frac{1}{2} - \frac{1}{2}\cos 6x\right) g(x) = \frac{1}{2}g(x) - \frac{1}{4}g(x)\, e^{6ix} - \frac{1}{4}g(x)\, e^{-6ix}.$$

If $\xi < 0$, by applying the residues Theorem and the Jordan Lemma to the function $g(z) = (z^2+2z+2)^{-1}$ in the half-plane $\operatorname{Im} z > 0$, we get

$$\widehat{g}(\xi) = 2\pi i \operatorname{Res}\big(g(z)e^{-i\xi z}, -1+i\big) = \pi e^{\xi(1+i)} \qquad \forall \xi < 0\,.$$

Analogously if $\xi > 0$ we consider the half-plane $\operatorname{Im} z < 0$

$$\widehat{g}(\xi) = -2\pi i \operatorname{Res}\big(g(z)e^{-i\xi z}, -1-i\big) = \pi e^{-\xi(1-i)} \qquad \forall \xi > 0\,.$$

From $g \in L^1$, we find that $\widehat{g}$ is continuous. We conclude

$$\widehat{g}(\xi) = \pi e^{+i\xi-|\xi|} \qquad \forall \xi \in \mathbb{R}$$

$$\widehat{f}(\xi) = \frac{\pi}{2} e^{+i\xi-|\xi|} - \frac{\pi}{4} e^{+i(\xi+6)-|\xi+6|} - \frac{\pi}{4} e^{+i(\xi-6)-|\xi-6|} \qquad \forall \xi \in \mathbb{R}.$$

□

Exercise 4.6. Convey f in terms of elementary functions, where

$$f(x) = \frac{1}{2\pi} \int_{-\infty}^{+\infty} e^{\,i\xi x-|\xi|} \cdot \frac{\sin\xi}{\xi}\, d\xi \qquad \forall x \in \mathbb{R}\,.$$

Solution - We can write

$$f(x) = \mathcal{F}^{-1}\left\{ e^{-|\xi|} \cdot \frac{\sin\xi}{\xi} \right\} = \mathcal{F}^{-1}\left\{ e^{-|\xi|} \right\} * \mathcal{F}^{-1}\left\{ \frac{\sin\xi}{\xi} \right\}$$

We recall the formulas

$$\mathcal{F}^{-1}\left\{ e^{-|\xi|} \right\} = \frac{1}{\pi} \frac{1}{x^2+1}$$

$$\mathcal{F}^{-1}\left\{ \frac{\sin\xi}{\xi} \right\} = \frac{1}{2}\, p(x)\,,$$

where $p(x) = 1$ se $|x| \le 1$, $p(x) = 0$ se $|x| > 1$. We deduce

$$f(x) = \frac{1}{2\pi} \frac{1}{x^2+1} * p(x) = \frac{1}{2\pi} \int_{-\infty}^{+\infty} \frac{p(x-t)}{t^2+1}\, dt = \frac{1}{2\pi} \int_{x-1}^{x+1} \frac{1}{t^2+1}\, dt =$$

$$= \frac{1}{2\pi} [\arctan(x+1) - \arctan(x-1)].$$

□

Exercise 4.7. Evaluate the Fourier transform of the function

$$f(x) = \frac{x e^{3-5ix}}{(3x-2i)^2} \qquad x \in \mathbb{R}.$$

Solution -

$$\widehat{f}(\xi) = e^3 \int_{-\infty}^{+\infty} e^{-ix(\xi+5)} \frac{x}{(3x-2i)^2}\, dx.$$

Since $|\psi(z)| \le C/|z|$ if $|z| \ge R_0$ with suitable R_0 and C, we can apply the Jordan Lemma to the function $\psi(z) = z/(3z-2i)^2$.
The function $\psi(z)\, e^{-iz(\xi+5)}$ has exactly one singularity: a double pole at $z = 2/3\, i$. The residue r is given by

$$r = \lim_{z\to\frac{2}{3}i} \frac{d}{dz} \left[e^{-iz(\xi+5)} \frac{z(z-2i/3)^2}{(3z-2i)^2} \right] = \frac{1}{9} \lim_{z\to\frac{2}{3}i} \frac{d}{dz} \left[z e^{-iz(\xi+5)} \right] =$$

$$= \frac{1}{9} \left[1 + \frac{2}{3}(\xi+5) \right] \; e^{2(\xi+5)/3}$$

Then choosing the loop in the half-plane $\operatorname{Im} z \ge 0$ or $\operatorname{Im} z \le 0$, according as $\xi < -5$ or $\xi > -5$, we obtain:

$$\widehat{f}(\xi) = 2\pi i e^3\, r = \frac{2\pi i e^3}{9} \left[1 + \frac{2}{3}(\xi+5) \right] \; e^{2(\xi+5)/3} \qquad \text{if } \xi < -5$$

$$\widehat{f}(\xi) = 0 \qquad \text{if } \xi > -5\,.$$

If H is the Heaviside function: $$H(x) = \begin{cases} 0 & \text{when } x < 0 \\ 1 & \text{when } x \geq 0 \end{cases}$$

we can write:
$$\widehat{f}(\xi) = \frac{2\pi i e^3}{27}(2\xi + 13)\, H(-\xi - 5)\, e^{2(\xi+5)/3}.$$

□

Exercise 4.8. Evaluate the Fourier transform of the function
$$f(x) = x(x^2+4)^{-2} \qquad x \in \mathbb{R}.$$

Solution - By setting
$$g(x) = (x^2+4)^{-1}$$
we obtain
$$f(x) = -\frac{1}{2}\, g'(x)$$
By exploiting (4.3), (4.5), $\mathcal{F}\{(x^2+1)^{-1}\} = \pi e^{-|\xi|}$ and
$$g(x) = \frac{1}{4}\left(\frac{x^2}{4}+1\right)^{-1}$$
we deduce
$$\widehat{g}(\xi) = \frac{\pi}{2} e^{-2|\xi|}$$
$$\widehat{f}(\xi) = -\frac{1}{2}\,\widehat{g'}(\xi) = -\frac{1}{2}\left(i\xi\,\widehat{g}(\xi)\right) = -\frac{\pi i}{4}\,\xi\, e^{-2|\xi|}.$$
Alternatively, we can compute the transform of f directly by (4.1), exploiting complex analysis techniques. □

Exercise 4.9. Evaluate the Fourier transform of the function
$$g(x) = \frac{7\sin(3x-2)}{x^2+25} - 2\,\exp(4 - 5|x-3|).$$

Solution - Setting
$$h_1(x) = \frac{1}{x^2+25} \qquad h_2(x) = e^{-5|x|}$$
we obtain
$$\begin{aligned} g(x) &= 7\, h_1(x)\, \sin(3x-2) - 2\, e^4\, h_2(x-3) = \\ &= \frac{7}{2i}\, h_1(x)\left[e^{i(3x-2)} - e^{-i(3x-2)}\right] - 2\, e^4\, h_2(x-3)\,, \end{aligned}$$
$$\widehat{g}(\xi) = \frac{7\, e^{-2i}}{2i}\,\widehat{h_1}(\xi-3) - \frac{7\, e^{2i}}{2i}\,\widehat{h_1}(\xi+3) - 2\, e^4\, e^{-3i\xi}\,\widehat{h_2}(\xi)\,.$$
Applying the Jordan Lemma and Residues Theorem we obtain
$$\widehat{h_1}(\xi) = 2\pi i \,\text{Res}\left(h_1(z)e^{-i\xi z}, 5i\right) = \frac{\pi}{5}\, e^{5\xi} \qquad \xi < 0\,.$$

Since h_1 is an even function, the same property holds true for $\widehat{h}_1$. The value at $\xi = 0$ is achieved by continuity of $\widehat{h_1}$ (a consequence of $h_1 \in L^1$). Then:
$$\widehat{h_1}(\xi) = \frac{\pi}{5}\, e^{-5|\xi|} \qquad \forall \xi \in \mathbb{R}.$$

Moreover

$$\widehat{h_2}(\xi) = \int_{-\infty}^{0} e^{5x-i\xi x}\, dx + \int_{0}^{+\infty} e^{-5x-i\xi x}\, dx = \frac{1}{5-i\xi} + \frac{1}{5+i\xi} = \frac{10}{\xi^2+25} \qquad \forall \xi \in \mathbb{R}.$$

Summarizing:

$$\widehat{g}(\xi) = \frac{7\, e^{-2i}}{2i}\,\frac{\pi}{5}\, e^{-5|\xi-3|} - \frac{7\, e^{2i}}{2i}\,\frac{\pi}{5}\, e^{-5|\xi+3|} + e^{4+3i\xi}\,\frac{20}{\xi^2+25}\,.$$

□

Exercise 4.10. Evaluate the Fourier transform of the function

$$f(x) = \frac{1-\cos 6x}{x} \qquad x \in \mathbb{R}\,.$$

Solution - We exploit the trigonometric formulas

$$\frac{1-\cos 6x}{x} = 2\frac{\sin^2 3x}{x} = 2\sin 3x\,\frac{\sin 3x}{x} = -i\left(e^{3ix} - e^{-3ix}\right)\,\frac{\sin 3x}{x}\,.$$

$x = 0$ is a removable singularity of f and the estimate $|f(x)| \leq 2/|x|$ hold true at infinity; then f belongs to $L^2(\mathbb{R})$, hence it is transformable.
After setting $p(\xi) = 1$ if $|\xi| < 1$ and $p(\xi) = 0$ if $|\xi| > 1$, the formula

$$\mathcal{F}\left\{\frac{\sin x}{x}\right\} = \pi\, p(\xi)$$

entails

$$\mathcal{F}\left\{\frac{\sin 3x}{x}\right\} = 3\mathcal{F}\left\{\frac{\sin 3x}{3x}\right\} = 3\frac{1}{3}\pi\, p\left(\frac{\xi}{3}\right) = \pi\, p\left(\frac{\xi}{3}\right)$$

$$\mathcal{F}\left\{\frac{1-\cos 6x}{x}\right\} = \pi i\left[p\left(\frac{\xi+3}{3}\right) - p\left(\frac{\xi-3}{3}\right)\right]$$

that is

$$\mathcal{F}\left\{\frac{1-\cos 6x}{x}\right\} = \begin{cases} +\pi i & \text{se } -6 < \xi < 0 \\ -\pi i & \text{se } 0 < \xi < 6 \\ 0 & \text{se } |\xi| > 6 \end{cases}$$

□

Exercise 4.11. Evaluate the following integral for any $\xi \in \mathbb{R}$:

$$f(\xi) = \int_{-\infty}^{+\infty} \frac{e^{ix(\xi^3-2\xi+6)}}{x^2+1}\, dx.$$

Solution - Set $\psi(x) = \dfrac{1}{x^2+1}$. Then $\widehat{\psi}(\xi) = \pi e^{-|\xi|}$. Hence

$$f(\xi) = \widehat{\psi}(\xi^3 - 2\xi + 6) = \pi\,\exp(-|\,\xi^3 - 2\xi + 6\,|).$$

□

Exercise 4.12. Perform a qualitative a priori study of the Fourier transform of

$$f(x) = \frac{x^2+x+1}{(x^2+1)^2}\,\sin(3x) \qquad x \in \mathbb{R}$$

without explicitly computing it.

Solution - Set $h(x) = (x^2 + x + 1)/(x^2 + 1)^2$.
From $h \in L^1 \cap L^2$ and $xh \in L^2$ we deduce that the transform $\widehat{h}$ is continuous, indeed absolutely continuous (Definition 2.6), it vanishes at infinity and $\widehat{h}, \widehat{h}' \in L^2$.
Moreover $h \in C^\infty$ with $h^{(n)} \in L^1 \cap L^2$ for any $n \geq 0$.
Then $\xi^n \widehat{h} \in L^2$ and tends to 0 at infinity for any n, that is $\widehat{h}$ has a fast decay.
It is not difficult to see that

$$\forall n \; \exists\, m \; : \; x^n h^{(m)} \in L^1.$$

Hence $\mathcal{F}\left\{x^n h^{(m)}\right\}$ is continuous, that is $\dfrac{d^n}{d\xi^n}(\xi^m \widehat{h})$ exists and is continuous. Summarizing

$$\forall n \; \exists\, m \; : \; \xi^m \widehat{h} \in C^n,$$

hence for any n, $\widehat{h} \in C^n$ for $\xi \neq 0$.
Hence $\widehat{h} \in C^\infty(\mathbb{R}\backslash\{0\})$. From the identity

$$\widehat{f}(\xi) = \frac{1}{2i}\left[\widehat{h}(\xi - 3) - \widehat{h}(\xi + 3)\right],$$

we obtain the conclusion: $\widehat{f}$ is absolutely continuous, $d\widehat{f}/d\xi \in L^2$, $\widehat{f}$ has fast decay at infinity, $\widehat{f}$ is C^∞ in the set $\{\xi \in \mathbb{R}\colon \xi \neq \pm 3\}$. □

Exercise 4.13. Evaluate the Fourier transform of the function

$$u(x) = 5x^2 e^{-2x} H(x),$$

where $H(x)$ is the Heaviside function ($H(x) = 0$ if $x < 0$, $H(x) = 1$ if $x \geq 0$).
Solution - Due to linearity property (4.2) and (4.4):

$$\widehat{u}(\xi) = 5\,\mathcal{F}\{x^2 e^{-2x} H(x)\} = 5\, i^2 \frac{d^2}{d\xi^2}\mathcal{F}\{e^{-2x} H(x)\}$$

By applying the definition (4.1) we get, for any $\xi \in \mathbb{R}$:

$$\mathcal{F}\{e^{-2x} H(x)\} = \int_{-\infty}^{+\infty} e^{-i\xi x}\, e^{-2x} H(x)\, dx = \int_0^{+\infty} e^{-i\xi x}\, e^{-2x}\, dx =$$

$$= \int_0^{+\infty} e^{-x(2+i\xi)}\, dx = -\frac{e^{-x(2+i\xi)}}{2+i\xi}\Bigg|_0^{+\infty} = \frac{1}{2+i\xi}$$

$$\widehat{u}(\xi) = 5\, i^2 \frac{d^2}{d\xi^2}\mathcal{F}\{e^{-2x} H(x)\} = -5\,\frac{d^2}{d\xi^2}\left(\frac{1}{2+i\xi}\right) = \frac{10}{(2+i\xi)^3}.$$

□

Exercise 4.14. Prove that

$$\mathcal{F}\{e^{-|x|}\} = \frac{2}{1+\xi^2} \qquad x,\, \xi \in \mathbb{R}.$$

Solution - By the definition (4.1), we obtain:

$$\mathcal{F}\{e^{-|x|}\} = \int_{-\infty}^{+\infty} e^{-i\xi x}\, e^{-|x|}\, dx = \int_{-\infty}^{+\infty} e^{-|x|}(\cos(\xi x) - i\sin(\xi x))\, dx =$$

;

$$= \int_{-\infty}^{+\infty} e^{-|x|}(\cos(\xi x))\, dx = 2\int_0^{+\infty} e^{-x}(\cos(\xi x))\, dx = \frac{2}{1+\xi^2}$$

integration by parts has been performed twice in the last step.

Alternative solution By taking into account (4.9), $\mathcal{F}\{1/(1+x^2)\} = \pi e^{-|\xi|}$ and the fact that $x \to e^{|x|}$ is an even function, we get:

$$\mathcal{F}\{e^{-|x|}\}(\xi) = 2\pi\,\mathcal{F}^{-1}\{e^{-|x|}\} = \frac{2}{1+\xi^2}\,.$$

□

Exercise 4.15. Evaluate the Fourier transform of the following function:

$$u(x) = 3xe^{-|x|}.$$

Solution - Due to the linearity property (4.2) and (4.4), we deduce:

$$\widehat{u}(\xi) = 3\,\mathcal{F}\{xe^{-|x|}\} = 3i\frac{d}{d\xi}\mathcal{F}\{e^{-|x|}\}\,.$$

We conclude by exploiting $\mathcal{F}\{e^{-|x|}\} = 2/(1+\xi^2)$:

$$\widehat{u}(\xi) = 3i\frac{d}{d\xi}\left(\frac{2}{\xi^2+1}\right) = \frac{-12i\xi}{(1+\xi^2)^2}$$

□

Exercise 4.16. Given $f \in L^1(\mathbb{R})$, find $u \in L^1(\mathbb{R})$ with u, u' absolutely continuous such that u is a solution of the differential equation

$$-\,u'' + u = f \qquad \text{in } \mathbb{R}\,.$$

Solution - If there exists such u, then $f, u, u', u'' \in L^1$, so we can transform both sides of the equation:

$$(1+\xi^2)\,\widehat{u}(\xi) = \widehat{f}(\xi) \qquad \xi \in \mathbb{R}.$$

The multiplier $1+\xi^2$ does not vanish for any $\xi \in \mathbb{R}$, so we can divide:

$$\widehat{u}(\xi) = \frac{\widehat{f}(\xi)}{1+\xi^2} \qquad \xi \in \mathbb{R}$$

and using Exercise 4.14 we obtain u by anti-transforming:

$$\mathcal{F}^{-1}\left\{\frac{1}{1+\xi^2}\right\} = \frac{1}{2}\,e^{-|x|}$$

$$u = \frac{1}{2}\,e^{-|x|} * f \tag{4.59}$$

that is $$u(x) = \frac{1}{2}\int_{-\infty}^{+\infty} \exp\left(-|x-y|\right) f(y)\,dy \qquad x \in \mathbb{R}\,.$$

we emphasize that the solution of the given equation

- exists for any integrable f,
- is unique due to the injectivity of $\mathcal{F}$,
- is explicitly computed explicitly up to an integration,
- $U = \frac{1}{2}\,e^{-|x|}$ is the fundamental solution of the differential operator $I - d^2/dx^2$.

The general integral of the homogeneous equation (dropping the request of L^1 solutions)

$$-v'' + v = 0 \qquad \text{in } \mathbb{R}$$

are linear combinations of $\exp(\pm x)$ and obviously they cannot be found by this technique since they do not belong to $L^1(\mathbb{R})$ (né ad $\mathcal{S}'(\mathbb{R})$).
The Fourier transform method allows to turn a differential problem into an algebraic equation, hence it helps in finding particular solutions of ordinary differential equations with constant coefficients and right-hand sides of general kind (possibly not continuous). If the polynomial multiplying the transformed solution has real zeroes then the transform of the solution may not be a function, in general it must be sought for in the space of tempered distributions (e.g. $v'' + v = f$ about whom see Exercises 4.32, 4.33). □

Exercise 4.17. Given $f,\ K \in L^1(\mathbb{R})$ study this integral equation in the unknown u.

$$\text{Find } u \in L^1(\mathbb{R}) : \qquad \int_{\mathbb{R}} K(x-y)\, u(y)\, dy \;=\; f(x) \qquad x \in \mathbb{R}. \tag{4.60}$$

Solution - Transforming both sides of the equation $K * u = f$, due to (4.50) (i), we obtain an algebraic problem

$$\widehat{K}(\xi)\,\widehat{u}(\xi) \;=\; \widehat{f}(\xi) \qquad \xi \in \mathbb{R}$$

If $\widehat{K}(\xi) \neq 0 \ \ \forall \xi \in \mathbb{R}$ and $(\widehat{f}/\widehat{K}) \in L^1{}_\xi$ then dividing times $\widehat{K}$ can be performed in an elementary way: $\widehat{u}(\xi) \;=\; K(\xi)^{-1}\,\widehat{f}(\xi)$. Anti-transforming we obtain the unique solution u:

$$u \;=\; \mathcal{F}^{-1}\,\{\widehat{f}/\widehat{K}\}\,.$$

Note - This exercise shows how the Fourier transform method provides a solving technique for integro-differential problems too, that is problems where the unknown appears under the action of both differential operators and convolution integrals.
In general the division problem and the inverse transform have to be faced and solved in the more general framework of distributions.
We notice that the integral equation (4.60) could have no solution in L^1: for instance consider the case of $K(x) = f(x) = \dfrac{1}{1+x^2}$; then $\widehat{f}(\xi) \neq 0 \ \ \forall \xi$ and possible solution should fulfil $\widehat{u}(\xi) \equiv 1 \ \forall \xi \in \mathbb{R}$, that is a contradiction with the Riemann-Lebesgue lemma. This remark also implies that there is no *"neutral element" with respect to the convolution product* in L^1, that is an element v fulfilling

$$f * v \;=\; f \qquad \forall f \in L^1\,.$$

Then any v fulfilling such equation cannot be an integrable function. Nevertheless (by defining the convolution in a wider framework than the one of functions (see Definition 4.15) the neutral element v is uniquely identified by the distribution "delta". □

Exercise 4.18. Find at least one solution of the following differential equation, by using the Fourier transform method:

$$u''(x) + 2xu'(x) + 4u(x) = 0 \qquad x \in \mathbb{R}\,.$$

Solution - After setting $y(\xi) = \widehat{u}(\xi)$, we obtain $-\xi^2 y(\xi) - 2(d/d\xi)\big(\xi y(\xi)\big) + 4y(\xi) = 0$ that is

$$2\,\xi\, y' \;=\; (2-\xi^2)\, y \qquad \xi \in \mathbb{R} \qquad (*)\,.$$

we look for a solution of (*) in the set $\{\xi \in \mathbb{R} :\ \xi > 0\}$:

$$\frac{y'}{y} = \frac{1}{\xi} - \frac{\xi}{2}$$
$$\ln y = \ln C + \ln \xi - \frac{\xi^2}{4}$$
$$y(\xi) = C\xi e^{-\xi^2/4} \qquad \xi > 0\,.$$

It easy to verify that one solution of (*) is given by

$$y_1(\xi) = C\xi e^{-\xi^2/4} \qquad \forall \xi \in \mathbb{R}\,.$$

Then one solution of the Exercise is given by the function u_1:

$$u_1(x) = \{\mathcal{F}\}^{-1}\{y_1\} = C\{\mathcal{F}\}^{-1}\{\xi e^{-\xi^2/4}\} = C_0\{\mathcal{F}\}^{-1}\{i\xi\sqrt{\pi}e^{-\xi^2/4}\} = C_0\frac{d}{dx}e^{-x^2} = C_1 x e^{-x^2}.$$

Remark - The equation (*) (where the coefficient of y' degenerates at $x = 0$) admits also the solution

$$y_2(\xi) = C|\xi|e^{-\xi^2/4} \qquad \xi \in \mathbb{R}\,.$$

Then
$$u_2(x) = \{\mathcal{F}\}^{-1}\{y_2\}$$

provides an integral of the given equation that is linearly independent from u_1. So the general integral is given by

$$u(x) \;=\; C_1 x e^{-x} + C_2\{\mathcal{F}\}^{-1}\{|\xi|e^{-\xi^2/4}\}(x) \qquad x \in \mathbb{R}\,.$$

The term $\{\mathcal{F}\}^{-1}\{|\xi|e^{-\xi^2/4}\}$ can be expressed as a convolution integral, anti-transforming y_2 in the framework of distributions by exploiting the relation $y_2(\xi) = \text{sign}(\xi)\, y_1(\xi)$. □

Exercise 4.19. (1) Evaluate the Fourier transform $\widehat{u}$ of the solution $u \in L^1(\mathbb{R})$ of

$$-\,4\,u'' \;+\; u * \frac{1}{2}e^{-|x|} \;=\; \frac{1}{2}e^{-|x|}\,.$$

(2) Once $\widehat{u}$ is known, study a priori the properties of u without computing it.
(3) Evaluate u and verify the a priori expected properties.

Solution - (1) Transforming the equation we have: $4\,\xi^2\,\widehat{u} + \dfrac{1}{1+\xi^2}\,\widehat{u} \;=\; \dfrac{1}{1+\xi^2}\,.$
Since $4\xi^4 + 4\xi^2 + 1 \neq 0 \neq 1 + \xi^2$ for any $\xi \in \mathbb{R}$, we obtain:

$$\widehat{u}(\xi) \;=\; \frac{1}{4\,\xi^4 + 4\,\xi^2 + 1} \;=\; \frac{1}{(2\,\xi^2 + 1)^2}$$

(2) A priori study of u :

$\widehat{u}$ real and even	$\Rightarrow$	u real and even ,
$\widehat{u} \in L^1 \cap L^2$	$\Rightarrow$	$u \in C^0 \cap L^\infty \cap L^2,\ \lim\limits_{\lvert x\rvert\to\infty} u(x) = 0,$,
$\xi\,\widehat{u},\ \xi^2\,\widehat{u} \in L^1 \cap L^2$	$\Rightarrow$	$u \in C^2,\ u', u'' \in L^\infty \cap L^2,\ \lim\limits_{\lvert x\rvert\to\infty} u^{(h)}(x) = 0,\ h = 1, 2,$
$\xi^3\,\widehat{u} \in L^2$	$\Rightarrow$	$u'' \in AC_{loc},\ u''' \in L^2,$
$\widehat{u} \in C^\infty,\ (\widehat{u})^{(k)} \in L^1 \cap L^2\ \forall k$	$\Rightarrow$	$x^k\,u \in L^2 \cap L^\infty\ ,\ \forall k,$
$\forall k\ \exists h\ :\ \xi^k\,(\widehat{u})^{(h)} \in L^1 \cap L^2$	$\Rightarrow$	$\forall k\ \exists h\ :\ \dfrac{d^k}{dx^k}(x^h u) \in C^0 \Rightarrow u \in C^\infty(\mathbb{R} \setminus \{0\}).$

(3) We consider the analytic extension $\widehat{u}(z)$ that has 2 double poles ($z = \pm i/\sqrt{2}$). Choose $x > 0$ and denote γ the oriented boundary of the half-disk centered at 0, with radius $R > 1/\sqrt{2}$ and contained in the upper half-plane. We apply the Jordan Lemma (Lemma 1.34 (i)) to the half-circumference (the multiplier of e^{izx} is an infinitesimal as $R \to +\infty$):

$$
\begin{aligned}
u(x) &= \frac{1}{2\pi}\int_{\mathbb{R}} \frac{e^{+i\xi x}}{(2\xi^2+1)^2}\,d\xi = \frac{1}{2\pi}\,\text{v.p.}\int_{\mathbb{R}} \frac{e^{+i\xi x}}{(2\xi^2+1)^2}\,d\xi = \frac{1}{2\pi}\oint_{\gamma} \frac{e^{+i\xi z}}{(2z^2+1)^2}\,dz = \\
&= \frac{2\pi i}{2\pi}\,\text{res}\left(\frac{e^{+izx}}{(2z^2+1)^2}\,,\, z = i/\sqrt{2}\right) = \frac{i}{4}\frac{d}{dz}\left.\frac{e^{ixz}}{(z+i/\sqrt{2})^2}\right|_{z=i/\sqrt{2}} \\
&= \frac{i}{4}\,\frac{e^{izx}}{(z+i/\sqrt{2})^4}\left.\left(ix(z+i/\sqrt{2})^2 - 2(z+i/\sqrt{2})\right)\right|_{z=i/\sqrt{2}} \\
&= \frac{i}{16}\,e^{-x\sqrt{2}}\,(-2ix - 2\sqrt{2}i) = \frac{e^{-x/\sqrt{2}}}{8}\,(x+\sqrt{2})\,, \qquad \forall x > 0.
\end{aligned}
$$

Since u is even e continuous, we conclude $u(x) = (|x| + \sqrt{2})\,\exp(-|x|/\sqrt{2})\,/\,8\,, \quad \forall x \in \mathbb{R}$.
The solution we have found fulfils the properties proven in (2), indeed we compute:
$u'(x) = -x\,\exp(-|x|/\sqrt{2})\,/\,(8\sqrt{2})\,, \quad u''(x) = (|x| - \sqrt{2})\,\exp(-|x|/\sqrt{2})\,/\,16\,.$ □

Exercise 4.20. Answer to the following questions, explaining your answers:

- a) study a priori the Fourier transform $\widehat{u}$ of $u(x) = \sin(x)\,\chi_{(-\pi,\pi)}(x) \quad x \in \mathbb{R}$.
- b) compute $\widehat{u}$.

Solution - a) u is a real and odd function $\Rightarrow$ $\widehat{u}$ imaginary and odd.

$\text{spt}\,u = [-\pi,\pi]$ is a compact set of $\mathbb{R}$ and $\|u\|_{L^1} = 4 \overset{Paley-WienerThm}{\Longrightarrow}$ $\widehat{u}$ admits an entire analytic extension that fulfils $|\widehat{u}(z)| \le 4\exp(\pi\,|\text{Im}z|)\ \forall z \in \mathbb{C}$.

$u \in L^1(\mathbb{R}) \cap L^2(\mathbb{R})$ (since u is continuous with compact support) $\Rightarrow$
$\widehat{u} \in C^0(\mathbb{R}) \cap L^\infty(\mathbb{R}) \cap L^2(\mathbb{R})$ e $\lim_{\xi\to\infty} \widehat{u}(\xi) = 0$.

$u \in C^0(\mathbb{R})$ with compact support and piece-wise C^1
(with classical derivative a.e. equal to $\cos(x)\,\chi_{(-\pi,\pi)}(x)$)

$\Rightarrow$ $u \in AC(\mathbb{R})$ with distributional derivative $u'(x) = \cos(x)\,\chi_{(-\pi,\pi)}(x) \in L^1(\mathbb{R})\cap L^2(\mathbb{R})$
$\Rightarrow$ $\xi\widehat{u} \in L^\infty(\mathbb{R}) \cap L^2(\mathbb{R})$ e $\lim_{\xi\to\infty} \xi\widehat{u}(\xi) = 0$.

b) Computation of $\widehat{u}$:

$$
\begin{aligned}
\widehat{u}(\xi) &= \int_{-\pi}^{\pi} \sin(x)\,e^{-i\xi x}\,dx = \frac{1}{2i}\int_{-\pi}^{\pi}\left(e^{ix-i\xi x} - e^{-ix-i\xi x}\right)dx \\
&= \frac{1}{2i}\left(\frac{e^{i\pi(1-\xi)} - e^{-i\pi(1-\xi)}}{i - i\xi} + \frac{e^{-i\pi(1+\xi)} - e^{i\pi(1+\xi)}}{i + i\xi}\right) \\
&= \frac{1}{2}\left(\frac{e^{i\pi\xi} - e^{-i\pi\xi}}{\xi - 1} - \frac{-e^{-i\pi\xi} + e^{i\pi\xi}}{\xi + 1}\right) \\
&= \frac{1}{2}\left(\frac{e^{i\pi\xi} - e^{-i\pi\xi}}{\xi - 1} + \frac{e^{-i\pi\xi} - e^{i\pi\xi}}{\xi + 1}\right) \\
&= \frac{2\,(e^{i\pi\xi} - e^{-i\pi\xi})}{2\,(\xi^2 - 1)} \\
&= 2\,i\,\frac{\sin(\pi\xi)}{\xi^2 - 1}\,.
\end{aligned}
$$

□

Exercise 4.21. Evaluate the Fourier transform of

$$f(t) = \begin{cases} \frac{1}{\sqrt{1-t^2}} & \text{se } |t| < 1 \\ 0 & \text{se } |t| \geq 1 \end{cases}$$

Hint: find a differential identity fulfilled by f, after removing its singularities with a polynomial multiplier.

Solution - $f \in L^1(\mathbb{R})$, hence and f is a transformable function: indeed it has bounded support and its two singularities (± 1) are both integrable since they behave like $(t-t_0)^{-1/2}$. After setting

$$g(t) = (1-t^2)f(t) \qquad t \in \mathbb{R}\,,$$

we have $g \in L^1$, moreover

$$g'(t) = -2tf(t) + (1-t^2)f'(t) = -tf(t) \qquad \forall t \neq \pm 1\,,$$

$$tf \in L^1(\mathbb{R}) \qquad \text{e} \qquad g(t) = \int_{-\infty}^{t} \big(-\tau f(\tau)\big)\, d\tau\,.$$

Hence: $g \in AC(\mathbb{R})$ and $g' = -tf \in L^1(\mathbb{R})$.

Taking into account of $\widehat{g} = \widehat{f} + \widehat{f}\,''$, we transform the equation $g' = -tf$:

$$\begin{aligned}
& i\xi\widehat{g} = i\widehat{(itf)} \\
& \xi(\widehat{f} + \widehat{f}\,'') = -\widehat{f}\,' \\
& \xi\widehat{f}\,'' + \widehat{f}\,' + \xi\widehat{f} = 0 \\
& \xi^2\widehat{f}\,'' + \xi\widehat{f}\,' + \xi^2\widehat{f} = 0 \qquad \text{Bessel equation of order 0.}
\end{aligned}$$

The general integral of the Bessel equation (Exercise 1.28) of order 0 is

$$c_1\, J_0(\xi) \,+\, c_2\, Y_0(\xi)$$

where J_0 is the Bessel function of first kind of order 0 (an entire analytic function) and Y_0 is the Bessel function of second kind of order 0 (an analytic function with a logarithmic singularity at the origin). Since $f \in L^1$ implies $\widehat{f} \in C^0$, we have to choose $c_2 = 0$. Eventually, taking into account of $J_0(0) = 1$,

$$c_1 = \widehat{f}(0) = \int_{-1}^{1} (1-t^2)^{-1/2} dt = \arcsin t \Big|_{t=-1}^{t=+1} = \pi\,,$$

$$\widehat{f}(\xi) \,=\, \pi\, J_0(\xi)\,, \qquad \xi \in \mathbb{R}\,.$$

Remark 1 - We had to introduce the function g, since $f \notin AC$: actually the distributional derivative of f does not belong to $L^1(\mathbb{R})$ (it is non integrable in any neighborhood ± 1).

Remark 2 - We compute the transform directly by using the definition:

$$\widehat{f}(\xi) \,=\, \int_{-1}^{1} \frac{e^{-it\xi}}{\sqrt{1-t^2}}\, dt \,=\, \int_{-1}^{1} \frac{\cos(t\xi)}{\sqrt{1-t^2}}\, dt$$

Then the solution of this exercise shows also a proof of this integral representation of J_0:

$$J_0(x) \,=\,=\, \frac{1}{\pi}\int_{-1}^{1} \frac{\cos(tx)}{\sqrt{1-t^2}}\, dt \,=\, \frac{1}{\pi}\int_{-1}^{1} \frac{e^{-itx}}{\sqrt{1-t^2}}\, dt\,. \tag{4.61}$$

Remark 3 - $J_0 \notin L^1(\mathbb{R})$.

Indeed, since f is even, by the inversion formula and Remark 2 we obtain

$$\widehat{J_0}\,(\xi) \,=\, 2\,\frac{p(\xi)}{(1-\xi^2)^{1/2}}\,, \quad \xi \in \mathbb{R} \tag{4.62}$$

that is J_0 has a discontinuous transform, then J_0 cannot belong to $L^1(\mathbb{R})$, otherwise there would be a contradiction with Theorem 4.3.
Remark 4 - $J_0' \notin L^1(\mathbb{R})$.
Indeed (4.62) entails $\mathcal{F}\{J_0'\}(\xi) = 2i\,\xi\,p(\xi)\,(1-\xi^2)^{-1/2}$ that is neither continuous, nor bounded, then it cannot be the Fourier transform of an L^1 function. □

Exercise 4.22. Consider the function $w(t) = \dfrac{\chi_{(-1,1)}}{\pi\sqrt{1-t^2}}$, $t \in L^1(\mathbb{R})$.
1) Study a priori the Fourier transform of w.
2) Evaluate the Fourier transform of w (without exploiting the previous exercise, but) using the formula (4.1). Hint: term-wise integration, after expanding in Taylor series the cosine function and taking into account of the dell'identity

$$\int_0^1 t^{2n}(1-t^2)^{-1/2}dt = \frac{\pi(2n)!}{2^{2n+1}(n!)^2}\,.$$

Solution - 1) *A priori study* -

- The function w is real and even, hence $\widehat{w}$ is real and even.
- $w \in L^1(\mathbb{R})$ since the singularities ($t = \pm 1$) are integrable, then $\widehat{w} \in C_0^0(\mathbb{R}) \cap L^\infty(\mathbb{R})$.
- $w \notin L^2(\mathbb{R})$, then $\widehat{w} \notin L^2(\mathbb{R})$.
- $w \in L^p(\mathbb{R})$ $\forall p \in [1,2)$ then $\widehat{w} \in L^q(\mathbb{R})$ $\forall q \in (2,+\infty]$.
- $\widehat{w} \notin L^1(\mathbb{R})$, since $w = \mathcal{F}^{-1}\widehat{w} = \frac{1}{2\pi}\mathcal{F}\widehat{w}(-x)$ ma $w \notin L^\infty$.
- $t^n w \in L^1(\mathbb{R})$ $\forall n \in \mathbf{N}$, then $\widehat{w} \in C^\infty(\mathbb{R})$ and all derivatives are bounded and infinitesimal at infinity.
- The support of w is $[-1,1]$ then, by the Paley-Wiener theorem $\widehat{w}$ has an entire extension in $\mathbf{C}$ fulfilling $|\widehat{w}(z)| \le \|w\|_{L^1(\mathbb{R})}\exp(|Im(z)|)$, $\forall z \in \mathbf{C}$.

2) *Transform computation* - By taking into account of the fact that w is even we obtain

$$\widehat{w}(x) = \int_{-\infty}^{+\infty} w(t)e^{-ixt}dt = \frac{1}{\pi}\int_{-1}^{+1}\frac{\cos(tx)}{\sqrt{1-t^2}}dt.$$

For any fixed $x \in \mathbb{R}$, by taking into account of the symmetries and expanding (Table 10.2) the cosine in Taylor series (that is uniformly convergent for $t \in [0,1]$) we get

$$\frac{1}{\pi}\int_{-1}^{+1}\frac{\cos(tx)}{\sqrt{1-t^2}}dt = \frac{2}{\pi}\int_0^{+1}(1-t^2)^{-1/2}\sum_{n=0}^{\infty}\frac{(-1)^n}{(2n)!}(tx)^{2n}dt =$$

$$= \frac{2}{\pi}\sum_{n=0}^{\infty}\frac{(-1)^n}{(2n)!}x^{2n}\int_0^{+1}(1-t^2)^{-1/2}t^{2n}dt = \sum_{n=0}^{\infty}\frac{(-1)^n}{(n!)^2}(x/2)^{2n} = J_0(x), \quad \forall x \in \mathbb{R}$$

where J_0 is the Bessel function of first kind of order zero (we refer to its Taylor series expansion: see Table 10.2). Due to the uniqueness of analytic extension $J_0(z) = \widehat{w}(z)$, $\forall z \in \mathbf{C}$. We notice that J_0 fulfils all the envisaged properties by the a priori study, in particular it is analytic entire, even, tends to zero as $x \to \pm\infty$, but $J_0 \notin L^1$ and $J_0 \notin L^2$.
□

Exercise 4.23. Prove that the function g depending on real variable defined by

$$g(x) \;=\; \begin{cases} \xi/e & \text{se } |\xi| < e\,, \\ \operatorname{sign}(\xi)/\ln|\xi| & \text{se } |\xi| > e\,, \end{cases}$$

does not belong to the space $\mathcal{V}$, then it is not the Fourier transform of any function in $L^1(\mathbb{R})$, though it is continuous, bounded and infinitesimal at infinity.

Solution - Obviously the function g belongs to $C^0 \cap L^\infty$, is infinitesimal at infinity, real and odd. If there existed an $f \in L^1(\mathbb{R})$ s.t. $g = \widehat{f}$ then f, should be imaginary and odd:

$$g(\xi) = \int_{\mathbb{R}} e^{-i\xi x} f(x)\, dx = -\int_{\mathbb{R}} \sin(\xi x)\ \mathrm{Im}\, f(x)\, dx = -i \int_{\mathbb{R}} \sin(\xi x)\, f(x)\, dx\,.$$

Notice that $g(\xi)/\xi$ belongs to $L^1_{loc}(\mathbb{R})$. Assuming by contradiction that $f \in L^1(\mathbb{R})$ and applying the Fubini-Tonelli Theorem we would find

$$\int_0^R \frac{g(\xi)}{\xi}\, d\xi \;=\; -i \int_0^R \frac{1}{\xi} \int_{\mathbb{R}} \sin(\xi x) f(x)\, dx\, d\xi \;=\; -i \int_0^R \int_{\mathbb{R}} \frac{\sin(\xi x)}{\xi} f(x)\, dx\, d\xi \;=$$
$$=\; -i \int_{\mathbb{R}} f(x) \int_0^{Rx} \frac{\sin(y)}{y}\, dy\, dx\,.$$

The above term is uniformly bounded for $R \in \mathbb{R}$ since, recalling (Exercise 1.22) that $\exists\ \lim_{T\to\pm\infty} \int_0^T \sin(x)/x\, dx = \pi/2$, from the previous equalities we get

$$\left| \int_0^R \frac{g(\xi)}{\xi}\, d\xi \right| \;\le\; \int_{\mathbb{R}} |f(x)| \sup_T \left| \int_0^T \frac{\sin(y)}{y}\, dy \right| dx < +\infty\,.$$

This last estimate is a contradiction with previous equalities since

$$\lim_{R\to+\infty} \int_0^R \frac{g(\xi)}{\xi}\, d\xi \;=\; \lim_{R\to+\infty} \int_0^R \frac{1}{\xi \ln \xi}\, d\xi \;=\; +\infty\,.$$

□

Exercise 4.24. Given $g \in L^1(\mathbb{R})$, write an integral representation of the solution u in $L^1(\mathbb{R})$ of

$$u(x) + \int_0^{+\infty} e^{-3t} u(x-t)\, dt = g(x)\,.$$

Then evaluate explicitly u with the choice $g(x) = e^{-|x|}$.

Solution - The equation can be rewritten as follows:

$$u(x) + \big(H(\cdot)e^{-3\cdot} * u(\cdot)\big)(x) = g(x)\,,$$

where H is the Heaviside function. Applying the Fourier transform to both sides of the equation we obtain

$$\widehat{u}(\xi) + \frac{1}{3+i\xi}\,\widehat{u}(\xi) \;=\; \widehat{g}(\xi)\,,$$

hence $(4 + i\xi \neq 0,\ \forall \xi \in \mathbb{R})$,

$$\widehat{u}(\xi) = \frac{3+i\xi}{4+i\xi}\,\widehat{g}(\xi)\,.$$

Actually the right-hand side is the product of two transforms:

$$\widehat{u}(\xi) \;=\; \mathcal{F}\left\{H(x)e^{-4x}\right\}(\xi)\ \left[3\widehat{g}(\xi) + \widehat{g'}\,(\xi)\right]\,.$$

Anti-transforming we obtain

$$u(x) = \left(H(\cdot)e^{-4\cdot}\right) * \left(3g + g'\right)(x) = \int_0^{+\infty} e^{-4t}\Big(3g(x-t) + g'(x-t)\Big)\, dt\,.$$

Now we face the case when $g(x) = e^{-|x|}$:

$$u(x) = \int_0^{+\infty} e^{-4t} e^{-|x-t|}\,[3 - \mathrm{sign}(x-t)]\, dt\,.$$

Setting apart the cases x positive vs negative and suitably splitting the integral, we find

$$u(x) = \begin{cases} \frac{4}{5}\, e^{x} & \text{se } x < 0\,, \\ \\ \frac{2}{3}\, e^{-x} + \frac{2}{15}\, e^{-4x} & \text{se } x > 0\,. \end{cases}$$

Eventually by setting $u(0) = 4/5$ we find $u \in C^0(\mathbb{R})$. □

Exercise 4.25. Find all the solutions $u \in L^2(\mathbb{R})$ of the equation

$$u''(x) + u(x) = (4x^2 - 1)e^{-x^2}\,, \qquad x \in \mathbb{R}\,.$$

Then find all the solutions $u \in C^2(\mathbb{R})$ of the same equation.

Solution - Transforming the equation we find

$$(1 - \xi^2)\widehat{u}(\xi) = -4\,\frac{d^2}{d\xi^2}\left(\sqrt{\pi}\, e^{-\xi^2/4}\right) - \sqrt{\pi}\, e^{-\xi^2/4} = \sqrt{\pi}(1 - \xi^2)\, e^{-\xi^2/4}\,,$$

then, $\widehat{u}(\xi) = \sqrt{\pi}e^{-\xi^2/4}$ for any $\xi \neq \pm 1$, hence $\widehat{u}(\xi) = \sqrt{\pi}e^{-\xi^2/4} \in L^2(\mathbb{R})$ and, anti-transforming, $u(x) = e^{-x^2} \in L^2(\mathbb{R})$. This function is the only solution in $L^2(\mathbb{R})$. Since the differential equation is linear and we know this particular solution, then the general integral n can be expressed by addition of this one with the general integral of the associated homogeneous equation. Then the solutions are $u(x) = e^{-x^2} + a\cos x + b\sin x$ for any $a, b \in \mathbb{R}$; only if $a = b = 0$ this function belongs to $L^2(\mathbb{R})$. □

Exercise 4.26. We denote by H the Heaviside function and set

$$f(x) = H(-x)xe^{3x} \qquad g(x) = H(2 - x)xe^{3x}.$$

Show that $f, g \in L^1(\mathbb{R})$ and evaluate the Fourier transform of f and g.

Solution - The functions f and g vanish respectively if $x > 0$ and $x > 2$; moreover, both decay exponentially to 0 as $x \to -\infty$, hence $f, g \in L^1(\mathbb{R})$. Setting $h(x) = H(-x)e^{3x}$, we have

$$\widehat{h}(\xi) = \int_{-\infty}^{+\infty} h(x)e^{-ix\xi}\, dx = \int_{-\infty}^{0} e^{(3-i\xi)x}\, dx = \frac{1}{3 - i\xi}\,.$$

Since $f(x) = xh(x)$, we deduce

$$\widehat{f}(\xi) = i\frac{d}{d\xi}\widehat{h}(\xi) = -\frac{1}{(3 - i\xi)^2}\,.$$

Moreover $g(x) = xe^6 h(x - 2)$, hence:

$$\widehat{g}(\xi) = i\, e^6\, \frac{d}{d\xi}\left(e^{-2i\xi}\, \widehat{h}(\xi)\right) = e^{6-2i\xi}\, \frac{5 - 2i\xi}{(3 - i\xi)^2}\,.$$

□

Exercise 4.27. Consider the following equation in the unknown $u \in L^1(\mathbb{R})$:

$$e^{-x^2} = \int_{-\infty}^{+\infty} e^{-|x-y|}u(y)\, dy \qquad (x \in \mathbb{R})$$

Find the Fourier transform $\widehat{u}$ and then the solution u.

Solution - The equation can be rewritten as $e^{-x^2} = e^{-|x|} * u(x)$. Transforming it, we obtain

$$\sqrt{\pi}\, e^{-\xi^2/4} = \frac{2}{1+\xi^2}\,\widehat{u}(\xi) \implies \widehat{u}(\xi) = \frac{\sqrt{\pi}}{2}(1+\xi^2)e^{-\xi^2/4}.$$

We notice that

$$\widehat{u}(\xi) = \frac{1}{2}\mathcal{F}\left\{e^{-x^2} - \frac{d^2}{dx^2}(e^{-x^2})\right\}(\xi);$$

hence $u(x) = (\frac{3}{2} - 2x^2)e^{-x^2}$. □

Exercise 4.28. Evaluate the Fourier transform of the functions

$$f_1(x) = \frac{xe^{3+i5x}}{(3x+2i)^2}, \quad f_2(x) = \chi_{[-1,1]}(x)\sin x, \quad f_3(x) = \chi_{[0,1]}(x)e^{-x},$$

$$f_4(x) = \frac{\sin^2(3x)}{x^2-2x+2}, \quad f_5(x) = H(1-x^2)\log|x|.$$

Solution - $f_1 \in L^1(\mathbb{R})$. Arguing as in Exercise 4.7 we find

$$\widehat{f_1}(\xi) = \frac{2\pi i e^3}{27}(2\xi - 13)H(\xi-5)e^{2(5-\xi)/3}.$$

By exploiting $\mathcal{F}\left\{\chi_{[-1,1]}(x)\right\}(\xi) = \frac{2\sin\xi}{\xi}$ and (4.7), we obtain

$$\widehat{f_2}(\xi) = i\frac{\sin(\xi+1)}{\xi+1} - i\frac{\sin(\xi-1)}{\xi-1}.$$

Applying the definition (4.1), a straightforward computation gives

$$\widehat{f_3}(\xi) = \int_0^1 e^{-(1+i\xi)x}dx = \frac{1-e^{-1-i\xi}}{1+i\xi}.$$

Arguing as in Exercise 4.5 we obtain

$$\widehat{f_4}(\xi) = \frac{\pi}{2}e^{-i\xi-|\xi|} - \frac{\pi}{4}e^{i(6-\xi)-|\xi-6|} - \frac{\pi}{4}e^{-i(\xi+6)-|\xi+6|}.$$

Transforming the equality $[xf_5(x)]' = f_5(x) + \chi_{[-1,1]}(x) \in L^1(\mathbb{R})$ we find

$$-\xi\widehat{f_5}'(\xi) = \widehat{f_6}(\xi) + \frac{2\sin\xi}{\xi};$$

the solution of the last equation is

$$\widehat{f_5}(\xi) = -\frac{2}{\xi}\int_0^{\xi}\frac{\sin x}{x}\,dx - \frac{C}{\xi}$$

where the integration constant C is 0 since $f_5 \in L^1(\mathbb{R}) \Rightarrow \widehat{f_5} \in C^0(\mathbb{R})$. □

4.2.2 Exercises on $\mathcal{F}$ transform in $\mathcal{S}'(\mathbb{R})$

Exercise 4.29. Evaluate the Fourier transform of the following distributions in $\mathcal{S}'(\mathbb{R})$.

1. 1, $\sin(x)$, $\cos(x)$
2. $\text{sign}(x)$ (sign function: $\text{sign}(x) = 1$ if $x > 0$, $sign(x) = -1$ se $x < 0$)
3. $H(x)$ (Heaviside function, $H(x) = 1$ if $x > 0$, $H(x) = 0$ se $x < 0$)

4. x
5. $|x|$
6. x^+
7. $(x^+)^2 \;=\; x^2\,H(x) \;=\; x\,(x^+)$
8. $\text{v.p.}\,\dfrac{1}{x}\,,\;\; \text{p.f.}\,\dfrac{1}{x^2}\,,\;\; \text{p.f.}\,\dfrac{1}{x^3}\,.$

Solution - 1. By $\widehat{\delta} = 1$, (4.43) and the fact that δ is even, we obtain $\widehat{1} \;=\; 2\,\pi\,\delta$. By the previous transform and $\mathcal{F}\{e^{i\xi_0 x}u(x)\} = \widehat{u}(\xi - \xi_0)$ we obtain:

$$\mathcal{F}\{\sin(x)\} = \mathcal{F}\{(e^{ix} - e^{-ix})\,1/(2i)\} = \pi\,i\,\big(\delta(\xi+1) - \delta(\xi-1)\big)$$

$$\mathcal{F}\{\cos(x)\} = \mathcal{F}\{(e^{ix} + e^{-ix})\,1/2\} = \;\pi\,\big(\delta(\xi+1) + \delta(\xi-1)\big)$$

2. After setting $v(x) = \text{sign}(x)$, we obtain $v' \;=\; 2\delta$. Transforming this identity, we have a division problem in $\mathcal{D}'(\mathbb{R})$ with unknown $\widehat{v}$:

$$i\,\xi\,\widehat{v}(\xi) \;=\; 2\,;$$

the division problem has the particular solution $\frac{2}{i}\,\text{v.p.}\,\frac{1}{\xi}$ (see Exercise 3.6), whereas the solutions of the associated homogeneous problem can be obtained by the formula (3.16); then the general solution is given by

$$\widehat{v}(\xi) = -2i\,\text{v.p.}\,\frac{1}{\xi} + C\delta(\xi);$$

with C arbitrary constant. We know that $\widehat{v}$ is imaginary and odd (since v is real and odd) and δ is even, therefore $C = 0$: $\mathcal{F}\{\text{sign}(x)\} \;=\; -\,2\,i\;\text{v.p.}\,\frac{1}{\xi}\,.$
3. Due to linearity of $\mathcal{F}$, the identity $H(x) = (1 + \text{sign}(x))/2$ and step 1, we obtain: $\widehat{H}(\xi) = \pi\delta(\xi) - i\,\text{v.p.}\,\frac{1}{\xi}$.
4. By (4.45) and $\widehat{1} = 2\pi\delta$ we obtain $\widehat{x} = \widehat{x\,1} \;=\; 2\,\pi\,\delta'(\xi)\,.$
5. Transforming $|x| = x\,\text{sign}(x)$ we obtain

$$\mathcal{F}\{\,|x|\} = \mathcal{F}\{x\,\text{sign}(x)\} = i\frac{d}{d\xi}\widehat{\text{sign}}(\xi) = 2\,\frac{d}{d\xi}\;\text{v.p.}\;\frac{1}{\xi} = -2\,\text{p.f.}\,\frac{1}{\xi^2}$$

6. Transforming $x^+ \;=\; x\,H(x)$ and taking into account of step 3, we obtain

$$\widehat{(x^+)} \;=\; \mathcal{F}\{x\,H\} \;=\; i\,\frac{d}{d\xi}\,\widehat{H} \;=\; i\,\pi\delta'(\xi) - \text{p.f.}\,\frac{1}{\xi^2}\,.$$

Alternatively transforming $x^+ = (x + |x|)/2$ and taking into account of steps 4 and 5:

$$\widehat{(x^+)} \;=\; \mathcal{F}\{x^+ = (x + |x|)/2\} \;=\; i\,\pi\delta'(\xi) - \text{p.f.}\,\frac{1}{\xi^2}\,.$$

7. After setting $\text{p.f.}\,1/x^3 \;=\; -\frac{1}{2}(d/dx)(\text{p.f.}\,1/x^2)$, we compute the transform, by exploiting point 6:

$$\mathcal{F}\{\,(x^+)^2\} = \mathcal{F}\{\,x^2 H(x)\} = \mathcal{F}\{\,x\,(x^+)\} \;=\; i\,\frac{d}{d\xi}\,\widehat{(x^+)} \;=\; i\,\frac{d}{d\xi}\left(i\,\pi\delta'(\xi) - \text{p.f.}\,\frac{1}{\xi^2}\right) =$$

$$= \;-\,\pi\,\delta''(\xi) \;+\; 2\,i\;\text{p.f.}\,\frac{1}{\xi^3}\,.$$

8. By the inversion formula (4.43):

$$\text{v.p.}\,\frac{1}{\xi} = \mathcal{F}\left\{\frac{i}{2}\,\text{sign}(x)\right\} \text{ is odd, hence } \mathcal{F}\left\{\text{v.p.}\,\frac{1}{x}\right\} = -\pi\, i\,\text{sign}(\xi)\,;$$

$$\text{p.f.}\,\frac{1}{\xi^2} = \mathcal{F}\left\{-\frac{|x|}{2}\right\} \text{ is even, hence } \mathcal{F}\left\{\text{p.f.}\,\frac{1}{x^2}\right\} = -\pi\,|\xi|\,.$$

By the knowledge of $\mathcal{F}\left\{\text{p.f.}\,\frac{1}{x^2}\right\}$, and $\text{p.f.}\,\frac{1}{x^3} = -\frac{1}{2}\frac{d}{dx}\,\text{p.f.}\,\frac{1}{x^2}$ we obtain:

$$\mathcal{F}\left\{\text{p.f.}\,\frac{1}{x^3}\right\} = -\frac{1}{2}\,i\xi(-\pi|\xi|) = \frac{\pi}{2}\,i\,\xi^2\,\text{sign}(\xi)\,.$$

□

Exercise 4.30. Given a distribution $f \in \mathcal{S}'(\mathbb{R})$ of order 0, find all the solutions u of

$$u \in \mathcal{S}'(\mathbb{R}): \qquad -u'' + u = f \qquad \text{in } \mathcal{S}'(\mathbb{R})\,. \tag{4.63}$$

- Analyze the regularity of the solutions if the datum f belongs to $L^2(\mathbb{R})$.
- Analyze the regularity of the solutions if f has a prescribed number of distributional derivatives in $L^2(\mathbb{R})$.

Solution - Transforming the equation and proceeding as in Exercise 4.16, we obtain a unique solution u fulfilling:

$$\widehat{u}(\xi) = \frac{\widehat{f}(\xi)}{1+\xi^2}\,, \tag{4.64}$$

$$u = \frac{1}{2}\,e^{-|x|} * f\,.$$

Notice that $U = \frac{1}{2}\exp(-|x|)$ is a fundamental solution of $I - d^2/dt^2$: $-U'' + U = \delta$ in $\mathbb{R}$. By (4.64) we obtain $|\widehat{u}(\xi)| \le |\widehat{f}(\xi)|/(1+\xi^2)$. The Plancherel identity (4.36) and the transformation rules (4.44) concerning derivatives entail

$$u^{(j+2)} \in L^2 \iff \xi^{j+2}\widehat{u} \in L^2 \iff \frac{\xi^{j+2}}{1+\xi^2}\,\widehat{f} \in L^2 \iff \xi^j\,\widehat{f} \in L^2 \iff f^{(j)} \in L^2;$$

this property together with the inclusion

$$\{v \in L^2(\mathbb{R}) : v' \in L^2(\mathbb{R})\} \subset C^0 \cap L^\infty(\mathbb{R})$$

allows to conclude:

- $f \in L^2(\mathbb{R}) \;\Rightarrow\; \xi^h\widehat{u} \in L^2(\mathbb{R}),\ h = 0,1,2, \;\Rightarrow$
  $\Rightarrow\; u, u', u'' \in L^2(\mathbb{R}) \;\Rightarrow\; u \in C^1,\ u, u' \in L^\infty;$
- $f, f', \dots, f^{(k)} \in L^2(\mathbb{R}) \quad\Rightarrow\quad \xi^h\widehat{u} \in L^2(\mathbb{R}),\ h = 0,1,\dots,k+2 \quad\Rightarrow$
  $\Rightarrow\; u^{(h)} \in L^2(\mathbb{R}),\ h = 0,1,\dots,k+2 \;\Rightarrow\; u \in C^{k+1},\ u^{(h)} \in L^\infty\ h = 0,\dots,k+1\,.$

This method for analyzing the regularity of the solution (in the L^2 framework) is very effective since and it can be applied also to partial differential equations (Exercise 4.46), in contrast with the more elementary result that holds in dimension $n = 1$ only for ordinary differential equations (in the C^k framework), and that we recall here for reader convenience:

$$u \in C^{k+2} \iff f \in C^k\,.$$

□

Exercise 4.31. Given a distribution $f \in \mathcal{E}'(\mathbb{R})$ of order 0, find the general integral u of

$$u \in \mathcal{D}'(\mathbb{R}): \qquad -u'' + u = f \qquad \text{in } \mathcal{D}'(\mathbb{R})\,. \tag{4.65}$$

Solution - From Exercise 4.30 we know a particular solution $(1/2\, e^{-|x|} * f)$ that is also the unique solution of (4.65) in $\mathcal{S}'(\mathbb{R})$. The convolution is well defined since f has compact support.
By summing with the solutions of the homogeneous equation $(-u'' + u = 0)$, that are not tempered distributions, we obtain the general integral in $\mathcal{D}'(\mathbb{R})$:

$$u(x) = 1/2\, e^{-|x|} * f + Ae^{x} + Be^{-x} \qquad A, B \in \mathbb{C}.$$

□

Exercise 4.32. Find a fundamental solution $U \in \mathcal{S}'(\mathbb{R})$ of $I + d^2/dt^2$ that is

$$U \in \mathcal{S}'(\mathbb{R}) : \qquad U'' + U = \delta \qquad \text{in } \mathcal{S}'(\mathbb{R}) \tag{4.66}$$

Solution - Transforming the equation we obtain the division problem in the unknown $\widehat{U}$

$$(1 - \xi^2)\, \widehat{U}(\xi) = 1\,,$$

whose solution (since the polynomial $1 - \xi^2$ has two simple zeroes at $\xi = \pm 1$) is given by

$$\widehat{U}(\xi) = \text{v.p.} \frac{1}{1 - \xi^2} + A\,\delta(\xi - 1) + B\,\delta(\xi + 1)\,.$$

Using the Fourier transform Table 10.8.2 we obtain:

$$\mathcal{F}\{\text{sign}(x)\} = -\,2\,i \text{ v.p.} \frac{1}{\xi}\,, \qquad \mathcal{F}\{-\,\text{sign}(x)\} = 2\,i \text{ v.p.} \frac{1}{\xi}\,,$$

$$\mathcal{F}\left\{-\,e^{ix}\,\text{sign}(x)\right\} = 2\,i \text{ v.p.} \frac{1}{\xi - 1}\,, \qquad \mathcal{F}\left\{-\,e^{-ix}\,\text{sign}(x)\right\} = 2\,i \text{ v.p.} \frac{1}{\xi + 1}\,,$$

$$\text{v.p.} \frac{1}{1 - \xi^2} = \frac{1}{2}\left(\text{v.p.} \frac{1}{1 + \xi} + \text{v.p.} \frac{1}{1 - \xi}\right) = \frac{1}{4i}\left(\text{v.p.} \frac{2i}{1 + \xi} - \text{v.p.} \frac{2i}{\xi - 1}\right) =$$

$$= \frac{1}{4i}\left(\mathcal{F}\left\{-\,e^{-ix}\,\text{sign}(x)\right\} + \mathcal{F}\left\{\,e^{ix}\,\text{sign}(x)\right\}\,\right) =$$

$$= \frac{1}{4i}\mathcal{F}\{\,2\,i\,\sin(x)\,\text{sign}(x)\} = \mathcal{F}\left\{\frac{1}{2}\,\sin(|x|)\,\right\}$$

$$\mathcal{F}^{-1}\left\{\text{v.p.} \frac{1}{1 - \xi^2}\right\} = \frac{1}{2}\,\sin(|x|) = \frac{1}{2}\,\sin(x)\,\text{sign}(x)$$

$$U(x) = \frac{1}{2}\,\sin(|x|) + a\,\cos(x) + b\,\sin(x)\,.$$

□

Exercise 4.33. Given $f \in L^1(\mathbb{R})$, find the general integral $u \in \mathcal{S}'(\mathbb{R})$ of

$$u'' + u = f \qquad \text{in } \mathcal{S}'(\mathbb{R}) \tag{4.67}$$

Solution - Due to Exercise 4.32 we know the fundamental solution $U(x) = 1/2\, \sin(|x|)$ that belongs to $L^\infty(\mathbb{R}) \cap C^0 \subset L^1_{loc}$. Then we can compute the convolution of U with $f \in L^1$ and obtain the general integral in $L^1_{loc} \cap C^0$:

$$u(x) = f * \frac{1}{2}\,\sin(|x|) + A \sin x + B \cos x \qquad A,\ B \in \mathbb{C}\,.$$

We notice that the previous formula provides the solution also if the datum f is a measure (or a distribution of order 0) with finite total mass (that is, $\exists C < \infty : \ |\int_{\mathcal{R}} f v\, dx| \leq C\|v\|_{L^\infty} \ \forall v \in C^0 \cap L^\infty(\mathbb{R})$), since U is continuous, integrable and bounded.

□

Exercise 4.34. 1. For any $k \in \mathbb{N}$, find the fundamental solution U_k of the operator $-d^2/dx^2 + 1/k^2\, I$, that is, solve the problem:

$$U_k \in \mathcal{S}'(\mathbb{R}) : \qquad -U_k'' + \frac{1}{k^2} U_k = \delta \, .$$

2. Find the fundamental solutions $V \in \mathcal{S}'(\mathbb{R})$ of $-d^2/dx^2$:

$$V \in \mathcal{S}'(\mathbb{R}) : \qquad -V'' = \delta \, .$$

3. After setting $V_k := U_k - k/2$, prove that $V_k \to V$ in $\mathcal{S}'(\mathbb{R})$ as $k \to \infty$, where V is the only one among solutions of item 2 that is even and vanish at 0.
Notice that $-V_k'' + \frac{1}{k^2} V_k = \delta - \frac{1}{2k}$.

Solution - 1. Proceeding as in Exercise 4.30 we obtain $U_k(x) = \frac{k}{2} e^{-|x|/k}$.
2. By Fourier transform, we obtain the division problem $\xi^2 \widehat{V}(\xi) = 1$, whose general solution (due to (3.16) and Exercise 3.6) is $\widehat{V}(\xi) = \text{p.f.} \frac{1}{\xi^2} + A\delta(\xi) + B\delta'(\xi)$. Anti-transforming the three terms we find (infinitely many) solutions V :

$$V(x) \; = \; -\frac{|x|}{2} + \frac{A}{2\pi} + \frac{B}{2\pi\, i} x \, .$$

3. In order that V is even and vanishes at 0, we choose $A = B = 0$: then

$$V(x) = -\frac{|x|}{2} , \qquad V_k(x) = \frac{k}{2} \left(e^{-|x|/k} - 1 \right)$$

and we are left to prove the convergence of V_k to V.

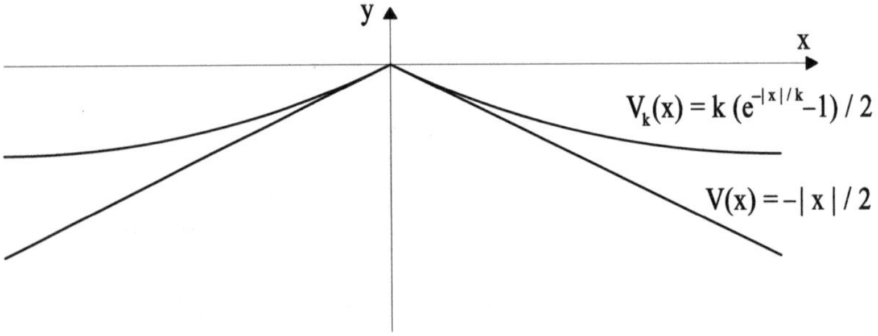

Figure 4.3 - Graphs of V and V_k (Exercise 4.34) .

Due to Taylor expansion of the exponential function, we can write:

$$V_k(x) \; = \; \left(\frac{k}{2} - \frac{|x|}{2} + \left(\frac{|x|}{k} \right) \right) - \frac{k}{2} \; = \; -\frac{|x|}{2} + O\left(\frac{|x|}{k} \right) .$$

Hence $V_k \to V$ in $L^1_{loc}(\mathbb{R})$. Then $V_k \to V$ in $\mathcal{D}'(\mathbb{R})$ and taking into account of the inequalities $-\frac{|x|}{2} = V(x) \leq V_k(x) \leq 0$, we deduce $V_k \to V$ in $\mathcal{S}'(\mathbb{R})$. □

Exercise 4.35. Prove these convergence properties for sequences of distributions:

(1) $$\sin(kx) \underset{k\to+\infty}{\overset{\mathcal{D}'(\mathbb{R})}{\longrightarrow}} 0$$

(2) $$\cos(kx) \underset{k\to+\infty}{\overset{\mathcal{D}'(\mathbb{R})}{\longrightarrow}} 0$$

(3) $$\big(\sin(kx)\big)^2 \underset{k\to+\infty}{\overset{\mathcal{D}'(\mathbb{R})}{\longrightarrow}} \frac{1}{2}$$

(4) $$\big(\cos(kx)\big)^2 \underset{k\to+\infty}{\overset{\mathcal{D}'(\mathbb{R})}{\longrightarrow}} \frac{1}{2}$$

Solution - (2) Given a function test $\varphi \in \mathcal{D}'(\mathbb{R})$, there exists a s.t. $\operatorname{spt}\varphi \subset [-a,+a]$ and

(∗) $$\frac{1}{k}\left|\int_{-a}^{+a} \sin(kx)\varphi'(x)\,dx\right| \le \frac{2a}{k}\sup|\varphi'|\,,$$

$$\int_{\mathbb{R}} \cos(kx)\varphi(x)\,dx = \int_{-a}^{a} \cos(kx)\varphi(x)\,dx = \int_{-a}^{a}\left(\frac{\sin(kx)}{k}\right)'\varphi(x)\,dx = \boxed{\text{int. by parts}}$$

$$= -\frac{1}{k}\int_{-a}^{a}\sin(kx)\,\varphi'(x)\,dx \underset{k\to+\infty}{\overset{(*)}{\longrightarrow}} 0\,.$$

In the integration by parts we have exploited $\varphi(\pm a) = 0$; the convergence (of the numerical sequence) in the last computation follows by the estimate (∗). The same conclusion holds true for any test function $\varphi \in \mathcal{D}'(\mathbb{R})$, so we have proved the convergence in the sense of distributions stated in item (1).
Item (2) can be proved in the same way. Item (3) and (4) follow by (1),(2) and trigonometry identities

$$\big(\sin(kx)\big)^2 = \frac{1}{2}\big(1-\cos(2kx)\big)\,, \qquad \big(\cos(kx)\big)^2 = \frac{1}{2}\big(1+\cos(2kx)\big)\,.$$

Alternative solution - Often the Fourier transform make easier to verify the the distributional convergence. In the present cases, the Fourier transform shows a stronger convergence than the one in $\mathcal{D}'$: all sequences (1) – (4) converge in $\mathcal{S}'$, the space of tempered distributions. Let us prove items (2) and (4) (items (1) and (3) can be dealt in the same way):

$$\mathcal{F}\{\cos(kx)\}(\xi) = \pi\big(\delta(\xi-k)+\delta(\xi+k)\big) \underset{k\to+\infty}{\overset{\mathcal{S}'(\mathbb{R})}{\longrightarrow}} 0$$

since $\int\delta(\xi\mp k)\psi(\xi) = \psi(\pm k\xi)\to 0\,\forall\psi\in\mathcal{S}(\mathbb{R})$. Item (1) follows from the continuity of $\mathcal{F}^{-1}$ in $\mathcal{S}'(\mathbb{R})$.

$$\mathcal{F}\left\{\big(\cos(kx)\big)^2\right\}(\xi) = \frac{1}{2\pi}\,\widehat{\cos(kx)}*\widehat{\cos(kx)} = \frac{\pi}{2}\big(\delta(\xi-k)+\delta(\xi+k)\big)*\big(\delta(\xi-k)+\delta(\xi+k)\big) =$$

$$= \frac{\pi}{2}\big(\delta(\xi-2k)+\delta(\xi+2k)+2\delta(\xi)\big) \underset{k\to+\infty}{\overset{\mathcal{S}'(\mathbb{R})}{\longrightarrow}} \pi\,\delta(\xi) = \mathcal{F}\left\{\frac{1}{2}\right\},$$

these convergence properties and the continuity of $\mathcal{F}^{-1}$ in $\mathcal{S}'(\mathbb{R})$ entail:

$$\cos(kx) \underset{k\to+\infty}{\overset{\mathcal{S}'(\mathbb{R})}{\longrightarrow}} 0\,, \qquad \big(\cos(kx)\big)^2 \underset{k\to+\infty}{\overset{\mathcal{S}'(\mathbb{R})}{\longrightarrow}} \frac{1}{2}\,.$$

□

Exercise 4.36. Find the limit in $\mathcal{S}'(\mathbb{R})$ of the sequence

$$v_k(x) = \frac{\sin(kx)}{\pi\, x} \qquad x \in \mathbb{R}\,.$$

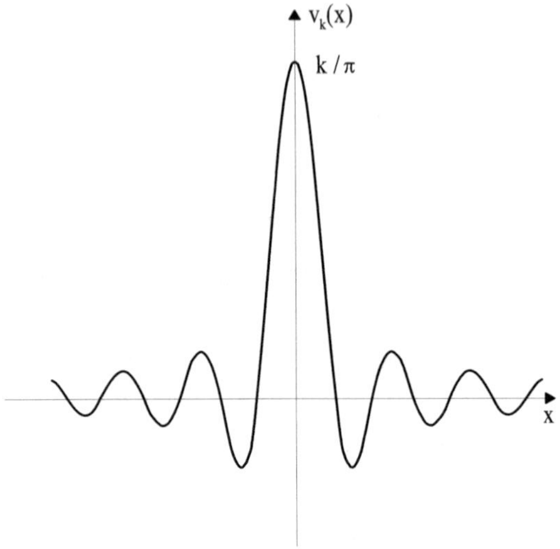

Figure 4.4 - Graph of v_k (Exercise 4.36).

Solution - After setting $u(x) = \sin(x)/x$, we obtain

$$v_k(x) = \frac{k}{\pi} u(kx)\,, \qquad \widehat{u}(\xi) = \pi p(\xi)\,, \qquad \widehat{v_k}(\xi) = p\left(\frac{\xi}{k}\right)\,.$$

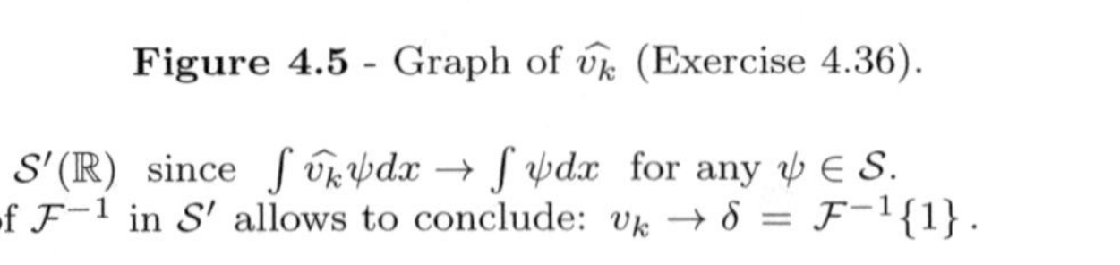

Figure 4.5 - Graph of $\widehat{v_k}$ (Exercise 4.36).

Then $\widehat{v_k} \to 1$ in $\mathcal{S}'(\mathbb{R})$ since $\int \widehat{v_k}\psi dx \to \int \psi dx$ for any $\psi \in \mathcal{S}$.
The continuity of $\mathcal{F}^{-1}$ in $\mathcal{S}'$ allows to conclude: $v_k \to \delta = \mathcal{F}^{-1}\{1\}$. □

Exercise 4.37. Replay to the following questions, by explaining your answers.

- (a) Set $f(x) = \arctan(2/x^2)$. Show that the Fourier transform of f is defined and a priori analyze its properties of integrability, regularity and behavior at infinity.

- (b) Show that $\widehat{f}(\xi) = 2\pi e^{-|\xi|} \frac{\sin\xi}{\xi}$ (hint: evaluate the transform of f').

- (c) Prove the identity

$$\arctan\frac{2}{x^2} = \arctan(x+1) - \arctan(x-1) \qquad (*)$$

and use it to show, independently from item (b), that

$$\mathcal{F}^{-1}\left\{2\pi e^{-|\xi|}\frac{\sin\xi}{\xi}\right\} = f(x)\,.$$

Solution - (a) We notice that $f \in C^\infty(\mathbb{R})$. The claim is obvious if $x \neq 0$; if we consider f near 0, we obtain

$$\lim_{x\to 0} f(x) = \lim_{t\to+\infty} \arctan t = \frac{\pi}{2}\,.$$

Then, setting $f(0) = \pi/2$, the function f is everywhere defined and continuous, also at $x = 0$. Applying the de l'Hôpital rule, we deduce the existence of

$$f'(0) = \lim_{x\to 0}\frac{f(x)-f(0)}{x} = \lim_{x\to 0}\frac{-4x}{x^4+4} = 0\,.$$

So f has classical derivative everywhere in $\mathbb{R}$ and

$$f'(x) = \frac{-4x}{x^4+4} \qquad \forall x \in \mathbb{R}\,.$$

Obviously: $f' \in C^\infty(\mathbb{R})$. Hence $f \in C^\infty(\mathbb{R})$.
Using de l'Hôpital again:

$$\lim_{|x|\to\infty}\frac{f(x)}{2/x^2} = \lim_{x\to 0_+}\frac{\arctan t}{t} = \lim_{t\to 0_+}\frac{1}{1+t^2} = 1\,.$$

Then $|f(x)| \leq C/x^2$ per $|x|$ grande. Hence $f \in L^1 \cap L^2$ e

1. $\widehat{f}(\xi)$ is defined for any ξ in $\mathbb{R}$,
2. $\widehat{f}(\xi)$ is continuous and vanishes at infinity,
3. $\widehat{f}(\xi)$ belongs to $L^2(\mathbb{R})$.

   The previous analysis entails $xf \in C^\infty(\mathbb{R})$ and it behaves as $2/x$ at infinity. Then $xf \in L^2$, $xf \notin L^1$. In particular, $f \in L^2$ and $xf \in L^2$ entail:
4. $\widehat{f}(\xi)$ is absolutely continuous e $\frac{d}{d\xi}\widehat{f}(\xi) \in L^2(\mathbb{R})$.

   Da $f'(x) = -4x/(x^4+4)$ segue $f^{(n)} \in L^1(\mathbb{R}) \cap L^2(\mathbb{R})\ \forall n$; hence
5. $\xi^n\widehat{f}(\xi)$ is nulla at infinity for any n.

   Items (4) and (5) imply
6. $\widehat{f} \in L^1(\mathbb{R})$.

   Eventually, since f is real and even, we obtain that
7. $\widehat{f}$ is real and even too.

(b) we set $\varphi(x) = f'(x) = -4x/(x^4+4)$. The function φ is real and odd. Hence $\widehat{\varphi}$ is imaginary and odd. We compute $\widehat{\varphi}(\xi)$ with $\xi < 0$.

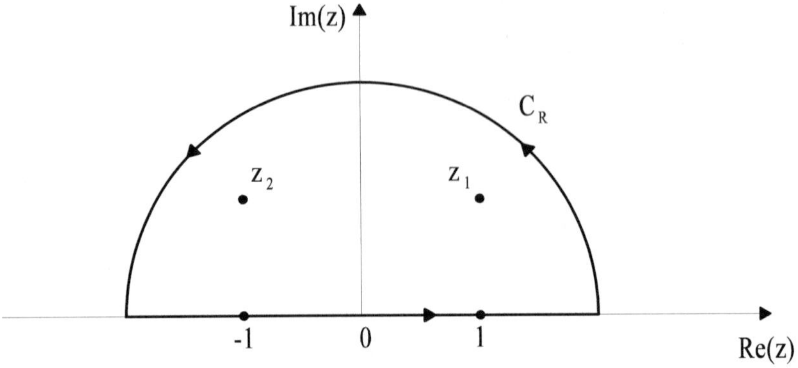

Figure 4.6 - Loop of the solution of Exercise 4.37(b).

$$\widehat{\varphi}(\xi) = \int_{-\infty}^{+\infty} \frac{-4xe^{-i\xi x}}{x^4+4}\,dx = \lim_{R\to+\infty} \int_{C_R} \frac{-4ze^{-i\xi z}}{z^4+4}\,dz = 2\pi i\,(r_1+r_2)$$

where r_j ($j=1,\,2$) is the residue of the integrand at z_j and $z_1 = 1+i$, $z_2 = -1+i$ are the simple poles in the upper half-plane:

$$r_j = \lim_{z\to z_j} (z-z_j)\frac{-4z\,e^{-i\xi z}}{z^4+4} = \frac{-4z_j\,e^{-i\xi z_j}}{4z_j^3} = -\frac{e^{-i\xi z_j}}{z_j^2},$$

$$r_1 = -\frac{e^{-i\xi}e^{\xi}}{2i}, \qquad r_2 = -\frac{e^{i\xi}e^{\xi}}{-2i},$$

$$\widehat{\varphi}(\xi) = 2\pi i\, e^{\xi}\,\frac{e^{i\xi}-e^{-i\xi}}{2i} = 2\pi i\, e^{\xi}\,\sin\xi \qquad \forall \xi < 0\,.$$

We deduce ($\widehat{\varphi}$ odd e continuous):

$$\widehat{\varphi}(\xi) = 2\pi i\, e^{-|\xi|}\,\sin\xi \qquad \forall\xi\in\mathbb{R},$$

$$\widehat{\varphi}(\xi) = \widehat{f'}(\xi) = i\xi\,\widehat{f}(\xi), \qquad \forall\xi\in\mathbb{R},$$

$$i\xi\,\widehat{f}(\xi) = 2\pi i\, e^{-|\xi|}\sin\xi, \qquad \forall\xi\in\mathbb{R},$$

$$\widehat{f}(\xi) = 2\,\pi\, e^{-|\xi|}\,\frac{\sin\xi}{\xi} + C\delta(\xi) \quad \text{in } \mathcal{S}'(\mathbb{R}).$$

Taking into account that f is real and even, we conclude

$$\widehat{f}(\xi) = 2\,\pi\, e^{-|\xi|}\,\frac{\sin\xi}{\xi}, \qquad \forall\xi\in\mathbb{R}.$$

(c) If $x=1$ the identity (*) is trivial; moreover both sides have exactly the same derivative; then (*) holds true for any real x. Recalling that $p(x) = \chi_{[-1,1]}(x)$, we find

$$\widehat{p}(\xi) = \frac{2\,\sin\xi}{\xi}, \qquad \mathcal{F}\left\{\frac{1}{x^2+1}\right\} = \pi\, e^{-|\xi|}.$$

Summarizing

$$\mathcal{F}^{-1}\left\{2\pi\, e^{-|\xi|}\,\frac{\sin\xi}{\xi}\right\} = \mathcal{F}^{-1}\left\{\pi\, e^{-|\xi|}\right\} * \left\{\frac{2\,\sin\xi}{\xi}\right\} = \frac{1}{x^2+1} * p(x) =$$

$$= \int_{-\infty}^{+\infty}\frac{p(x-s)}{s^2+1}\,ds = \int_{x-1}^{x+1}\frac{ds}{s^2+1} = \arctan(x+1)-\arctan(x-1) = \arctan(2/x^2).$$

□

Exercise 4.38. Evaluate the Fourier transform of the function $f(x) = \dfrac{\arctan x}{x}$

Solution - We notice that $f \in L^2(\mathbb{R})$ is even and fulfils the differential equation

$$x f'(x) = \frac{1}{1+x^2} - f(x);$$

by transforming this equation we are lead to a division problem in the unknown $\widehat{f}\,'$:

$$-\xi \widehat{f}\,'(\xi) - \widehat{f}(\xi) = \pi e^{-|\xi|} - \widehat{f}(\xi) \;\Rightarrow\; \widehat{f}'(\xi) = -\pi\frac{e^{-|\xi|}}{\xi} + C\delta(\xi)\,.$$

f is even, hence $C = 0$. By looking separately at the two cases $\xi > 0$ and $\xi < 0$, we get

$$\widehat{f}(\xi) = \pi\int_{|\xi|}^{+\infty}\frac{dy}{ye^y},$$

so $\widehat{f}(\xi)$ has a logarithmic singularity as $\xi \to 0$. □

Exercise 4.39. Prove that

$$\mathcal{F}\left\{\frac{1}{\sqrt{|x|}}\right\} = \frac{\sqrt{2\pi}}{\sqrt{|\xi|}} \qquad x,\,\xi \in \mathbb{R}\,.$$

Solution - After setting $u(x) = |x|^{-1/2}$, we observe that u is even, $u \in L^1_{loc}(\mathbb{R})$ and $u\,\chi_{\{|x|>1\}} \in L^\infty(\mathbb{R})$. Then $u \in \mathcal{S}'(\mathbb{R})$, there exists the limit in $\mathcal{S}'(\mathbb{R})$ of $\mathcal{F}\left\{\dfrac{\chi_{\{|x|<R\}}}{|x|^{1/2}}\right\}(\xi)$ as $R \to +\infty$ and such limit coincides with $\widehat{u}(\xi)$:

$$\widehat{u}(\xi) = \text{v.p.}\int_{\mathbb{R}} e^{-i\xi x}u(x)\,dx = \text{v.p.}\int_{\mathbb{R}}\cos(\xi x)\,|x|^{-1/2}\,dx =$$

$$= 2\lim_{R\to+\infty}\int_0^R \cos(\xi x)\,|x|^{-1/2}\,dx = \qquad \boxed{\xi x = s^2,\ dx = 2s\,ds/\xi\,,\ \xi>0}$$

$$= 2\lim_{R\to+\infty}\int_0^{\sqrt{\xi R}}\cos(s^2)\,2\,|\xi|^{-1/2}ds = \frac{4}{|\xi|^{1/2}}\lim_{T\to+\infty}\int_0^T\cos(s^2)\,ds = \frac{\sqrt{2\pi}}{\sqrt{\xi}}\,,\ \forall\xi>0.$$

In the last computation we have inserted the the Fresnel's integral value (Exercise 1.23):

$$\lim_{T\to+\infty}\int_0^T\cos(s^2)\,ds = \frac{1}{2}\sqrt{\frac{\pi}{2}}$$

Due to symmetry we get $\widehat{u}(\xi) = \sqrt{2\pi}/\sqrt{|\xi|}\ \forall\xi\in\mathbb{R}\backslash\{0\}$. □

Exercise 4.40. Evaluate $\mathcal{F}\left\{\dfrac{1}{|x|^a}\right\} \qquad x,\,\xi \in \mathbb{R},\ 0 < a < 1.$

Solution - We proceed as in the previous exercise. $|x|^{-a}$ belongs to $\mathcal{S}'$ and is even. If $\xi > 0$ we obtain:

$$\mathcal{F}\left\{\frac{1}{|x|^a}\right\}(\xi) = \text{v.p.}\int_{\mathbb{R}} e^{-i\xi x}|x|^{-a}\,dx = \text{v.p.}\int_{\mathbb{R}} \cos(\xi x)\,|x|^{-a}\,dx =$$

$$= 2\lim_{R\to+\infty}\int_0^R \cos(\xi x)\,|x|^{-a}\,dx = \boxed{\xi x = s^2,\ dx = 2s\,ds/\xi\ ,\ \xi>0}$$

$$= 2\lim_{R\to+\infty}\int_0^{\sqrt{\xi R}} \cos(s^2)\,\frac{|\xi|^a}{s^{2a}}\,\frac{2s}{\xi}\,ds = \left(4\lim_{T\to+\infty}\int_0^T \frac{\cos(s^2)}{s^{2a-1}}\,ds\right)\frac{1}{|\xi|^{1-a}}$$

Due to symmetry the result holds $\forall \xi \in \mathbb{R}\backslash\{0\}$ too.
The numerical value of the constant that appears in front of $1/|\xi|^{1-a}$ in the transform can be written as: $2^{1-a}\sqrt{\pi}\,\Gamma\left(\frac{1-a}{2}\right)/\,\Gamma\left(\frac{a}{2}\right)$; in this respect see also Exercise 4.59. □

Exercise 4.41. **(Difference quotient in L^2)** Given $u \in \mathcal{D}'(\mathbb{R})$, for any $h \in \mathbb{R}$ with $h \neq 0$ we define

$$w_h(x) = \frac{1}{h}\,(u(x+h) - u(x))\,.$$

We know (by Exercise 3.22) that $w_h(x) \to u'(x)$ in $\mathcal{D}'(\mathbb{R})$, as $h \to 0$.
Prove that, if $u \in L^2(\mathbb{R})$ and $u' \in L^2(\mathbb{R})$ (where u' denotes the distributional derivative) then

$$\lim_{h\to 0} w_h(x) = u'(x) \quad \text{in } L^2(\mathbb{R})\,,$$

Solution - We prove the convergence of the Fourier transform of w_h.

$$(*) \qquad \widehat{w_h}(\xi) = \frac{e^{ih\xi}-1}{h}\,\widehat{u}(\xi) \overset{L^2(\mathbb{R})}{\longrightarrow} i\xi\,\widehat{u}(\xi) = \widehat{u'}(\xi)\,,$$

the convergence in L^2 in the above computation coincides with the convergence of integrals:

$$\int_{\mathbb{R}}\left|\frac{e^{ih\xi}-1}{h}\,\widehat{u}(\xi)\right|^2 d\xi \overset{L^2(\mathbb{R})}{\longrightarrow} \int_{\mathbb{R}} |\,i\xi\,\widehat{u}(\xi)\,|^2\,d\xi\,;$$

this last convergence follows by Lebesgue's dominated convergence theorem (Theorem 2.5), since the integrands a.e. point-wise converge to $|i\xi\widehat{u}(\xi)|^2$ and are dominated by an integrable function

$$\left|\frac{e^{ih\xi}-1}{h}\,\widehat{u}(\xi)\right|^2 \leq |\,\xi\,\widehat{u}(\xi)\,|^2 \in L^1(\mathbb{R})\,,$$

due to

$$|e^{i\lambda}-1| = \left((\cos\lambda - 1)^2 + (\sin\lambda)^2\right)^{1/2} = \left(2(1-\cos\lambda)\right)^{1/2} = \left(4\left(\sin\frac{\lambda}{2}\right)^2\right)^{1/2} =$$

$$= 2\left|\sin\left(\frac{\lambda}{2}\right)\right| \leq |\lambda|\,.$$

The convergence in $(*)$ and the Plancherel identity (Theorem 4.8) entail the convergence in $L^2(\mathbb{R})$ of $w_h(x) = \mathcal{F}^{-1}\{\frac{e^{ih\xi}-1}{h}\,\widehat{u}(\xi)\}$ ad $u'(x) = \mathcal{F}^{-1}\{i\xi\widehat{u}\}$ in $L^2(\mathbb{R})$. □

Exercise 4.42. Under the conditions $u,\ u' \in L^1(\mathbb{R})$, prove:

$$1)\qquad \mathcal{F}\left\{\int_{-\infty}^{x} u(t)\,dt\right\} = \pi\,\widehat{u}(0)\,\delta(\xi) - i\,\widehat{u}(\xi)\text{ v.p.}\frac{1}{\xi},$$

$$2)\qquad \mathcal{F}\left\{\int_{x}^{\infty} u(t)\,dt\right\} = \pi\,\widehat{u}(0)\,\delta(\xi) + i\,\widehat{u}(\xi)\text{ v.p.}\frac{1}{\xi}$$

$$3)\qquad \mathcal{F}\left\{\int_{0}^{x} u(t)\,dt\right\} = \pi\left(\int_{0}^{\infty} u(t)\,dt - \int_{-\infty}^{0} u(t)\,dt\right)\delta(\xi) - i\,\widehat{u}(\xi)\text{ v.p.}\frac{1}{\xi}$$

Solution - First we notice that v.p. $\frac{1}{\xi}$ is a distribution of order 1, then it can be multiplied by $C^1(\mathbb{R})$ functions. Moreover $u,\ u' \in L^1(\mathbb{R})$ entail $\widehat{u} \in C^1(\mathbb{R})$.

$$1)\qquad \int_{-\infty}^{x} u(t)\,dt = \int_{-\infty}^{+\infty} u(t)\,H(x-t)\,dt = \big(u * H\big)(x),$$

using the formula $\mathcal{F}\{H\}(\xi) = \pi\delta(\xi) - i\text{ v.p.}\frac{1}{\xi}$ (Table 10.8.2):

$$\mathcal{F}\left\{\int_{-\infty}^{x} u(t)\,dt\right\} = \mathcal{F}\left\{\,\big(u*H\big)(x)\,\right\} = \widehat{u}(\xi)\left(\pi\delta(\xi) - i\text{ v.p.}\frac{1}{\xi}\right) = \pi\widehat{u}(0)\delta(\xi) - i\widehat{u}(\xi)\text{ v.p.}\frac{1}{\xi}.$$

$$2)\qquad \int_{x}^{+\infty} u(t)\,dt = \int_{\mathbb{R}} u(t)\,dt - \int_{-\infty}^{x} u(t)\,dt = \widehat{u}(0) - \int_{-\infty}^{x} u(t)\,dt,$$

using the formula $\mathcal{F}\{1\}(\xi) = 2\pi\delta(\xi)$ (Tabella 10.8.2):

$$\mathcal{F}\left\{\int_{x}^{+\infty} u(t)\,dt\right\} = (2\pi-\pi)\,\widehat{u}(0)\,\delta(\xi) + i\,\widehat{u}(\xi)\text{ v.p.}\frac{1}{\xi} = \pi\,\widehat{u}(0)\,\delta(\xi) + i\,\widehat{u}(\xi)\text{ v.p.}\frac{1}{\xi}.$$

3) Exploiting item 1) and $\int_0^x u(t)\,dt = \int_{-\infty}^x u(t)\,dt - \int_{-\infty}^0 u(t)\,dt$ we obtain:

$$\begin{aligned}\mathcal{F}\left\{\int_{0}^{x} u(t)\,dt\right\} &= +\pi\widehat{u}(0)\delta(\xi) - i\widehat{u}(\xi)\text{ v.p.}\frac{1}{\xi} - 2\pi\left(\int_{-\infty}^{0} u(t)dt\right)\delta(\xi) = \\ &= \left(\pi\left(\int_{-\infty}^{+\infty} u(t)dt\right) - 2\pi\left(\int_{-\infty}^{0} u(t)dt\right)\right)\delta(\xi) - i\widehat{u}(\xi)\text{ v.p.}\frac{1}{\xi}.\end{aligned}$$ □

Exercise 4.43. 1) Find $u \in L^1(\mathbb{R})$ such that $u * u = \sqrt{\frac{\pi}{2}}\,e^{-x^2/2}$, $x \in \mathbb{R}$.
2) If we face the same problem of the previous item but looking for solutions in $L^2(\mathbb{R})$ instead of $L^1(\mathbb{R})$, do we find the same solutions?

Solution - 1) By $\mathcal{F}$ transform: $\big(\widehat{u}(\xi)\big)^2 = \pi\,e^{-\xi^2/2}$: the fractional powers denote arithmetical roots; $\widehat{u}(\xi) = \pm\sqrt{\pi}\,e^{-\xi^2/4}$ $\forall \xi \in \mathbb{R}$, that is there are 2 choices for any ξ, corresponding to infinitely many possibilities for $\widehat{u}$. Nevertheless $\widehat{u}$ must belong to C^0 due to Riemann-Lebesgue Lemma (Theorem 4.3). Summarizing there are only two admissible choices for the function $\widehat{u}$. Accordingly there are two solutions in $L^1(\mathbb{R})$: $u_1(x) = e^{-x^2}$ and $u_2(x) = -\,e^{-x^2}$.
2) In this case $\widehat{u}$ is not necessarily continuous, then everyone among the infinitely many choices with values of $\widehat{u}(\xi)$ equal to $\pm\sqrt{\pi}\,e^{-\xi^2/4}$ are admissible, provided they give a measurable function. Every choice of this kind gives a solution. Then the problem has infinitely many solutions (in addition to u_1 and u_2, the only two solutions in L^1): the anti-transform of all the choices described above. For instance the function below is an example of admissible solution, different from u_1 and u_2:

$$u(x) = \mathcal{F}^{-1}\left\{\text{sign}(\xi)\,\sqrt{\pi}\,e^{-\xi^2/4}\right\} = -\frac{1}{\pi}\text{ v.p.}\frac{1}{x} * e^{-x^2}.$$ □

4.2.3 Exercises on $\mathcal{F}$ transform in several variables

Exercise 4.44. Show that the transform of Gaussian function in $\mathbb{R}^n$ fulfils for any $a > 0$:

- $\mathcal{F}\{\exp(-ax^2)\} = \sqrt{\frac{\pi}{a}}\exp(-\xi^2/4a) \qquad x, \xi \in \mathbb{R}\,,$
- $\mathcal{F}\{\exp(-a|\mathbf{x}|^2)\} = \left(\frac{\pi}{a}\right)^{\frac{n}{2}}\exp(-|\boldsymbol{\xi}|^2/4a) \qquad \mathbf{x}, \boldsymbol{\xi} \in \mathbb{R}^n\,.$

Solution - They are straightforward consequences of $\mathcal{F}\{\exp(-x^2)\}(\xi) = \sqrt{\pi}\exp(-\xi^2/4)$ (see Exercise 4.2), of the transform scaling properties (4.5) and (4.18), and of the identity

$$\int_{\mathbb{R}^n} \exp(-i\,\boldsymbol{\xi}\cdot\mathbf{x})\exp(-|\mathbf{x}|^2)\,d\mathbf{x} = \prod_{j=1}^{n}\int_{\mathbb{R}} \exp(-i\,\xi_j x_j)\exp(-x_j^2)\,dx_j\,.$$

□

Exercise 4.45. Prove that, if u_j depend on one real variable and are $\mathcal{F}$ trasformable for $j = 1,\ldots,n$, then

$$\mathcal{F}\{\,u_1(x_1)\,u_2(x_2)\ldots u_n(x_n)\,\} = \widehat{u_1}(\xi_1)\,\widehat{u_2}(\xi_2)\ldots\widehat{u_n}(\xi_n)$$

where the products stand for tensor product when the u_j's are distributions.

Solution - If $u_j \in L^1(\mathbb{R})$ $\forall j = 1,\ldots,n$, then by taking into account of $\exp(-i\boldsymbol{\xi}\cdot\mathbf{x}) = \exp(-i\xi_1 x_1)\exp(-i\xi_2 x_2)\ldots\exp(-i\xi_n x_n)$ and applying the Fubini-Tonelli Theorem,

$$\int_{\mathbb{R}^n} \exp(-i\,\boldsymbol{\xi}\cdot\mathbf{x})\big(u_1(x_1)\,u_2(x_2)\ldots u_n(x_n)\big)\,d\mathbf{x} = \prod_{j=1}^{n}\int_{\mathbb{R}} \exp(-i\,\xi_j x_j)\,u_j(x_j)\,dx_j\,.$$

If $u_j \in \mathcal{S}'(\mathbb{R})$ $\forall j = 1,\ldots,n$, then each u_j can be approximated by a sequence of test functions in $\mathcal{D}(\mathbb{R})$: $\forall j\ \exists\, \varphi_{k,j} \in \mathcal{D}(\mathbb{R})$ s.t. $\varphi_{k,j} \to u_j$ in $\mathcal{S}'(\mathbb{R})$. Then $\big(\varphi_{k,1}(x_1)\,\varphi_{k,2}(x_2)\ldots\varphi_{k,n}(x_n)\big) \to \big(u_1(x_1)\,u_2(x_2)\ldots u_n(x_n)\big)$ in $\mathcal{S}'(\mathbb{R}^n)$ and the formula to be proved follows in the general case since it holds true for L^1 functions due to the fact that the $\mathcal{F}$ transform is continuous with respect to the convergence in $\mathcal{S}'(\mathbb{R}^n)$. □

Exercise 4.46. Given a compactly supported distribution f find all solutions u of

$$u \in \mathcal{S}'(\mathbb{R}^n)\ : \qquad -\Delta u + u = f \qquad \text{in } \mathcal{S}'(\mathbb{R}^n)\,, \tag{4.68}$$

where $\Delta = \sum_{j=1}^{n}\dfrac{\partial^2}{\partial x_j^2}$.

- Analyze the regularity of the solutions when the datum f belongs to $L^2(\mathbb{R}^n)$.
- Analyze the regularity of the solutions when the datum f has all distributional derivatives up to the order k in $L^2(\mathbb{R}^n)$.

Solution - f ha compact support, hence $f \in \mathcal{S}'(\mathbb{R}^n)$.
Transforming the equation and taking into account of (4.44) we obtain

$$(1 + |\boldsymbol{\xi}|^2)\,\widehat{u}(\boldsymbol{\xi}) = \widehat{f}(\boldsymbol{\xi})\,.$$

Since $(1 + |\boldsymbol{\xi}|^2) \neq 0$ $\forall\boldsymbol{\xi} \in \mathbb{R}^n$ we deduce the existence and uniqueness of the solution $u \in \mathcal{S}'(\mathbb{R}^n)$ fulfilling:

$$\widehat{u}(\boldsymbol{\xi}) = \frac{\widehat{f}(\boldsymbol{\xi})}{1 + |\boldsymbol{\xi}|^2}\,, \tag{4.69}$$

$$u = \mathcal{F}^{-1}\left\{\frac{1}{1 + |\boldsymbol{\xi}|^2}\right\} * f\,.$$

Notice that $U=\mathcal{F}^{-1}\left\{\frac{1}{1+|\boldsymbol{\xi}|^2}\right\}$ is the fundamental solution of $I-\Delta$, for any dimension $n\geq 1$, that is U solves

$$-\Delta U+U=\delta \qquad \text{in } \mathcal{S}'(\mathbb{R}^n).$$

Except in the case $n=1$ (see Exercise 4.30) computing the anti-transform $\mathcal{F}^{-1}\left\{1/(1+|\boldsymbol{\xi}|^2)\right\}$ is a tricky issue and the resulting analytic form of U is different according to the dimension n. See Exercise 4.48 for the case $n=3$ and the subsequent Note for $n=2$.
Anyway knowing only the transform of the solution (4.69) allows a careful regularity analysis of the solution itself.
By (4.64) we deduce $|\widehat{u}(\boldsymbol{\xi})|\leq|\widehat{f}(\boldsymbol{\xi})|/(1+|\boldsymbol{\xi}|^2)$. The Plancherel identity (4.36) and the rules for transforming derivatives (4.44) lead to the statement:

$$u\in H^{k+2}(\mathbb{R}^n) \qquad \Longleftrightarrow \qquad f\in H^k(\mathbb{R}^n)\,, \tag{4.70}$$

where $H^k(\mathbb{R}^n)$ denotes the *Sobolev space* of functions in $L^2(\mathbb{R}^n)$ whose derivatives up to the order k belong to $L^2(\mathbb{R}^n)$. $H^k(\mathbb{R}^n)$ is an Hilbert space when endowed with
the norm $\|v\|_{H^k(\mathbb{R}^n)}=\left(\sum_{|\boldsymbol{\alpha}|\leq k}\|D^{\boldsymbol{\alpha}}v\|^2_{L^2(\mathbb{R}^n)}\right)^{1/2}$ and
the scalar product $(u,v)_{H^k(\mathbb{R}^n)}=\sum_{|\boldsymbol{\alpha}|\leq k}(D^{\boldsymbol{\alpha}}u,D^{\boldsymbol{\alpha}}v)_{L^2(\mathbb{R}^n)}$.
On the other hand we notice that $f\in C^k$ does not entails $u\in C^{k+2}$, if $n>1$. □

Exercise 4.47. Prove the next statement.
Sobolev Theorem (Sergej L'vovič Sobolev, 1908-1989) - If $k>m+\dfrac{n}{2}$ with $k,m,n\in\mathbb{N}$ then

$$H^k(\mathbb{R}^n)\subset C^m(\mathbb{R}^n) \tag{4.71}$$

and there exists C such that, for any $u\in H^k(\mathbb{R}^n)$ and any multi-index $\boldsymbol{\alpha}$, $|\boldsymbol{\alpha}|\leq m$,

$$\|D^{\boldsymbol{\alpha}}u\|_{L^\infty(\mathbb{R}^n)}\leq C\,\|u\|_{H^k(\mathbb{R}^n)}\,.$$

Solution - Considering the case $m=0$ is enough: indeed the general case is an easy consequence of this one by induction.
Due to Plancherel identity $u\in H^k(\mathbb{R}^n)$ if and only if $(1+|\boldsymbol{\xi}|^2)^{k/2}\,\widehat{u}(\boldsymbol{\xi})\in L^2(\mathbb{R}^n)$. Moreover $\|u\|_{H^k(\mathbb{R}^n)}$ and $\|(1+|\boldsymbol{\xi}|^2)^{k/2}\widehat{u}(\boldsymbol{\xi})\|_{L^2(\mathbb{R}^n)}$ are equivalent norms in $H^k(\mathbb{R}^n)$.
The function $\dfrac{1}{(1+|\boldsymbol{\xi}|^2)^{k/2}}$ belongs to $L^2(\mathbb{R}^n)$ if and only if $k>n/2$. In fact, if this is the case and referring to polar coordinates, we have $\int_0^{+\infty}\varrho^{n-1}/(1+\varrho^2)^k\,d\varrho<+\infty$.
Therefore if $u\in H^k(\mathbb{R}^n)$ and $k>n/2$, then the function

$$\widehat{u}(\boldsymbol{\xi})=\left((1+|\boldsymbol{\xi}|^2)^{k/2}\,\widehat{u}(\boldsymbol{\xi})\right)\frac{1}{(1+|\boldsymbol{\xi}|^2)^{k/2}}$$

belongs to $L^1(\mathbb{R}^n)$ since it is the product of two $L^2(\mathbb{R}^n)$ functions.
By Riemann-Lebesgue Lemma and the Schwarz inequality $u=\mathcal{F}^{-1}\{\widehat{u}\}$ is continuous and fulfils the estimate

$$2\pi\,\|u\|_{L^\infty(\mathbb{R}^n)}\leq\|\widehat{u}\|_{L^1(\mathbb{R}^n)}\leq\|(1+|\boldsymbol{\xi}|^2)^{k/2}\widehat{u}(\boldsymbol{\xi})\|_{L^2(\mathbb{R}^n)}\left\|\frac{1}{(1+|\boldsymbol{\xi}|^2)^{k/2}}\right\|_{L^2(\mathbb{R}^n)}.$$ □

Exercise 4.48. Prove that, if $a>0$ and $n=3$, then the fundamental solution of the operator $a^2I-\Delta$, that is the unique solution of

$$U\in\mathcal{S}'(\mathbb{R}^3) \qquad -\Delta U+a^2U=\delta \qquad \text{in } \mathcal{S}'(\mathbb{R}^3)\,,$$

fulfils

$$U(\mathbf{x})=\frac{1}{4\pi}\frac{e^{-a|\mathbf{x}|}}{|\mathbf{x}|} \qquad \mathbf{x}\in\mathbb{R}^3\setminus\{\mathbf{0}\}\,. \tag{4.72}$$

By the previous formula we deduce $U\in L^p(\mathbb{R}^3)$ $\forall p\in[1,3)$.

Solution - Proceeding as in Exercise 4.46 we obtain $\widehat{U}(\boldsymbol{\xi}) = 1/(a^2 + |\boldsymbol{\xi}|^2)$.
Then $\widehat{U}$ is radially symmetric and real, moreover $\widehat{U} \in L^2(\mathbb{R}^3)\backslash L^1(\mathbb{R}^3)$.
We have to evaluate $U = \mathcal{F}^{-1}\{\widehat{U}\} \in L^2(\mathbb{R}^3)$ (due to Plancherel identity, Theorem 4.8): we must pay attention since point-wise values are only defined almost everywhere.
U must be radially symmetric and real, that is $U(\mathbf{x}) = v(r)$, with $r = |\mathbf{x}|$, for a suitable function v :

$$a^2 v - r^{-2}\partial_r^2(r^2 v) = a^2 U - \Delta U = 0 \qquad \text{in } \mathbb{R}^3 \setminus \{\mathbf{0}\}$$
$$v_{rr} + \frac{2}{r} v_r - a^2 v = 0 \qquad r \in (0,\infty).$$

The ordinary differential equation of the second order (singular at 0) has two independent solutions ($e^{\pm ar}/r$), then the general integral is $v(r) = B\, e^{-ar}/r + C\, e^{+ar}/r$. We must choose $C = 0$ since $U \in \mathcal{S}'$. At this level we know that

$$U(\mathbf{x}) = B\, e^{-a|\mathbf{x}|}/|\mathbf{x}| \qquad \text{in } \mathbb{R}^3 \setminus \{\mathbf{0}\},$$
$$\text{spt}\left(U - B\, e^{-a|\mathbf{x}|}/|\mathbf{x}|\right) \subset \{\mathbf{0}\}.$$

The knowledge of the support entails (by formula (3.17)) that

$$U(\mathbf{x}) - B\, e^{-a|\mathbf{x}|}/|\mathbf{x}| = \sum_{\boldsymbol{\alpha}} c_{\boldsymbol{\alpha}} D^{\boldsymbol{\alpha}} \delta(\mathbf{x}) \qquad \text{where a finite number of summands is present.}$$

On the other hand every coefficient $c_{\boldsymbol{\alpha}}$ must vanish since the left-hand side belongs to L^2.
We conclude by computing B :

$$\frac{1}{a^2} = \widehat{U}(\mathbf{0}) = \int_{\mathbb{R}^3} U(\mathbf{x})\, d\mathbf{x} \overset{d\mathbf{x}=r^2 d\sigma\, dr}{=} \int_{\partial B_1} d\sigma \int_0^\infty B\,(e^{-ar}/r)\, r^2 dr =$$
$$= 4\pi\, B \int_0^\infty e^{-ar}\, r\, dr = \frac{4\pi\, B}{a^2}.$$

By (4.72), integrating in polar coordinates, we obtain $U \in L^p(\mathbb{R}^3)$ if and only if $e^{-ap\,r} r^2/r^p$ belongs to $L^1(0,\infty)$, that is if and only if $1 \le p < 3$.
Remark - The technique we have used can be exploited in any dimension n too. Nevertheless the solution of the ordinary equation in v that one meets has no elementary solution contrarily to the case $n = 3$. For instance, if $n = 2$ and $a = 1$, then solving

$$U \in \mathcal{S}'(\mathbb{R}^2) \qquad -\Delta U + U = \delta \qquad \text{in } \mathcal{S}'(\mathbb{R}^2),$$

one finds the modified Bessel equation of order 0 (see Exercise 1.29):

$$v_{rr} + \frac{1}{r} v_r - v = 0 \qquad r \in (0,\infty);$$

picking out its solution $\mathcal{K}_0$ (with a logarithmic singularity at the origin) one achieves $U(\mathbf{x}) = C_2\, \mathcal{K}_0(|\mathbf{x}|)$, with $C_2 = 1/\left(2\pi \int_0^{+\infty} \mathcal{K}_0(r) r\, dr\right)$. □

Exercise 4.49. By using the results of the previous exercise prove that, if $n = 3$, then

$$U(\mathbf{x}) = \frac{1}{4\pi\, |\mathbf{x}|} \qquad \mathbf{x} \in \mathbb{R}^3 \setminus \{\mathbf{0}\} \tag{4.73}$$

is a fundamental solution of the operator $-\Delta$, that is a solution of

$$U \in \mathcal{S}'(\mathbb{R}^3), \qquad -\Delta U = \delta \qquad \text{in } \mathcal{S}'(\mathbb{R}^3). \tag{4.74}$$

Find all the solutions of (4.74)

Solution - For any $k \in \mathbb{N}$, we denote by U_k the fundamental solution of $(1/k^2)I - \Delta$:

$$U_k \in \mathcal{S}'(\mathbb{R}^3), \qquad \frac{1}{k^2}U_k \; - \; \Delta U_k \; = \; \delta\,. \tag{4.75}$$

By Exercise 4.48 we know that $U_k(\mathbf{x}) = \frac{1}{4\,\pi}\frac{e^{-|\mathbf{x}|/k}}{|\mathbf{x}|}$. Hence $U_k \to U$ uniformly in any fixed compact of $\mathbb{R}^3\backslash\{\mathbf{0}\}$ (not in $\mathbb{R}^3$). Then: $U_k \to U$ in $\mathcal{D}'(\mathbb{R}^3)$, $\Delta U_k \to \Delta U$ in $\mathcal{D}'(\mathbb{R}^3)$, $U_k/k^2 \to 0$ in $\mathcal{D}'(\mathbb{R}^3)$ and we can pass to the limit in the equation (4.75) that marks out U_k, hence obtaining that U solves (4.74).
Since $0 \le U_k(\mathbf{x}) \le U(\mathbf{x})$ and $U \in \mathcal{S}'(\mathbb{R}^3)$, we have $U_k \to U$ in $\mathcal{S}'(\mathbb{R}^3)$ too.
Notice that $U_k \in L^1(\mathbb{R}^3)$ (indeed $U_k \in L^p \; \forall p \in [1,3)$) ma $U \notin L^1(\mathbb{R}^3)$, nevertheless $U \in L^1_{loc}(\mathbb{R}^3)$.
Then U is a particular solution of (4.74).
In order to find the general solution we are left only to solve the homogeneous problem:

$$W \in \mathcal{S}'(\mathbb{R}^3)\,, \qquad -\,\Delta W \; = \; 0 \quad \text{in } \mathcal{S}'(\mathbb{R}^3)\,.$$

The Fourier transform of the equation in W leads to a division problem in the distributional framework with unknown $\widehat{W}$:

$$\widehat{W} \in \mathcal{S}'(\mathbb{R}^3)\,, \qquad |\boldsymbol{\xi}|^2\, \widehat{W}(\boldsymbol{\xi}) \; = \; 0\,.$$

Thanks to the characterization (3.17) we know that its solution is $\widehat{W}(\boldsymbol{\xi}) = \sum_{\boldsymbol{\alpha}} c_{\boldsymbol{\alpha}}\, D^{\boldsymbol{\alpha}}\, \delta$.
The correspondent anti-transform is $W = \sum_{\boldsymbol{\alpha}} c_{\boldsymbol{\alpha}}\, (-i)^{|\boldsymbol{\alpha}|}\, \mathbf{x}^{\boldsymbol{\alpha}}$: a real-valued polynomial in 3 real variables, that must be a harmonic function too.
Summarizing the general solution W of the homogeneous equation is any harmonic polynomial in 3 real variables (examples: $x^2 - y^2\,;\; x^3 - 3xy^2\,;\; x^2 - y^2 + z\,;\; xz$).
The general solution of (4.73) is of the kind $W \, + \, 1/(4\,\pi\,|\mathbf{x}|)$.
Note - When solving this exercise, we have implicitly proved that:

- the only harmonic functions in $\mathbb{R}^3$ that are also tempered distributions are the polynomials;
- **(Liouville Theorem)** all bounded harmonic functions in $\mathbb{R}^3$ are constants.

□

Exercise 4.50. (1) Prove that every harmonic tempered distributions in $\mathbb{R}^n$ are polynomials, that is every solution $u \in \mathcal{S}'(\mathbb{R}^n)$ of $-\Delta u = 0$ is a polynomial in n real variables.
(2) Prove that, if $L = \sum_{\boldsymbol{\alpha}} c_{\boldsymbol{\alpha}} D^{\boldsymbol{\alpha}}$ and $\sum c_{\boldsymbol{\alpha}} (i)^{|\boldsymbol{\alpha}|} \boldsymbol{\xi}^{\boldsymbol{\alpha}} \neq 0 \;\forall \boldsymbol{\xi} \neq \mathbf{0}$, then every solution of

$$u \in \mathcal{S}'(\mathbb{R}^n) \qquad -\,Lu = 0 \qquad \mathbb{R}^n$$

is a polynomial in n real variables. In particular if $n = 2$, it is a function of x, y that can be expressed as the real part of a complex polynomial in one complex variable $z = x + iy$.
(3) Prove that, if $L = \sum_{\boldsymbol{\alpha}} c_{\boldsymbol{\alpha}} D^{\boldsymbol{\alpha}}$ and there exists $\widetilde{\boldsymbol{\xi}} \in \mathbb{R}^n$ s.t. $\sum c_{\boldsymbol{\alpha}} (i)^{|\boldsymbol{\alpha}|} \boldsymbol{\xi}^{\boldsymbol{\alpha}} \neq 0 \;\forall \boldsymbol{\xi} \neq \widetilde{\boldsymbol{\xi}}$, then every solution of

$$u \in \mathcal{S}'(\mathbb{R}^n) \qquad -\,Lu = 0 \qquad \mathbb{R}^n$$

is a polynomial in n real variables multiplied by the complex modulation $e^{i\,\widetilde{\boldsymbol{\xi}}\cdot\mathbf{x}}$.

Solution - Hint: use the information about $\operatorname{spt}\widehat{u}$ and apply the characterization (3.17) to the Fourier transform of the solutions. □

Exercise 4.51. (Fourier transform of radially symmetric functions in $L^1(\mathbb{R}^2)$)
Prove that if $u \in L^1(\mathbb{R}^2)$ with $u(\mathbf{x}) = \psi(|\mathbf{x}|)$ then $\widehat{u}(\boldsymbol{\xi}) = \varphi(|\boldsymbol{\xi}|)$ with

$$\varphi(|\boldsymbol{\xi}|) \; = \; \mathcal{H}_0\{\psi\} \;\stackrel{\text{def}}{=}\; \int_0^\infty \; J_0(\varrho\,|\boldsymbol{\xi}|)\; \psi(\varrho)\; \varrho\; d\varrho \tag{4.76}$$

where the integrand belongs to $L^1(0,\infty)$, J_0 is the Bessel function of first kind of order 0 (see Exercise 1.28) and $\mathcal{H}_0$ is called Hankel transform (Hermann Hankel, 1839-1873).

Solution - By (4.20) we know that $\widehat{u}$ is radially symmetric, that is $\widehat{u}(\boldsymbol{\xi}) = \varphi(|\boldsymbol{\xi}|)$. Moreover $u \in L^1$ entails $\int_0^\infty \psi(\varrho)\,\varrho\,d\varrho < +\infty$, Then the integral in (4.76) is convergent. We set

$$(2k)!! = (2k)(2k-2)\cdots 4\times 2\,, \qquad (2k-1)!! = (2k-1)(2k-3)\cdots 3\times 1$$

and recall:

$$(2k)! \;=\; (2k)!!\,(2k-1)!!\,, \qquad \int_0^{2\pi} (\cos\vartheta)^{2k}\,d\vartheta \;=\; \frac{(2k-1)!!}{(2k)!!}\,,$$

$$\cos(z) \;=\; \sum_{k=0}^{\infty} (-1)^k\,\frac{z^{2k}}{(2k)!}\,, \qquad J_0(z) \;=\; \sum_{k=0}^{\infty} \frac{(-1)^k}{(k!)^2}\left(\frac{z}{2}\right)^{2k} \qquad \forall z\in\mathbb{C}\,.$$

In order to prove (4.76) we exploit the previous identities together with the fact that $\widehat{u}$ is radially symmetric and ψ is even:

$$\begin{aligned}
\varphi(|\boldsymbol{\xi}|) \;=\; \widehat{u}(\boldsymbol{\xi}) \;=\; \widehat{u}(|\boldsymbol{\xi}|,0) \;&=\; \int_{\mathbb{R}^2} e^{-i|\boldsymbol{\xi}|x_1}\,u(\mathbf{x})\,d\mathbf{x} \;=\\
&=\; \int_0^\infty\!\!\int_0^{2\pi} e^{-i|\boldsymbol{\xi}|\varrho\,\cos\vartheta}\,\psi(\varrho)\,\varrho\,d\varrho\,d\vartheta \;=\\
&=\; \int_0^\infty\!\!\int_0^{2\pi} \cos\left(\,|\boldsymbol{\xi}|\varrho\,\cos\vartheta\,\right)\,\psi(\varrho)\,\varrho\,d\varrho\,d\vartheta \;=\\
&=\; \int_0^\infty\!\!\int_0^{2\pi} \sum_{k=0}^{\infty} (-1)^k\,\frac{|\boldsymbol{\xi}|^{2k}\,\varrho^{2k}\,(\cos\vartheta)^{2k}}{(2k)!}\,\psi(\varrho)\,\varrho\,d\varrho\,d\vartheta \;=\\
&=\; \int_0^\infty \sum_{k=0}^{\infty} (-1)^k\,\frac{|\boldsymbol{\xi}|^{2k}\,\varrho^{2k}}{(k!)^2\,2^{2k}}\,\psi(\varrho)\,\varrho\,d\varrho \;=\\
&=\; \int_0^\infty J_0(|\boldsymbol{\xi}|\,\varrho)\,\psi(\varrho)\,\varrho\,d\varrho\,.
\end{aligned}$$

□

Exercise 4.52. (**Fourier transform of radially symmetric functions in** $L^2(\mathbb{R}^2)$)
Prove that if $u \in L^2(\mathbb{R}^2)$ with $u(\mathbf{x}) = \psi(|\mathbf{x}|)$ then $\widehat{u}(\boldsymbol{\xi}) = \varphi(|\boldsymbol{\xi}|)$ with

$$\varphi(|\boldsymbol{\xi}|) \;=\; \lim_{R\to+\infty} \int_0^R J_0(\varrho\,|\boldsymbol{\xi}|)\,\psi(\varrho)\,\varrho\,d\varrho\,. \tag{4.77}$$

where the limit denotes the strong convergence in $L^2(\mathbb{R}^2_{\boldsymbol{\xi}})$.
Notice that, under the assumption that u is radially symmetric and $u \in L^2(\mathbb{R}^2)$, the integrand $J_0(\varrho\,|\boldsymbol{\xi}|)\,\psi(\varrho)\,\varrho$ may not belong to $L^1(0,\infty)$.

Solution - We can repeat the computations in the solution of Exercise 4.77 starting from (4.31) rather than (4.15). □

Exercise 4.53. Using the $\mathcal{F}$ transform formula for radially symmetric functions in $L^2(\mathbb{R}^2)$, that we proved in Exercise 4.52, write an integral representation of the fundamental solution of the operator $a^2\,I - \Delta$, with $a > 0$ and $n = 2$, that is the unique solution of

$$U \in \mathcal{S}'(\mathbb{R}^2) \qquad -\,\Delta U + a^2 U \;=\; \delta \qquad \text{in } \mathcal{S}'(\mathbb{R}^2)\,.$$

Solution - We notice that $1/(a^2+|\boldsymbol{\xi}|^2) = \mathcal{F}\{U\}$ belongs to $L^2(\mathbb{R}^2)$, then using the inversion formula for $\mathcal{F}$ in $L^2(\mathbb{R}^2)$, $1/(a^2+|\boldsymbol{\xi}|^2)$ radially symmetric and (4.77), we obtain:

$$U(\mathbf{x}) = \mathcal{F}^{-1}\left\{\frac{1}{a^2+|\boldsymbol{\xi}|^2}\right\} = \frac{1}{(2\pi)^2}\,\mathcal{F}\left\{\frac{1}{a^2+|\boldsymbol{\xi}|^2}\right\} = \frac{1}{(2\pi)^2}\lim_{R\to\infty}\int_0^R \frac{J_0(\varrho\,|\mathbf{x}|)}{a^2+\varrho^2}\,\varrho\,d\varrho\,.$$

□

Exercise 4.54. Solve the Cauchy problem for the **heat equation**:
given $u_0 \in \mathcal{S}'(\mathbb{R}^n)$ and $f \in C^0\big((0,+\infty),\mathcal{S}'(\mathbb{R}^n_{\mathbf{x}})\big)$, find the solution u of:

$$\begin{cases} u \in C^1\big((0,+\infty),\mathcal{S}'(\mathbb{R}^n)\big)\ : & \\ u_t - \Delta_{\mathbf{x}} u \;=\; f & \mathbf{x}\in\mathbb{R}^n,\ t>0\,, \\ u(\mathbf{x},0) \;=\; u_0(\mathbf{x}) & \mathbf{x}\in\mathbb{R}^n\,. \end{cases} \tag{4.78}$$

Solution - We consider the partial Fourier transform (with respect to the variables $\mathbf{x}$ only):

$$\mathcal{F}_{\mathbf{x}}\{u\} \;=\; \widehat{u}(\mathbf{x},t) \;=\; \int_{\mathbb{R}^n} e^{-i\,\boldsymbol{\xi}\cdot\mathbf{x}}\,u(\mathbf{x},t)\,d\mathbf{x} \tag{4.79}$$

and we proceed formally, by assuming the commutation property

$$\mathcal{F}_{\mathbf{x}}\left\{\partial_t u\right\} \;=\; \partial_t\left(\mathcal{F}_{\mathbf{x}}\{u\}\right) \tag{4.80}$$

whose validity is verified later, due to the regularity of the solution that we will compute. Transforming the partial differential equation, we obtain the one-parameter ($\boldsymbol{\xi}\in\mathbb{R}^n$) family of ordinary differential equations

$$\begin{cases} \widehat{u}_t + |\boldsymbol{\xi}|^2\widehat{u} \;=\; \widehat{f}(\boldsymbol{\xi},t) & t>0 \\ \widehat{u}(\boldsymbol{\xi},0) \;=\; \widehat{u_0}(\boldsymbol{\xi})\,, & \end{cases}$$

whose solutions are

$$\widehat{u}(\boldsymbol{\xi},t) = \widehat{u_0}(\boldsymbol{\xi})\,e^{-|\boldsymbol{\xi}|^2\,t} + \int_0^t f(\boldsymbol{\xi},t-\tau)\,e^{-|\boldsymbol{\xi}|^2\,\tau}\,d\tau = \widehat{u_0}(\boldsymbol{\xi})\,e^{-|\boldsymbol{\xi}|^2\,t} + f(\boldsymbol{\xi},t)*_t\left(e^{-|\boldsymbol{\xi}|^2\,t}H(t)\right).$$

In the last summand, the notation $*_t$ is used to specify the convolution with respect to the only variable t, H denotes the Heaviside function and the trivial extension of f to $\mathbb{R}$ is understood: $f(\boldsymbol{\xi},t)\equiv 0,\ t<0$.
We anti-transform $\widehat{u}$, recalling how $\mathcal{F}_{\mathbf{x}}$ acts on the convolution (with respect to the transform variables $\mathbf{x}$ only: formula (4.51)) and

$$\mathcal{F}_{\mathbf{x}}^{-1}\{e^{-|\boldsymbol{\xi}|^2\,t}\} = \frac{1}{(4\,\pi\,t)^{n/2}}e^{-|\mathbf{x}|^2/(4t)}\,.$$

After setting

$$U(\mathbf{x},t) \;=\; \frac{1}{(4\,\pi\,t)^{n/2}}e^{-|\mathbf{x}|^2/(4t)}\,H(t) \tag{4.81}$$

and noticing that $U(\mathbf{x},t)\in\mathcal{S}'(\mathbb{R}^n)$ for any t, we can write the solution:

$$u \;=\; u_0 *_{\mathbf{x}} U \;+\; f *_{\mathbf{x},t} U \tag{4.82}$$

where $*_{\mathbf{x}}$ denotes again the convolution in $\mathbf{x}$ and $*_{\mathbf{x},t}$ denotes the convolution in $\mathbf{x}$ and t.
We make it explicit:

$$u(\mathbf{x},t) =$$
$$= \frac{1}{(4\,\pi\,t)^{n/2}} \int_{\mathbb{R}^n} u_0(\mathbf{y}) e^{-|\mathbf{x}-\mathbf{y}|^2/(4t)} d\mathbf{y} + \int_0^t \int_{\mathbb{R}^n_{\mathbf{x}}} f(\mathbf{y}, t-\tau) \frac{e^{-|\mathbf{x}-\mathbf{y}|^2/(4\tau)}}{(4\,\pi\,\tau)^{n/2}} d\mathbf{y}\, dt.$$

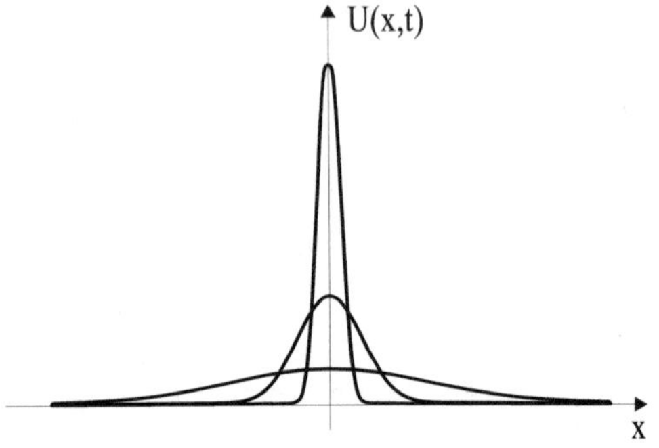

Figure 4.7 - Exercise 4.54: graph of $U(x,t)$ for $n = 1$ and $t = 1,\ 1/10,\ 1/100$.

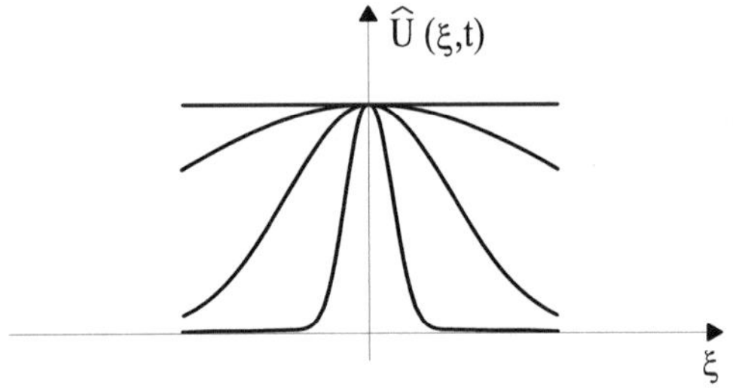

Figure 4.8 - Exercise 4.54: graph of $\widehat{U}(\xi, t)$ for $n = 1$ and $t = 1,\ 1/10,\ 1/100$.

The formula we have found for the solution u entails some remarkable facts.

- The regularizing effect of the heat operator:
  due to the $\mathbf{x}$-convolution of $U(\cdot, t) \in \mathcal{S}(\mathbb{R}^n_{\mathbf{x}})$ $\forall t > 0$ with the data, we obtain
  $$\mathbf{x} \mapsto u(\mathbf{x}, t) \in C^\infty(\mathbb{R}^n) \quad \forall t > 0\,,$$
  even if $\mathbf{x} \mapsto u_0(\mathbf{x}) = u(\mathbf{x}, 0)$ is a function discontinuous!
- The infinite velocity of propagation associated to the heat operator:
  initial data u_0 and u_1 with compact support and source term f vanishing everywhere, lead to a solution u with $\operatorname{spt} u(\cdot, t) = \mathbb{R}^n_{\mathbf{x}}$ $\forall t > 0$.
  This property is a consequence of the fact that $\mathbf{x} \mapsto u(\mathbf{x}, t)$ is an entire analytic function of n real variables that is not identically zero, $\forall t > 0$.
- The solution (4.82) of the Cauchy problem for the heat equation (4.78) is computed under the assumption that the initial temperature u_0 is a tempered distribution (in order to exploit the Fourier transform). Nevertheless (4.82) solves (4.78) in far more general cases: for instance concerning the initial datum u_0, it is enough that the convolution $U *_{\mathbf{x}} u_0$ is well defined. The smooth functions with exponential growth at infinity belong to this class: e.g $n = 1,\ u_0(x) = e^x$. □

Exercise 4.55. Prove that the function defined by (4.81), is the fundamental solution U of the heat operator $\partial_t - \Delta_{\mathbf{x}}$ in n space dimensions, that is U solves:

$$U \in \mathcal{D}'(\mathbb{R}^n), \qquad U_t - \Delta_{\mathbf{x}} U = \delta(\mathbf{x}, t) \quad \text{in } \mathcal{D}'(\mathbb{R}^{n+1}).$$

Moreover prove:

$$\begin{aligned} \operatorname{spt} U &\subset \{(\mathbf{x}, t) \in \mathbb{R}^{n+1} : \ t \geq 0\}, \\ \widehat{U}(\boldsymbol{\xi}, t) &= H(t)\, e^{-|\boldsymbol{\xi}|^2 t}. \end{aligned}$$

Solution - Hint: exploit the solutions of Exercises 4.54 and 6.31. □

Exercise 4.56. Prove that, if $f \equiv 0$ and $u_0 \geq 0$, then the solution u of the Cauchy problem (4.78) for the heat equation in n space dimensions is nonnegative and fulfils:

1) if $u_0 \in L^1(\mathbb{R}^n)$, then $u(\cdot, t) \in L^1(\mathbb{R}^n)$, $\forall t > 0$ and $\displaystyle\int_{\mathbb{R}^n} u(\mathbf{x}, t)\, d\mathbf{x} = \int_{\mathbb{R}^n} u_0(\mathbf{x})\, d\mathbf{x}\ \forall t > 0$.

2) if $u_0 \in L^2(\mathbb{R}^n)$, then $u(\cdot, t) \in L^2(\mathbb{R}^n)$, $\forall t > 0$ and $\|u(\cdot, t)\|_{L^2(\mathbb{R}^n)}$ is monotone decreasing and

$$\lim_{t \to +\infty} \|u(\cdot, t)\|_{L^2(\mathbb{R}^n)} = 0.$$

Solution - Nonnegativity follows from the fact that u is the convolution of two nonnegative functions. By taking into account of $u_t = \Delta_{\mathbf{x}} u$ and the 1st Green's identity (Table 10.4, formula 5.), we prove the *conservation of the total heat quantity* (that is $\|u\|_{L^1(\mathbb{R}^n)} = \int_{\mathbb{R}^n} u\, d\mathbf{x}$ is constant) and the *energy dissipation* ($\frac{1}{2}\|u\|^2_{L^2(\mathbb{R}^n)} = \frac{1}{2}\int_{\mathbb{R}^n} u^2\, d\mathbf{x} \downarrow 0_+$ if $t \to +\infty$).

1) $$\frac{d}{dt}\int_{\mathbb{R}^n} u(\mathbf{x}, t)\, d\mathbf{x} = \int_{\mathbb{R}^n} u_t(\mathbf{x}, t)\, d\mathbf{x} = \int_{\mathbb{R}^n} \Delta_{\mathbf{x}} u(\mathbf{x}, t)\, d\mathbf{x} = 0.$$

2) $$\frac{d}{dt}\|u(\cdot, t)\|^2_{L^2(\mathbb{R}^n)} = \frac{d}{dt}\int_{\mathbb{R}^n} u(\mathbf{x}, t)^2\, d\mathbf{x} = 2\int_{\mathbb{R}^n} u_t(\mathbf{x}, t)\, u(\mathbf{x}, t)\, d\mathbf{x} =$$

$$= 2\int_{\mathbb{R}^n} \Delta_{\mathbf{x}} u(\mathbf{x}, t)\, u(\mathbf{x}, t)\, d\mathbf{x} = -2\int_{\mathbb{R}^n} |\nabla_{\mathbf{x}} u(\mathbf{x}, t)|^2\, d\mathbf{x} \leq 0,$$

moreover this inequality is strict if $u(\mathbf{x}, t)$ does not vanish identically in $\mathbb{R}^n$.
In both cases, due to the regularizing properties of convolution, the differentiation with respect to t in the integral with respect to $\mathbf{x}$ is allowed, moreover the boundary terms appearing in the Green identity are negligible, due to fast decay of u at infinity. □

Exercise 4.57. Solve the Cauchy problem for the **wave equation**:
given $u_0, u_1 \in \mathcal{S}'(\mathbb{R}^n)$, $f \in C^0\big((0, +\infty), \mathcal{S}'(\mathbb{R}^n_{\mathbf{x}})\big)$ and $c > 0$, find the solution u of

$$\begin{cases} u \in C^2\big((0, +\infty), \mathcal{S}'(\mathbb{R}^n)\big) \text{ s.t.} & \\ u_{tt} - c^2 \Delta_{\mathbf{x}} u = f & \mathbf{x} \in \mathbb{R}^n,\ t > 0, \\ u(\mathbf{x}, 0) = u_0(\mathbf{x}) & \mathbf{x} \in \mathbb{R}^n, \\ u_t(\mathbf{x}, 0) = u_1(\mathbf{x}) & \mathbf{x} \in \mathbb{R}^n. \end{cases} \tag{4.83}$$

Solution - We perform the partial Fourier transform (4.79) only with respect to the space variables $\mathbf{x}$ on both sides of the equation, due to (4.28) and (4.80) we obtain a family of Cauchy problems for second order ordinary differential equations (with the parameter $\boldsymbol{\xi}$),

$$\begin{cases} \widehat{u}_{tt}(\boldsymbol{\xi}, t) + c^2 |\boldsymbol{\xi}|^2 \widehat{u}(\boldsymbol{\xi}, t) = \widehat{f}(\boldsymbol{\xi}, t) \\ \widehat{u}(\boldsymbol{\xi}, 0) = \widehat{u_0}(\boldsymbol{\xi}) \\ \partial_t \widehat{u}(\boldsymbol{\xi}, 0) = \widehat{u_1}(\boldsymbol{\xi}), \end{cases}$$

whose solutions are

$$\widehat{u}(\boldsymbol{\xi},t)=\widehat{u_0}(\boldsymbol{\xi})\,\cos(ct|\boldsymbol{\xi}|)+\widehat{u_1}(\boldsymbol{\xi})\,\frac{\sin(ct|\boldsymbol{\xi}|)}{c|\boldsymbol{\xi}|}+\int_0^t\widehat{f}(\boldsymbol{\xi},t-\tau)\,\frac{\sin(c\tau|\boldsymbol{\xi}|)}{c|\boldsymbol{\xi}|}\,d\tau. \tag{4.84}$$

The transform $\widehat{u}$ of the solution is expressed by the same analytic formula (4.84) in any space dimension n. The last summand in (4.84) corresponds to $\left(\frac{\sin(ct|\boldsymbol{\xi}|)}{|\boldsymbol{\xi}|}\,H(t)\right)*_t f(\boldsymbol{\xi},t)$, where $*_t$ denotes the convolution in the variable t only (the trivial extension of f to $\mathbb{R}$ is understood: $f(\boldsymbol{\xi},t)\equiv 0,\ t<0$) and H denotes the Heaviside function.
Moreover, introducing the notation $\Box_c = \partial_{tt} - c^2\Delta$ for the *d'Alembertian operator* or *wave operator* we can compute the transform of fundamental solution U for $\Box_c$ in $\mathbb{R}^{n+1}$. Indeed, after setting

$$\widehat{U}(\boldsymbol{\xi},t)\ \stackrel{def}{=}\ \frac{\sin(ct|\boldsymbol{\xi}|)}{c|\boldsymbol{\xi}|}\,H(t)\,, \tag{4.85}$$

$$U(\mathbf{x},t)\ =\ \mathcal{F}^{-1}\left\{\frac{\sin(ct|\boldsymbol{\xi}|)}{c|\boldsymbol{\xi}|}\,H(t)\right\}, \tag{4.86}$$

exploiting Exercise 6.32, we obtain

$$\begin{cases} U\in\mathcal{S}'(\mathbb{R}^{n+1})\ : \\ \Box_c U\ =\ \delta(\mathbf{x},t)\ \text{ in }\ \mathcal{S}'(\mathbb{R}^{n+1})\,, \\ \operatorname{spt} U\ \subset\ \{\,(\mathbf{x},t)\in\mathbb{R}^{n+1}\ :\ t\geq 0\,\}\,. \end{cases}$$

Since

$$\partial_t\,\widehat{U}(\boldsymbol{\xi},t)\ =\ \cos(ct|\boldsymbol{\xi}|)\,H(t) \tag{4.87}$$

we get

$$\widehat{u}(\boldsymbol{\xi},t)\ =\ \widehat{u_0}(\boldsymbol{\xi})\ \partial_t\,\widehat{U}(\boldsymbol{\xi},t)\ +\ \widehat{u_1}(\boldsymbol{\xi})\ \widehat{U}(\boldsymbol{\xi},t)\ +\ \widehat{f}(\boldsymbol{\xi},t)\ *_t\ \widehat{U}(\boldsymbol{\xi},t)\,. \tag{4.88}$$

Anti-transforming $\widehat{U}$ we obtain the fundamental solution U of the wave operator, that makes possible writing the solution u of (4.83) in term of a convolution.

$$u\ =\ u_0 *_{\mathbf{x}} \partial_t U\ +\ u_1 *_{\mathbf{x}} U\ +\ f *_{\mathbf{x},t} U \tag{4.89}$$

where $U=\mathcal{F}^{-1}\widehat{U}$ is given by (4.86). Nevertheless the explicit computation of $\mathcal{F}_{\mathbf{x}}^{-1}\left\{\widehat{U}\right\}$ requires different techniques according to the space dimension n.

- $n=1$, **wave equation in** $\mathbb{R}$.
  Cosine and sine are respectively even and odd: (4.84) simplifies (ξ is scalar):

$$\widehat{u}(\xi,t)\ =\ \widehat{u_0}(\xi)\,\cos(ct\,\xi)\ +\ \widehat{u_1}(\xi)\,\frac{\sin(ct\,\xi)}{c\,\xi}\ +\ \int_0^t\widehat{f}(\xi,t-\tau)\,\frac{\sin(c\tau\,\xi)}{c\,\xi}\,d\tau\,;$$

  by $\cos(ct\,\xi)\ =\ (e^{ict\,\xi}+e^{-ict\,\xi})/2$, shifting formula (4.5) and scaling formula (4.6), convolution transform (4.50) and $\sin(\xi)/\xi\ =\ \frac{1}{2}\widehat{p}(\xi)$, anti-transformig we obtain the fundamental solution $U(x,t)\ =\ H(ct-|x|)/(2c)$ and

  the solution of (4.83) for $n=1$ (**D'Alembert's formula**):

$$u(x,t)=\frac{1}{2}(u_0(x-ct)+u_0(x+ct))+\frac{1}{2c}\int_{x-ct}^{x+ct}u_1(y)\,dy+\frac{1}{2c}\int_0^t\int_{x-c(t-\tau)}^{x+c(t-\tau)}f(y,\tau)\,dy\,d\tau.$$

We notice that, if $n = 1$, the initial datum u_0 alone produces two half-split copies of u_0 propagating in opposite directions with the same scalar velocity, whose value is exactly c. Whereas u_1 produces effects propagating with velocity less than to c and equal to c. Finally the third summand in d'Alembert's formula is half the integral of the source term f in the dependence domain for the signal at x and time t, that is triangular set $\{(y,\tau) \colon\ 0 \le \tau \le t - |y - x|/c\}$, bounded by characteristic lines passing by (x,t).

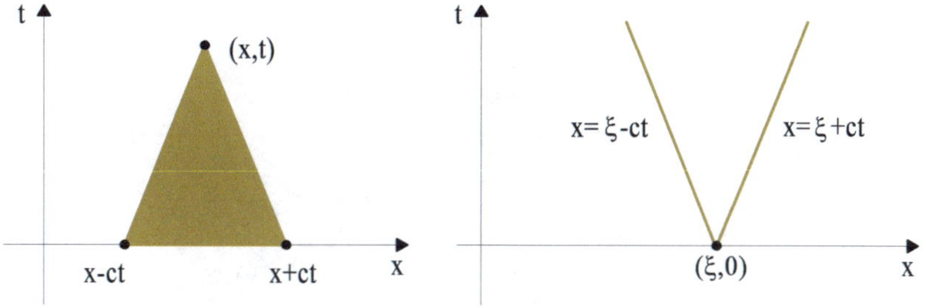

Figure 4.9 - Waves, $n = 1$. Dependence domain on f and initial data at (x,t). characteristic lines passing by $(\xi,0)$: half-lines.

- **general n, wave equation in $\mathbb{R}^n$.**
  In any dimension $n \ge 1$ the transform (4.85) of the fundamental solution $U(\cdot,t)$ has an entire analytic extension from $\mathbb{R}^n_{\mathbf{x}}$ to $\mathbb{C}^n$ (for any time $t > 0$), since $\widehat{U}(\cdot,t)$ is the sum of a power series with convergence radius equal to $+\infty$:

$$\widehat{U}(\boldsymbol{\xi},t) = \frac{\sin(ct\,|\boldsymbol{\xi}|)}{|\boldsymbol{\xi}|}\,H(t) = \sum_{k=0}^{\infty} (-1)^k \frac{(ct)^{2k+1}}{(2k+1)!}\,|\boldsymbol{\xi}|^{2k}\,H(t) \qquad \boldsymbol{\xi} \in \mathbb{R}^n,\ t \in \mathbb{R}.$$

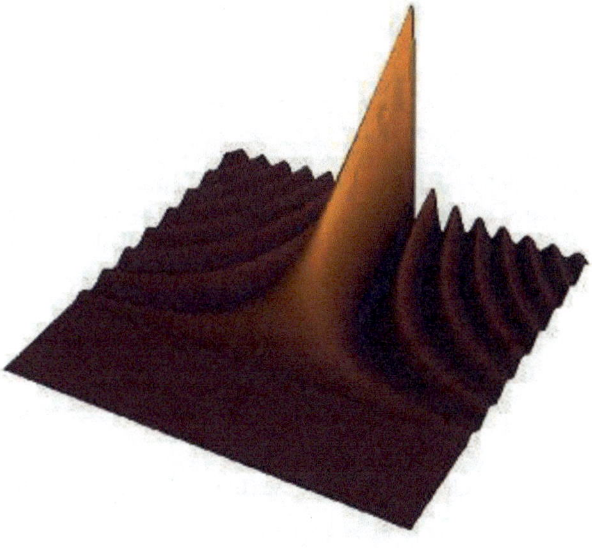

Figure 4.10 - Wave equation, $n = 1$: graph of $\widehat{U}(\xi,t) = H(t)\,\sin(ct\xi)/(c\xi)$, Fourier transform of fundamental solution of the wave equation.

Moreover we can prove that $|\widehat{U}(\boldsymbol{\xi},t)| \leq C_n \exp(ct\,|\operatorname{Im}\boldsymbol{\xi}|)$, where $\operatorname{Im}\boldsymbol{\xi} = (\operatorname{Im}\xi_1,\ldots,\operatorname{Im}\xi_n)$. Hence, due to the Paley-Wiener Theorem (Theorem 4.13),

$$\operatorname{spt} U(\mathbf{x},t) \subset \overline{B_{ct}(\mathbf{0})} = \{\mathbf{x}\in\mathbb{R}^n : |\mathbf{x}|\leq ct\} \qquad \forall t>0\,, \tag{4.90}$$

that is (in any space dimension n) the wave equation rules phenomena with *finite velocity of propagation* (less than or equal to c).

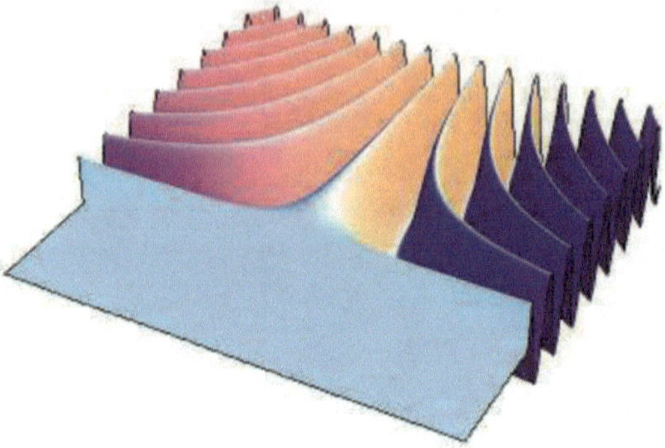

Figure 4.11 - Wave equation, $n=1$: graph of $\partial_t\widehat{U}(\xi,t) = \cos(ct\xi)\,H(t)$.

The velocity is exactly c when the space dimension n is odd and fulfils $n>1$. We specify precisely these properties in the physically relevant case of dimension 3.

- $n=3$, **wave equation in $\mathbb{R}^3$**.

Huygens' Principle - If $n=3$ and U is the fundamental solution (4.86) of $\square_c$, then

$$\operatorname{spt} U(\cdot,t) = \left\{\mathbf{x}\in\mathbb{R}^3 : |\mathbf{x}| = ct\right\} \qquad \forall t>0\,. \tag{4.91}$$

Formula (4.91) entails that the propagation velocity is exactly c, since a signal concentrated at the origin of $\mathbb{R}^3$ at time $t=0$ produces a solution $U(\mathbf{x},t)$ that for any time $t>0$ is supported on the boundary of the 3-dimensional sphere with radius ct.
We prove Huygens' principle in dimension $n=3$: the property $\widehat{U}$ radially symmetric in $\boldsymbol{\xi}$ implies U radially symmetric in $\mathbf{x}$; using the formula of the Laplacian operator in polar coordinates and recalling that $\widehat{U}$ is radially symmetric, we obtain an identity (in $|\boldsymbol{\xi}|\neq 0$), that can be analytically extended to the whole $\mathbb{R}^3_{\boldsymbol{\xi}}$ since 0 is a removable singularity:

$$\left(c^2t^2 I + \Delta_{\boldsymbol{\xi}}\right)\widehat{U}(\boldsymbol{\xi},t) = \left(c^2t^2 I + \partial_{\varrho\varrho} + \frac{n-1}{\varrho}\partial_\varrho\right)\widehat{U}(\boldsymbol{\xi},t) =$$

$$= (n-3)\,\frac{ct\,|\boldsymbol{\xi}|\,\cos(ct\,|\boldsymbol{\xi}|) - \sin(ct\,\boldsymbol{\xi})}{|\boldsymbol{\xi}|^3}\,H(t) \overset{\boxed{n=3}}{=} 0\,.$$

Anti-transforming this identity (by $\mathcal{F}_{\mathbf{x}}^{-1}$) we find a division problem in the unknown U:

$$\left(c^2t^2 - |\mathbf{x}|^2\right)U(\mathbf{x},t) = 0 \qquad \text{in } \mathbb{R}^4 = \mathbb{R}^3_{\mathbf{x}}\times\mathbb{R}_t\,,$$

whose solutions are radially symmetric, non trivial and fulfil (4.91).

Starting from (4.91) we evaluate U: we know that it is radially symmetric and its support is $\partial B_{ct}(\mathbf{0})$ for any $t > 0$. Then are left only to identify exactly the constant (with respect to $\mathbf{x}$, possibly depending on t) that multiplies a Dirac delta concentrated on $\partial B_{ct}(\mathbf{0})$.

Such constant is made explicit in the box below; subsequently we will verify that this choice is the right one.

Fundamental solution of the wave equation ($n = 3$)

$$U(\mathbf{x}, t) \;=\; \frac{H(t)}{4\,\pi\, c^2 t}\,\delta_{\partial B_{ct}(\mathbf{0})}(\mathbf{x})\,. \tag{4.92}$$

Explicitly the fundamental solution U of $\Box_c$ in 3 space dimensions acts on test functions as follows, we explicit three formulations of duality $\langle U, \varphi\rangle$, $\forall \varphi \in \mathcal{D}(\mathbb{R}^3_{\mathbf{x}})$, at time $t > 0$:

$$\begin{aligned}\iiint_{\mathbb{R}^3} U(\mathbf{x}, t)\,\varphi(\mathbf{x})\,d\mathbf{x} &= \frac{1}{4\,\pi\, c^2 t}\iiint_{\mathbb{R}^3} \delta_{\partial B_{ct}(\mathbf{0})}(\mathbf{x})\,\varphi(\mathbf{x})\,d\sigma(\mathbf{x}) \;=\\ &= \frac{1}{4\,\pi\, c^2 t}\iint_{\partial B_{ct}(\mathbf{0})} \varphi(\mathbf{x})\,d\sigma(\mathbf{x}) \;=\\ &= t \;⨏\!\!\!\!\iint_{\partial B_{ct}(\mathbf{0})} \varphi(\mathbf{x})\,d\sigma(\mathbf{x})\,.\end{aligned}$$

We prove (4.92) by showing that the $\mathcal{F}$ transform of U is $H(t)\sin(c|\boldsymbol{\xi}|t)/(c|\boldsymbol{\xi}|)$.
$U = (1/4\pi c^2 t)\delta_{\partial B_{ct}(\mathbf{0})}(\mathbf{x})$ is radially symmetric and has compact support, therefore its transform is entire analytic and can be evaluated with the duality $\langle U, e^{-i\boldsymbol{\xi}\cdot\mathbf{x}}\rangle$:

$$\begin{aligned}\widehat{U}(\boldsymbol{\xi}, t) &= \widehat{U}(0, 0, |\boldsymbol{\xi}|, t) =\\ &= \iiint_{\mathbb{R}^3} U(\mathbf{x}, t)\, e^{-i\,(0,0,|\boldsymbol{\xi}|)\cdot\mathbf{x}}\,d\mathbf{x}\\ &= \frac{1}{4\,\pi\, c^2 t}\iint_{\partial B_{ct}(\mathbf{0})} e^{-i\,|\boldsymbol{\xi}|\,x_3}\,d\sigma(\mathbf{x}) \;=\\ &= \frac{1}{4\,\pi\, c^2 t}\int_0^{2\pi} d\vartheta \int_0^{\pi} e^{-i|\boldsymbol{\xi}|ct\cos\varphi}\,c^2 t^2\,\sin\varphi\,d\varphi \;=\\ &= \frac{t}{2}\,\frac{\exp(-\,i\,|\boldsymbol{\xi}|\,ct\,\cos\varphi)}{i\,|\boldsymbol{\xi}|\,ct}\Bigg|_{\varphi=0}^{\varphi=\pi} \;=\\ &= \frac{1}{2ic|\boldsymbol{\xi}|}\left(e^{i\,|\boldsymbol{\xi}|\,ct} - e^{-i\,|\boldsymbol{\xi}|\,ct}\right) \;=\\ &= \frac{\sin(|\boldsymbol{\xi}|ct)}{c\,|\boldsymbol{\xi}|} \qquad \forall t > 0\,.\end{aligned}$$

We come back to problem (4.83) with $n = 3$: summarizing, we can rewrite (4.83) as

$$\Box_c u(\mathbf{x}, t) \;=\; u_0(\mathbf{x}) \otimes \delta'(t) + u_1(\mathbf{x}) \otimes \delta(t) + f(\mathbf{x}, t) \qquad \text{in } \mathcal{D}'(\mathbb{R}^4)$$

where the trivial extensions of u and f are denoted by the same symbols (u and f): $u(\mathbf{x}, t) \equiv 0,\; t < 0,\; f(\mathbf{x}, t) \equiv 0,\; t < 0$; moreover, referring to the fundamental solution (4.86), via $\mathcal{F}_{\mathbf{x}}$ transform in the space variables only and recalling $\widehat{\delta} = 1$:

$$\widehat{u}_{tt} + c^2|\boldsymbol{\xi}|^2\widehat{u} = \widehat{f}(\boldsymbol{\xi}, t) \qquad \text{in } \mathcal{D}'(\mathbb{R}^4)$$

$$\Box_c U = \delta(\mathbf{x}) \otimes \delta(t) \qquad \text{in } \mathcal{D}'(\mathbb{R}^4)$$

$$\widehat{U}_{tt} + c^2|\boldsymbol{\xi}|^2\widehat{U} = \widehat{\delta}(\boldsymbol{\xi}) \otimes \delta(t) = \delta(t) \qquad \text{in } \mathcal{D}'(\mathbb{R}^4)$$

Therefore, in 3 space dimensions, the solution (4.89) of the problem (4.57) can be made explicit as follows.

Kirchhoff's formula

$$\begin{aligned} u(\mathbf{x},t) \;=\; & \int_0^t \frac{1}{4\pi\, c^2\tau} \iint_{\partial B_{c\tau}(\mathbf{x})} f(\mathbf{y},t-\tau)\, d\sigma(\mathbf{y})\, d\tau \;+ \\ & + \iint\!\!\!\!\!\!-_{\partial B_{ct}(\mathbf{x})} \Big(u_0(\mathbf{y}) + \nabla u_0(\mathbf{y})\cdot(\mathbf{y}-\mathbf{x}) \Big)\, d\sigma(\mathbf{y}) \;+ \\ & + \frac{1}{4\pi\, c^2 t} \iint_{\partial B_{ct}(\mathbf{x})} u_1(\mathbf{y})\, d\sigma(\mathbf{y}) \,, \qquad \forall \mathbf{x}\in\mathbb{R}^3,\ t>0\,. \end{aligned}$$

The second summand in the Kirchhoff formula is deduced by following computation:

$$\begin{aligned} \partial_t \left(t \iint\!\!\!\!\!\!-_{\partial B_{ct}(\mathbf{x})} u_0(\mathbf{y})\, d\sigma(\mathbf{y}) \right) \;=\; & \partial_t \left(t \iint\!\!\!\!\!\!-_{\partial B_{ct}(\mathbf{0})} u_0(\mathbf{x}+t\mathbf{z})\, d\sigma(\mathbf{z}) \right) \;= \\ = \; & \iint\!\!\!\!\!\!-_{\partial B_{ct}(\mathbf{x})} u_0(\mathbf{y})\, d\sigma(\mathbf{y}) \;+\; \iint\!\!\!\!\!\!-_{\partial B_{ct}(\mathbf{x})} \nabla u_0(\mathbf{y})\cdot(\mathbf{y}-\mathbf{x})\, d\sigma(\mathbf{y})\,. \end{aligned}$$

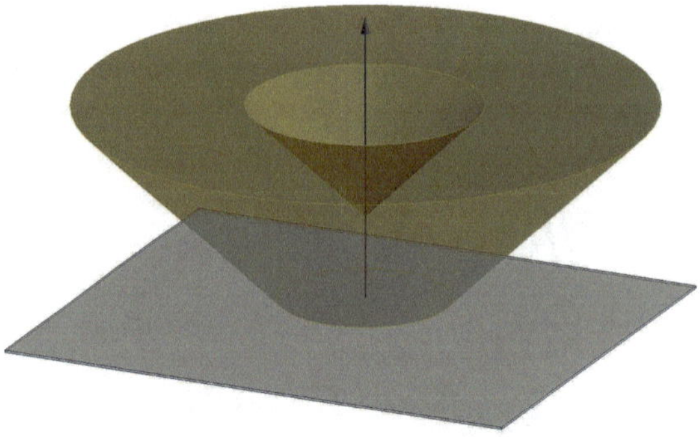

Figure 4.12 - Waves, $n = 3$. Influence domain of initial data supported in a disk.

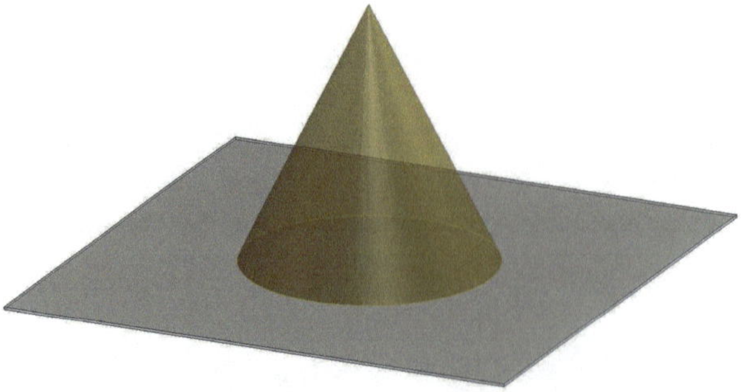

Figure 4.13 - waves, $n = 3$. Dependence domain of the solution at one point on initial data and sources: conical surface in $\mathbb{R}^4$.

- $n = 2$, **wave equation in $\mathbb{R}^2$.**
  The 2-dimensional case, can be lead back to the 3-dimensional one, supposing that data and solutions (depending on x_1, x_2) are extended to $\mathbb{R}^3$ in such a way that they are (constant) independent of x_3. The fundamental solution U_2 solves then the problem:

$$\Box_c U_2 \;=\; \delta(x_1, x_2) \otimes 1(x_3) \otimes \delta(t) \qquad \mathcal{S}'(\mathbb{R}_{x_1,x_2} \times \mathbb{R}_t)\,. \tag{4.93}$$

Then by denoting U_3 the fundamental solution in 3 space dimensions, $\mathbf{x} = (x_1, x_2, x_3)$, $\mathbf{x}' = (x_1, x_2)$ and α the angle between $\mathbf{x}$ and the $\mathbf{x}'$ plane, we find:

$$U_2(\mathbf{x}', t) \;=\; U_3 *_{\mathbf{x},t} \Big(\delta(x_1, x_2) \otimes 1(x_3) \otimes \delta(t)\Big) = U_3 *_{x_3} 1(x_3) = \int_{\mathbb{R}} U_3(\mathbf{x}', x_3, t)\, dx_3 =$$
$$= 2\,\frac{1}{4\pi c^2 t}\,\frac{1}{|\sin\alpha|} = \frac{1}{2\pi c^2 t}\,\frac{ct}{\sqrt{c^2t^2 - |\mathbf{x}'|^2}} = \frac{1}{2\,\pi\,c}\,\frac{\chi_{B_{ct}(\mathbf{0})}}{\sqrt{c^2t^2 - |\mathbf{x}'|^2}}\,H(t)\,,$$

so U_2 is a function with compact support in $\mathbb{R}^2_{\mathbf{x}}$ and belongs to $L^1(\mathbb{R}^2_{\mathbf{x}})$ for every t.

□

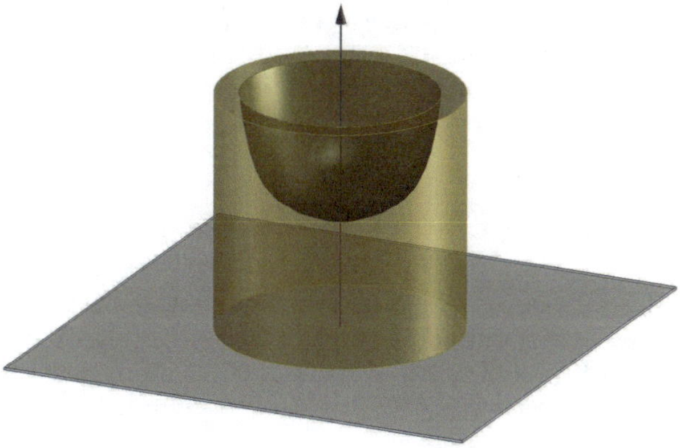

Figure 4.14 - $n = 2$. Graph of the fundamental solution of the wave equation.

Exercise 4.58. Prove that, if $u_0 \in H^1(\mathbb{R}^n)$, $u_1 \in L^2(\mathbb{R}^n)$, $f \equiv 0$, then the solution u of the Cauchy problem (4.83) for the wave equation in any space dimension n fulfils:
1) *conservation of total mechanical energy*: $E(t) = E(0)\ \forall t > 0$ where:

$$E(t) \;=\; \frac{1}{2}\int_{\mathbb{R}^n} \left(|u_t(\mathbf{x}, t)|^2 + c^2\, |\nabla_{\mathbf{x}} u(\mathbf{x}, t)|^2 \right) d\mathbf{x}\,.$$

2) *conservation of total energy separately for any value of the frequency*:

$$\widehat{e}(\boldsymbol{\xi}, t) \;=\; \widehat{e}(\boldsymbol{\xi}, 0) \qquad \forall t > 0,\ \forall \boldsymbol{\xi} \in \mathbb{R}^n\,,$$

where:

$$\widehat{e}(\boldsymbol{\xi}, t) \;=\; \frac{1}{2}\left(|\widehat{u}_t(\boldsymbol{\xi}, t)|^2 + c^2\, |\boldsymbol{\xi}|^2\, |\widehat{u}(\boldsymbol{\xi}, t)|^2 \right).$$

Solution - 1) The conditions $u_0 \in L^2$, $\nabla u_0 \in L^2$, $u_1 \in L^2$, $f \equiv 0$ and the formula (4.84) imply $\nabla u \in L^2$, $u_t \in L^2$ for any $t > 0$. Differentiating with respect to t, integrating by parts in $\mathbb{R}^n_{\mathbf{x}}$ and taking into account of $u_{tt} = \Delta_{\mathbf{x}} u$, we obtain

$$\frac{d}{dt}\,E(t) = \int_{\mathbb{R}^n} \left(u_t\, u_{tt} + c^2\, \nabla_{\mathbf{x}} u \cdot \nabla_{\mathbf{x}} u_t \right) d\mathbf{x} \;=\; \int_{\mathbb{R}^n} u_t \left(u_{tt} - c^2\, \Delta_{\mathbf{x}} u \right) d\mathbf{x} \;=\; 0\,.$$

2) Taking into account of $\widehat{u}_{tt}(\boldsymbol{\xi},t)+|\boldsymbol{\xi}|^2\widehat{u}=0$, we differentiate $\widehat{e}$ with respect to t:

$$\frac{d}{dt}\,\widehat{e} = \left(\widehat{u}_{tt} + |\boldsymbol{\xi}|^2\widehat{u}\right)\widehat{u}_t = 0\,.$$

□

Exercise 4.59. Prove these formulas for the Fourier transform in $\mathcal{S}'(\mathbb{R}^n)$:

$$\mathcal{F}\left\{\frac{1}{|\mathbf{x}|^a}\right\} = \frac{C_{n,a}}{|\boldsymbol{\xi}|^{n-a}} \qquad \text{where } C_{n,a} = 2^{n-a}\,\pi^{n/2}\,\frac{\Gamma(\frac{n-a}{2})}{\Gamma(\frac{a}{2})}\,,\quad n\geq 1,\ \ 0<a<n\,. \tag{4.94}$$

Solution - I) First we prove that, under the assumptions made on a and n, there exists a constant $C=C(n,a)$ s.t.

$$(*)\qquad \mathcal{F}\left\{\frac{1}{|\mathbf{x}|^a}\right\} = \frac{C(n,a)}{|\boldsymbol{\xi}|^{n-a}}\,,$$

subsequently we compute the constant C.
We set $u(\mathbf{x})=1/|\mathbf{x}|^a$. Then u belongs to $L^1_{loc}(\mathbb{R}^n)$, as it can be easily verified by using polar coordinates since $0<a<n$ entails $\varrho^{n-a-1}\in L^1(0,1)$.
u is the sum of an L^1 function and an L^∞ function: $u(\mathbf{x})=\chi_{B_1(\mathbf{0})}/|\mathbf{x}|^a+\chi_{\mathbb{R}^n\setminus B_1(\mathbf{0})}/|\mathbf{x}|^a$, so $u\in\mathcal{S}'(\mathbb{R}^n)$ and u is the limit in $\mathcal{S}'(\mathbb{R}^n)$ of $u_R=\chi_{B_R(\mathbf{0})}/|\mathbf{x}|^a$ as $R\to+\infty$.
due to the continuity of $\mathcal{F}$ in $\mathcal{S}'(\mathbb{R}^n)$ we obtain

$$\mathcal{F}\left\{\frac{1}{|\mathbf{x}|^a}\right\}(\boldsymbol{\xi}) = \lim_R \int_{B_R(\mathbf{0})}\frac{e^{-i\,\boldsymbol{\xi}\cdot\mathbf{x}}}{|\mathbf{x}|^a}\,d\mathbf{x} = \text{v.p.}\int_{\mathbb{R}^n}\frac{e^{-i\,\boldsymbol{\xi}\cdot\mathbf{x}}}{|\mathbf{x}|^a}\,d\mathbf{x}\,.$$

In the sequel the principal value is always implied: every integral in $\mathbb{R}^n$ related to Fourier transform must be read as its principal value.
Taking into account of (4.18) and of the fact that u is radially symmetric, for any $n\times n$ real matrix $\mathbb{M}$ that is orthogonal (i.e. $\mathbb{M}^{-1}=\mathbb{M}^T$, $|\det\mathbb{M}|=1$) and for any $\psi\in\mathcal{S}(\mathbb{R}^n)$, we obtain

$$\int\widehat{u}(\mathbb{M}\boldsymbol{\xi})\,\psi(\boldsymbol{\xi})\,d\boldsymbol{\xi} = |\text{det}\mathbb{M}|^{-1}\int u(\mathbb{M}^{-T}\mathbf{x})\,\widehat{\psi}(\mathbf{x})\,d\mathbf{x} = \int u(\mathbf{x})\,\widehat{\psi}(\mathbf{x})\,d\mathbf{x} = \int\widehat{u}(\boldsymbol{\xi})\,\psi(\boldsymbol{\xi})\,d\boldsymbol{\xi}$$

that is u is radially symmetric too.
For any $t>0$, $\mathbf{x},\mathbf{y}\in\mathbb{R}^n$, we set $\mathbf{x}=t\mathbf{y}$. Then $d\mathbf{x}=t^n d\mathbf{y}$ and

$$\widehat{u}(\boldsymbol{\xi}) = \int|\mathbf{x}|^{-a}e^{-i\,\boldsymbol{\xi}\cdot\mathbf{x}}d\mathbf{x} = \int|t\mathbf{y}|^{-a}e^{-i\boldsymbol{\xi}\cdot(t\mathbf{y})}t^n d\mathbf{x} = t^{n-a}\int|\mathbf{y}|^{-a}e^{-i(t\boldsymbol{\xi})\cdot\mathbf{y}}d\mathbf{x} = t^{n-a}\,\widehat{u}(t\boldsymbol{\xi}).$$

Therefore $\widehat{u}(t\boldsymbol{\xi})=t^{a-n}\widehat{u}(\boldsymbol{\xi})$, that is on any open half-line from the origin of $\mathbb{R}^n$ we have $\widehat{u}(t\boldsymbol{\xi}) = Ct^{a-n}$ for a suitable constant associated to the half-line. Since $\widehat{u}$ is radially symmetric, the constant must be independent of the half-line:

$$(**)\qquad \widehat{u}(\boldsymbol{\xi}) = C|\boldsymbol{\xi}|^{a-n}\,.$$

To be thorough we remark that $\widehat{u}$ cannot have singular part concentrated at 0 : indeed it is an L^1_{loc} function outside the origin, moreover $\widehat{u}$ is an $(a-n)$ homogeneous distribution; hence spt $\left(\widehat{u}(\boldsymbol{\xi})-C|\boldsymbol{\xi}|^{a-n}\right)\subset\{\mathbf{0}\}$; this restriction on the support entails (formula (3.17) for the division problem in $\mathbb{R}^n$) that $\widehat{u}(\boldsymbol{\xi})-C|\boldsymbol{\xi}|^{a-n}=\sum_{\boldsymbol{\alpha}}c_{\boldsymbol{\alpha}}D^{\boldsymbol{\alpha}}\delta$ where finitely many summands are present; then every coefficient $c_{\boldsymbol{\alpha}}$ must vanish since δ is $(-n)$ homogeneous and $D^{\boldsymbol{\alpha}}\delta$ is $(-n-|\boldsymbol{\alpha}|)$omogenea. Then there are no singular terms at the origin.

II) We evaluate the constant $C(n,a)$ in $(*)$ by using known identities for Gaussian function. we set: $u(\mathbf{x}) = 1/|\mathbf{x}|^a \in \mathcal{S}'(\mathbb{R}^n)$, $G(\mathbf{x}) = e^{-|\mathbf{x}|^2}$. By formula (4.41) we obtain:

$$\int_{\mathbb{R}^n} u\,\widehat{G} = \int_{\mathbb{R}^n} \widehat{u}\,G\,;$$

taking into account of $(**)$ and of $\widehat{G}(\boldsymbol{\xi}) = \pi^{n/2}\,e^{-|\boldsymbol{\xi}|^2/4}$ (Exercise 4.44), the last identity can be rewritten as

$$\pi^{n/2}\int_{\mathbb{R}^n} e^{-|\mathbf{x}|^2/4}\,\frac{1}{|\mathbf{x}|^a}\,d\mathbf{x} = \int_{\mathbb{R}^n} e^{-|\boldsymbol{\xi}|^2}\,\frac{C_{n,a}}{|\boldsymbol{\xi}|^{n-a}}\,d\boldsymbol{\xi}\,;$$

introducing the polar coordinates and using (2.13) we have:

$$(\pi)^{n/2}\,\sigma_{n-1}\int_0^\infty e^{-\varrho^2/4}\,\varrho^{n-1-a}\,d\varrho = \sigma_{n-1}\,C_{n,a}\int_0^\infty e^{-r^2}\,r^{a-n+n-1}\,dr\,;$$

performing the change of variables $\left\{\varrho^2 = 4t,\ d\varrho = dt/\sqrt{t}\right\}$ and $\left\{r^2 = \tau,\ 2r\,dr = d\tau\right\}$, we get:

$$2^{n-a-1}\,(\pi)^{n/2}\int_0^\infty e^{-t}\,t^{(n-a)/2\,-1}\,dt = \frac{1}{2}\,C_{n,a}\int_0^\infty e^{-\tau}\,\tau^{(a-1)/2}\,\tau^{-1/2}\,d\tau\,;$$

recalling the definition of the Euler Gamma function ($\Gamma(x) = \int_0^\infty e^{-t}t^{x-1},\ x>0$), we conclude:

$$C_{n,a} = 2^{n-a}\,\pi^{n/2}\,\frac{\Gamma(\frac{n-a}{2})}{\Gamma(\frac{a}{2})}\,.$$

□

Exercise 4.60. Evaluate

$$-\Delta\,\frac{1}{|\mathbf{x}|} \qquad \text{in } \mathcal{S}'(\mathbb{R}^n)\,,\quad n\geq 2$$

if possible or at least express it in terms of its Fourier transform.

Solution - First we remark that $1/|\mathbf{x}\,| \in L^1_{loc}(\mathbb{R}^n) \subset \mathcal{S}'(\mathbb{R}^n)$ if and only if $n \geq 2$. Then the distributional Laplacian we have to evaluate is well defined.
$1/|\mathbf{x}|$ is a function C^∞ outside the origin of $\mathbb{R}^n$, then we can evaluate the classical Laplacian by the formula for smooth radially symmetric functions (Table 10.6.4):

$$(-\Delta)\,\frac{1}{|\mathbf{x}|} = -\left(\partial_{\varrho\varrho} + \frac{n-1}{\varrho}\partial_\varrho\right)\frac{1}{\varrho} = \frac{n-3}{\varrho^3} = \frac{n-3}{|\mathbf{x}|^3} \qquad \mathbf{x}\in\mathbb{R}^n\setminus\{\mathbf{0}\},\ \varrho = |\mathbf{x}|\,.$$

To shed light on the presence or absence of singular terms at the origin it is necessary to inspect this issue separately in various dimensions $n \geq 2$.
By formula (4.94), we know that $\mathcal{F}\{1/|\mathbf{x}|\} = C_{n,1}/|\boldsymbol{\xi}|^{n-1}$ if $n \geq 2$.

- $n=3$. Then $-\Delta(1/|\mathbf{x}|) = 0$ in $\mathbb{R}^3\setminus\{\mathbf{0}\}$. Nevertheless $-\Delta(1/|\mathbf{x}|) \not\equiv 0$ due to a singular term at the origin. In fact, through $\mathcal{F}$ transform and subsequent anti-transform:

$$\mathcal{F}\left\{-\Delta\frac{1}{|\mathbf{x}|}\right\} = |\boldsymbol{\xi}|^2\,\frac{C_{3,1}}{|\boldsymbol{\xi}|^2} = 4\pi = \sigma_2\,,$$

$$-\Delta\,\frac{1}{|\mathbf{x}|} = 4\,\pi\,\delta \qquad \text{in } \mathbb{R}^3.$$

$\dfrac{-1}{4\pi|\mathbf{x}|}$ is the **fundamental solution of the Laplace equation** in $\mathbb{R}^3$:

$$(-\Delta)\,\frac{1}{4\pi|\mathbf{x}|} = \delta \qquad \text{in } \mathbb{R}^3.$$

- $n \geq 4$. Then $-\Delta(1/|\mathbf{x}|) = (n-3)/|\mathbf{x}|^3$ in $\mathbb{R}^n \setminus \{\mathbf{0}\}$, moreover $1/|\mathbf{x}|$ is a function locally integrable $\forall n \geq 4$. We prove that $-\Delta(1/|\mathbf{x}|)$ has no singular terms at the origin.

$$\mathcal{F}\left\{-\Delta\frac{1}{|\mathbf{x}|}\right\} = |\boldsymbol{\xi}|^2\,\frac{C_{n,1}}{|\boldsymbol{\xi}|^{n-1}} = \frac{C_{n,1}}{|\boldsymbol{\xi}|^{n-3}} \in L^1_{loc}(\mathbb{R}^n),\ n \geq 4\,.$$

We anti-transform this radially symmetric function with (4.94):

$$-\Delta\,\frac{1}{|\mathbf{x}|} = \mathcal{F}^{-1}\left\{\frac{C_{n,1}}{|\boldsymbol{\xi}|^{n-3}}\right\} = \frac{C_{n,1}}{(2\pi)^n}\,\mathcal{F}\left\{\frac{1}{|\boldsymbol{\xi}|^{n-3}}\right\} \in L^1_{loc}(\mathbb{R}^n),\ n \geq 4\,.$$

$$-\Delta\,\frac{1}{|\mathbf{x}|} = \frac{C_{n,1}\,C_{n,n-3}}{(2\pi)^n}\,\frac{1}{|\mathbf{x}|^3} \in L^1_{loc}(\mathbb{R}^n),\ n \geq 4\,.$$

- $n = 2$. It is the most tricky case: indeed $-\Delta(1/|\mathbf{x}|) = -1/|\mathbf{x}|^3$ in $\mathbb{R}^2 \setminus \{\mathbf{0}\}$, but the function $-1/|\mathbf{x}|^3$ is not even locally integrable in $\mathbb{R}^2$. Transforming we find:

$$\mathcal{F}\left\{-\Delta\frac{1}{|\mathbf{x}|}\right\} = |\boldsymbol{\xi}|^2\,\frac{C_{2,1}}{|\boldsymbol{\xi}|} = 2\pi\,|\boldsymbol{\xi}|.$$

We cannot make explicit the anti-transform of this term.
Nevertheless, recalling the identity $\mathcal{F}\{-\Delta\,\delta\} = |\boldsymbol{\xi}|^2$, it is natural to set conventionally

$$-\Delta\frac{1}{|\mathbf{x}|} = 2\pi\,(-\Delta)^{1/2}\,\delta(x) \in \mathcal{S}'(\mathbb{R}^2)\,.$$

□

Exercise 4.61. Evaluate

$$-\Delta\,\frac{1}{|\mathbf{x}|^{n-2}} \qquad \text{in } \mathcal{S}'(\mathbb{R}^n)\,,\quad n \geq 3\,.$$

Gather that the fundamental solution W of the *Laplace equation* in $\mathbb{R}^n$, $n \geq 3$:

$$-\Delta\,W = \delta(\mathbf{x}) \qquad \mathcal{S}'(\mathbb{R}^n)\,,\ n \geq 3\,. \tag{4.95}$$

Infer the fundamental solution U of the Laplacian Δ in $\mathbb{R}^n$, $n \geq 3$.

Solution - First we remark that $1/|\mathbf{x}|^{n-2} \in L^1_{loc}(\mathbb{R}^n) \subset \mathcal{S}'(\mathbb{R}^n)$ $\forall n \geq 3$. Then the distributional Laplacian we have to evaluate is well defined.
$1/|\mathbf{x}|$ is a function C^∞ except at the origin of $\mathbb{R}^n$, so we can compute the Laplacian (referring to classical derivatives) with the formula for radially symmetric functions (Table 10.6.4):

$$(-\Delta)\,\frac{1}{|\mathbf{x}|^{n-2}} = -\left(\partial_{\varrho\varrho} + \frac{n-1}{\varrho}\partial_\varrho\right)\frac{1}{\varrho^{n-2}} = \frac{-(n-2)(n-1)+(n-2)(n-1)}{\varrho^n} = 0$$

$$\forall\,\mathbf{x} \in \mathbb{R}^n \setminus \{\mathbf{0}\},\ \varrho = |\mathbf{x}|\,.$$

Then $-\Delta \frac{1}{|\mathbf{x}|^{n-2}}$ reduces to a singular term supported by the origin; to evaluate this term, we use the Fourier transform and the identities (2.13),(4.94) and $\Gamma(z+1) = z\Gamma(z)$:

$$\mathcal{F}\left\{-\Delta \frac{1}{|\mathbf{x}|^{n-2}}\right\} = |\boldsymbol{\xi}|^2 \frac{C_{n,n-2}}{|\boldsymbol{\xi}|^2} = (n-2)\,\sigma_{n-1} \qquad n \geq 3,$$

$$-\Delta \frac{1}{|\mathbf{x}|^{n-2}} = (n-2)\,\sigma_{n-1}\,\delta(\mathbf{x}) \qquad n \geq 3.$$

So we have proved the following relations.

$$W(\mathbf{x}) = \frac{1}{(n-2)\,\sigma_{n-1}\,|\mathbf{x}|^{n-2}}$$

is the **fundamental solution of the Laplace equation** in $\mathbb{R}^n$, $n \geq 3$:

$$(-\Delta)\,\frac{1}{(n-2)\,\sigma_{n-1}\,|\mathbf{x}|^{n-2}} = \delta \qquad \text{in } \mathbb{R}^n,\ n \geq 3.$$

Sign-changed, $U = -W$ is the **fundamental solution of the Laplace operator** Δ **.**

Note - Explicitly, referring also to (3.26), we have $\Delta_{\mathbf{x}} U = \delta$ in $\mathbb{R}^n$, $n \geq 3$. □

Exercise 4.62. (**Helmholtz equation**) Given a parameter $a > 0$ and a datum f in $\mathcal{E}'(\mathbb{R}^n)$, this equation in the unknown u

$$u \in \mathcal{S}'(\mathbb{R}^n): \qquad \Delta u + a^2\,u = f \quad \mathcal{S}'(\mathbb{R}^n) \tag{4.96}$$

is called *Helmholtz equation* (Hermann Ludwig Ferdinand von Helmholtz, 1821–1894).

1) Prove that, if $\omega > 0$ and $v(\mathbf{x},t) = u(\mathbf{x})e^{i\omega t}$ is a solution of this wave equation in $\mathbb{R}^n$, $v_{tt}(\mathbf{x},t) - c^2\Delta v(\mathbf{x},t) = -\,c^2\,e^{i\omega t}\,f(\mathbf{x})$, then u solves the Helmholtz equation with $a = \omega/c$.

2) Find the symmetric C^2 solutions, if they exist, of the homogeneous Helmholtz equation ($f \equiv 0$) for $n = 2, 3$ and analyze the uniqueness of the solution.

3) Find the radially symmetric solutions of the homogeneous Helmholtz equation in all the Euclidean space but the origin, if $n = 2, 3$:

$$z \in \mathcal{S}'(\mathbb{R}^n): \qquad \Delta z + a^2\,z = 0 \qquad \mathcal{S}'(\mathbb{R}^n) \setminus \{\mathbf{0}\}$$

and use the result to describe the solutions of the kind $w(\mathbf{x},t) = z(\mathbf{x})e^{i\omega t}$ of the equation

$$w_{tt}(\mathbf{x},t) - c^2\Delta w(\mathbf{x},t) = -\,c^2\,e^{i\omega t} \qquad \mathcal{S}'(\mathbb{R}^3 \setminus \{\mathbf{0}\}).$$

4) Find the fundamental solution of the Helmholtz equation in the physical case $n = 3$.

Solution - We compute

1) $-\,c^2\,e^{i\omega t}\,f = \Box_c\,v = \Box_c\big(u(\mathbf{x})e^{i\omega t}\big) = e^{i\omega t}\big(-\omega^2\,u - c^2\Delta u\big)$ hence (4.96) with $a = \omega/c$.

2) Applying the Fourier transform to the homogeneous differential equation $\Delta u + a^2 u = 0$ we get $(a^2 - |\boldsymbol{\xi}|^2)\widehat{u}(\boldsymbol{\xi}) = 0$, that is a division problem in the distributional framework. There is no uniqueness of solution, since the polynomial multiplier vanishes on $\partial B_a(\mathbf{0})$. By the analysis done in Exercise 4.60 we know that $\widehat{u}(\boldsymbol{\xi}) = C\,\delta_{\partial B_a(\mathbf{0})}(\boldsymbol{\xi})$ is a solution of the division problem for any constant C, then at least in dimension $n = 3$ we expect to find the multiples of $\sin(a|\mathbf{x}|)/|\mathbf{x}|$ among the solutions of the homogeneous Helmholtz equation.

In order to achieve complete description of radially symmetric solutions we proceed in a different way, using the polar coordinates (Tables 10.6.1 and 10.6.2), and we obtain additional tips useful for the subsequent item 3).

- $n=3$. The differential equation confined to $\mathbb{R}^3 \setminus \{\mathbf{0}\}$ is equivalent to

$$u_{\varrho\varrho} + \frac{2}{\varrho}\, u_\varrho + a^2\, u \;=\; 0 \qquad \varrho > 0\,,$$

whose solutions are linear combinations of $\sin(a|\mathbf{x}|)/|\mathbf{x}|$ and $\cos(a|\mathbf{x}|)/|\mathbf{x}|$. Nevertheless the solution must be regular at the origin too. Then all radially symmetric solution are of the kind

$$u(\mathbf{x}) \;=\; C\sin(a|\mathbf{x}|)/|\mathbf{x}|\,, \qquad \text{se } n=3\,,$$

with arbitrary constant C, and there is no uniqueness. Notice that these u are entire analytic functions.

- $n=2$. The differential equation in $\mathbb{R}^2 \setminus \{\mathbf{0}\}$ reads

$$u_{\varrho\varrho} + \frac{1}{\varrho}\, u_\varrho + a^2\, u \;=\; 0 \qquad \varrho > 0\,,$$

whose solutions are linear combinations of $J_0(a\varrho)$ and $Y_0(a\,\varrho)$, with J_0 and Y_0 Bessel functions of order 0 respectively of first and second kind (see Exercise 1.28): we recall that J_0 is a function $C^\infty(\mathbb{R})$, whereas $Y_0 \in C^\infty(\mathbb{R} \setminus \{0\})$ and Y_0 has a logarithmic singularity at the origin. Nevertheless the solution u must be regular also at the origin. Then every radially symmetric solution is (with arbitrary constant C) of the kind

$$u(\mathbf{x}) \;=C\, J_0(a\,|\mathbf{x}|)\,, \qquad \text{se } n=2\,.$$

3) We exploit the polar coordinates as in item 2), but now it is not necessary to rule out the singularities at the origin. Moreover we express the solutions of the ordinary differential also in the equivalent complex formulation (more suitable for the applications) and we obtain

$$z(\mathbf{x}) \;=\; A\,\frac{\sin(a|\mathbf{x}|)}{|\mathbf{x}|} + B\,\frac{\cos(a|\mathbf{x}|)}{|\mathbf{x}|} \;=\; C\,\frac{e^{i\,a\,|\mathbf{x}|}}{|\mathbf{x}|} + D\,\frac{e^{-\,i\,a\,|\mathbf{x}|}}{|\mathbf{x}|} \qquad \text{se } n=3\,.$$

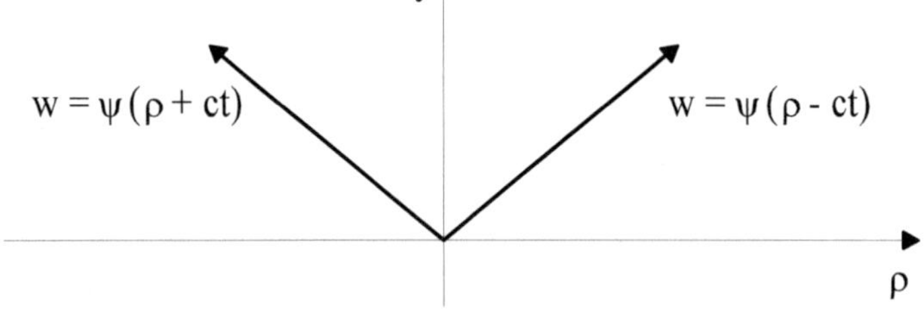

Figure 4.15 - Exercise 4.62. Antenna radiation $w(\mathbf{x},t) = \psi(\varrho - ct)$, with $\psi(s) = \frac{1}{s}\,\exp(-\frac{i\omega}{c}\,s)$, in the physical space $\varrho > 0$.

We notice that in the present case ($n=3$) we can, either explain w as the radiation emitted by an antenna in the surrounding physical environment and that propagates toward infinity

$$w(\mathbf{x},t) \;=\; \frac{e^{i\,\frac{\omega}{c}\,(ct-|\mathbf{x}|)}}{|\mathbf{x}|} \qquad \text{corresponding to } z(\mathbf{x}) = \frac{e^{-\,i\,a\,|\mathbf{x}|}}{|\mathbf{x}|}\,,$$

or explain it as the received radiation at the antenna from the surrounding physical space

$$w(\mathbf{x},t) \;=\; \frac{e^{i\,\frac{\omega}{c}\,(ct+|\mathbf{x}|)}}{|\mathbf{x}|} \qquad \text{corresponding to } z(\mathbf{x}) = \frac{e^{i\,a\,|\mathbf{x}|}}{|\mathbf{x}|}$$

that is a radiation coming to the origin.
The second one has no physical meaning if we are interested in the description of wave sources contained in bounded domains. Notice that the wave $w(\mathbf{x},t) = \dfrac{e^{i\,\omega t - |\mathbf{x}|/c}}{|\mathbf{x}|}$, corresponding to the radiation emitted by the antenna, has temporal frequency of oscillation equal to $\omega/2\pi$ and oscillation amplitude at $|\mathbf{x}|$ equal to $1/|\mathbf{x}|$.
In the 2-dimensional case we find:

$$z(\mathbf{x}) \;=\; A\,J_0(a\,|\mathbf{x}|) \,+\, B\,Y_0(a\,|\mathbf{x}|) \qquad \text{se } n=2\,.$$

4) Items 2) and 3) entail that $\cos(a|\mathbf{x}|)/|\mathbf{x}|$ solves the homogeneous Helmholtz equation in $\mathbb{R}^3 \setminus \{\mathbf{0}\}$. Moreover $\cos(a|\mathbf{x}|)/|\mathbf{x}| \;\sim\; 1/|\mathbf{x}|$ in a neighborhood of the origin, whereas the perturbation $\cos(a|\mathbf{x}|) \in C^\infty(\mathbb{R}^3)$. Then the singular part of the Laplacian computed on $\cos(a|\mathbf{x}|)/|\mathbf{x}|$ is identical to the one of Laplacian computed on $1/|\mathbf{x}|$. Thank to the analysis done by Exercise 4.60, we can conclude with the explicit formula of the fundamental solution in 3 space dimensions.

Fundamental solution of the Helmholtz equation in $\mathbb{R}^3$

$$-\,\frac{1}{4\,\pi}\,\frac{\cos(a|\mathbf{x}|)}{|\mathbf{x}|}\,. \tag{4.97}$$

Remark - The Helmoltz equation exhibits a remarkably different nature with respect to the equation $-\Delta u + a^2\,u \;=\; f$ (see Exercise 4.48) that displays existence and uniqueness of the solution for any $f \in \mathcal{S}'$. The difference is due to the fact that the symbol (see Section 4.1.6) of the Helmholtz differential operator is $a^2 - |\boldsymbol{\xi}|^2$ which vanish in $\partial B_a(\mathbf{0})$, whereas the symbol of $-\Delta u + a^2\,I$ is $a^2 + |\boldsymbol{\xi}|^2 \neq 0$ for any $\boldsymbol{\xi} \in \mathbb{R}^n$. This property is referred to by saying that $-\Delta u + a^2\,I$ is an *elliptic operator.*

□

Exercise 4.63. (Estimate of the diffusion coefficient in the heat equation)
Consider the homogeneous heat equation $u_t - a\,\Delta_{\mathbf{x}} u = 0$ in $\mathbb{R}^n \times (0,\infty)$.
Prove that, if $u(\mathbf{x},0) = u_0(\mathbf{x}) \in L^1(\mathbb{R}^n)$, then

$$a \;=\; \frac{\displaystyle\int_{\mathbb{R}^n} |\mathbf{x}|^2 \Big(u(\mathbf{x},T) - u_0(\mathbf{x})\Big)\,d\mathbf{x}}{\displaystyle 2\,n\,T \int_{\mathbb{R}^n} u_0(\mathbf{x})\,d\mathbf{x}} \qquad \forall T > 0\,. \tag{4.98}$$

Nota - This formula (4.98) may be helpful when looking for an estimate of the diffusion coefficient a, by performing measurements on the solution of the Cauchy problem for the heat equation.
We notice that the right-hand side in formula (4.98) provides the exact value of the diffusion coefficient a only if we know the temperature $u(\mathbf{x},T)$ at time T for every $\mathbf{x}$. Nevertheless we can exploit the formula for finding an approximation of a since we can perform the measurements at several different times T and for many different initial values u_0, by suitable repetition of the experiment.

Solution - By the explicit formula for the solution of the Cauchy problem (Exercise 4.54): $u(\mathbf{x},t) = u_0 *_{\mathbf{x}} U \ \forall t > 0$, where $U = U(\mathbf{x},t)$ is the fundamental solution of the heat operator.
Due to the decay of U at infinity and the properties of convolution $u(\mathbf{x},t) = o(1)$ is an infinitesimal of exponential order as $|\mathbf{x}| \to \infty$, for any fixed $t > 0$. This fact entails the cancelation "at infinity" of these boundary integrals (that subsequently appear when we use the divergence Theorem):

$$\lim_{R\to+\infty} \int_{\partial B_R(\mathbf{0})} |\mathbf{x}|^2 \frac{\partial u}{\partial \nu}\, d\sigma(\mathbf{x}) = \lim_{R\to+\infty} \int_{\partial B_R(\mathbf{0})} |\mathbf{x}|\, u\, d\sigma(\mathbf{x}) = 0 .$$

We multiply the equation $u_t - a\,\Delta_{\mathbf{x}} u = 0$ times $|\mathbf{x}|^2$ and integrate over $\mathbb{R}^n \times (0,t)$:

$$\begin{aligned}
\int_{\mathbb{R}^n} |\mathbf{x}|^2 u(\mathbf{x},t)\, d\mathbf{x} - \int_{\mathbb{R}^n} |\mathbf{x}|^2 u_0(\mathbf{x})\, d\mathbf{x} &= a \int_0^t \int_{\mathbb{R}^n} |\mathbf{x}|^2\, \Delta u(\mathbf{x},\tau)\, d\mathbf{x}\, d\tau \\
&= a \int_0^t \lim_{R\to+\infty} \int_{B_R(\mathbf{0})} |\mathbf{x}|^2\, \Delta u(\mathbf{x},\tau)\, d\mathbf{x}\, d\tau \\
\boxed{\text{divergence thm}} \quad &= -a \int_0^t \lim_{R\to+\infty} \int_{B_R(\mathbf{0})} 2\,\mathbf{x} \cdot \nabla u(\mathbf{x},t)\, d\mathbf{x}\, d\tau \\
\boxed{\text{divergence thm}} \quad &= a \int_0^t \lim_{R\to+\infty} \int_{B_R(\mathbf{0})} 2n\, u(\mathbf{x},\tau)\, d\mathbf{x}\, d\tau \\
&= 2na \int_0^t \int_{\mathbb{R}^n} u(\mathbf{x},\tau)\, d\mathbf{x}\, d\tau \\
\boxed{(*)} \quad &= 2na \int_0^t \int_{\mathbb{R}^n} u_0(\mathbf{x})\, d\mathbf{x}\, d\tau \\
&= 2nat \int_{\mathbb{R}^n} u_0(\mathbf{x})\, d\mathbf{x}
\end{aligned}$$

Hence (4.98) holds true for any choice of $t = T > 0$.
In the step labeled by (*) we used the *conservation of the total quantity of heat* $\left(\int_{\mathbb{R}^n} u(\mathbf{x},t)\, d\mathbf{x} = \int_{\mathbb{R}^n} u_0(\mathbf{x})\, d\mathbf{x} \ \forall t > 0 \right)$, whose proof can be found in Exercises 4.56 (1) and 8.9.

□

Chapter 5
Fourier Series

Fourier series is a classical tool to analyze and approximate periodic functions. The method can be easily adapted to the case of functions defined on a bounded interval, considering its periodic extension with period equal to the length of the interval. In this chapter we introduce the Fourier series of periodic functions and distributions and we consider the problem of finding periodic solutions of ordinary differential equations.
There is a precise connection between the Fourier series of a signal and its Fourier transform, as described in Theorem 5.8.
The application of the Fourier series to the method of separation of variables for solving partial differential equation is postponed to Chapters 7 and 8.
The basic ideas of the method of the Fourier series have been clarified along the years, rising subtle theoretical questions and suggesting elegant solutions. Recently they fostered important developments in numerical analysis, signal processing and image compression: Discrete Fourier Transform (DFT), Fast Fourier Transform (FFT), wavelets, JPG and JPG2000 protocols. We refer the interested reader to ([16],[20],[22],[23],[25]).

5.1 Definitions and properties of the Fourier Series

Let us first consider periodic functions that are locally integrable in $\mathbb{R}$ and, subsequently, the general case of periodic distributions in $\mathbb{R}$.

5.1.1 Fourier series of periodic functions

A suitable functional framework to study the Fourier series is the Hilbert space $L^2(0, 2\pi)$ endowed with the scalar product

$$(u, v)_{L^2} := \int_0^{2\pi} u(x)\,\overline{v(x)}\,dx \qquad \forall u,\, v \in L^2(0, 2\pi) \tag{5.1}$$

and with the induced norm

$$\|u\|_{L^2} = \left(\int_0^{2\pi} |u(x)|^2\,dx\right)^{1/2} \qquad \forall u \in L^2(0, 2\pi).$$

We will consider the (countable) set $\mathcal{B}$ of functions

$$\left\{\frac{1}{\sqrt{2\pi}}, \frac{\cos x}{\sqrt{\pi}}, \frac{\sin x}{\sqrt{\pi}}, \frac{\cos(2x)}{\sqrt{\pi}}, \frac{\sin(2x)}{\sqrt{\pi}}, \ldots, \frac{\cos(nx)}{\sqrt{\pi}}, \frac{\sin(nx)}{\sqrt{\pi}}, \ldots\right\} \tag{5.2}$$

referring to Exercise 5.1 to verify its elementary properties. Of each function $u \in L^2(0, 2\pi)$ we can compute the Fourier coefficients:

$$a_n = \frac{1}{\pi}\int_0^{2\pi} u(x)\cos(nx)\,dx, \quad b_n = \frac{1}{\pi}\int_0^{2\pi} u(x)\sin(nx)\,dx, \quad \forall\, n \geq 0. \tag{5.3}$$

In particular notice that $b_0 = 0$. With these coefficients we can write

$$u(x) = \frac{a_0}{2} + \sum_{n=1}^{\infty}\Big(a_n\cos(nx) + b_n\sin(nx)\Big), \tag{5.4}$$

reading u as the limit in $L^2(0, 2\pi)$ (and point-wise a.e. in $(0, 2\pi)$) of the sequence of partial sums defined as follows:

$$S_N(x) = \frac{a_0}{2} + \sum_{n=1}^{N}\Big(a_n\cos(nx) + b_n\sin(nx)\Big). \tag{5.5}$$

Explicitly this means that $\{S_N\}$ converges in square mean to u, that is

$$\lim_{N\to\infty}\int_0^{2\pi} |u(x) - S_N(x)|^2\,dx = 0\,, \tag{5.6}$$

which is equivalent to

$$\|u - S_N\|_{L^2(0,2\pi)} \overset{N\to+\infty}{\longrightarrow} 0 \qquad \forall u \in L^2(0, 2\pi)\,. \tag{5.7}$$

The series in the right hand side of (5.4) is called *Fourier series* of u.
The corresponding coefficients a_n, b_n are called Fourier coefficients of u.
After extending periodically the function from $(0, 2\pi)$ to the whole $\mathbb{R}$, formulas (5.3) provide the same values of the coefficients also changing the integration interval with any other integration interval of length 2π. In particular, referring to an interval $(-\pi, \pi)$, we deduce for any 2π periodic function u:

$$\text{if } u \text{ is even} \quad \text{then } b_n = 0 \ \forall n\,, \tag{5.8}$$

$$\text{if } u \text{ is odd} \quad \text{then } a_n = 0 \ \forall n\,. \tag{5.9}$$

Observe that the computation of the coefficients a_n and b_n with (5.3) is possible under the only hypothesis $u \in L^1(0, 2\pi)$. Nevertheless, in this space important properties of convergence are missing and they can be recovered in the smaller space $L^2(0, 2\pi)$. In particular, in $L^2(0, 2\pi)$ the following property, which corresponds to the infinite-dimensional version of the Pythagorean Theorem, holds

Theorem 5.1. (Parseval's Identity). *Let $u \in L^2(0, 2\pi)$; then*

$$\int_0^{2\pi} |u(x)|^2\, dx = \|u\|_{L^2}^2 = \pi \left\{ \frac{|a_0|^2}{2} + \sum_{n=1}^{\infty} (|a_n|^2 + |b_n|^2) \right\}$$

where a_n and b_n are defined by (5.3).

Clearly it is possible to proceed as above also in intervals $(0, T)$ with a generic $T > 0$. In such a case, a function $u \in L^2(0, T)$ has Fourier coefficients

$$a_n = \frac{2}{T} \int_0^T u(x) \cos \frac{2\pi n x}{T}\, dx\,, \ b_n = \frac{2}{T} \int_0^T u(x) \sin \frac{2\pi n x}{T}\, dx\,, \ n \geq 0, \tag{5.10}$$

and the corresponding *Fourier expansion* of u is given by

$$u(x) = \frac{a_0}{2} + \sum_{n=1}^{\infty} \left(a_n \cos \frac{2\pi n x}{T} + b_n \sin \frac{2\pi n x}{T} \right). \tag{5.11}$$

Also observe that, besides the real form (5.11), the Fourier series can be expressed in the complex exponential form, that is as a series of a appropriate complex exponential functions. Let $u \in L^2(-T/2, T/2)$, we define

$$c_k = \frac{1}{T} \int_{-T/2}^{T/2} e^{-i\,k\,\frac{2\pi}{T}\,x} u(x)\, dx\,, \qquad \forall k \in \mathbb{Z}\,. \tag{5.12}$$

Thus the following relations hold:

$$a_n = c_n + c_n\,, \qquad b_n = i(c_n - c_{-n*})\,, \qquad \forall n \in \mathbb{N}\,, \tag{5.13}$$

$$c_n = \frac{a_n - i b_n}{2}\,, \qquad c_{-n} = \frac{a_n + i b_n}{2}\,, \qquad \forall n \in \mathbb{N}\,, \tag{5.14}$$

and we have the equality with convergence in $L^2(0, T)$ on the right hand side:

$$u(x) = \sum_{k=-\infty}^{+\infty} c_k \, e^{+i\,k\,\frac{2\pi}{T}\,x} \,. \tag{5.15}$$

Equation (5.15) is the *exponential Fourier series in complex form* of u.
We can interpret the coefficients as scalar products with pairwise orthogonal functions in L^2: the Fourier series corresponds to the sum (possibly infinite sum, or series) of the "components" of u along the directions of such functions.
Set

$$v_k(x) = \frac{e^{+i\,k\,(2\pi/T)x}}{\sqrt{T}}, \qquad \forall k \in \mathbb{Z},$$

we have: $\|v_k\|_{L^2(0,2\pi)} = 1\,, \ (v_k, v_h)_{L^2(0,2\pi)} = 0 \ \ \forall k \in \mathbb{Z}$ and

$$c_k = (u\,,\, v_k)_{L^2(0,2\pi)} \,/\sqrt{T}\,, \qquad \forall k \in \mathbb{Z},$$

$$u(x) = \sum_{k=-\infty}^{+\infty} (u\,,\, v_k)_{L^2} \, v_k(x)\,.$$

Referring to the expansion in the exponential form, Parseval's Identity (Theorem 5.1) reads as follows.

(Parseval's Identity). If $u \in L^2(0,T)$ and c_k are defined by (5.12), then

$$\|u\|^2_{L^2(0,T)} = T \sum_{k=-\infty}^{+\infty} |c_k|^2 \,. \tag{5.16}$$

Besides the square mean convergence, there are other types of convergence of a certain interest in Fourier series analysis.
We now state some results connected with the point-wise and the uniform convergence of the Fourier series. By Weierstrass criterion if

$$\sum_{n=0}^{\infty} (|a_n| + |b_n|) < \infty$$

then the sequence of partial sums S_N in (5.5) uniformly converges in $[0, 2\pi]$ to u and thus, in particular, $u \in C^0([0, 2\pi])$. By differentiating term by term the Fourier series, this argument can be iterated (observe that the derivative of a periodic function is periodic with the same period): the m-th formal derivative of u is a Fourier series with, up to a sign, coefficients of the form $n^m a_n$ and $n^m b_n$ that multiply functions sine/cosine. Hence, the following convergence and regularity criterion holds true.

Theorem 5.2. (Uniform convergence of Fourier series).
Let $u \in C^0[0, 2\pi]$ and let a_n and b_n be defined by (5.3). Then

$$\sum_{n=0}^{\infty} n^m (|a_n| + |b_n|) < \infty \implies u \in C^m([0, 2\pi])$$

and the j-th derivative of the Fourier series of u uniformly converges to the j-th derivative of u on $[0, 2\pi]$ for every $j = 0, ..., m$.

In particular, if $a_n(v)$ and $b_n(v)$ are the Fourier coefficients of a T periodic function v and $\sum_{n=0}^{\infty} n(|a_n(u)| + |b_n(u)|) < \infty$, then the following relations hold true

$$a_n(u') = n \frac{2\pi}{T} b_n(u), \qquad b_n(u') = -n \frac{2\pi}{T} a_n(u), \qquad \forall n \geq 1. \tag{5.17}$$

Let us observe that, if a T periodic function u is given, then its antiderivative is T periodic if and only if u has zero mean value on the period, that is if

$$⨍_0^T u(x)\, dx := \frac{1}{T} \int_0^T u(x)\, dx = 0. \tag{5.18}$$

If $u \in L^2_{loc}(\mathbb{R})$ is T periodic and (5.18) holds then, setting $U(x) = \int_0^x u(y)\, dy$, the following relations hold true

$$a_n(U) = -\frac{1}{n} \frac{T}{2\pi} b_n(u), \qquad b_n(U) = \frac{1}{n} \frac{T}{2\pi} a_n(u), \qquad \forall n \geq 1. \tag{5.19}$$

Definition 5.3. *A real-valued functions defined on an interval $I \subset \mathbb{R}$ is said to be a function of bounded variation on I if it is equal to the difference of two bounded monotone functions.*
The space of bounded variation functions in I is denoted by $BV(I)$.

Observe that $BV(I)$ is neither contained in $C^0(I)$ and nor it contains $C^0(I)$. Indeed, the function w, defined by

$$w(x) = 0 \text{ if } x \in [0, \pi], \quad w(x) = 1 \text{ if } x \in (\pi, 2\pi],$$

is monotone and thus in $BV([0, 2\pi])$, but it is not a continuous function.
On the other hand, the function v, defined by

$$v(x) = x \sin \frac{1}{x} \text{ if } x \in (0, 2\pi], \quad v(0) = 0 \text{ if } x = 0,$$

cannot be represented as difference of monotone functions in spite of being a continuous function on $[0, 2\pi]$. Nevertheless, by the properties of monotone functions, we have

$$u \in BV(I) \;\Rightarrow\; \lim_{x\to\widetilde{x}_+} u(x) \text{ and } \lim_{x\to\widetilde{x}_-} u(x) \text{ exist finite } \forall\, \widetilde{x} \in I\,. \tag{5.20}$$

Theorem 5.4. (Point-wise convergence of the Fourier series). *Let $u \in BV([0,2\pi])$ and let S_N be the sequence in (5.5) with a_n and b_n defined by (5.3). Then the following limit exists and*

$$\lim_{N\to\infty} S_N(\widetilde{x}) \;=\; \frac{1}{2}\left(\lim_{x\to\widetilde{x}_-} u(x) + \lim_{x\to\widetilde{x}_+} u(x)\right) \qquad \forall \widetilde{x} \in [0,2\pi]\,.$$

Remark 5.5. Analogous versions of Theorems 5.2 and 5.4 for Fourier series in complex form and the corresponding coefficients c_k easily follow, from the identity

$$S_N(x) = \sum_{k=-N}^{k=+N} c_k\, e^{i\,k\,\frac{2\pi}{T}\,x}\,. \tag{5.21}$$

Remark 5.6. The Fourier coefficients of a function in $L^2(0,T)$ are infinitesimal as the index diverges:

$$\lim_{n\to+\infty} a_n \;=\; \lim_{n\to+\infty} b_n \;=\; \lim_{k\to+\infty} c_k \;=\; \lim_{k\to-\infty} c_k \;=\; 0\,. \tag{5.22}$$

5.1.2 Fourier series of periodic distributions

It is also possible to define the Fourier series of periodic distributions. For the notations and properties of distributions we refer to Chapter 3.

Definizione 5.7. (Periodic tempered distributions).
A distribution $u \in \mathcal{D}'(\mathbb{R})$ is said to be periodic of period $T > 0$ if $u(x) = u(x+T)$ in $\mathbb{R}$, that is

$$\int_{\mathbb{R}} u(x)\varphi(x-T)\,dx = \int_{\mathbb{R}} u(x)\varphi(x)\,dx \qquad \forall \varphi \in \mathcal{D}(\mathbb{R}).$$

Any periodic distribution is tempered.
We denote by $\mathcal{S}'_T(\mathbb{R})$ the space of T periodic distributions in $\mathbb{R}$.

We now give a result that characterizes the periodic distributions through the properties of their Fourier transform.

Theorem 5.8. *$u \in \mathcal{S}'_T(\mathbb{R})$ if and only if its Fourier transform $\widehat{u}$ can be written as a convergent series in $\mathcal{S}'(\mathbb{R})$ of the form*

$$\widehat{u}(\xi) = \sum_{k=-\infty}^{+\infty} \gamma_k \, \delta\Big(\xi - \frac{2\pi k}{T}\Big) \tag{5.23}$$

with tempered coefficients $\gamma_k \in \mathbb{C}$, that is

$$\exists\, A,\, B,: \qquad |\gamma_k| \le A(1+|k|)^B \qquad \forall k \in \mathbb{Z}. \tag{5.24}$$

A series as (5.23) is called *delta train* or *impulse train.*
Theorem 5.8 guarantees the existence and uniqueness of the *Fourier series of a periodic distribution.* Indeed, for every $u \in \mathcal{S}'_T$ we have (5.23), (5.24) with suitable coefficients γ_k depending on u; setting $\widehat{u}_k = \gamma_k/(2\pi)$ $\forall k$ and antitransforming (5.23) we obtain explicitly the Fourier series of u:

$$u(x) = \sum_{k=-\infty}^{+\infty} \widehat{u}_k \, e^{i\,k\,\frac{2\pi}{T}\,x}\,, \tag{5.25}$$

a series that converges to u in $\mathcal{S}'(\mathbb{R})$ and that has coefficients $\widehat{u}_k$ that are tempered, that is, they verify (5.24).

Remark 5.9. The Fourier transform of any periodic signal of period $T > 0$ is a train of impulses, whose support is contained in the set of the multiples of $2\pi/T$ and whose corresponding intensity are expressed by tempered coefficients. In particular these coefficients may be neither infinitesimal (thus (5.22) is not guaranteed) nor bounded, nevertheless they can grow polynomially at most, according to (5.24).
In Signal Theory this fact is usually expressed by saying that periodic signals have a *discrete spectrum*, in opposition with a *continuous spectrum* that characterizes integrable functions in $\mathbb{R}$: in this context *spectrum* stands for support of the Fourier transform.
It is important to notice that even if we have a discrete spectrum (that is if $\operatorname{spt} u$ is made of isolated points) the signal may not be periodic. Consider for instance $\widehat{u}(\xi) = \delta(\xi - 1) + \delta(\xi - \sqrt{7})$, where the spectrum consists of two points (thus isolated) but it is not contained in a set of integer multiples of $2\pi/T$ for no choice of T, because of the irrationality of the ratio 1 over $\sqrt{7}$.

Remark 5.10. Observe that both series at the right hand side of (5.23) and (5.25) converge in the sense of tempered distributions (that is in $\mathcal{S}'(\mathbb{R})$, see Definition 3.17) respectively to $\widehat{u}$ and u. Therefore the operations $\mathcal{F}$ and $\mathcal{F}^{-1}$ (Fourier transform and antitransform, we refer to Chapter 4) are correct, not only formally, when they are computed term by term.

Remark 5.11. The fact that the Fourier series (5.25) converges in $\mathcal{S}'(\mathbb{R})$, allows to perform term by term differentiation (of any order) and obtaining a series that converges in $\mathcal{S}'(\mathbb{R})$ to the corresponding derivative of the sum (this would be false if we refer to a more elementary kind of convergence, unless adding assumptions on the regularity of sum function of the Fourier series, e.g. Theorem 5.2). The possibility to differentiate term by term the series is very useful when looking for periodic solutions of ordinary differential equations. Indeed expressing the data and the unknown through their Fourier series makes possible to reduce differential problems to a system of uncoupled algebraic equations that easily allows to find the coefficients of the solution, when it exists: see Exercises 5.8, 5.9, 5.15, 5.16, 5.17.
The relations between the coefficients of the distribution and of its distributional derivative remain true (without any assumption on the decay of the coefficients, provided they are tempered). Precisely, referring to the complex form of the Fourier series, we denote by $\widehat{v}_k$ the kth Fourier coefficient of a distribution $v \in \mathcal{S}'_T$ and by v' the distributional derivative of v, then the following relations, analogous to (5.17), hold true:

$$\widehat{u'}_k = k\left(i\frac{2\pi}{T}\right)\widehat{u}_k\,, \qquad \forall k \in \mathbb{Z}\,. \tag{5.26}$$

Moreover, if U is the antiderivative of u and (5.18) holds, then we have the analogous of (5.19):

$$\widehat{U}_k = -\frac{1}{k}\left(i\frac{T}{2\pi}\right)\widehat{u}_k\,, \qquad \forall k \in \mathbb{Z}\,. \tag{5.27}$$

In this context the integral over the period (in (5.18)) of the distribution $u \in \mathcal{S}'_T$ has to be properly interpreted: the following quantity must vanish

$$\int_{\mathbb{R}} u(x)\psi(x)\,dx := \langle\, u\,,\,\psi\,\rangle \qquad \text{whenever } \psi \in \mathcal{S} \text{ and } \sum_{k\in\mathbb{Z}}\psi(x-kT) \equiv 1 \text{ in } \mathbb{R}\,. \tag{5.28}$$

Note that the mean value of the Tperiodic distribution u defined in (5.28) does not depend on the choice of ψ provided that ψ belongs to the Schwartz space and verifies the identity $\sum_k \psi(x-kT) \equiv 1$ in $\mathbb{R}$.

Remark 5.12. If $u \in \mathcal{S}'_T(\mathbb{R}) \cap L^2_{loc}$, that is the distribution u coincides with the Tperiodic extension of an $L^2(0,T)$ function, then the coefficients c_k computed through (5.12) coincide with the $\widehat{u}_k$ of (5.25), by uniqueness of the Fourier series.
On the other hand, the coefficients of a Fourier series of a periodic distribution cannot be computed through (5.12), if it is not locally integrable.
Nevertheless every periodic distribution u has finite order (since all distributions have locally finite order, see Theorem 3.11). Thus every periodic distribution is, up to a polynomial correction, the derivative of finite order

of a function $w \in L^1_{loc}$ and periodic of the same order: thus it is possible to compute the Fourier coefficients of the function w and from them deduce those of u by applying iteratively (5.26).
About a different computation method for the coefficients of $u \in \mathcal{S}'_T$ see Exercise 5.23.

Let us now consider an example that plays a fundamental role in the analysis of signals: in such a context it is natural referring to the variable t (time) for the signal and to the variable ω (frequency) for its Fourier transform. If a signal has period $T > 0$ then its frequency is $2\pi/T$.

Example 5.13. (**comb function** or **Dirac Comb**)
We call *Dirac Comb* of step $T > 0$ and we denote by comb_T the following train of impulses spaced by intervals of length T and with all coefficients equal to 1:

$$\text{comb}_T(t) = \sum_{k=-\infty}^{+\infty} \delta(t - kT). \tag{5.29}$$

comb_T is a T periodic distribution, thus it is possible to compute its Fourier series, its Fourier transform and (since its transform turns out to be $2\pi/T$ periodic: the Fourier transform of comb_T is another Dirac comb of step $2\pi/T$ and amplitude $2\pi/T$ instead of 1) also the Fourier series of its transform. Precisely (we refer to Exercise 5.12 for the proof) we have:

$$\text{comb}_T(t) = \sum_{k=-\infty}^{+\infty} \delta(t - kT) = \frac{1}{T} \sum_{k=-\infty}^{+\infty} e^{i\,k\,\frac{2\pi}{T}\,t} \qquad t \in \mathbb{R}, \tag{5.30}$$

$$\widehat{\text{comb}_T}(\omega) = \frac{2\pi}{T} \sum_{k=-\infty}^{+\infty} \delta\Big(\omega - k\frac{2\pi}{T}\Big) = \sum_{k=-\infty}^{+\infty} e^{i\,k\,T\,\omega} \qquad \omega \in \mathbb{R}. \tag{5.31}$$

Observe that the Fourier series in (5.30) and (5.31) do not converge pointwise at any point of $\mathbb{R}$ (the general terms are not infinitesimal). Nonetheless they converge in $\mathcal{S}'(\mathbb{R})$ to the respective first terms (Definition 3.17): in the duality with functions of $\mathcal{S}(\mathbb{R})$ there is a balance between positive, negative and imaginary terms annihilating every contribution at every $t \in \mathbb{R}$, except at $t = kT$ and $t = 2k\pi/T$ respectively in (5.30) and (5.31), $\forall k \in \mathbb{Z}$.

Let us define, for $T > 0$:

$$L^2_T(\mathbb{R}) = L^2_{loc}(\mathbb{R}) \cap \mathcal{S}'_T(\mathbb{R}). \tag{5.32}$$

Denoting by $\widehat{u}_k$ the k th Fourier coefficient of some $u \in \mathcal{S}'_T$, we have

$$u(t) = \sum_{k \in \mathbb{Z}} \widehat{u}_k \, \exp\big(i\,k\,(2\pi/T)\,t\big).$$

Moreover by Parseval identity we obtain a necessary and sufficient condition for u to belong to L^2_T and an estimate through the norm series and of the scalar product in L^2_T. Precisely:

$$u \in L^2_T \iff \sum_{k\in\mathbb{Z}} |\widehat{u}_k|^2 < +\infty\,, \tag{5.33}$$

$$\|u\|^2_{L^2_T} := \int_0^T |u(t)|^2\,dt = T\sum_{k\in\mathbb{Z}} |\widehat{u}_k|^2 \qquad \forall u \in L^2_T\,, \tag{5.34}$$

$$(u, v)_{L^2_T} := \int_0^T u(t)\,\overline{v(t)}\,dt = T\sum_{k\in\mathbb{Z}} \widehat{u}_k\,\overline{\widehat{v}_k} \qquad \forall u \in L^2_T\,. \tag{5.35}$$

It is very useful in the analysis of differential equations to consider functions that, besides being in L^2_T, have one or more distributional derivatives in L^2_T. For this purpose we introduce the Sobolev spaces of T periodic functions.

Definizione 5.14. *(Sobolev spaces of periodic functions)*

$$H^1_T := \left\{u \in L^2_T \,:\, u' \in L^2_T\right\}, \tag{5.36}$$

$$H^k_T := \left\{u \in L^2_T \,:\, u', u'', \dots u^{(k)} \in L^2_T\right\}, \qquad k \in \mathbb{N}\,. \tag{5.37}$$

Theorem 5.15. *Let $u \in \mathcal{S}'_T$ be given. Then*

$$u \in H^1_T \iff \sum_{h\in\mathbb{Z}} |\,h\,\widehat{u}_h\,|^2 < +\infty\,, \tag{5.38}$$

$$u \in H^k_T \iff \sum_{h\in\mathbb{Z}} |\,h^k\,\widehat{u}_h\,|^2 < +\infty\,. \tag{5.39}$$

The previous theorem suggests a definition for the property of having *fractional derivatives* in L^2_T.

Definizione 5.16. *(Sobolev spaces with non integer index) Let $s \in \mathbb{R}$ and $s > 0$, then we set*

$$H^s_T := \left\{u \in L^2_T \,:\, \sum_{h\in\mathbb{Z}} |\,h^s\,\widehat{u}_h\,|^2 < +\infty\right\}. \tag{5.40}$$

Belonging to H^s_T with a non integer s is a regularity property between having exactly $[s]$ derivatives in L^2_T and having $[s]+1$ derivatives, where $[s]$ denotes the integer part of s.
There are interesting relations between the Sobolev spaces of periodic functions and the spaces C^m, as we can see in the following statement.

Theorem 5.17. *If* $u \in H^1_T$ *then* $u \in C^0(\mathbb{R}) \cap L^\infty(\mathbb{R})$.
If $u \in H^s_T$, $s > m+1/2$ *and* $m \in \mathbb{N}$, *then* $u^{(m)} \in C^0(\mathbb{R}) \cap L^\infty(\mathbb{R})$ $\forall m \leq [s]$.

Summarizing, there is a precise correspondence between the regularity of a periodic function and the decay of its Fourier coefficients (as $|k| \to \infty$). Theorem 5.17 and formulas (5.22), (5.24) describe quantitatively this correspondence.

5.2 Exercises

5.2.1 *Exercises on Fourier series of periodic functions*

Exercise 5.1. Verify that the family of functions $\mathcal{B}$ defined in (5.2) is orthonormal with respect to the scalar product defined in (5.1).

Solution - We have to prove that with the scalar product defined in (5.1) and the associated Hilbertian norm $\|\cdot\|_{L^2(0,2\pi)}$ it holds true that:

$$\|v\|_{L^2} = 1 \ \ \forall v \in \mathcal{B}; \qquad (v,w)_{L^2} = 0 \ \ \forall v,\, w \in \mathcal{B}:\ v \neq w\,.$$

We have:

$$\left\|\frac{1}{\sqrt{\pi}}\cos(nx)\right\|_{L^2}^2 = \frac{1}{\pi}\int_0^{2\pi} |\cos(nx)|^2 dx = \frac{1}{\pi}\int_0^{2\pi}(1+\cos(2nx)\,dx = 1$$

$$\left\|\frac{1}{\sqrt{\pi}}\sin(nx)\right\|_{L^2}^2 = \frac{1}{\pi}\int_0^{2\pi} |\sin(nx)|^2 dx = \frac{1}{\pi}\int_0^{2\pi}(1-\cos(2nx)\,dx = 1\,,$$

Moreover, using the formulas $\cos\theta = (e^{i\theta}+e^{-i\theta})/2$ and $\sin\theta = (e^{i\theta}-e^{-i\theta})/(2i)$, it is possible to prove that

$$\begin{aligned}
\big(\cos(nx),\cos(mx)\big)_{L^2} &= 0\,, & \forall n,m \in \mathbb{N},\ n\neq m\,,\\
\big(\sin(nx),\sin(mx)\big)_{L^2} &= 0\,, & \forall n,m \in \mathbb{N},\ n\neq m\,,\\
\big(\sin(nx),\cos(mx)\big)_{L^2} &= 0\,, & \forall n,m \in \mathbb{N}\,.
\end{aligned} \tag{5.41}$$

□

Exercise 5.2. Determine a sequence of coefficients $\{b_n\}_{n\geq 1} \subset \mathbb{R}$ such that

$$\sum_{n=1}^{+\infty} b_n \sin(nx) = 1 \quad \text{for any } x \in (0,\pi).$$

Deduce the sum of the series

$$\sum_{m=0}^{+\infty}\frac{1}{(2m+1)^2}\,, \qquad \sum_{m=0}^{+\infty}\frac{(-1)^m}{2m+1}\,.$$

Was it possible, before calculating the coefficients b_n, to expect that

$$\sum_{n=1}^{+\infty}|b_n| = +\infty \ ?$$

Solution - A Fourier series of only sines corresponds to an odd function u. By assumptions, it must be $u(x) = 1$ in $(0,\pi)$. The series, we are looking for, corresponds to a 2π periodic function u such that

$$u(x) := \begin{cases} 1 & x \in (0,\pi)\\ -1 & x\in(-\pi,0). \end{cases}$$

Some calculations (see Exercise 5.3) allow us to conclude that $b_n = 0$ if n even and $b_n = 4/(n\pi)$ if n odd.

The sum of the second series is $\pi/4$ and it is possible to obtain it by inserting $x = \pi/2$ in the associated Fourier series. Observe that since u is a bounded variation function, the series converges point-wise. In particular, for $x = \pi/2$, it converges to $u(\pi/2) = 1$.
The sum of the first series is $\pi^2/8$ and it is possible to obtain it by using Parseval identity:

$$\sum_{n=1}^{+\infty} |b_n|^2 = \frac{1}{\pi} \|u\|^2_{L^2(0,2\pi)} \quad \Longleftrightarrow \quad \frac{16}{\pi^2} \sum_{m=0}^{+\infty} \frac{1}{(2m+1)^2} = \frac{1}{\pi} \int_0^{2\pi} 1\,dx = 2.$$

It cannot be $\sum_{n=0}^{+\infty} |b_n| < +\infty$, since otherwise by Weirstrass criterion the Fourier series would converge uniformly to a continuous function on $\mathbb{R}$ and this is a contradiction since u is not a continuous function.

□

Exercise 5.3. Compute the Fourier series of $f(x) = \operatorname{sign}(\sin x)$, $x \in \mathbb{R}$, and specify its convergence properties.

Solution - f is an odd function: $f(-x) = -f(x)$. Thus its Fourier series is of the form:

$$\sum_{n=1}^{\infty} b_n \sin(nx).$$

$$b_n = \frac{1}{\pi} \int_{-\pi}^{+\pi} f(x) \sin(nx) dx = \frac{2}{\pi} \int_0^{+\pi} \sin(nx) dx = \frac{2}{n\pi} (1 - (-1)^n) = \begin{cases} \dfrac{4}{n\,\pi} & \text{if } n = 2m+1 \\ 0 & \text{if } n = 2m \end{cases}$$

thus the Fourier series of f is

$$(*) \qquad f(x) = \frac{4}{\pi} \sum_{m=0}^{\infty} \frac{1}{2m+1} \sin\left((2m+1)\, x \right).$$

The series $(*)$ converges in $L^2(-\pi, \pi)$ to f because $f \in L^2(-\pi, \pi)$, hence it converges in $\mathcal{S}'(\mathbb{R})$. Moreover, f is a function of bounded variation, thus the series converges point-wise at any real value x to $(f(x_+) + f(x_-))/2$, that is to 0 if $x = k\pi,\ k \in \mathbb{Z}$, and to $\operatorname{sign}(\sin(x))$ if $x \neq k\pi,\ k \in \mathbb{Z}$.

□

Exercise 5.4. Consider the function

$$f(x) = \begin{cases} |x| & x \in [-\pi, \pi] \\ 2\pi \text{ periodic}. \end{cases}$$

1) Draw the graph of the function $y = f(x)$ in the interval $[-2\pi, 2\pi]$.
2) Determine the Fourier series associated to f.
3) By using the convergence of the Fourier series at $x = 0$, compute

$$\sum_{m=0}^{\infty} \frac{1}{(2m+1)^2}.$$

Solution - 1) The graph of f is the following

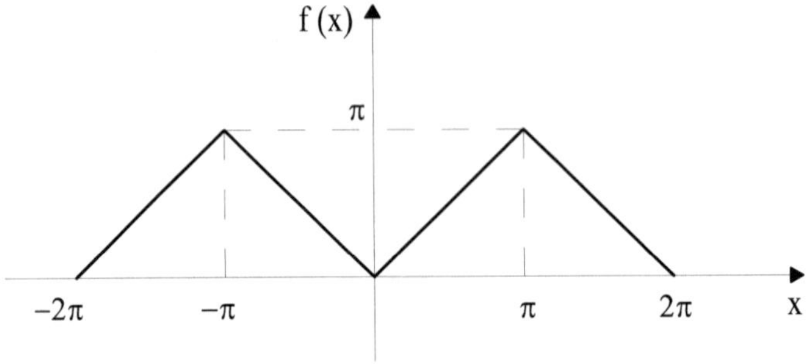

Figure 5.1 - Graph of the function f (Exercise 5.4).

2) The function f is even, thus its Fourier series consists of cosines only. We have

$$a_0 = \frac{1}{\pi}\int_{-\pi}^{\pi} |x|\,dx = \frac{2}{\pi}\int_0^{\pi} x\,dx = \pi\ ,$$

$$a_n = \frac{2}{\pi}\int_0^{\pi} x\cos(nx)\,dx = \frac{2}{\pi}\left[x\,\frac{\sin(nx)}{n}\right]_0^{\pi} - \frac{2}{\pi}\int_0^{\pi}\frac{\sin(nx)}{n}\,dx = \frac{2[(-1)^n-1]}{\pi\,n^2}\ .$$

Therefore $a_n = 0$ if n is even and different from 0. Hence, the Fourier series associated to f is

$$f(x) = \frac{\pi}{2} - \frac{4}{\pi}\sum_{m=0}^{\infty}\frac{\cos[(2m+1)x]}{(2m+1)^2}\ .$$

3) Since $f \in C^0$ and $f \in BV$, we have point-wise convergence at any x. The convergence at $x = 0$ entails

$$0 = f(0) = \frac{\pi}{2} - \frac{4}{\pi}\sum_{m=0}^{\infty}\frac{1}{(2m+1)^2} \qquad \Longrightarrow \qquad \sum_{m=0}^{\infty}\frac{1}{(2m+1)^2} = \frac{\pi^2}{8}\ .$$

□

Exercise 5.5. Consider the odd, 2π periodic function $f(x)$, defined by setting

$$f(x) = \begin{cases} \frac{\pi-x}{2} & \text{if } 0 < x \le \pi \\ 0 & \text{if } x = 0. \end{cases}$$

1) Draw the graph of f in the interval $[-2\pi, 2\pi]$.
2) Determine the Fourier series associated to f.
3) Study the point-wise and the square mean convergence of the obtained series.
4) Write the Parseval identity related to the obtained series and deduce the value of

$$\sum_{n=1}^{\infty}\frac{1}{n^2}\ .$$

Solution - 1) The graph of f is the following

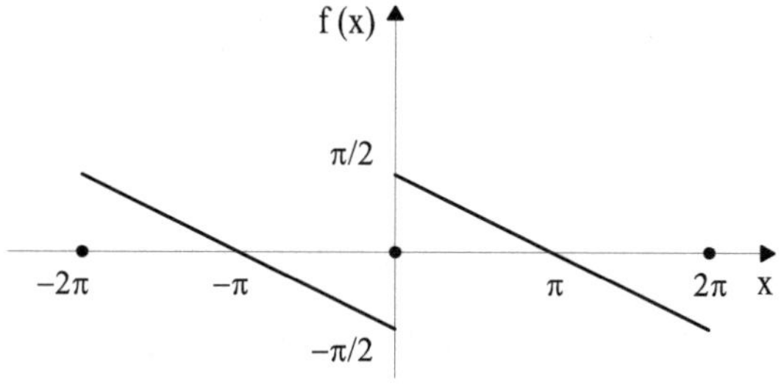

Figure 5.2 - Graph of the function f (Exercise 5.5).

2) f has a Fourier series which consists of sines only: $b_n = \dfrac{2}{\pi}\displaystyle\int_0^\pi \frac{\pi - x}{2}\sin(nx)\,dx = \frac{1}{n}$,
$f(x) = \displaystyle\sum_{n=1}^{\infty} \frac{\sin\, nx}{n}$.
3) Since $f \in L^2(0,2\pi)$ we have square mean convergence of the partial sums of the Fourier series: $S_n \to f$ in $L^2(0,2\pi)$. Moreover, f is monotone and bounded in $(0,2\pi)$, thus $f \in BV(0,2\pi)$ and (by Theorem 5.4) we have point-wise convergence of $S_n(x)$ to $f(x)$ in every point $x \in [0,2\pi]$, also in the point of discontinuity $x = 0$, because $f(0) = 0 = \big(f(0_+) + f(0_-)\big)/2$.

4) We have:
$$2\int_0^\pi f(x)^2\,dx = \pi \sum_{n=1}^{\infty} \frac{1}{n^2} \implies \sum_{n=1}^{\infty} \frac{1}{n^2} = \frac{\pi^2}{6}.$$

□

Exercise 5.6. Consider the function defined by

$$f(x) = \begin{cases} \pi^2 - x^2 & \text{if } x \in [-\pi,\pi) \\ 2\pi - \text{periodic.} \end{cases}$$

1) Draw the graph of f in the interval $[-2\pi, 2\pi]$.
2) Determine the Fourier series associated to f and determine the point-wise limit.
3) Using item 2), compute $\displaystyle\sum_{n=1}^{\infty} \frac{(-1)^{n+1}}{n^2}$.

Solution - 1) The graph of f is the following

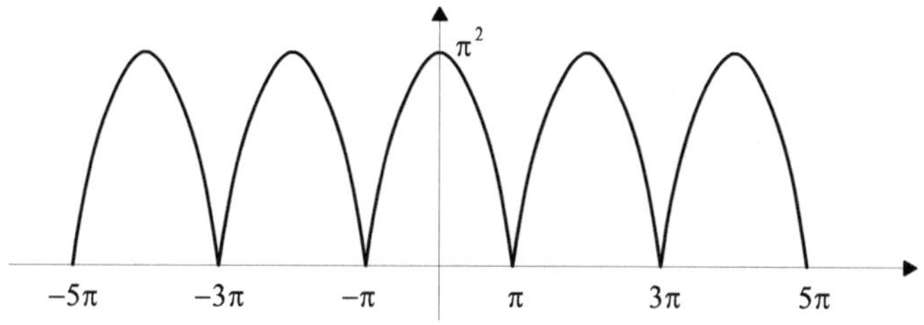

Figure 5.3 - Graph of the function f (Exercise 5.6).

2) The function is even. Thus we look for an expansion of cosines only ($b_n = 0$, $\forall n \in \mathbb{N}$). We get $a_0 = 4\pi^3/3$ and

$$a_n = \frac{2}{\pi}\int_0^{\pi}(\pi^2 - x^2)\cos(nx)dx = \frac{2}{\pi}\int \pi^2\cos(nx)dx - \frac{2}{\pi}\int x^2\cos(nx)dx =$$

$$= 2\pi\left[\frac{\sin(nx)}{n}\right]_0^{\pi} - \frac{2}{\pi}\left[\frac{2x\cos(nx)}{n^2} + \frac{(n^2x^2-2)\sin(nx)}{n^3}\right]_0^{\pi} = (-1)^{n+1}\frac{4}{n^2},$$

hence the Fourier series: $$f(x) = \frac{2}{3}\pi^2 + 4\sum_{n=1}^{+\infty}\frac{(-1)^{n+1}}{n^2}\cos(nx).$$

The function f belongs to $BV(0,2\pi)\cap C^0(\mathbb{R})$, thus its Fourier series converges point-wise to f at every point of $\mathbb{R}$. Analyzing the coefficients ($\sum|a_n| < +\infty$) we have that the convergence is uniform too, by Weierstrass criterion.

3) By the point-wise convergence at $x = 0$ we have

$$\pi^2 = f(0) = \frac{2}{3}\pi^2 + 4\sum_{n=1}^{+\infty}\frac{(-1)^{n+1}}{n^2}\cos(0)$$

$$\sum_{n=1}^{\infty}\frac{(-1)^{n+1}}{n^2} = \frac{1}{12}\pi^2.$$

□

Exercise 5.7. Expand the function $g(x) = \pi x - x^2$, $x \in (0,\pi)$, in a Fourier series of sines only.

Solution - Let us consider the odd extension of g to $(-\pi,\pi)$ and apply formulas (5.3). By exploiting the cancelations due to the symmetry we compute the coefficients a_n (that are all zero) and b_n, so we get

$$\alpha_m = \frac{2}{\pi}\int_0^{\pi} x\,\sin(mx)\,dx = \frac{2}{\pi}\left[\frac{\sin(mx)}{m^2} - \frac{x\,\cos(mx)}{m}\right]_0^{\pi} = \frac{2(-1)^{m+1}}{m}$$

$$\beta_m = \frac{2}{\pi}\int_0^{\pi} x^2\,\sin(mx)\,dx = \frac{2}{\pi}\left[-\frac{x^2\,\cos(mx)}{m} + \frac{2x\sin(mx)}{m^2} + \frac{2\cos(mx)}{m^3}\right]_0^{\pi} =$$

$$= \frac{2\pi(-1)^{m+1}}{m} + \left[\frac{2(-1)^m}{m^3} - \frac{2}{m^3}\right]\frac{2}{\pi}.$$

Thus

$$b_n = \frac{2}{\pi}\int_0^{\pi}(\pi x - x^2)\,\sin(mx)\,dx = \pi\,\alpha_m - \beta_m = \frac{4}{\pi\,m^3}[1 + (-1)^{m+1}].$$

□

Exercise 5.8. Show that the equation

$$u''(x) + u(x) = \sin(x) \quad x \in \mathbb{R}$$

has no 2π periodic solution of class $C^3(\mathbb{R})$.

Solution - A $C^3(\mathbb{R})$ solution should be expandable, with suitable coefficients a_n, b_n, in a Fourier series uniformly converging to u:

$$u(x) = \frac{1}{2}a_0 + \sum_{n=1}^{+\infty}(a_n \cos(nx) + b_n \sin(nx)).$$

By substitution of the series in the differential equation, after a formal derivation, we obtain:

$$\frac{a_0}{2} + \sum_{n=1}^{+\infty}(1-n^2)\,(a_n \cos(nx) + b_n \sin(nx)) = \sin(x).$$

By uniqueness of the Fourier series we obtain an infinite (but uncoupled) system of algebraic equations that is solved by the coefficients:

$$a_0 = 0, \qquad (1-n^2)\,a_n = 0 \ \forall n > 1,$$

$$0 \cdot b_1 = 1, \quad (1-n^2)\,b_n = 0 \ \forall n > 1.$$

The equation in the unknown b_1 has no solution, thus there exists no 2π periodic function that solves the given differential equation.

Let us check that the use of the series of derivatives (that is the use of formulas (5.17)) is not formal only.
If $u \in C^3(\mathbb{R})$ it follows that $u''' \in C^0 \subset L^2_{loc}$, thus $\sum(|n^3 a_n|^2 + |n^3 b_n|^2) < +\infty$.
It is known that the modulus of a product can be estimated from above by the sum of the square of the moduli ($|\alpha\beta| \leq |\alpha|^2 + |\beta|^2$). Applying such an inequality to the product $n^2 a_n = (n^3 a_n)\,\frac{1}{n}$ we get:

$$\left| n^2\, a_n \right| \leq \left| n^3\, a_n \right|^2 + \frac{1}{n^2}.$$

Similarly, for b_n we obtain $\left| n^2\, b_n \right| \leq \left|n^3 b_n\right|^2 + 1/n^2$.
Thus

$$\sum_{n=1}^{+\infty}\left(|n^2\, a_n| + |n^2\, b_n|\right) < +\infty,$$

since the general term of this series, whose terms are all positive, can be bounded from above by term of convergent series. Applying Theorem 5.2 with $m = 2$, we can conclude that the series of derivatives provides a Fourier series (uniformly convergent) of the derivative of the first one. Similar conclusions hold true for the second derivative too.

Remark - The differential equation $u''(x) + u(x) = \sin x,\ x \in \mathbb{R}$ has infinitely many solutions, all in $C^\infty(\mathbb{R})$, but none of them is 2π periodic. We might wonder whether there are less regular solutions: the answer is negative too. There are no 2π periodic distributions that solve the given equation (see Exercise 5.15). □

Exercise 5.9. For all $\lambda \in \mathbb{R}$ find, if they exist, the 2π periodic solutions of the equation:

$$u''(x) + \lambda u(x) = |\sin(x)|.$$

Find also the π periodic solutions.

Solution - First, let us observe that $|\sin(x)|$ is a even, π periodic function, thus it is also 2π periodic. It is convenient to use the real form of the expansion since the datum is in real form. Through some calculation we can deduce the Fourier expansion (of cosine only!) convergent in $\mathcal{S}'(\mathbb{R})$:

$$\frac{2}{\pi}\left(1 - 2\sum_{m=1}^{+\infty}\frac{1}{4m^2-1}\cos(2mx)\right).$$

Notice that, since $\sum_{m=1}^{+\infty}|\frac{1}{4m^2-1}\cos(2mx)| \leq \sum_{m=1}^{+\infty}\frac{1}{4m^2-1} < +\infty$, the Weierstrass criterion ensures the uniform (and thus point-wise) convergence of the Fourier series to $|\sin(x)|$.
Suppose that u is a 2π periodic solution (the computation in the π periodic case is similar)

of the equation, then it admits a Fourier series expansion:

$$u(x)=\frac{a_0}{2}+\sum_{n=1}^{+\infty}\left(a_n\cos(nx)+b_n\sin(nx)\right).$$

After formal term-by term differentiation and substitution in the equation, we obtain the identity

$$\lambda\,\frac{a_0}{2}+\sum_{n=1}^{+\infty}(\lambda-n^2)\left(a_n\cos(nx)+b_n\sin(nx)\right)=\frac{2}{\pi}\left(1-2\sum_{m=1}^{+\infty}\frac{1}{4m^2-1}\cos(2mx)\right).$$

By the uniqueness of the coefficients, we obtain the following system of infinite uncoupled equations:

$$\begin{cases}\lambda\,a_0=\frac{4}{\pi} & \\ (\lambda-n^2)\,a_n=\begin{cases}0 & \text{if } n=2m+1\\ \frac{-4}{\pi(4m^2-1)} & \text{if } n=2m\end{cases} & \text{if } n\geq 1\\ (\lambda-n^2)\,b_n=0 & \text{if } n\geq 1.\end{cases}$$

By summarizing we have:

- if $\lambda=0$ there are no solutions (the first equation is not solvable);
- if $\lambda\neq n^2$ for every $n\in\mathbb{N}$, then there exists a unique solution:

$$u(x)=\frac{2}{\pi\lambda}-\frac{4}{\pi}\sum_{m=1}^{+\infty}\frac{1}{(4m^2-1)(\lambda-4m^2)}\cos(2mx);$$

  observe that the series that define u and u'' (via term-by-term differentation) converge uniformly by Weierstrass criterion, this justifies the formal operations performed above;
- if $\lambda=\overline{n}^2$, for $\overline{n}\in\mathbb{N}$ fixed, then
  - the problem has no solution if $\overline{n}$ is even (the equation corresponding to $a_{\overline{n}}$ has no solution),
  - there are infinitely many solutions if $\overline{n}$ is odd ($a_{\overline{n}}$ and $b_{\overline{n}}$ turn out to be arbitrary).

Both the datum $|\sin(x)|$ and the obtained expansion, expressed in terms of $\cos(2mx)$, are π periodic. Therefore in the case of π periodic solutions, we have the same situation and the same answer as in the case of 2π periodic solutions. □

Exercise 5.10. In the interval $[-2\pi,4\pi]$ draw the graph of the function $f(x)$, periodic of period 2π, defined by

$$f(x)=\begin{cases}2 & \text{if } 0<x\leq\frac{\pi}{2}\\ 0 & \text{if } \frac{\pi}{2}<x\leq\pi\\ 1 & \text{if } \pi<x\leq 2\pi\,.\end{cases}$$

Write the Fourier expansion associated to f and identify the point-wise limit.

Solution - The graph of f is

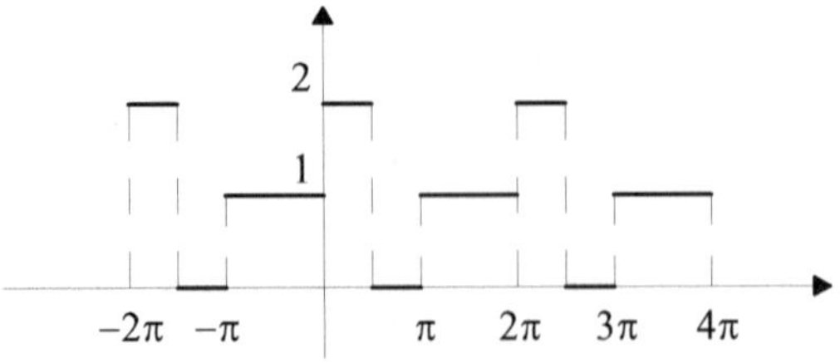

Figure 5.4 - Graph of the function f (Exercise 5.10).

The conclusion of the exercise is left to the reader. □

5.2.2 *Exercises on the Fourier series of periodic distributions*

Exercise 5.11. By exploiting the solution of Exercise 5.3, write the Fourier series of

$$v(x) \;=\; \sum_{k=-\infty}^{+\infty} (-1)^k\,\delta(x-k) \qquad x\in\mathbb{R}\,.$$

Solution - The Fourier series of $f(x)=\operatorname{sign}(\sin x)$ is

$$(*)\qquad f(x) \;=\; \frac{4}{\pi}\sum_{m=0}^{\infty}\frac{1}{2m+1}\,\sin\left((2m+1)x\right)$$

The series $(*)$ converges in $L^2(-\pi,\pi)$ to f because $f\in L^2(-\pi,\pi)$. This implies the convergence in $\mathcal{S}'(\mathbb{R})$. Observe that $v=f'/2$ in $\mathcal{D}'(\mathbb{R})$. Due to the convergence in $\mathcal{S}'(\mathbb{R})$ of $(*)$, the derivative can be obtained through term by term differentiation of the series:

$$(**)\qquad v(x) \;=\; \frac{4}{\pi}\sum_{m=0}^{\infty}\cos\left((2m+1)x\right) \;=\; \frac{2}{\pi}\sum_{k=-\infty}^{+\infty}\exp\Big(i(2k+1)x\Big)\,.$$

The series $(**)$ converges in $\mathcal{S}'(\mathbb{R})$ to v.

□

Exercise 5.12. Write the complex exponential form of the Fourier series of the Dirac comb of length T (defined by (5.29)) and its Fourier transform, thus proving the formulas (5.30) (5.31).
Write the real form of the Fourier series appearing in the two formulas (5.30) (5.31).

Solution - It suffices to calculate the Fourier series of comb_T

$$(*)\qquad \operatorname{comb}_T(t) \;=\; \sum_{k=-\infty}^{+\infty}\delta(t-kT) \;=\; \frac{1}{T}\sum_{k=-\infty}^{+\infty} e^{i\,k\,\frac{2\pi}{T}\,t}\,,$$

since the second formula

$$\widehat{\operatorname{comb}_T}(\omega) \;=\; \frac{2\pi}{T}\sum_{k=-\infty}^{+\infty}\delta\Big(\omega-k\frac{2\pi}{T}\Big) \;=\; \sum_{k=-\infty}^{+\infty} e^{i\,k\,T\,\omega}$$

is a straightforward consequence of the first one via Fourier transformation, by using the fact that both the series in (*) converge in $\mathcal{S}'(\mathbb{R})$ and the identities

$$\mathcal{F}\{\delta(t-a)\} \;=\; e^{i\,a\,\omega}\,, \qquad \mathcal{F}\{e^{i\,a\,t}\} \;=\; 2\pi\,\delta\Big(\omega-a\Big)\,, \qquad \forall a\in\mathbb{R}\,.$$

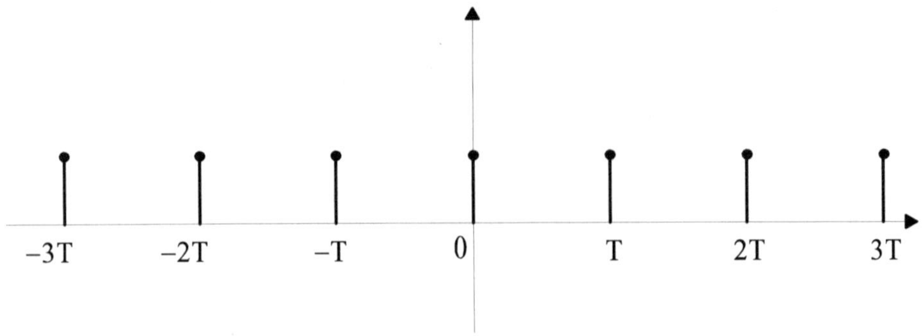

Figure 5.5 - Graphic representation of the Dirac comb_T .

We are left to prove (∗), that is to compute the Fourier coefficients (in the exponential complex form) of the distribution $\mathrm{comb}_T \in \mathcal{S}'_T$. We cannot use (5.12) because u is not in L^1_{loc}. Nevertheless, we can take advantage of Remark 5.11 and (5.26).

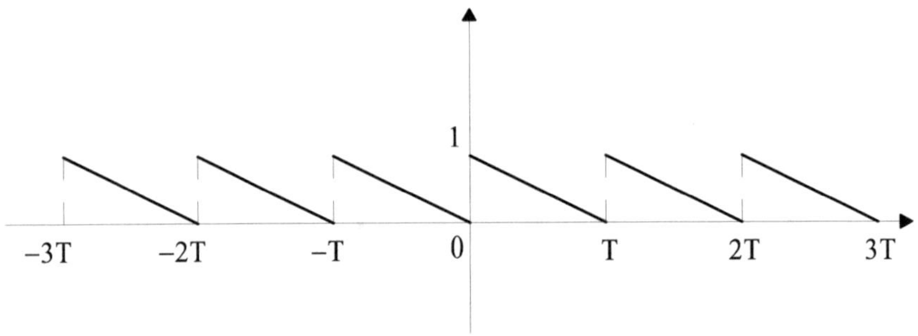

Figure 5.6 - Graph of the function g in Exercise 5.12.

The T periodic function $g : \mathbb{R} \to \mathbb{R}$, defined by $g(t) = 1 - t/T$ if $t \in [0, T)$, belongs to L^2_{loc}. Thus it is possible to use (5.12) to compute the Fourier coefficients of g :

$$c_0 = \frac{1}{T}\int_0^T (1 - t/T)\,dt = \frac{1}{2}\ ;$$

$$\text{if } k \neq 0 \quad \text{then } c_k = \frac{1}{T}\int_0^T (1 - t/T)\, e^{-i\,k\,\frac{2\pi}{T}\,t}\,dt \ \overset{\boxed{\text{int. by parts}}}{=}\ \frac{1}{k\,2\pi\,i}\ ;$$

$$g(t) = \sum_{k=-\infty}^{+\infty} c_k\, e^{i\,k\,\frac{2\pi}{T}\,t} = \frac{1}{2} + \frac{1}{2\pi\,i}\sum_{k\neq 0}\frac{e^{ik\frac{2\pi}{T}\,t}}{k} = \frac{1}{2} + \sum_{n=1}^{+\infty}\frac{1}{n\,\pi}\sin\left(n\frac{2\pi}{T}t\right).$$

By differentiating g in the sense of distributions, and using the result proved in Exercise 3.11:

$$g'(t) = -\frac{1}{T} + \sum_{k=-\infty}^{+\infty}\delta(t - kT) = \mathrm{comb}_T(t) - \frac{1}{T}\,.$$

Computing g' in a different way, namely by differentiating term by term the Fourier series (Remark 5.11 and formula (5.26)):

$$g'(t) = \frac{1}{T}\sum_{k\neq 0} e^{ik\frac{2\pi}{T}t} = \frac{2}{T}\sum_{n=1}^{+\infty}\cos\left(n\frac{2\pi}{T}t\right).$$

By equating we get the following identities, where all the series are convergent in $\mathcal{S}'(\mathbb{R})$:

$$\mathrm{comb}_T(t) = \sum_{k=-\infty}^{+\infty}\delta(t-kT) = \frac{1}{T}\sum_{k=-\infty}^{+\infty} e^{ik\frac{2\pi}{T}t} = \frac{1}{T}+\frac{2}{T}\sum_{n=1}^{+\infty}\cos\left(n\frac{2\pi}{T}t\right),$$

from which, through the $\mathcal{F}$ transform:

$$\widehat{\mathrm{comb}_T}(\omega) = \frac{2\pi}{T}\sum_{k=-\infty}^{+\infty}\delta\Big(\omega-k\frac{2\pi}{T}\Big) = \sum_{k=-\infty}^{+\infty} e^{ikT\omega} = 1+2\sum_{n=1}^{+\infty}\cos(nT\omega).$$

□

Exercise 5.13. 1) Prove the following formula.

Poisson summation formula

$$\sum_{k=-\infty}^{+\infty}\psi(k) = \sum_{k=-\infty}^{+\infty}\widehat{\psi}(2k\pi) \qquad \forall\psi\in\mathcal{S}(\mathbb{R}). \tag{5.42}$$

2) Show that both numeric series in the Poisson summation formula are absolutely convergent.
3) Rescale the formula in such a way to obtain the identity

$$h\sum_{k=-\infty}^{+\infty}\psi(kh) = \sum_{k=-\infty}^{+\infty}\widehat{\psi}(2k\pi/h) \qquad \forall\psi\in\mathcal{S}(\mathbb{R}),\ h>0. \tag{5.43}$$

4) Use item 3) to obtain information on: the numerical approximation of the integrals over $\mathbb{R}$ of a Schwartz function and of its Fourier transform; approximation of the numerical values of $\psi(0)$ and $\widehat{\psi}(0)$.

Solution - 1) We evaluate the duality pairing $\langle \mathrm{comb}_1, \psi\rangle$ where ψ is a Schwartz function, starting from the Fourier series of the Dirac comb (formula (5.30)) that allows to express comb_1 through two different series converging in $\mathcal{S}'(\mathbb{R})$. Formula (5.42) follows, by recalling that $\widehat{\psi}(2k\pi) = \int_{\mathbb{R}}\psi(t)e^{-i\,2k\pi\,t}\,dt$, by definition of $\mathcal{F}$ transform.

2) The numerical series in the Poisson summation formula are absolutely convergent due to the fast decay at infinity of ψ and $\widehat{\psi}$ (both in $\mathcal{S}(\mathbb{R})$): at infinity we have $\psi(t) = o(|t|^{-n})$ and $\widehat{\psi}(\omega) = o(|\omega|^{-n})$ for any $n\in\mathbb{N}$.

3) We can proceed as in item 1) using comb_h instead of comb_1.

4) The identity (5.43) has two remarkable consequences. On the left hand side of (5.43) we recognize a Riemann sum relative to a uniform partition of $\mathbb{R}$ in intervals of length h that approximates the integral of $\int_{\mathbb{R}}\psi\,dt = \widehat{\psi}(0)$; this Riemann sum is equal to the sum on the right hand side that converges very rapidly, as $h\to 0$, because $2\pi/h\to+\infty$ and $\psi\in\mathcal{S}$. Thus the difference between the Riemann sum $h\sum_{k=-\infty}^{+\infty}\psi(kh)$ of a Schwartz function ψ and its integral $\int_{\mathbb{R}}\psi = \widehat{\psi}(0)$ is $o(h^m)$ for any $m\in\mathbb{N}$ as $h\to 0_+$.

This surprising numerical estimate is due to a balance of positive and negative errors when we substitute $\int_{kh}^{(k+1)h} \psi$ with $h\psi(kh)$.
On the right hand side of (5.43) there is a Riemann sum relative to a uniform partition of $\mathbb{R}$ in intervals of length $2\pi/h$ that approximates the integral of $\widehat{\psi}$ over $\mathbb{R}$.

By letting $h \to +\infty$ in (5.43), we get the well known inversion formula (evaluated at the origin)

$$\psi(0) = \frac{1}{2\pi} \int_{-\infty}^{+\infty} \widehat{\psi}(\omega)\, d\omega \,,$$

and, before passing to the limit we can read:

$$\psi(0) + \text{infinitesimals} = \frac{1}{2\pi} \left(\text{Riemann sum related to the integral over } \mathbb{R} \text{ of } \widehat{\psi}\right),$$

where, with the same arguments used for $h \to 0$, the infinitesimals are $o(h^{-m})\, \forall m \in \mathbb{N}$, and the Riemann sum differs from the integral of ψ by an infinitesimal of the same order.
□

Exercise 5.14. Prove the following formulation of the Poisson summation formula (5.42), (5.43), which is more general and with weaker assumptions.

Assume that $T > 0$ and a complex valued function $u \in C^2(\mathbb{R})$ are given, such that $u, u', u'' \in L^1(\mathbb{R})$ and $|u(t)| \leq \dfrac{1}{1+t^2}$. Define the fundamental mode $\omega_0 = 2\pi/T$.
Then the following identity holds true:

$$\sum_{k=-\infty}^{+\infty} u(t-kT) = \frac{1}{T} \sum_{k=-\infty}^{+\infty} \widehat{u}(k\omega_0)\, e^{ik\omega_0 t} \tag{5.44}$$

The function on the left hand side of (5.44) is called *periodic repetition* of u with period $T > 0$ and it is denoted by u_T:

$$u_T(t) := \sum_{k=-\infty}^{+\infty} u(t-kT)\,. \tag{5.45}$$

Hint: 1) Prove that the series in (5.45) converges in L^1_{loc}, uniformly and in $\mathcal{S}'(\mathbb{R})$ to a well defined $u_T \in L^1_{loc} \cap C^0(\mathbb{R})$. 2) Show that u_T is a T-periodic distribution which belongs to $L^1_{loc}(\mathbb{R})$ and also to $L^2(0,T)$.
3) Determine the Fourier expansion of u_T in the complex form and deduce (5.44).

Solution - 1) We have:

$$\int_0^T |u_T(t)|\, dt = \int_0^T \left|\sum_k u(t-kT)\right| dt \leq \int_0^T \sum_k |u(t-kT)|\, dt =$$
$$= \sum_k \int_0^T |u(t-kT)|\, dt = \sum_k \int_{kT}^{(k+1)T} |u(t)|\, dt = \|u\|_{L^1(\mathbb{R})} < +\infty\,.$$

Thus the series that defines u_T converges in $L^1(0,T)$ and, consequently, in $\mathcal{S}'(\mathbb{R})$. By completeness of L^1 the sum u_T belongs to $L^1(0,T)$.

Observe that switching the summation and the integral is correct since we are dealing with positive quantities that are uniformly bounded from above.
In view of the estimate $|u(t)| \leq \dfrac{1}{1+t^2}$, the series converges also uniformly.

2) We have: $u_T(t+T) = \sum_k u(t-(k-1)T) = \sum_k u(t-kT) = u_T(t)\, \forall t \in \mathbb{R}$, thus u_T is T-periodic.
By our assumptions ($u, u'' \in L^1(\mathbb{R})$) and the Riemann-Lebesgue Lemma, we know that $(1+\omega^2)\widehat{u}(\omega) \in L^\infty(\mathbb{R})$. Therefore

$$(*) \qquad |\widehat{u}(\omega)| \leq C(1+\omega^2)^{-1}$$

hence we infer $\widehat{u} \in L^2(\mathbb{R})$ and $u \in L^2(\mathbb{R})$, by Plancherel identity.

3) We can compute the Fourier coefficients of the function $u_T \in L^2(0,T)$ with formulas (5.12)

$$c_k = \frac{1}{T}\int_{-T/2}^{T/2} \sum_{m=-\infty}^{+\infty} u(t-mT)\, e^{-i2\pi kt/T}\, dt = \frac{1}{T}\sum_{m=-\infty}^{+\infty} \int_{-T/2-mT}^{T/2+mT} u(t)\, e^{-i2\pi kt/T}\, dt$$

where we have performed the change of variable $t - mT \mapsto t$ and we have exploited the periodicity of the complex exponential function; switching the summation and the integration is allowed due to the L^1 convergence in bounded intervals.
Joining all integration intervals, we can rewrite the last integral emphasizing the Fourier transform of u, hence, by recalling that we have set $\omega_0 = 2\pi/T$:

$$c_k \;=\; \frac{1}{T}\int_{\mathbb{R}} u(x)\, e^{-i2\pi kx/T}\, dx \;=\; \frac{1}{T}\,\widehat{u}\,(k\omega_0)\,.$$

Inserting these coefficients in the Fourier series of u_T we get (5.44). Notice that estimate (*), valid under our assumptions, ensures that $|c_k|$ is the general term of a numeric converging series, therefore the right hand side of (5.44) (that is the Fourier series if u_T) converges uniformly.
Moreover u is an absolutely continuous function (Definition 2.6). Thus u is a well defined continuous function, in particular computing the point-wise values that appears in (5.44) is correct.

Remark - An alternative way, quicker but less trivial, to deduce the Fourier series of the periodic repetition u_T is the following that makes use of the convolution.
Observe that the relation $u_T \;=\; \text{comb}_T * u$ holds true. Its Fourier transform is

$$\widehat{u_T}(\omega) \;=\; \frac{2\pi}{T}\,\text{comb}_{2\pi/T}(\omega)\,\widehat{u}(\omega) \;=\; \frac{2\pi}{T}\sum_{k\in\mathbb{Z}} \widehat{u}(2k\pi/T)\,\delta(\omega - 2k\pi/T)\,,$$

by antitransforming we get $u_T(t) \;=\; \sum_{k\in\mathbb{Z}} \widehat{u}(2k\pi/T)\,\exp(2k\pi i t/T)\,.$ □

Exercise 5.15. Show that the equation

$$u''(x) + u(x) = \sin(x) \quad x \in \mathbb{R}$$

has no 2π periodic solution (compare the solution with the one of exercise 5.8).

Solution - A possible solution in $\mathcal{S}'_{2\pi}$ would be expandable, for suitable coefficients c_k, in a Fourier series convergent in $\mathcal{S}'(\mathbb{R})$ to u:

$$u(x) \;=\; \sum_{k=-\infty}^{+\infty} c_k\,\exp(i\,kx)\,.$$

By performing term by term differentiation (formulas (5.26)) and substituting in the differential equation, we get:

$$\sum_{k=-\infty}^{+\infty} (1-k^2)\, c_k \, \exp(i\, kx) \;=\; \frac{1}{2i}\left(\exp(ix) - \exp(-ix)\right).$$

By uniqueness of the Fourier series we obtain an uncoupled infinite system of algebraic equations that is solved by the coefficients:

$$0 \cdot c_1 \;=\; 1/(2i)\,, \quad 0 \cdot c_{-1} \;=\; -1/(2i)\,, \quad (1-k^2)\, c_k \;=\; 0 \;\forall k \neq 1\,.$$

The first two algebraic equations have no solution, therefore there exits no 2π periodic distribution that solves the given differential equation. □

Exercise 5.16. Rephrase and solve Exercise 5.9 in the framework of 2π periodic and π periodic distributions.

Solution - Use Remark 5.11, to achieve the conclusion: the answers are equal to those of Exercise 5.9. In spite of the fact that the admissible setting is much larger the possible solutions are $C^2(\mathbb{R})$ functions. □

Exercise 5.17. 1) Specify under which conditions on the datum w, the following problem in the unknown $u \in S'(\mathbb{R})$ admits 2π periodic solutions and solve it when such conditions are met:

$$u(t) - u''(t) = w(t), \quad t \in \mathbb{R}.$$

2) Solve the particular case $w(t) = \sum_{k=-\infty}^{+\infty} \delta(t - k\pi)$.

3) Does the solution of item 2) belong to some periodic Sobolev space?

Solution - 1) A necessary and sufficient condition to guarantee the existence of 2π periodic solutions is that the datum w is a 2π periodic distribution.
Indeed, if u is periodic, then u'' is also periodic and, accordingly also $w = u - u''$.
Conversely the adjoint homogeneous problem $\{$find 2π periodic $z\,:\, z - z'' = 0$ in $\mathbb{R}\}$ has only the trivial solution $z = 0$. Hence if w is 2π periodic, then u can be obtained applying the Fourier series method.
Since w is 2π periodic, it can be expanded as a series:

$$w(t) = \sum_{k=-\infty}^{+\infty} \gamma_k e^{ikt} \text{ in } \; S'(\mathbb{R}),$$

for suitable tempered coefficients that are uniquely determined and can be computed if w is given. Proceeding formally, let us write the series expansion of u

$$u(t) = \sum_{k=-\infty}^{+\infty} c_k e^{ikt} \text{ in } \; S'(\mathbb{R}),$$

where the c_k are not known.
The convergence of the Fourier series in $S'(\mathbb{R})$ allows to differentiate term by term the series. By substitution, we get

$$\sum_{k=-\infty}^{+\infty} (1+k^2) c_k e^{ikt} \;=\; \sum_{k=-\infty}^{+\infty} \gamma_k e^{ikt} \text{ in } \; S'(\mathbb{R}).$$

The uniqueness of the Fourier series implies the equivalence of the last identity with the following system of infinite uncoupled equations:

$$(1+k^2)c_k = \gamma_k, \quad k \in \mathbb{Z}$$

that admits the unique solution $c_k = \dfrac{\gamma_k}{(1+k^2)}$ for $k \in \mathbb{Z}$.
Summarizing, for every 2π periodic w (whose Fourier coefficient in complex form are the γ_ks), the problem has a unique solution

$$u(t) = \sum_{k=-\infty}^{+\infty} \frac{\gamma_k}{1+k^2} e^{ikt}.$$

2) From the Table 10.8.5 we infer $w(t) = \displaystyle\sum_{k=-\infty}^{+\infty} \delta(t-k\pi) = \frac{1}{\pi}\sum_{h=-\infty}^{+\infty} e^{i2ht}$, that is

$$\gamma_k = \frac{1}{\pi} \quad \text{if } k = 2h, \quad h \in \mathbb{Z} \quad \text{and } \gamma_k = 0 \text{ if } k \text{ is not a multiple of 2.}$$

It follows that

$$u(t) = \frac{1}{\pi}\sum_{h=-\infty}^{+\infty} \frac{1}{1+4h^2} e^{i2ht} = \frac{1}{\pi} + \frac{2}{\pi}\sum_{h=1}^{+\infty} \frac{\cos(2ht)}{1+4h^2}.$$

Observe that $c_h = \dfrac{1}{\pi}\dfrac{1}{(1+4h^2)} \sim \dfrac{1}{4\pi h^2}$, therefore $hc_h \sim \dfrac{1}{4\pi h}$ that is $u \in H^1_T(\mathbb{R}) \subset C^0(\mathbb{R})$ (Sobolev space of periodic functions) with period $T=1$.

3) $\displaystyle\sum_{h=-\infty}^{+\infty} |h^{3/2-\varepsilon} c_h|^2 < +\infty$, thus $u \in H_1^{3/2-\varepsilon}(\mathbb{R})$, for every $\varepsilon > 0$. □

Exercise 5.18. Establish which one among the following distributions in $\mathbb{R}$ (dependent on the real variable ξ) is a Fourier transform of a periodic signals and, when the answer is affirmative, determine the corresponding signal.

1) $\sin(\xi)$,
2) $\sum_{k\in\mathbb{Z}} e^k\, \delta(\xi - 2k\pi)$,
3) $\pi\left(\delta(\xi-1)+\delta(\xi+1)\right)$,
4) $3\pi\left(\delta(\xi-3)+\delta(\xi-\sqrt{3})\right)$,
5) $\delta(\xi-1)+\delta(\xi-\pi)$,
6) $\delta(\xi)+\delta(\xi-\pi)$,
7) $\pi\, i\left(\delta(\xi+31)-\delta(\xi-31)+\delta(\xi+30)-\delta(\xi-30)\right)$.

Solution - We refer to the necessary and sufficient conditions stated by Theorem 5.8.

1) $\sin(\xi)$ is not the Fourier transform of any periodic signal, because its support coincides with $\mathbb{R}$ (it is a continuous spectrum).

2) $\sum_{k\in\mathbb{Z}} e^k\, \delta(\xi - 2k\pi)$ is not the Fourier transform of a periodic signal, though its spectrum is contained in the 2π multiple of integers, because the coefficients of the delta train grow exponentially with respect to the index (they are not tempered).

3) $\pi\left(\delta(\xi-1)+\delta(\xi+1)\right)$ is the Fourier transform of a periodic signal, because it is has a discrete spectrum: $\operatorname{spt}\widehat{u} \subset \{1,-1\}$, therefore it is a subset of integer multiples of 1; hence u is a 2π periodic function.
By antitransforming: $\pi\left(\delta(\xi-1)+\delta(\xi+1)\right) = \mathcal{F}\{\cos t\}$.

4) $3\pi\big(\,\delta(\xi-3)+\delta(\xi-\sqrt{3})\,\big)$ is not the Fourier transform of any periodic signal, because its support $\{3,\sqrt{3}\}$ is not contained in the set $\{2k\pi/T,\ k\in\mathbb{Z}\}$ for any choice of $T>0$, due to the irrational ratio between 3 and $\sqrt{3}$.
It is the Fourier transform of the sum of two periodic signals with periods that are irrational between them: $3e^{3\,i\,t}$ and $3e^{\sqrt{3}\,i\,t}$: their sum is not periodic of any real period: *chaotic sum*.

5) $\delta(\xi-1)+\delta(\xi-\pi)$ is not the Fourier transform of a periodic signal, for the same reasons of item 4): here $\operatorname{spt}\widehat{u}\ =\ \{1,\pi\}$.

6) $\delta(\xi)+\delta(\xi-\pi)$ is the Fourier transform of a periodic signal because it is a discrete spectrum: $\operatorname{spt}\widehat{u}=\{0,\pi\}$, thus $\operatorname{spt}\widehat{u}$ is contained in the set of integer multiples of π; hence u is 2 periodic.
By antitransforming: $\delta(\xi)+\delta(\xi-\pi)\ =\ \mathcal{F}\left\{\dfrac{1+e^{i\pi x}}{2\pi}\right\}$.

7) It is the Fourier transform of a periodic signal (it verifies the necessary and sufficient conditions): the minimum period of such a signal is $2\pi/\big(GCD(30,31)\big)\ =2\pi$, where we have denoted by GCD the largest common divisor. By antitransforming:

$$\pi\, i\,\big(\,\delta(\xi+31)-\delta(\xi-31)+\delta(\xi+30)-\delta(\xi-30)\,\big)\ =\mathcal{F}\{\sin(31t)+\sin(30t)\}\,.$$

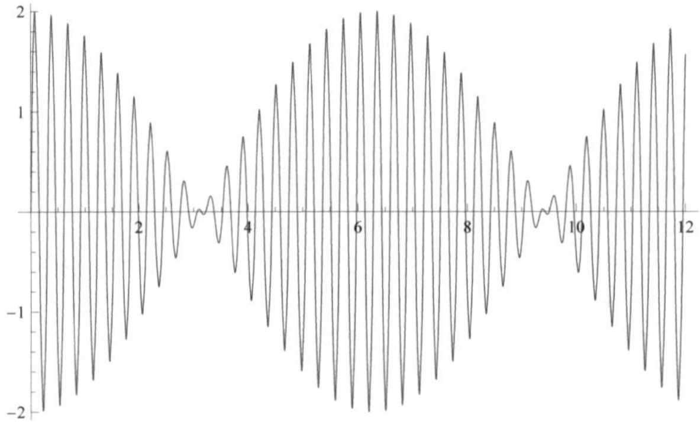

Figure 5.7 - Beats (Exercise 5.18).

This signal has period $T\ =\ 1$ and we observe that it is the sum of two signals with much shorter periods, equal to $1/30$ and $1/31$. Moreover it shows the phenomena of *beats*, corresponding to signals whose frequencies that are large and very close but different: in such cases there is no interference, neither constructive nor destructive, but the resulting signal has period much larger than those of the addenda and it has a frequency much smaller than those of the addenda. Here they have frequencies 60π and 62π respectively.
In the case considered here the frequency is $\nu=2\pi/T=1$ and we may further specify the analysis, using the trigonometric prostaphaeresis formulas (Table 10.1.5): we get

$$\sin(31t)+\sin(30t)\ =\ 2\,\sin\left(\frac{61}{2}\,t\right)\,\cos\left(\frac{t}{2}\right)\,,$$

so that the signal is the product of two functions, one (primary) has period 4π and frequency $1/2$, the other (that modulates the primary) has period $4\pi/61$ and much larger frequency: $61/2$. □

Exercise 5.19. Expand in Fourier series the distribution p.v. $\big(\cot(x/2)\big)$.

Solution - We will prove that

$$\frac{1}{2}\ \text{p.v.}\ \big(\cot(x/2)\big)\ =\ \sum_{n=1}^{+\infty}\sin(nx)\ . \tag{5.46}$$

Setting $f(x)\ =\ -\dfrac{1}{2}\ \text{v.p.}\ \big(\cot(x/2)\big)$ and

$$\langle\, f\,,\, v\,\rangle\ =\ -\frac{1}{2}\ \text{v.p.}\int_{\mathbb{R}}\cot(x/2)\,v(x)\,dx\qquad \forall v\in\mathcal{S}(\mathbb{R})\,,$$

we observe that f is an odd 2π periodic distribution of order 1 in the open set $\mathbb{R}$. Indeed f is C^{∞} except at the singularities ($x=2k\pi,\ k\in\mathbb{Z}$) and it is a locally integrable function except in the neighborhoods of the points $2k\pi,\ k\in\mathbb{Z}$; moreover it is odd with respect to such points, where it has a non integrable singularity such that $f(x-k\pi/2)\sim\frac{2}{(x-k\pi/2)}$; therefore we can argue as for p.v. $\frac{1}{x}$ of item 1) of Exercise 3.4.
We define

$$g(x)\ =\ \ln\frac{1}{|\sin(x/2)|}\ =\ -\ \ln|\sin(x/2)|\,.$$

Thus: g is 2π periodic and even, $g\in L^2_{loc}(\mathbb{R})$. We can compute the Fourier coefficients of g in the real form with (5.3). It turns out that $b_n=0,\ n=1,2,3,\dots$. By computing (details are postponed at the end of the solution) we obtain $a_n=1/n,\ n=1,2,3,\dots$ and the Fourier series of g:

$$g(x)\ =\ \frac{a_0}{2}\ +\ \sum_{n=1}^{+\infty}\frac{\cos(nx)}{n}.$$

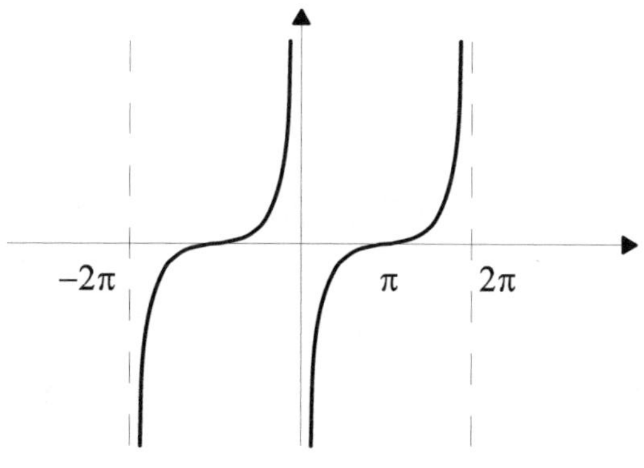

Figure 5.8 - Graph of the function $f(x)=-(1/2)\cot(x/2)$ of Exercise 5.19.

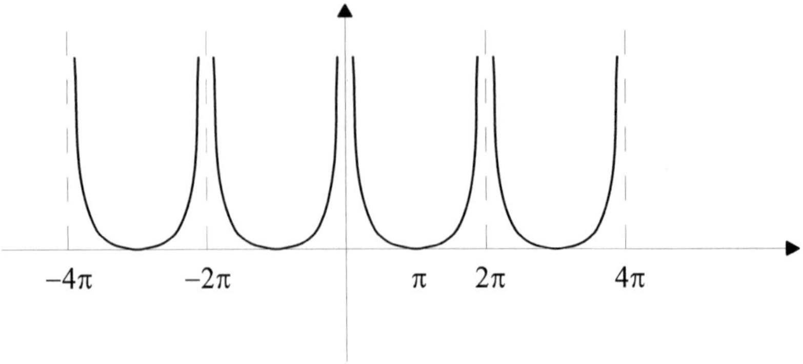

Figure 5.9 - Graph of the function $g(x) = -\ln|\sin(x/2)|$ of Exercise 5.19.

By differentiationg g in the distributional sense (analogously to item 2) of Exercise 3.4) we get $g' = f$. Furthermore the Fourier series of g converges in the distributional sense, therefore it can be differentiated term by term. By equating the two just obtained values of g', we get the expression (5.46) of the Fourier series of f.

Eventually, we compute the coefficients $a_n = a_n(g)$, $n > 0$. It is known (Table 10.1.7) that

$$(*)\qquad \frac{\sin\left(\left(n+\frac{1}{2}\right)x\right)}{2\,\sin(x/2)} = \frac{1}{2} + \cos(x) + \cos(2x) + \ \ldots\ + \cos(nx)\,.$$

We deduce

$$\begin{aligned}
a_n &= -\frac{1}{\pi}\int_{-\pi}^{\pi} \ln|\sin(x/2)|\,\cos(nx)\,dx \overset{\boxed{\text{by parts}}}{=} \\
&= \frac{1}{n\pi}\int_{-\pi}^{\pi}\frac{1}{2}\cot(x/2)\sin(nx)\,dx - \frac{1}{n\pi}\left[\ln|\sin(x/2)|\,\sin(nx)\right]_{-\pi}^{\pi} \overset{\boxed{\cot(x/2)\text{ even}}}{=} \\
&= \frac{1}{n\pi}\int_0^{\pi}\cos(x/2)\sin(nx)\,\frac{1}{\sin(x/2)}\,dx = \\
&= \frac{1}{n\pi}\int_0^{\pi}\frac{\sin\left(\left(n+1/2\right)x\right) - \sin(x/2)\,\cos(nx)}{\sin(x/2)}\,dx \overset{\boxed{(*)}}{=} \\
&= \frac{1}{n\pi}\int_0^{\pi}\Big(1 + 2\,\cos(x) + 2\,\cos(2x) + \ldots + 2\,\cos(nx) - \cos(nx)\Big)\,dx = \\
&= \frac{1}{n}\,.
\end{aligned}$$

□

Exercise 5.20. Prove Theorem 5.8.

Solution - By taking the Fourier transform of the periodicity condition we obtain

$$u \in \mathcal{S}'_T \iff u(t) = u(t+T) \iff \left\{\widehat{u} \in \mathcal{S}' \text{ e } \left(e^{i\xi T} - 1\right)\widehat{u}(\xi) = 0\right\}.$$

Therefore $\widehat{u}$ must be a tempered distribution that solves a division problem in the distributional setting; the multiplier $(e^{i\xi T}-1)$ has infinitely many zeros, $\xi = 2k\pi/T,\ k\in\mathbb{Z}$, and all of them are simple zeros (the derivative $iTe^{i\xi T}$ does not vanish); by (3.16) we infer that the solution of the division problem is a delta train of the form (5.23).
We conclude with the following technical remark: if $a>0$ and ξ denotes the real variable, then the series $\sum_{k\in\mathbb{Z}}\gamma_k\delta(\xi-ka)$ converges to a tempered distribution in $\mathcal{S}'(\mathbb{R})$ if and only if the γ_k-s verify (5.24). Therefore the coefficients of the delta train that represents $\widehat{u}$ must be tempered. □

Exercise 5.21. Prove Theorem 5.15.

Solution - Let us prove (5.38), by applying (5.34) to u and u':

$$u\in H^1_T \iff \left\{u,\,u'\in L^2_T\right\} \iff \left\{\sum_h |\widehat{u}_h|^2<+\infty\,,\ \sum_h |h\widehat{u}_h|^2<+\infty\right\}$$
$$\iff \left\{\sum_h |h\widehat{u}_h|^2<+\infty\right\}.$$

The equivalence (5.39) can be proved by iterating (5.38). □

Exercise 5.22. Prove Theorem 5.17.

Solution - For any real number s such that $1/2<s\le 1$ we have

$$u\in H^s_T \iff \sum_{h\in\mathbb{Z}} |\,h^s\,\widehat{u}_h\,|^2 \ <\ +\infty\,.$$

Using *Young's inequality*: $|ab|\ \le\ \dfrac{1}{2}\left(|a|^2+|b|^2\right)\ \ \forall a,b\in\mathbb{C}$, we have

$$|\widehat{u}_h|\ \le\ \frac{1}{h^s}\,|\,h^s\,\widehat{u}_h\,|\ \le\ \frac{1}{h^{2s}}+|\,h^s\,\widehat{u}_h\,|^2\,.$$

Thus $\sum_{h\in\mathbb{Z}}|\widehat{u}_h|\ <\ +\infty$, since its general term $|\widehat{u}_h|$ is bounded by the general terms of two converging series. The Weierstrass criterion entails the uniform convergence of the Fourier series of u, and hence the continuity and boundedness of u.

If $s>m+1/2$ with integer $m\ge 1$ then it is possible to repeat the same analysis for the n-th derivative, for all $n\le m$ by recalling that due to (5.26): $\widehat{u^{(n)}}_h\ =\ k^n(2\pi i/T)^n\,\widehat{u}_h\,,\ \forall h$; thus we conclude $u\in C^m(\mathbb{R})\cap L^\infty(\mathbb{R})$. □

Exercise 5.23. Prove that it is possible to compute the Fourier coefficients $\widehat{u}_k$ of a periodic distribution u in the following way too.

$$\widehat{u}_k\ =\ \int_{\mathbb{R}} u(t)\,e^{-\,i\,k\,\frac{2\pi}{T}\,t}\,\psi(t)\,dt \tag{5.47}$$

where $\psi\in\mathcal{S}(\mathbb{R})$ verifies

$$\psi_T(t) := \sum_{k\in\mathbb{Z}}\psi(t-kT)\ \equiv\ 1 \qquad \forall t\in\mathbb{R} \tag{5.48}$$

and the integral in the (5.47) denotes the pairing $\langle\, u\,,\,\exp(-2k\pi x/T)\,\psi\,\rangle$.

Prove that the values of $\widehat{u}_k$ computed by (5.47) do not depend on the Schwartz function ψ chosen, provided ψ fulfils (5.48).

Solution - 1) If $u \in L^1_{loc}$ then the $\widehat{u}_k$-s computed with (5.47) are equal to the $c_k(u)$-s computed with (5.12).
Indeed in such a case the duality is also a Lebesgue integral of an integrable function, therefore it is possible to write it as the sum of integrals computed over a decomposition of the domain

$$\begin{aligned}\langle u, \exp(-2k\pi it/T)\psi\rangle &= \int_{\mathbb{R}} u(t)\, e^{-ik\frac{2\pi}{T}t}\psi(t)\, dt = \sum_{h\in\mathbb{Z}} \int_{hT}^{(h+1)T} u(t)\, e^{-ik\frac{2\pi}{T}t}\psi(t)\, dt = \\ &= \sum_{h\in\mathbb{Z}} \int_{hT}^{(h+1)T} u(t+hT)\, e^{-ik\frac{2\pi}{T}t}\psi(t)\, dt = \\ &= \sum_{h\in\mathbb{Z}} \int_0^T u(t)\, e^{-ik\frac{2\pi}{T}t}\psi(t-hT)\, dt = \\ &= \int_0^T u(t)\, e^{-ik\frac{2\pi}{T}t}\left(\sum_{k\in\mathbb{Z}} \psi(t-hT)\right) dt = \\ &= \int_0^T u(t)\, e^{-ik\frac{2\pi}{T}t}\, dt \;=\; c_k\ .\end{aligned}$$

2) If $u_n,\ u \in \mathcal{S}'_T$ and u_n converges to u in $\mathcal{S}'(\mathbb{R})$, then all Fourier coefficients converge: that is the real sequences $\widehat{u_n}_k$ converge to $\widehat{u}_k$, $\forall k \in \mathbb{Z}$.
In fact, applying (5.47) and Definition 3.17, we obtain

$$\widehat{u_n}_k \;=\; \langle u_n, \exp(-2k\pi it/T)\psi\rangle \longrightarrow \langle u, \exp(-2k\pi it/T)\psi\rangle \;=\; \widehat{u}_k \qquad \forall k \in \mathbb{Z}\,.$$

3) Every periodic distribution can be approximated by smooth periodic functions.
Indeed, given $u \in \mathcal{S}'_T$, it is possible to construct an approximating sequence (T periodic and C^∞) by the convolution: $u_n \;=\; u * \varphi_n$, where $\varphi_n(t) \;=\; n\varphi(nt)$ and $\varphi \in \mathcal{D}(\mathbb{R})$ is a fixed test function such that $\int_{\mathbb{R}} \varphi(x)\, dx = 1$.
Summarizing, due to to 1), 2), 3), given a generic $u \in \mathcal{S}'_T$, there exists u_n such that

$$\widehat{u_n}_k \;=\; c_k(u_n) \;=\; \int_{\mathbb{R}} u_n(t)\, \psi(t)\, exp(-2k\pi it/T)\, dt \;=\; \int_{\mathbb{R}} \psi(t)\, \exp(-2k\pi it/T)\, dt\,,$$

which yields (5.47) since we have $\widehat{u_n}_k \to \widehat{u}_k$, as $n \to +\infty$ and
$\int_{\mathbb{R}} u_n(t)\, \psi(t)\, \exp(-2k\pi it/T)\, dt \;\to\; \int_{\mathbb{R}} u(t)\, \psi(t)\, \exp(-2k\pi it/T)\, dt$.
The independence of ψ for the value of $\widehat{u}_k$ obtained with (5.47) is a consequence of the uniqueness of the Fourier coefficients.

Remark - As noticed in Remark 5.12 it is meaningful to compute the coefficients with formulas (5.12) only if $u \in L^1_{loc}$, since (5.12) "see u through a rectangular window",

$$\text{explicitely } \widehat{u}_k = \int_{\mathbb{R}} u(t) e^{-ik\frac{2\pi}{T}t}\chi_{[-T/2,T/2)}dt\ ,$$

because it is not possible to give a good definition of the product $u\,\chi_{[-T/2,T/2)}$ if $u \in \mathcal{S}'_T$. Whereas formula (5.47) "see u through a window with rounded edges" so that the coefficients are well defined for any $u \in \mathcal{S}'_T$ □

Chapter 6
Laplace transform

In this chapter we introduce the Laplace transform (Pierre-Simon Laplace, 1749-1827) together with its main properties.
This transform is a classical tool: it was introduced for the study of differential equations and it proved very useful also in the analysis of integral equations, differential equations with delay, electrical circuits and signals.
The Fourier transform is a powerful tool to solve a wide class of problems which are set in $\mathbb{R}$ or in the Euclidean space $\mathbb{R}^n$ (see Chapter 4), while the Laplace transform is focussed on the study of problems that are set in the half-line of positive real numbers with prescribed initial data: a remarkable example is the Cauchy problem for a differential equation or for a system of differential equations (Section 6.1.4).
Another important property of the Laplace transform is the fact that it works also on functions with exponential growth, that is objects that cannot undergo the Fourier transform, neither in the functional nor in the distributional setting. Nevertheless there is a close connection between the two transforms as it will be precisely stated in Section 6.1.6.

6.1 The Laplace transform method

The Laplace transform is an effective tool to solve a wide range of problems. This motivates the extension of its definition to a more general functional setting: the one of distributions that are Laplace transformable.
In the next sections we introduce the transform first in the functional setting and then in the distributional setting, together with an outline of its properties.
For further readings we refer to [5] and [25].

6.1.1 Laplace transform of $\mathcal{L}$ transformable functions

Definition 6.1. *$f : \mathbb{R} \to \mathbb{C}$ is called* ***Laplace transformable function*** *(or $\mathcal{L}$ transformable function) if:*

$$\operatorname{spt} f \subset [0, \infty)\,, \tag{6.1}$$

$$\text{there exists } \lambda \in \mathbb{R} \text{ such that } e^{-\lambda t} f(t) \in L^1(\mathbb{R})\,. \tag{6.2}$$

Definition 6.2. *We call* ***abscissa of convergence*** *of f the real number*

$$\lambda_f = \inf\{\, \lambda \in \mathbb{R} : \; e^{-\lambda t} f(t) \in L^1(\mathbb{R})\}\,. \tag{6.3}$$

Definition 6.3. *Let f be a $\mathcal{L}$ transformable function, then the function F defined by*

$$F(s) = \int_0^{\infty} e^{-st} f(t)\, dt = \int_{\mathbb{R}} e^{-st} f(t)\, dt \quad \forall s \in \mathbb{C} : \operatorname{Re} s > \lambda_f\,. \tag{6.4}$$

is called ***Laplace transform*** *of f. We use the notation: $F = \mathcal{L}\{f\}$.*

The Laplace transform of f is a complex valued function whose domain is a half-plane contained in the complex plane: $\{s \in \mathbb{C} : \operatorname{Re} s > \lambda_f\}$.
We use the convention of indicating transformable functions with lowercase letters and their transforms with the corresponding capital letters; let us mention that in some contexts the opposite convention is used.
Often it is convenient to write explicitly the variables, which the function and its transform depend on, by adopting the following notation:

$$F(s) = \mathcal{L}\{f(t)\}\,.$$

Example 6.4. The Heaviside function H, defined by $H(t) = 1$ if $t \geq 0$, $H(t) = 0$ if $t < 0$, is an $\mathcal{L}$ transformable function and it plays a central role in the Laplace transform method. The following identity is a straightforward consequence of (6.4):

$$\mathcal{L}\{H(t)\} = \frac{1}{s} \qquad \forall s \in \mathbb{C} : \operatorname{Re} s > \lambda_H = 0\,.$$

6.1.2 Properties of the Laplace transform of $\mathcal{L}$ transformable functions

We list some properties that are useful when operating with the Laplace transform.

If f, g are $\mathcal{L}$ transformable functions, we denote by $F = \mathcal{L}\{f\}$, $G = \mathcal{L}\{g\}$ and by λ_f, λ_g their Laplace transforms and abscissas of convergence respectively, then:

L1. The Laplace transform is a linear and injective operator:

$$\begin{gathered}\mathcal{L}\{\alpha f + \beta g\}(s) \;=\; \alpha\,\mathcal{L}\{f\}(s) \,+\, \beta\,\mathcal{L}\{g\}(s) \\ \forall \alpha, \beta \in \mathbb{C},\ \forall s \in \mathbb{C} :\ \mathrm{Re}(s) > \max(\lambda_f, \lambda_g).\end{gathered} \tag{6.5}$$

$$\text{If } \exists \lambda \in \mathbb{R} :\ F(s) = G(s)\ \forall s \in \mathbb{C} :\ \mathrm{Re}\, s > \lambda\,, \quad \text{then } f = g \ \ a.e. \tag{6.6}$$

L2. The domain of definition of the transform depends on the values of λ fulfilling the estimate (6.2), precisely the domain of the transform F (of a $\mathcal{L}$ transformable function f) is the nonempty half-plane

$$\{ s \in \mathbb{C} :\ \mathrm{Re}\, s > \lambda_f \}\,, \tag{6.7}$$

such a half-plane is said **half-plane of convergence** of f because in this set the Lebesgue integral in (6.4) is well defined.

L3. The Laplace transform F of an $\mathcal{L}$ transformable function f is an analytic function (that is to say, it is differentiable in complex sense) in the half-plane of convergence and

$$F'(s) \;=\; -\,\mathcal{L}\{\, t\, f\,\} \tag{6.8}$$

$$F^{(k)}(s) \;=\; (-1)^k\, \mathcal{L}\{\, t^k\, f\,\} \qquad \forall k \in \mathbb{N}\,. \tag{6.9}$$

L4. If $f \in C^1([0, +\infty))$ and $\dfrac{df}{dt}$ is a $\mathcal{L}$ transformable function (df/dt denotes the derivative of f in $\mathbb{R}^+$ extended with 0 value in $\mathbb{R}^-$), then

$$\mathcal{L}\left\{\frac{df}{dt}\right\}(s) \;=\; s\, F(s) - f(0_+)\,. \tag{6.10}$$

The assumption "$f \in C^1([0, +\infty))$" can be weakened: if it is replaced by "$f \in AC_{loc}([0, +\infty))$" then (6.10) is still true.

L5. If $\exists k \geq 2 :\ f \in C^k([0, +\infty))$ and $f, \dfrac{df}{dt}, \dots, \dfrac{d^k f}{dt^k}$ are $\mathcal{L}$ transformable functions (here $\dfrac{d^k f}{dt^k}$ denotes k-th derivative of f in $\mathbb{R}^+$ extended with zero value in the left half-line), then in addition to (6.10) we have

$$\mathcal{L}\left\{\frac{d^2 f}{dt^2}\right\}(s) = s^2\,F(s) - s\,f(0_+) - \frac{df}{dt}(0_+)\,, \tag{6.11}$$

$$\begin{aligned}\mathcal{L}\left\{\frac{d^h f}{dt^h}\right\}(s) &= s^h F(s) - s^{h-1} f(0_+) - \cdots \\ &\quad - s\,\frac{d^{h-2} f}{dt^{h-2}}(0_+) - \frac{d^{h-1} f}{dt^{h-1}}(0_+) \quad h \in \mathbb{N},\, 2 \le h \le k.\end{aligned} \tag{6.12}$$

L6. The antiderivative of an $\mathcal{L}$ transformable function f is an $\mathcal{L}$ transformable function that satisfies

$$\mathcal{L}\left\{\int_0^t f(\tau)\,d\tau\right\} = \frac{1}{s}\mathcal{L}\{f\}(s) \qquad \forall s:\ \operatorname{Re} s > \max(\lambda_f, 0)\,. \tag{6.13}$$

L7. If $f(t)/t$ is an $\mathcal{L}$ transformable function, then also f is an $\mathcal{L}$ transformable function and

$$\mathcal{L}\left\{f(t)/t\right\}(s) = \int_s^{\infty} F(\sigma)\,d\sigma\,. \tag{6.14}$$

L8. If $F = \mathcal{L}\{f\}$, where f is an $\mathcal{L}$ transformable function, then

$$\lim_{\operatorname{Re} s \to +\infty} F(s) = 0\,. \tag{6.15}$$

Let us recall some **algebraic transformation rules**.
If $f = f(t)$ is an $\mathcal{L}$ transformable function and $F = \mathcal{L}\{f\}$, then also $f(at)$, $f(t-a)$ and $e^{\alpha t} f(t)$ are $\mathcal{L}$ transformable functions, for any real number $a > 0$ and any complex number α. Moreover we have:

$$\mathcal{L}\{f(a\,t)\} = \frac{1}{a}\,F\left(\frac{s}{a}\right), \qquad \operatorname{Re} s > a\,\lambda_f \qquad \text{(rescaling)}, \tag{6.16}$$

$$\mathcal{L}\{f(t-a)\} = e^{-a\,s}\,F(s), \qquad \operatorname{Re} s > \lambda_f \qquad \text{(delay)}, \tag{6.17}$$

$$\mathcal{L}\{e^{\alpha t} f(t)\} = F(s-\alpha), \quad \operatorname{Re} s > \lambda_f + \operatorname{Re}\alpha \ \text{(product with exponential)}. \tag{6.18}$$

Theorem 6.5. ***(Laplace transform of the convolution)***
If f and g are $\mathcal{L}$ transformable functions, then:
1) it is possible to define the convolution of f and g, denoted by $f * g$:

$$f * g\,(t) = \int_0^t f(\tau)\,g(t-\tau)\,d\tau \qquad t \in \mathbb{R}\,, \tag{6.19}$$

2) the convolution $f * g$ is an $\mathcal{L}$ transformable function and we have:

$$\mathcal{L}\{f * g\}(s) = F(s)\,G(s) \qquad \forall s \in \mathbb{C}:\ \operatorname{Re} s > \max(\lambda_f, \lambda_g)\,, \tag{6.20}$$

$$e^{-\lambda t}\,(f * g) = \left(e^{-\lambda t} f\right) * \left(e^{-\lambda t} g\right) \qquad \forall \lambda \in \mathbb{R}:\ \lambda > \max(\lambda_f, \lambda_g)\,. \tag{6.21}$$

Remark 6.6. Notice the (apparent) difference with respect to the product of convolution introduced in the study of the Fourier transform: in Definition 4.14 the integral domain is the whole real line $\mathbb{R}$ whereas in (6.19) is the interval $[0, t]$, Nevertheless the two definitions coincide when applied to real valued functions that are **identically zero** in the half-line $(-\infty, 0)$, as in the present case of $\mathcal{L}$ transformable functions f and g verifying the support constraint (6.1).

Theorem 6.7. ***(Initial value theorem)*** *If*

$$f \in AC_{loc}([0, +\infty)), \ f \ \textit{and} \ \frac{df}{dt} \ \textit{are } \mathcal{L} \textit{ transformable functions} \tag{6.22}$$

(here $\frac{df}{dt}$ *stands for the derivative of* f *in* $\mathbb{R}^+$ *extended zero in* $\mathbb{R}^-$*), then:*

$$f \ \textit{tends to a finite limit} \ f(0_+) \ \textit{as} \ t \ \textit{goes to} \ 0_+ ,$$

$$\lim_{\mathrm{Re}\, s \to +\infty} s\, F(s) \ = \ f(0_+) .$$

Theorem 6.8. ***(Final value theorem)*** *If*

$$f \in AC\big((0, +\infty)\big)$$

and we denote by f *also the trivial extension of* f *to the whole* $\mathbb{R}$ *(that is to say* $f(t) = 0$ *if* $t \in (-\infty, 0)$ *), then:*

$$f \ \textit{is } \mathcal{L} \textit{ transformable,}$$

$$f \ \textit{tends to a finite limit} \ f(\infty) \ \textit{as} \ t \ \textit{goes to} \ +\infty ,$$

$$\lambda_f \leq 0 ,$$

$$\lim_{s \to 0,\, \mathrm{Re}\, s > 0} s\, F(s) \ = \ f(\infty) .$$

Theorem 6.9. *We say that a sequence of* $\mathcal{L}$ *transformable functions* f_k ***converges to*** f ***in the space of*** $\mathcal{L}$ ***transformable functions*** *if*

$$\exists \lambda \in \mathbb{R} : \ e^{-\lambda t} f_k(t) \to e^{-\lambda t} f(t) \ \textit{in} \ L^1(\mathbb{R}) , \quad \mathrm{spt}\, f_k \subset [0, +\infty) \ \forall k . \tag{6.23}$$

If f_k *converges to* f *in the space of* $\mathcal{L}$ *transformable functions (that is* (6.23) *holds true), then* f *is an* $\mathcal{L}$ *transformable function,* $\lambda_f, \lambda_{f_k} \leq \lambda$ *and the* $\mathcal{L}$ *transforms converge:*

$$F_k(s) \to F(s) \ \textit{uniformly in the half-plane} \ \{ \mathrm{Re}\, s \geq \lambda \} . \tag{6.24}$$

6.1.3 Inverse transform in the framework of $\mathcal{L}$ transformable functions

Given a complex variable function F whose domain contains a half-plane of the form $\{s \in \mathbb{C} : Re(s) > \lambda\}$, we may ask whether F is the Laplace transform of an $\mathcal{L}$ transformable function f and, in the affirmative case, to determine f. If such f exists, then it is unique by the injectivity of the Laplace transform.
We recall two necessary conditions in order that F is the Laplace transform of an $\mathcal{L}$ transformable function (see properties L3 and L8): F must be analytic in a right half-plane and satisfy the existence of $\lim_{x\to+\infty} F(x+iy)=0$, $\forall y \in \mathbb{R}$.
Let us state a sufficient (but not necessary) condition which implies that F is a Laplace transform of an $\mathcal{L}$ transformable function:
if there exist $M \geq 0$, $\alpha > 1$ and $\lambda \in \mathbb{R}$, such that

$$F \text{ is analytic and } \quad |F(s)| \leq \frac{M}{1+|s|^\alpha} \quad \forall s \in \mathbb{C} : Re(s) > \lambda \tag{6.25}$$

then we have $F = \mathcal{L}\{f\}$, where f is an $\mathcal{L}$ transformable function and $\lambda_f \leq \lambda$.
In the effective calculation of the of the inverse transform (or anti-transform), it is convenient to try first the use of properties listed in the Table 10.7 and the transforms itemized in the Table 10.7.1 (Chapter 10).
For instance, if $F(s)$ is a rational function of the variable s with the degree of the denominator strictly larger than the degree of the numerator, then F is a Laplace transform of an $\mathcal{L}$ transformable function and its inverse transform can be achieved by using the decomposition of F as a sum of simple ratios and, subsequently anti-transforming the summand with the Table 10.7.1. In this regard, we recall the following useful property.

Proposition 6.10. Let

$$F(s) = \frac{a_m\, s^m + a_{m-1}\, s^{m-1} + \cdots + a_0}{b_n\, s^n + b_{n-1}\, s^{n-1} + \cdots + b_0} \qquad (s \in \mathbb{C})$$

with $n > m$, and denote by $z_1, ..., z_N \in \mathbb{C}$ the distinct roots of the denominator, respectively with algebraic multiplicity $r_1, ..., r_N$ respectively. Then there exist coefficients $c_{kj} \in \mathbb{C}$ (with $k = 1, ..., N$ and $j = 1, ..., r_k$) such that

$$F(s) = \sum_{j=1}^{r_1} \frac{c_{1j}}{(s-z_1)^j} + \sum_{j=1}^{r_2} \frac{c_{2j}}{(s-z_2)^j} + ... + \sum_{j=1}^{r_N} \frac{c_{Nj}}{(s-z_N)^j},$$

$$f(t) = \mathcal{L}^{-1}\{F\} = \left(\sum_{j=1}^{r_1} \frac{c_{1j}\, t^{j-1}\, e^{z_1 t}}{(j-1)!} + ... + \sum_{j=1}^{r_N} \frac{c_{Nj}\, t^{j-1}\, e^{z_N t}}{(j-1)!} \right) H(t)\,.$$

If n is large then the decomposition in simple ratios may turn out hardworking or unattainable, since it requires first the exact solution of high order algebraic equation and second the solution of a high-dimensional linear system. Nevertheless, in the particular case in which the denominator $Q(s)$ of the rational function $f = P/Q$ has only simple zeros ($r_k = 1, \forall k$) and we are able to explicit them all, the anti-transforming procedure is simplified by the following formula that provides the coefficients c_{k1} explicitly.

Theorem 6.11. ***(Heaviside expansion Formula)*** *If $F = P/Q$ is a lowest terms rational function, where the degree of polynomial P is strictly less than the degree of polynomial Q and all zeros z_k of Q are simple ($r_k = 1\,\forall k$), then $F = \mathcal{L}\{f\}$ where f is an $\mathcal{L}$ transformable function and*

$$f(t) = \mathcal{L}^{-1}\left\{\frac{P(s)}{Q(s)}\right\} = \sum_k \frac{P(z_k)}{Q'(z_k)}\, e^{z_k t}\, H(t)\,. \tag{6.26}$$

A further technique to find the inverse transform is given by the following statement that is an immediate consequence of Theorem 6.9 applied to the sequence of partial sums of a series.

Theorem 6.12. ***(Term-wise inverse transformation of series)***
If f_k are $\mathcal{L}$ transformable functions, $F_k = \mathcal{L}\{f_k\}$, and there is $\lambda \in \mathbb{R}$ s.t.

$$\sum_{k=0}^{+\infty} e^{-\lambda t} f_k(t) \qquad \textit{converges in } L^1(\mathbb{R}), \tag{6.27}$$

then $f(t) := \sum_0^{+\infty} f_k(t)$ is an $\mathcal{L}$ transformable function,
the series on the right hand side of the definition of f converges in $L^1_{loc}(\mathbb{R})$,
$F(s) := \sum_0^{+\infty} F_k(s)$ is the transform of the $\mathcal{L}$ transformable function f,
the series on the right hand side of the definition of F converges uniformly in $\{s :\ \mathrm{Re}\, s \geq \lambda\}$ and it is possible to anti-transform term-wise

$$f(t) = \mathcal{L}^{-1}\{F\} = \sum_{k=0}^{+\infty} \mathcal{L}^{-1}\{F_k\} = \sum_{k=0}^{+\infty} f_k(t)\,. \tag{6.28}$$

The previous theorem is stated for series (a handy formulation for computations: see Exercises 6.17 and 6.27), but it can be easily adapted to sequences. The next theorem provides a general formula for the inverse transform.

Theorem 6.13. ***(Riemann-Fourier inversion formula)***
If F is the Laplace transform of an $\mathcal{L}$ transformable function f and there exists $x > \lambda_f$ such that the function $y \to F(x+iy)$ belongs to $L^1(\mathbb{R})$, then the following Riemann-Fourier inversion formula holds true:

$$f(t) = \frac{1}{2\pi i} \int_{\gamma_x} e^{st}\, F(s)\, ds \tag{6.29}$$

where the integral path γ_x is a vertical line which passes through x and is swept in the increasing ordinates direction: $\{\,\gamma_x(y) = x + iy,\ y \in \mathbb{R}\,\}$.

Remark 6.14. Let us observe that formula (6.29) holds for any $t \in \mathbb{R}$, thus the right-hand side is zero if $t < 0$ (by (6.1)) and it can have a nonzero value only if $t \geq 0$. In the actual computation of the right hand side in formula (6.29) (in the case $t > 0$) we refer to the (unique) analytic extension of F, therefore it might be helpful Jordan Lemma, precisely its version related to the half-circumference in the left half-plane: Lemma 1.34 (iii).
In this regard notice that starting from the knowledge of F in the half-plane of convergence (property L2) often it is possible to perform an analytic extension of F to a subset of the complex plane which is wider than the half-plane of convergence; in this process isolated singularities and/or branching points in the half-plane $\operatorname{Re} s \leq \lambda_f$ may appear. When branching points are present, in order to compute the integral (6.29) it is necessary to introduce cuts in the complex plane and precisely define the analytic branch of the multi-valued function which is coincident with F in the convergence half-plane.

Remark 6.15. In the literature ([1],[7]) another (different from (6.4)) definition of Laplace transform is available: this alternative definition does not require Lebesgue integrability of $e^{-st}f(t)$, but it requires only existence of its generalized integral on $[0, +\infty)$, that is: given a function locally integrable $f : \mathbb{R} \to \mathbb{C}$ one defines the transform of f equal to the function (dependent on the variable s)

$$\lim_{T \to +\infty} \int_0^T e^{-st} F(t)\, dt \tag{6.30}$$

whose domain is the set of $s \in \mathbb{C}$ so that such a limit exists and is finite.
In the case of Definition (6.30) one speaks of Laplace transform of *simply transformable functions*, to mark the distinction from $\mathcal{L}$ transformable functions (fulfilling (6.1),(6.2)) that, in such context are called *absolutely transformable functions*. Observe that absolute transformability implies simple transformability, whereas the opposite does not hold in general.
The definition of simple transformability is only slightly more general of the one we have given and it verifies much less properties of that listed above. Therefore we will always refer to Definitions 6.1, 6.3 or, in the distributional setting to the subsequent Definitions 6.16, 6.17.

6.1.4 Solving differential equations with initial conditions by $\mathcal{L}$ transform

The method of the Laplace transform allows to solve initial problems for differential equations. Let us consider, for instance, the following Cauchy problem.

$$\begin{cases} \text{Given: } \mathcal{L} \text{ transformable function } f,\ v_j \in \mathbb{R},\ j = 0, \dots, n-1\,; \\ \text{find the } \mathcal{L} \text{ transformable function } v \text{ that solves :} \\ Lv = f \text{ in } (0, +\infty)\,, \\ v(0) = v_0,\ v'(0) = v_1, \dots, v^{(n-1)} = v_{n-1}\,. \end{cases} \tag{6.31}$$

where $L = \sum_{j=0}^{n} c_j\, d^j/dt^j$ is a linear differential operator with constant coefficients of order n ($c_j \in \mathbb{R}$, $j = 0, \dots, n$, are given) defined in $C^n\big((0, +\infty)\big)$. The differential equation $Lv = f$ in problem (6.31) has the explicit form

$$c_n\, v^{(n)}(t) + \cdots + c_1 v'(t) + c_0 v(t) = f(t) \qquad \text{in } (0, +\infty)\,.$$

We denote by $\mathcal{P}$ the polynomial associated to the differential operator L, that is the polynomial of degree n which has the following expression when evaluated in the variable s : $\mathcal{P}(s) = \sum_{j=0}^{n} c_j\, s^j$.
We set $V = \mathcal{L}\{v\}$ and we transform both sides of the differential equation, by taking into account (6.12) and the initial conditions. We obtain an algebraic equation whose unknown is the transform V of the solution v:

$$\begin{aligned} \mathcal{P}(s)\, V(s) = \ & F(s) + \ v_0\big(c_n s^{n-1} + c_{n-1} s^{n-2} + \cdots + + c_2 s + c_1\big) + \\ & + \ \ldots + \ v_{n-2}(c_n s + c_{n-1}) \ + \ v_{n-1} c_n\,. \end{aligned}$$

The polynomial $\mathcal{P}$ has a finite number of zeros s_k, therefore dividing by $\mathcal{P}(s)$ is allowed in the half-plane $\operatorname{Re} s > \max_k \operatorname{Re} s_k$:

$$V(s) \ = \ \frac{F(s) \ + \ \mathcal{R}(s)}{\mathcal{P}(s)}\,,$$

with $\mathcal{R}$ is a polynomial of degree smaller or equal to $n-1$, whose coefficients depend only on c_j, v_j : $\mathcal{R}(s) = \sum_{h=0}^{n} \sum_{j=1}^{h} c_h v_{j-1} s^{h-j}$. Thus V is the transform of an $\mathcal{L}$ transformable function, due to sufficient condition (6.25) and the fact that $\mathcal{R}/\mathcal{P}$ is a proper ratio. Eventually, we set $u = \mathcal{L}^{-1}\{1/\mathcal{P}\}$, and anti-transform V to get the unique solution v of the Cauchy problem (6.31):

$$v = u * f + \mathcal{L}^{-1}\left\{\frac{\mathcal{R}}{\mathcal{P}}\right\} = u * f + \sum_{h=0}^{n} \sum_{j=1}^{h} c_h v_{j-1} u^{(h-j)} \qquad t \in (0, +\infty)\,.$$

The method can be applied also to:

- systems of ordinary differential equations with initial conditions (that are transformed into systems of algebraic equations),
- integro-differential equations and differential equations with delay (that are transformed into transcendental equations),
- linear difference equations.

About the first and second item, we refer to exercises in Section 6.2.2, about the third one we mention that difference equations can be solved with the $\mathcal{Z}$ transform, the discrete analogous of the $\mathcal{L}$ transform ([16]).

6.1.5 *Laplace transform of $\mathcal{L}$ transformable distributions*

Definition 6.16. *A* ***Laplace transformable distribution*** *(briefly denoted by $\mathcal{L}$ transformable distribution) is a distribution $f \in \mathcal{D}'(\mathbb{R})$ that fulfils:*

$$\operatorname{spt} f \ \subset \ [0,\infty) \tag{6.32}$$

$$\text{there exists } \lambda \in \mathbb{R} \text{ such that } e^{-\lambda t}\, f(t) \ \in \ \mathcal{S}'(\mathbb{R})\,. \tag{6.33}$$

We keep the convention of denoting transformable distributions with lower case letters and their transforms with capital letters.

Definition 6.17. *Let f be an $\mathcal{L}$ transformable distribution, then the function F defined by*

$$F(s) = \int_{\mathbb{R}} e^{-st}\, f(t)\, dt \qquad \forall s \in \mathbb{C}:\ \operatorname{Re} s > \lambda_f\,. \tag{6.34}$$

is called ***Laplace transform*** *of f.*
Notation: $F(s) = \mathcal{L}\{f(t)\}\,,\ \ \lambda_f = \inf\{\lambda :\, e^{-\lambda t} f \in \mathcal{S}'\}\,.$

The integral $\int_{\mathbb{R}} e^{-st} f(t)\, dt$ in formula (6.34) stands for the pairing between $\mathcal{S}'(\mathbb{R})$ and $\mathcal{S}(\mathbb{R})$ that is:

$$F(s) \ = \ \langle\, e^{-\lambda t} f(t)\,,\, e^{-(s-\lambda)t}\,\rangle \tag{6.35}$$

where $\lambda \in \mathbb{R}$ is such that $\operatorname{Re} s > \lambda > \lambda_f$ and (6.33) holds true: note that if $\operatorname{Re} s > \lambda_f$ then this λ exists and $e^{-(s-\lambda)t}$ decays exponentially as $t \to +\infty$.
The Laplace transform of a distribution is a complex valued function whose domain is a right half-plane contained in the complex plane.
Every $\mathcal{L}$ transformable function is an $\mathcal{L}$ transformable distribution too: Definitions (6.4) and (6.34)-(6.35) coincide for an $\mathcal{L}$ transformable function.

Example 6.18. A fundamental example is the Dirac δ: δ is a compact supported distribution ($\operatorname{spt} \delta = \{0\}$); therefore δ is an $\mathcal{L}$ transformable distribution and its transform is an entire analytic function:

$$\mathcal{L}\{\delta\} \ = \ 1\,, \qquad \lambda_\delta = -\infty\,.$$

About the Laplace transform of an $\mathcal{L}$ transformable distributions, we emphasize that property L8 (Section 6.1.2) cannot hold true: for instance, the transform of δ is not an infinitesimal at infinity. Theorem 6.23 specifies the behavior of the transform of a distribution in the convergence half-plane.

Remark 6.19. Every $\mathcal{L}$ transformable distribution satisfies all algebraic properties (6.16),(6.17),(6.18) and all functional properties L1-L7 (but not L8 !). Nevertheless properties L4 and L5 have to be suitably understood, as it is clarified by the next statement.

Theorem 6.20. *If f is an $\mathcal{L}$ transformable distribution and we denote by f' and $f^{(k)}$ respectively the first distributional derivative of f in $\mathbb{R}$ and the distributional derivative of order k of f in $\mathbb{R}$, $k \in \mathbb{N}$, then*

$$\mathcal{L}\{f'\} = s\,F(s), \qquad \mathcal{L}\{f^{(k)}\} = s^k\,F(s)\,. \tag{6.36}$$

There is no contradiction between Theorem 6.20 and properties L4, L5. Actually, if we denote by df/dt and $d^k f/dt^k$ the trivial extensions of the first and of the k-th distributional derivative in $\mathbb{R}^+$ and denote by f', $f^{(k)}$ the distributional derivatives in $\mathbb{R}$, then the following identities hold true in $\mathcal{D}'(\mathbb{R})$:

$$\begin{aligned} f' &= \frac{df}{dt} + f(0_+)\,\delta\,, \\ f^{(k)} &= \frac{d^k f}{dt^k} + f(0_+)\,\delta^{(k-1)} + f'(0_+)\,\delta^{(k-2)} + \ldots + f^{(k-1)}(0_+)\,\delta\,. \end{aligned}$$

Theorem 6.21. ***(*$\mathcal{L}$ *transform of distributions convolution)***
If f and g are $\mathcal{L}$ transformable distributions, then:
1) we can define the convolution of f and g, denoted by $f * g$, by

$$\langle\,(f * g)\,,\,\varphi\,\rangle = \langle\,f\,,\,(\check{g} * \varphi)\,\rangle \qquad \forall \varphi \in \mathcal{D}(\mathbb{R})\,, \tag{6.37}$$

where $\check{g}(t) = g(-t)$.
2) the convolution $f * g$ is an $\mathcal{L}$ transformable distribution that satisfies:

$$\mathcal{L}\{\,f * g\,\}(s) = F(s)\,G(s) \qquad \forall s \in \mathbb{C}:\ \operatorname{Re} s > \max(\lambda_f, \lambda_g)\,. \tag{6.38}$$

Theorem 6.22. *Given a sequence f_k of distributions, we will say that f_k* ***converges to f in the space of $\mathcal{L}$ transformable distributions*** *if:*

$$\exists \lambda \in \mathbb{R}:\ e^{-\lambda t}\,f_k(t) \to e^{-\lambda t}\,f(t)\ \text{ in } \mathcal{S}'(\mathbb{R})\,, \quad \operatorname{spt} f_k \subset [0, +\infty)\ \forall k\,. \tag{6.39}$$

If f_k converges to f in the space of $\mathcal{L}$ transformable distributions, that is (6.39) holds true, then f, f_k are $\mathcal{L}$ transformable distributions, $\lambda_f, \lambda_{f_k} \le \lambda$ and their transforms converge, precisely

$$\begin{aligned} F_k \to F \ &\text{pointwise in the open half-plane } \{\operatorname{Re} s > \lambda\} \text{ and} \\ &\text{uniformly in every compact set contained in } \{\operatorname{Re} s > \lambda\}. \end{aligned} \tag{6.40}$$

6.1.6 Relationship between Laplace transform and Fourier transform

If F is the Laplace transform of an $\mathcal{L}$ transformable function (or of a distribution) f, then the restriction of F to a vertical line contained in the half-plane of convergence is the Fourier transform of an exponential correction of f. Precisely, for any fixed $x > \lambda_f$ by setting $s = x + iy,\ y \in \mathbb{R}$, we find

$$\mathcal{L}\{f\}(s) = F(s) = F(x+iy) = \mathcal{F}_{t,y}\left\{e^{-xt}f(t)\right\} \tag{6.41}$$

where $\mathcal{F}_{t,y}$, denotes the Fourier transform which acts on integrable functions (or tempered distributions) that depend on variable t, by transforming them in functions or distributions dependent on y.
We emphasize that the function $y \to F(x+iy)$ admits an analytic extension to the half-plane $\{y \in \mathbb{C} : \ \operatorname{Im} y < x - \lambda_f\}$.

6.1.7 Inverse Laplace transform in the framework of $\mathcal{L}$ transformable distributions

Paley-Wiener Theorem and Section 6.1.6 (Theorem 4.13) provide a sharp condition to establish whether a function of complex variable is the Laplace transform of an $\mathcal{L}$ transformable distribution.

Theorem 6.23. *A function F, whose domain is an open set of the complex plane, is the Laplace transform of an $\mathcal{L}$ transformable distribution if and only if F is defined (or at least it can be extended) in a right half-plane s.t. there exist λ, A, $B \in \mathbb{R}$ such that*

$$F \text{ is analytic in } \{\, s \in \mathbb{C} : \ \operatorname{Re} s > \lambda \,\}, \tag{6.42}$$

$$|F(s)| \leq A(1+|s|)^B \text{ in } \{\, s \in \mathbb{C} : \ \operatorname{Re} s > \lambda \,\}. \tag{6.43}$$

We notice that, if (6.42), (6.43) hold true and we set $G(s) = F(s)/s^m$ (where m is an integer: $m \geq B+2$), then G verifies (6.25), thus $g = \{\mathcal{L}\}^{-1}\{G\}$ can be computed with the Riemann-Fourier formula (6.29). Thus we get $f = g^{(m)}$: distributional derivative in $\mathbb{R}$ of order m of g.
Nevertheless, simpler (than the general Riemann-Fourier formula) techniques should be preferred to evaluate inverse $\mathcal{L}$ transform in practice, whenever it is possible: a good alternative is provided by the next statement.

Theorem 6.24. *(term-wise inverse transformation of series in the framework of $\mathcal{L}$ transformable distributions)*
If f_k are $\mathcal{L}$ transformable distributions, $F_k = \mathcal{L}\{f_k\}$ and there is $\lambda \in \mathbb{R}$ s.t.

$$\sum_{k=0}^{+\infty} e^{-\lambda t} f_k(t) \quad \text{converges in } \mathcal{S}'(\mathbb{R}),$$

then:
$f(t) := \sum_0^{+\infty} f_k(t)$ is an $\mathcal{L}$ transformable distribution,
the series on the right hand side of the definition of f converges in $\mathcal{D}'(\mathbb{R})$,
$F(s) := \sum_0^{+\infty} F_k(s)$ is the Laplace transform of f,
the series in the definition of F converges point-wise if $\mathrm{Re}\, s > \lambda$
and it is possible to anti-transform term-wise:

$$f(t) = \mathcal{L}^{-1}\{F\} = \sum_{k=0}^{+\infty} \mathcal{L}^{-1}\{F_k\} = \sum_{k=0}^{+\infty} f_k(t)\,. \tag{6.44}$$

6.1.8 Differential equations with support conditions

The Laplace transform in the distributional framework allows to find the solution with support contained in $[0, +\infty)$ of a linear ordinary differential equation in $\mathbb{R}$. For instance consider the following problem.

$$\begin{cases} \text{Given a } \mathcal{L} \text{ transformable distribution } g\,, \\ \text{find the } \mathcal{L} \text{ transformable distribution } w \text{ that solves} \\ Lw = g \text{ in } \mathcal{D}'(\mathbb{R})\,. \end{cases} \tag{6.45}$$

where $L = \sum_{j=0}^{k} c_j\, d^j/dt^j$ is a constant coefficients linear differential operator of order k ($c_k \neq 0$) and the derivatives have to be understood in the distributional sense.
The differential equation $Lw = g$ in the problem (6.45) has the explicit form

$$c_k\, w^{(k)}(t) + \cdots + c_1 w'(t) + c_0 w(t) = g(t) \qquad \text{in } \mathcal{D}'(\mathbb{R})\,.$$

Unlike (6.31), the problem (6.45) considered here has no initial values: actually they are implicitly required by the support condition together with the fact that the differential equation holds in the whole $\mathbb{R}$.
We denote by $\mathcal{P}$ the polynomial of degree k associated to the differential operator L, that is $\mathcal{P}(s) = \sum_{j=0}^{k} c_j\, s^j$, then we transform both sides of the differential equation. By setting $W = \mathcal{L}\{w\}$, due to (6.36), we obtain an algebraic equation whose unknown is the transform W of the solution w:

$$\mathcal{P}(s)\, W(s) = G(s)\,.$$

Polynomial $\mathcal{P}$ has a finite number of zeros s_m, Therefore it is possible to divide by $\mathcal{P}(s)$ in the half-plane $\{\, \mathrm{Re}\, s > \max_m \mathrm{Re}\, s_m \,\}$:

$$W(s) = G(s)/\mathcal{P}(s)\,.$$

W is the transform of an $\mathcal{L}$ transformable distribution due to (6.42), (6.43). By anti-transforming W we get the unique solution w of the problem (6.45):

$$w = u * g \qquad \text{where } u = \mathcal{L}^{-1}\{1/\mathcal{P}\}\,.$$

This u is the fundamental solution (Section 3.1.4) of L with support in $[0, +\infty)$.

This technique can be applied also to systems of ordinary differential equations whenever one looks for solutions with support in $\mathbb{R}^+$.

6.2 Exercises

6.2.1 *Computation of Laplace transforms and anti-transforms*

Exercise 6.1. Ascertain which one of these functions are $\mathcal{L}$ transformable and, in the affirmative cases, compute the abscissa of convergence and the $\mathcal{L}$ transform:

$$f_1(t) = H(t)\cos(\omega t)\,, \qquad f_2(t) = H(t)\sin(\omega t)\,, \qquad f_3(t) = H(t)\sum_{k=0}^{m} \alpha_k t^k \ (\alpha_k \in \mathbb{R})\,,$$

$$f_4(t) = H(t)\cosh(\omega t)\,, \qquad f_5(t) = H(t)\sinh(\omega t)\,,$$

where ω is a real parameter such that $\omega > 0$.

Solution - All functions are $\mathcal{L}$ transformable. Indeed they all have support contained in $[0, +\infty)$; moreover, f_1, f_2, f_3 have the abscissa of convergence equal to 0 (multiplication by $e^{-\lambda t}$ makes them integrable if $\lambda > 0$; not integrable if $\lambda \leq 0$), whereas f_4, f_5 have abscissa of convergence equal to ω (multiplication by $e^{-\lambda t}$ makes them integrable if $\lambda > \omega$, not integrable if $\lambda \leq \omega$).

By (6.18) and the identities $\cos(\omega t) = \frac{e^{i\omega t}+e^{-i\omega t}}{2}$, $\sin(\omega t) = \frac{e^{i\omega t}-e^{-i\omega t}}{2i}$ we get, for any $s \in \mathbb{C}$ such that $\mathrm{Re}\, s > 0$,

$$\mathcal{L}\{H(t)\cos(\omega t)\} = \frac{1}{2}\mathcal{L}\{e^{i\omega t} + e^{-i\omega t}\} = \frac{1}{2}\,\frac{1}{s - i\omega} + \frac{1}{2}\,\frac{1}{s + i\omega} = \frac{s}{s^2 + \omega^2}\,,$$

$$\mathcal{L}\{H(t)\sin(\omega t)\} = \frac{1}{2i}\mathcal{L}\{e^{i\omega t} - e^{-i\omega t}\} = \frac{1}{2i}\,\frac{1}{s - i\omega} - \frac{1}{2i}\,\frac{1}{s + i\omega} = \frac{\omega}{s^2 + \omega^2}\,.$$

By exploiting (6.5), (6.9) we obtain, for every $s \in \mathbb{C}$ such that $\mathrm{Re}\, s > 0$,

$$\mathcal{L}\left\{\sum_{k=0}^{m} \alpha_k t^k\right\} = \sum_{k=0}^{m} \alpha_k\, \mathcal{L}\{t^k\} = \sum_{k=0}^{m} \alpha_k\, \frac{k!}{s^{k+1}}\,.$$

By (6.18) and the identities $\cosh(\omega t) = \frac{e^{\omega t}+e^{-\omega t}}{2}$, $\sinh(\omega t) = \frac{e^{\omega t}-e^{-\omega t}}{2}$ we get, for every $s \in \mathbb{C}$ s.t. $\mathrm{Re}\, s > 0$,

$$\mathcal{L}\{\cosh(\omega t)\} = \frac{1}{2}\mathcal{L}\{e^{\omega t} + e^{-\omega t}\} = \frac{1}{2}\,\frac{1}{s - \omega} + \frac{1}{2}\,\frac{1}{s + \omega} = \frac{s}{s^2 - \omega^2};$$

$$\mathcal{L}\{\sinh(\omega t)\} = \frac{1}{2}\mathcal{L}\{e^{\omega t} - e^{-\omega t}\} = \frac{1}{2}\,\frac{1}{s - \omega} - \frac{1}{2}\,\frac{1}{s + \omega} = \frac{\omega}{s^2 - \omega^2}.$$

Remark - A wide range of properties and transforms can be found in the Tables 10.7 and 10.7.1.
□

Exercise 6.2. Establish which one of the functions listed below are $\mathcal{L}$ transformable functions and, in the affirmative cases, determine the abscissa of convergence:

$$f_1(t) = H(t)\,\arctan t\,, \qquad f_2(t) = H(t)\,\ln t\,, \qquad f_3(t) = H(t)\,\frac{1}{t}\,,$$

$$f_4(t) = H(t)\,e^{-t^2}\,, \qquad f_5(t) = H(t)\,e^{t^2 - t}\,, \qquad f_6(t) = H(t)\,\frac{\sinh t}{t}\,.$$

Solution - f_1, f_2, f_4, ed f_6 are $\mathcal{L}$ transformable functions: $\lambda_{f_1} = \lambda_{f_2} = 0$, $\lambda_{f_4} = -\infty$, $\lambda_{f_6} = 1$. Whereas f_3 and f_5 are not $\mathcal{L}$ transformable functions. □

Exercise 6.3. Find the real values α that make the function

$$f_\alpha(t) = e^{\alpha(t^\alpha + \ln t)}$$

an $\mathcal{L}$ transformable function and compute the corresponding abscissae of convergence.

Solution - We have

$$f_\alpha(t)\, e^{-st} = t^\alpha\, e^{\alpha t^\alpha - s\,t}\,.$$

The integral over $[0,1]$ is convergent for every α (and every s). The integral over $[1,+\infty]$ is convergent (for some s) if and only if $\alpha \leq 1$.
The values of α that render f is an $\mathcal{L}$ transformable function are the ones fulfilling the condition: $\alpha \leq 1$. The related abscissae of convergence are: $\lambda_{f_1} = 1$, $\lambda_{f_\alpha} = 0$ if $\alpha < 1$. □

Exercise 6.4. For each one functions listed below say whether it is the Laplace transform of an $\mathcal{L}$ transformable function:

$$\frac{1}{s^2}\,, \quad \frac{1}{|s|}\,, \quad \frac{1}{\operatorname{Re} s}\,, \quad \frac{1}{\cos s}\,.$$

Solution - $1/s^2$ is the Laplace transform of the $\mathcal{L}$ transformable function $t^+ = t\,H(t)$ that is $1/s^2 = \mathcal{L}\{t\,H(t)\}$. None among the others satisfies the necessary conditions (F must be analytic in a right half-plane and fulfil $\lim_{\operatorname{Re} s\to+\infty} F(s)=0$, whenever $F = \mathcal{L}\{f\}$ for an $\mathcal{L}$ transformable function f): $1/|s|$ and $1/(\operatorname{Re} s)$ are not analytic functions (since they do not fulfil the Cauchy-Riemann conditions: they are real-valued and nonconstant); eventually $1/\cos s$ is not analytic in any right half-plane because of the infinitely many simple poles at $s = \pi/2 + k\pi$, $k \in \mathbb{N}$. □

Exercise 6.5. Prove that none of the following functions is the Laplace transform of an $\mathcal{L}$ transformable function:

$$e^{1/s}\,, \qquad e^{-|s|}\,, \qquad e^{-s^2}\,, \qquad \frac{e^{-s}}{\sin s}\,, \qquad e^{-s}\,.$$

Solution - $e^{1/s}$ is not an $\mathcal{L}$ transform since

$$\lim_{\operatorname{Re} s\to+\infty} e^{1/s} = 1 \neq 0$$

contradicts property (6.15).
$e^{-|s|}$ is not an $\mathcal{L}$ transform since it is a real-valued nonconstant function. Therefore it is not analytic in any right half-plane: a contradiction with the necessary conditions.
e^{-s^2} is not a transform because it is not infinitesimal as $\operatorname{Re} s$ goes $+\infty$: for instance, take into account the half-line $s = x\,(1+i)$ with $x > 0$, then

$$|e^{-s^2}| = |e^{-x^2(1+i)^2}| = |e^{-2ix^2}| = 1 \qquad \forall x \in \mathbb{R}.$$

$e^{-s}/\sin s$ is not a transform because for any $\lambda \in \mathbb{R}$, it has infinitely many simple poles on $\mathbb{R}$, that are bigger than λ ($s = k\pi$, $k \in \mathbb{N}$, $k\pi > \lambda$), therefore it can not be analytic in any right half-plane.
Though e^{-s} verifies the necessary conditions, e^{-s} is not the Laplace transform of an $\mathcal{L}$ transformable function. Indeed arguing by contradiction: if it were, then by setting $F(s) = e^{-s}$, $f = \mathcal{L}^{-1}\{f\}$, due to (6.8) we would have $\mathcal{L}\{(1-t)f(t)\} = F + F' \equiv 0$, that is $(1-t)f(t) = 0$ a.e. (by (6.29)), hence $f(t) = 0$ a.e., a contradiction with the injectivity of $\mathcal{L}$. □

Exercise 6.6. Compute $\mathcal{L}\left\{e^{3t}\int_0^t \frac{\sin(2x)}{x}\,dx\right\}$.

Solution - It is known that $\mathcal{L}\{\sin\,t\} = \frac{1}{s^2+1}$; by (6.16), (6.14), (6.13), (6.18) we get:

$$\mathcal{L}\{\sin(2t)\} = \frac{2}{s^2+4}$$

$$\mathcal{L}\left\{\frac{\sin(2t)}{t}\right\} = \int_s^\infty \frac{2}{\sigma^2+4}\,d\sigma = \left[\arctan\left(\frac{\sigma}{2}\right)\right]_s^\infty = \frac{\pi}{2} - \arctan\frac{s}{2}$$

$$\mathcal{L}\left\{\int_0^t \frac{\sin(2x)}{x}\,dx\right\} = \frac{1}{s}\,\mathcal{L}\left\{\frac{\sin(2t)}{t}\right\} = \frac{1}{s}\left(\frac{\pi}{2} - \arctan\frac{s}{2}\right)$$

$$\mathcal{L}\left\{e^{3t}\int_0^t \frac{\sin(2x)}{x}\,dx\right\} = \frac{1}{s-3}\left(\frac{\pi}{2} - \arctan\frac{s-3}{2}\right)\,.$$

□

Exercise 6.7. Compute $\mathcal{L}\left\{e^{-t^2}\right\}$.

Solution - Let s be real.

$$\mathcal{L}\left\{e^{-t^2}\right\} = \int_0^\infty e^{-st}e^{-t^2}\,dt = e^{s^2/4}\int_0^\infty e^{-(t+s/2)^2}\,dt \overset{\begin{cases} t+s/2=\tau \\ dt = d\tau\end{cases}}{=}$$

$$= e^{s^2/4}\int_{s/2}^\infty e^{-\tau^2}\,d\tau = \frac{\sqrt{\pi}}{2}e^{s^2/4}\,\operatorname{erfc}\left(\frac{s}{2}\right)$$

where erfc stands for the complementary error function.

Consider the entire analytic extension $\operatorname{erfc}(z)$ of $\operatorname{erfc}(x)$. Since $\mathcal{L}\left\{e^{-t^2}\right\}$ is defined for every real s, it is an entire analytic function. But $\frac{\sqrt{\pi}}{2}\,e^{s^2/4}\,\operatorname{erfc}\left(\frac{s}{2}\right)$ is an entire function too and it coincides with $\mathcal{L}\left\{e^{-t^2}\right\}$ for real s. The uniqueness of analytic extension entails

$$\mathcal{L}\left\{e^{-t^2}\right\} = \frac{\sqrt{\pi}}{2}e^{\frac{s^2}{4}}\,\operatorname{erfc}\left(\frac{s}{2}\right) \qquad \forall s\in\mathbb{C}.$$

We can write explicitly the Taylor expansion of $\operatorname{erfc}(z)$. For every $x\in\mathbb{R}$ we have:

$$\operatorname{erfc}(x) = 1-\operatorname{erf}(x) = 1-\frac{2}{\sqrt{\pi}}\int_0^x e^{-t^2}\,dt = 1-\frac{2}{\sqrt{\pi}}\int_0^x\sum_{n=0}^\infty\frac{(-1)^n t^{2n}}{n!}\,dt =$$

$$= 1-\frac{2}{\sqrt{\pi}}\sum_{n=0}^\infty\frac{(-1)^n}{n!}\int_0^x t^{2n}\,dt = 1-\frac{2}{\sqrt{\pi}}\sum_{n=0}^\infty\frac{(-1)^n}{n!}\,\frac{x^{2n+1}}{2n+1}\,.$$

The above computations are justified by the fact that the exponential series has infinite radius of convergence. Also the last series has infinite radius of convergence. Then

$$\operatorname{erfc}(z) = 1-\frac{2}{\sqrt{\pi}}\sum_{n=0}^\infty\frac{(-1)^n}{n!}\,\frac{z^{2n+1}}{2n+1} \qquad \forall z\in\mathbb{C}. \tag{6.46}$$

An alternative approach to find an explicit representation of erfc can be obtained starting from Theorem 1.20: if $h(z)$ is an entire function then the integral $\int_0^z h(w)\,dw$ over any path joining 0 to z does not depend on the path and defines an entire function $H(z)$, with $H'(z) = h(z)$. So the map

$$z \mapsto 1 - \frac{2}{\sqrt{\pi}} \int_0^z e^{-w^2}\, dw \qquad z \in \mathbb{C}$$

is an entire function of complex variable that coincides with $\mathrm{erfc}(z)$ for every $z \in \mathbb{R}$, thus

$$\mathrm{erfc}(z) = 1 - \frac{2}{\sqrt{\pi}} \int_0^z e^{-w^2}\, dw\,. \tag{6.47}$$

Remark - We introduce a symbol for the normalized antiderivative of the gaussian: it is labeled erf and it is called **error function**, due to the important role it plays in the error measurement theory. The definition is:

$$\mathrm{erf}(x) = \frac{2}{\sqrt{\pi}} \int_0^x e^{-t^2}\, dt \qquad \forall x \in \mathbb{R}\,. \tag{6.48}$$

erf fulfils $\mathrm{erf}(x) + \mathrm{erfc}(x) \equiv 1$, it has an entire analytic extension, with Taylor expansion:

$$\mathrm{erf}(z) = \frac{2}{\sqrt{\pi}} \sum_{n=0}^{\infty} \frac{(-1)^n}{n!} \frac{z^{2n+1}}{2n+1} \qquad \forall z \in \mathbb{C}.$$

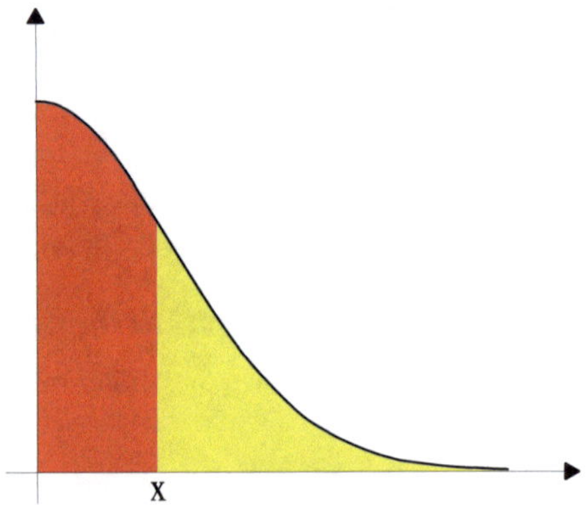

Figure 6.1 - The numerical values $\mathrm{erf}(x)$ and $\mathrm{erfc}(x)$ provide respectively the areas of the red and yellow region in the subgraph of $2e^{-x^2}/\sqrt{\pi}$.

□

Exercise 6.8. Compute $\mathcal{L}\left\{e^{-t^2} \sin\, t\right\}$ and $\mathcal{L}\{\mathrm{erf}(t)\}$.

Solution -

$$\mathcal{L}\left\{e^{-t^2} \sin\, t\right\} = \frac{1}{2i}\mathcal{L}\left\{e^{-t^2} e^{it}\right\} - \frac{1}{2i}\mathcal{L}\left\{e^{-t^2} e^{-it}\right\} =$$
$$= \frac{\sqrt{\pi}}{4i}\left[e^{(s-i)^2/4}\, \mathrm{erfc}\left(\frac{s-i}{2}\right) - e^{(s+i)^2/4}\, \mathrm{erfc}\left(\frac{s+i}{2}\right)\right]$$

where erf and erfc are defined by (6.48) and (6.46).

$$\mathcal{L}\{\mathrm{erf}(t)\} = \frac{2}{\sqrt{\pi}}\, \frac{1}{s}\, \mathcal{L}\left\{e^{-t^2}\right\} = \frac{1}{s}\, e^{s^2/4}\, \mathrm{erfc}\left(\frac{s}{2}\right).$$

□

Exercise 6.9. Let f be an $\mathcal{L}$ transformable function with $f(0) = 1$ and $f'(0) = 0$. Set

$$g(t) = \int_0^t \frac{\sin[2(t-\tau)]}{t-\tau}\, e^{-\tau}\, f''(\tau)\, d\tau,$$

and compute $G = \mathcal{L}\{g\}$ in terms of $F = \mathcal{L}\{f\}$.

Solution - g is the convolution of two $\mathcal{L}$ transformable functions:

$$g(t) = \frac{\sin(2t)}{t} * [e^{-t}\, f''(t)]$$

Therefore, by (6.20) we obtain:

$$G = \mathcal{L}\left\{\frac{\sin(2t)}{t}\right\} \cdot \mathcal{L}\left\{e^{-t}\, f''(t)\right\}$$

We make the two factors of G explicit:

$$\mathcal{L}\left\{\frac{\sin(2t)}{t}\right\} = \int_s^\infty \frac{2}{\sigma^2+4}\, d\sigma = \frac{\pi}{2} - \arctan\frac{s}{2}$$

$$\mathcal{L}\{f''\} = s^2\, F(s) - s\, f(0) - f'(0) = s^2\, F(s) - s$$

$$\mathcal{L}\{e^{-t}\, f''\} = (s+1)^2\, F(s+1) - (s+1)$$

$$G(s) = \left(\frac{\pi}{2} - \arctan\frac{s}{2}\right)(s+1)[(s+1)\, F(s+1) - 1].$$

□

Exercise 6.10. Set

$$f(t) = \begin{cases} 1 & 0 \le t < 1 \\ 2-t & 1 \le t \le 2 \\ 0 & 2 < t \end{cases} \qquad g(t) = \begin{cases} t & 0 \le t < 1 \\ 2-t & 1 \le t < 2 \\ t-2 & 2 \le t\,. \end{cases}$$

Compute $\mathcal{L}\{f\}$ and $\mathcal{L}\{g\}$ and their respective abscissae of convergence.

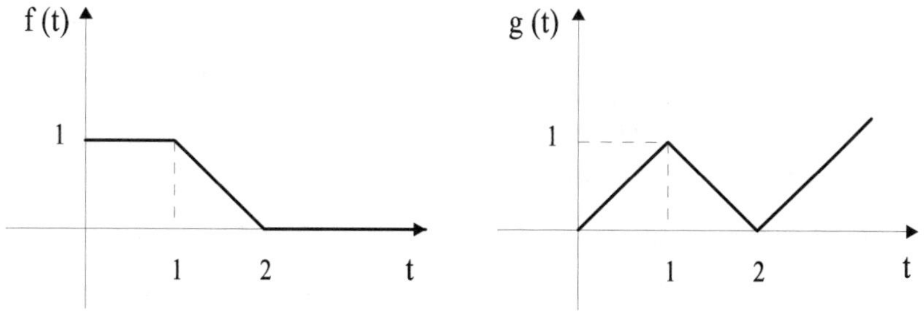

Figure 6.2 - Graphs of f e g : Exercise 6.10 .

Solution - In these cases we can use the so called graphic addition method:

$$f(t) = H(t) - H(t-1) + (2-t)[H(t-1) - H(t-2)] = H(t) - (t-1)\, H(t-1) + (t-2)\, H(t-2)$$

where $H(t)$ denotes the Heaviside function. Therefore:

$$\mathcal{L}\{f\} = \frac{1}{s} - \frac{e^{-s}}{s^2} + \frac{e^{-2s}}{s^2}, \qquad \lambda_f = 0\,.$$

When dealing with $g(t)$, we can apply the same method used for f:

$$g(t) = t[H(t) - H(t-1)] + (2-t)[H(t-1) - H(t-2)] + (t-2)\, H(t-2) =$$
$$= t - 2(t-1)\, H(t-1) + 2(t-2)\, H(t-2)\,.$$

Therefore: $$\mathcal{L}\{g\} = \frac{1}{s^2} - 2\frac{e^{-s}}{s^2} + 2\frac{e^{-2s}}{s^2}\,, \qquad \lambda_g = 0\,.$$

□

Exercise 6.11. Compute

$$\mathcal{L}\left\{\left(1+t^4\right)\, e^{-3t+1}\, \frac{\cos(2t)}{\sqrt{t}}\right\}\,.$$

Solution - We set

$$f(t) = e \cdot e^{-3t}\, \cos(2t)\left(\frac{1}{\sqrt{t}} + t^{7/2}\right) = \frac{e}{2}\, \left[e^{(-3+2i)t} + e^{(-3-2i)t}\right]\left(\frac{1}{\sqrt{t}} + t^{7/2}\right)$$

and denote the Euler Gamma function by Γ (see Exercise 1.26), we have

$$\mathcal{L}\{t^\alpha\} = \frac{\Gamma(\alpha+1)}{s^{\alpha+1}} \qquad \operatorname{Re}\alpha > -1$$
$$\Gamma\left(\frac{1}{2}\right) = \sqrt{\pi} \qquad \Gamma(\alpha+1) = \alpha\, \Gamma(\alpha)$$
$$\Gamma\left(\frac{9}{2}\right) = \frac{7}{2}\, \Gamma\left(\frac{7}{2}\right) = \frac{7}{2}\,\frac{5}{2}\, \Gamma\left(\frac{5}{2}\right) = \frac{7}{2}\,\frac{5}{2}\,\frac{3}{2}\,\frac{1}{2}\, \Gamma\left(\frac{1}{2}\right) = \frac{105}{16}\, \sqrt{\pi}$$
$$\mathcal{L}\left\{\frac{1}{\sqrt{t}} + t^{7/2}\right\} = \frac{\Gamma\left(\frac{1}{2}\right)}{\sqrt{s}} + \frac{\Gamma\left(\frac{9}{2}\right)}{s^{9/2}} = \sqrt{\frac{\pi}{s}} + \frac{105}{16}\,\frac{\sqrt{\pi}}{s^{9/2}}$$

$$\mathcal{L}\{f\} = \frac{e\sqrt{\pi}}{2}\left[\frac{1}{\sqrt{s+3-2i}} + \frac{1}{\sqrt{s+3+2i}}\right] + \frac{105 e\sqrt{\pi}}{32}\left[(s+3-2i)^{-9/2} + (s+3+2i)^{-9/2}\right].$$

□

Exercise 6.12. Compute $\underbrace{H * H * \cdots * H}_{k \text{ factors}}$, $k \in \mathbb{N}$, where H is the Heaviside function.

Solution - Using the definition of convolution (6.19) we get:

$$H * H = t^+\,, \qquad H * H * H = (H * H) * H = t^+ * H(t) = \frac{t^2}{2}\, H(t)\,,$$

however, if k is large, the direct computation is nontrivial (it would be anyway easy to guess the general expression and proving it by induction).
An efficient method to obtain directly the general result consists in computing first the $\mathcal{L}$ transform and afterwards the inverse transform:

$$\mathcal{L}\{\, H * H * \cdots * H\,\} = \frac{1}{s^k}$$

$$\underbrace{H * H * \cdots * H}_{k \text{ factors}} = \mathcal{L}^{-1}\left\{\frac{1}{s^k}\right\} = \frac{t^{k-1}}{(k-1)!}\, H(t)\,.$$

□

Exercise 6.13. Compute

$$(a)\,\mathcal{L}^{-1}\left\{\frac{1}{(s+2)^3}\right\},\quad (b)\,\mathcal{L}^{-1}\left\{\frac{e^{-4s}}{s^2+1}\right\},\quad (c)\,\mathcal{L}^{-1}\left\{\frac{1}{s(s+5)^2}\right\},\quad (d)\,\mathcal{L}^{-1}\left\{\frac{s}{(s+3)^4}\right\}.$$

Solution - They are all transforms of $\mathcal{L}$ transformable functions, because they are proper rational functions whose denominator degree is larger than the numerator degree.
(a) By $\mathcal{L}\{(t^2/2)H(t)\} = 1/s^3$ and $\mathcal{L}\{e^{\alpha t}f(t)\} = F(s-\alpha)$ we obtain

$$\mathcal{L}\{e^{-2t}(t^2/2)H(t)\} = 1/(s+2)^3\,,$$

$$\mathcal{L}^{-1}\{1/(s+2)^3\} = e^{-2t}(t^2/2)H(t)\,.$$

(b) By (6.17) and $\mathcal{L}\{\sin(t)H(t)\} = 1/(s^2+1)$ we obtain

$$\mathcal{L}\{\,\sin(t-4)H(t-4)\,\} = \frac{e^{-4s}}{s^2+1}\,,$$

$$\mathcal{L}^{-1}\left\{\frac{e^{-4s}}{s^2+1}\right\} = \sin(t-4)H(t-4)\,.$$

(c) By (6.8),(6.13),(6.18), we obtain

$$\mathcal{L}\left\{e^{-5t}tH(t)\right\} = \frac{1}{(s+5)^2}\,,$$

$$\mathcal{L}\left\{\int_0^t \tau e^{-5\tau}\,d\tau\right\} = \frac{1}{s(s+5)^2}\,,$$

$$\mathcal{L}^{-1}\left\{\frac{1}{s(s+5)^2}\right\} = H(t)\int_0^t \tau e^{-5\tau}\,d\tau\,.$$

(d) By (6.18),(6.10) we obtain

$$\mathcal{L}\left\{e^{-3t}\,\frac{t^3}{3!}\,H(t)\right\} = \frac{1}{(s+3)^4}\,,$$

$$\mathcal{L}\left\{H(t)\,\frac{d}{dt}\left(e^{-3t}\frac{t^3}{3!}\right)\right\} = \frac{s}{(s+3)^4} - \left[e^{-3t}\frac{t^3}{3!}\right]_{t=0}\,,$$

$$\mathcal{L}^{-1}\left\{\frac{s}{(s+3)^4}\right\} = \frac{e^{-3t}}{2}\,(t^2-t^3)\,H(t)\,.$$

□

Exercise 6.14. We define the *(unnormalized) cardinal sine* (denoted by the symbol Sinc), the *normalized cardinal sine* (denoted by the symbol sinc) and the *integral sine* (denoted with the symbol Si) as follows:

$$\mathrm{Sinc}(t) := \frac{\sin t}{t}\,,\qquad \mathrm{sinc}(t) := \frac{\sin(\pi\,t)}{\pi\,t}\,,\qquad \mathrm{Si}(t) := \int_0^t \frac{\sin x}{x}\,dx = \int_0^t \mathrm{Sinc}(x)\,dx\,.$$

Prove that

a) $\mathrm{Si}(t) = \dfrac{\pi}{2} - \displaystyle\int_t^{+\infty} \frac{\sin(x)}{x}\,dx$ where $\int$ denotes the generalized integral.

b) $\mathrm{Si}(t) = \displaystyle\sum_{k=0}^{+\infty} \frac{(-1)^k\,t^{2k+1}}{(2k+1)\,(2k+1)!}\,.$

c) $\mathcal{L}\left\{\mathrm{Sinc}(t)\,H(t)\right\}(s) = \dfrac{\pi}{2} - \arctan(s) = \arctan\left(\dfrac{1}{s}\right).$

d) $\mathcal{L}\{\mathrm{Si}(t)\,H(t)\}(s) = \frac{1}{s}\left(\frac{\pi}{2} - \arctan(s)\right) = \frac{1}{s}\arctan\left(\frac{1}{s}\right).$

e) $\int_{\mathbb{R}} \mathrm{Sinc}\,t\,dt = \pi\,, \qquad \int_{\mathbb{R}} \mathrm{sinc}\,t\,dt = 1\,.$

Solution - a) It follows from the identity p.v. $\int_{\mathbb{R}} \mathrm{Sinc}(x)\,dx = \pi$ (see Exercise 1.22) and by the fact that Sinc is an even function.

b) The Taylor series of the sine function converges uniformly in compact sets, therefore it is possible to integrate the series term-wise:

$$\mathrm{Si}(t) = \int_0^t \frac{\sin x}{x}\,dx = \int_0^t \sum_{k=0}^{+\infty}(-1)^k \frac{x^{2k}}{(2k+1)!}\,dx = \sum_{k=0}^{+\infty}(-1)^k \int_0^t \frac{x^{2k}}{(2k+1)!}\,dx =$$
$$= \sum_{k=0}^{+\infty} \frac{(-1)^k t^{2k+1}}{(2k+1)(2k+1)!}$$

c) $\mathcal{L}\{H(t)\sin(t)\} = 1/(1+s^2)$ e (6.14) $\Rightarrow \mathcal{L}\left\{H(t)\frac{\sin(t)}{t}\right\} = \int_s^{\infty}(1+\sigma^2) = \frac{\pi}{2} - \arctan(s).$

d) From c) and (6.13) we deduce $\mathcal{L}\{H(t)\,\mathrm{Si}(t)\} = \frac{1}{s}\left(\frac{\pi}{2} - \arctan(s)\right).$

e) Straightforward consequences of Exercise 1.22. □

Exercise 6.15. Compute the inverse Laplace transform of

$$F(s) = \frac{2\,e^{-3s}\,\cosh s}{s^2 + s^3}$$

Solution - We have

$$F(s) = \left(\frac{1}{s+1} - \frac{1}{s} + \frac{1}{s^2}\right)\left(e^{-2s} + e^{-4s}\right)$$

$$\mathcal{L}^{-1}\{F\} = e^{-(t-2)}\,H(t-2) - H(t-2) + (t-2)\,H(t-2) + e^{-(t-4)}\,H(t-4) +$$
$$-H(t-4) + (t-4)\,H(t-4) = \begin{cases} 0 & 0 < t < 2 \\ e^{-(t-2)} + t - 3 & 2 < t < 4 \\ e^{-t}(e^2 + e^4) + 2t - 8 & 4 < t\,. \end{cases}$$

□

Exercise 6.16. By exploiting $\mathcal{L}\{\ln t\} = \frac{\Gamma'(1) - \ln s}{s}$ (see the previous exercise), compute

$$\mathcal{L}^{-1}\left\{\frac{\ln s}{s^2}\right\} \qquad \text{e} \qquad \mathcal{L}^{-1}\left\{\frac{\ln(s+2)}{(s+1)^2}\right\}.$$

Solution - Since $\mathcal{L}\{\Gamma'(1)\} = \frac{\Gamma'(1)}{s}$, we have that $\mathcal{L}\{\Gamma'(1) - \ln t\} = \frac{\ln s}{s}$

$$\mathcal{L}^{-1}\left\{\frac{\ln s}{s^2}\right\} = \int_0^t (\Gamma'(1) - \ln\tau)\,d\tau = \Gamma'(1)\,t - (t\,\ln t - t) = (1 + \Gamma'(1))\,t - t\,\ln t$$

To compute $\mathcal{L}^{-1}\left\{\frac{\ln(s+2)}{(s+1)^2}\right\}$ we can use the first part of the exercise:

$$\mathcal{L}^{-1}\left\{\frac{\ln(s+2)}{(s+1)^2}\right\} = \mathcal{L}^{-1}\left\{\frac{\ln(s+2)}{(s+2)}\,\frac{(s+2)}{(s+1)^2}\right\} = \mathcal{L}^{-1}\left\{\frac{\ln(s+2)}{(s+2)}\right\} * \mathcal{L}^{-1}\left\{\frac{(s+2)}{(s+1)^2}\right\} =$$

$$= \left\{e^{-2t}\,[\Gamma'(1) - \ln t]\right\} * \left(\mathcal{L}^{-1}\left\{\frac{1}{(s+1)}\right\} + \mathcal{L}^{-1}\left\{\frac{1}{(s+1)^2}\right\}\right) =$$

$$= \left\{e^{-2t}\,[\Gamma'(1) - \ln t]\right\} * \left[e^{-t}\,(1+t)\right] = \int_0^t e^{-2\tau}\,[\Gamma'(1) - \ln \tau]\, e^{-(t-\tau)}\,(1+t-\tau)\,d\tau =$$

$$= \Gamma'(1)\,e^{-t}\int_0^t e^{-\tau}\,(1+t-\tau)\,d\tau - e^{-t}\int_0^t e^{-\tau}\,\ln\tau\,(1+t-\tau)\,d\tau =$$

$$= \Gamma'(1)\,t\,e^{-t} - e^{-t}\int_0^t e^{-\tau}\,\ln\tau\,(1+t-\tau)\,d\tau\,.$$

The second integral can not be expressed elementary. □

Exercise 6.17. Prove

$$\mathcal{L}^{-1}\left\{\frac{1}{\sqrt{1+s^2}}\right\} \quad = \quad J_0(t)\,H(t)\;.$$

Solution - Notice that, due to the identity

$$\frac{1}{\sqrt{1+s^2}} - \frac{1}{s} \quad = \quad \frac{-1}{s\sqrt{1+s^2}\,(s+\sqrt{1+s^2})}\,,$$

the function $F(s) = (1+s^2)^{-1/2} - 1/s$ verifies the sufficient condition (6.25) to be the Laplace transform of an $\mathcal{L}$ transformable function; since $1/s = \mathcal{L}\{H\}$, we conclude that also $(1+s^2)^{-1/2}$ is the Laplace transform of an $\mathcal{L}$ transformable function.
By recalling the binomial series:

$$(1+z)^\alpha \;=\; \sum_0^\infty \binom{\alpha}{n} z^n \qquad \text{where} \quad \binom{\alpha}{n} \overset{\text{def}}{=} \frac{1}{n!}\alpha(\alpha-1)\cdots(\alpha-n+1)\,,$$

$$F(s) = \left(1+s^2\right)^{-1/2} = \frac{1}{s}\left(1+\frac{1}{s^2}\right)^{-1/2} = \sum_0^\infty \binom{-1/2}{n}\frac{1}{s^{2n+1}}\,,$$

then, by formally term-wise anti-transforming

$$\mathcal{L}^{-1}\{F\} \overset{?}{=} \sum_0^\infty \binom{-1/2}{n}\mathcal{L}^{-1}\left\{\frac{(2n)!}{s^{2n+1}}\right\} = \sum_0^\infty \binom{-1/2}{n} t^{2n} H(t) =$$
$$= \sum_0^\infty \frac{(-1)^n}{n!2^n\left((2n)!/(2n)!!\right)}\,t^{2n}H(t) = \sum_0^\infty \frac{(-1)^n}{(n!)^2}\left(\frac{t}{2}\right)^{2n} H(t) = J_0(t)H(t)$$

where, for any integer $m \geq -1$, the double factorial $m!!$ denotes: the product of the positive integers less or equal to m with the same parity of m, if $m > 0$; by convention $(-1)!! = 0!! = 1$. The equality denoted by $\overset{?}{=}$ is justified by Theorem 6.12 (with $\lambda > 1$) and

$$\sum_0^\infty \frac{1}{(n!)^2}\left(\frac{t}{2}\right)^{2n} = \sum_0^\infty \frac{1}{2^n n!}\,\frac{t^{2n}}{(2n)!!} \leq \sum_0^\infty \frac{t^{2n}}{((2n)!!)^2} \leq \sum_0^\infty \frac{t^{2n}}{(2n)!} \leq \sum_0^\infty \frac{t^n}{n!} = e^t.$$

□

Exercise 6.18. By using the Riemann-Fourier formula evaluate

$$\mathcal{L}^{-1}\left\{\frac{e^{-1/s}}{s}\right\}$$

Solution - For any fixed $\gamma > 0$, we have

$$\mathcal{L}^{-1}\left\{\frac{e^{-1/s}}{s}\right\} = \frac{1}{2\pi i}\int_{\gamma-i\infty}^{\gamma+i\infty}\frac{1}{s}\,e^{st-1/s}\,ds\,.$$

The analytic function (dependent on the complex variable s) that appears in the integral has an essential singularity in the origin:

$$\frac{1}{s}\,e^{st-1/s} = \frac{1}{s}\sum_{n=0}^{+\infty}\frac{s^n t^n}{n!}\sum_{m=0}^{+\infty}\frac{(-1)^m}{s^m m!} = \sum_{n,m=0}^{+\infty}\frac{(-1)^m\,t^n}{n!\,m!}\,s^{n-m-1}\,,$$

by selecting $n = m$ we obtain the residual in the origin

$$\text{res}\left(\frac{1}{s}\,e^{st-1/s}\,,\,0\right) = \sum_{n=0}^{+\infty}\frac{(-1)^n\,t^n}{(n!)^2} = J_0(2\sqrt{t})\,.$$

By setting $s = x + iy$ for every $s \in \mathbb{C}$ such that $|s| > 1$, we obtain: $|x| \le |s| \le |s|^2$ and

$$\left|\frac{1}{s}\,e^{-1/s}\right| = \frac{1}{|s|}\left|\exp\left(\frac{-x+iy}{x^2+y^2}\right)\right| = \frac{1}{|s|}\,e^{-x/|s|^2} \le \frac{e}{|s|}\,.$$

Therefore it is possible to apply the estimate (1.4) and (if $t > 0$) the Jordan Lemma (Lemma 1.34 (iii)) along the path $ABCDE$:

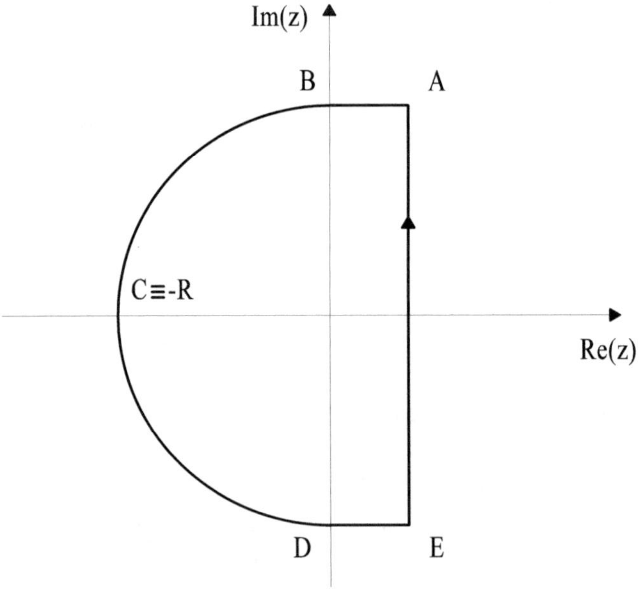

Figure 6.3 - Path $ABCDE$ in Exercise 6.18.

we get

$$\lim_{R\to+\infty}\int_{ABCDE}\frac{1}{s}\,e^{st-1/s}\,ds = 0\,.$$

By Cauchy Theorem (Theorem 1.20) we conclude

$$\mathcal{L}^{-1}\left\{\frac{e^{-1/s}}{s}\right\} = \operatorname{res}\left(\frac{1}{s}\, e^{st-1/s}\,,\, 0\right) = J_0(2\sqrt{t})\, H(t)\,.$$

□

Exercise 6.19. By exploiting the Riemann-Fourier formula, evaluate

$$\mathcal{L}^{-1}\left\{\frac{\Gamma'(1) - \ln s}{s}\right\}$$

where Γ is the Euler Gamma function and $\ln s$ denotes the real-valued natural logarithm if $s \in \mathbb{R}^+$.

Solution - We consider the complex plane with a cut along the half-line of the negative real numbers, we choose the following analytic branch of the logarithm

$$s = \rho\, e^{i\theta}, \ -\pi < \theta < \pi, \ \ln s = \ln\rho + i\theta$$

and we consider the path in Figure 6.4, where the convergence radius of the semi circle CDHI is R and ε is the radius of EFG.

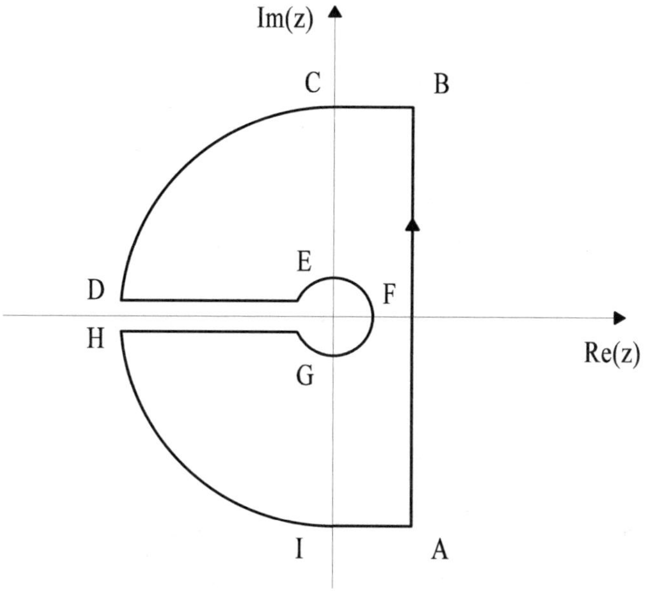

Figure 6.4 - Path $ABCDEFGHI$ in Exercise 6.19.

We denote by γ the Euler-Mascheroni constant: $\gamma = \Gamma'(1)$.
We set $f = \mathcal{L}^{-1}\left\{\frac{\gamma - \ln s}{s}\right\}$. By the Cauchy Theorem, we get

$$\begin{aligned} f(t) &= \frac{1}{2\pi i} \lim_{R\to\infty} \int_{AB} e^{st}\, \frac{\gamma - \ln s}{s}\, ds = \\ &= -\frac{1}{2\pi i} \lim_{\substack{R\to\infty \\ \varepsilon\to 0}} \int_{BCDEFGHIA} e^{st}\, \frac{\gamma - \ln s}{s}\, ds\,. \end{aligned} \tag{6.49}$$

Since $\left|\dfrac{\gamma - \ln s}{s}\right| \le C\,\dfrac{\ln\rho}{\rho}$, by applying the inequality (1.4) to the paths BC and IA and the Jordan Lemma (Lemma 1.34 (iii)) to the paths CD and HI with $t > 0$, we obtain

$$\lim_{R\to\infty}\left(\int_{BCD} + \int_{HIA}\right) e^{st}\,\frac{\gamma - \ln s}{s}\,ds \,=\, 0\,. \tag{6.50}$$

Along DE we have: $s = -\rho$, $\theta = \pi$, $\ln s = \ln\rho + i\pi$ and $ds = -d\rho$, therefore

$$\int_{DE} e^{st}\,\frac{\gamma - \ln s}{s}\,ds = \int_R^{\varepsilon} e^{-\rho t}\,\frac{\gamma - \ln\rho - i\pi}{-\rho}\,(-d\rho)\,. \tag{6.51}$$

Along GH we have: $s = -\rho$, $\theta = -\pi$, $\ln s = \ln\rho - i\pi$ and $ds = -d\rho$, thus

$$\int_{GH} e^{st}\,\frac{\gamma - \ln s}{s}\,ds = \int_{\varepsilon}^{R} e^{-\rho t}\,\frac{\gamma - \ln\rho + i\pi}{-\rho}\,(-d\rho). \tag{6.52}$$

By adding (6.51) and (6.52):

$$\begin{aligned}
\int_{DE\bigcup GH} e^{st}\frac{\gamma - \ln s}{s}\,ds \,&=\, 2\pi i\int_{\varepsilon}^{R}\frac{e^{-\rho t}}{\rho}\,d\rho \,= \\
&= 2\pi i\left(e^{-Rt}\,\ln R - e^{-\varepsilon t}\,\ln\varepsilon + t\int_{\varepsilon}^{R} e^{-\rho t}\,\ln\rho\,d\rho\right) = \boxed{t\rho = y\,,\ d\rho = dy/t} \\
&= 2\pi i\left(e^{-Rt}\,\ln R - e^{-\varepsilon t}\,\ln\varepsilon + \int_{t\varepsilon}^{tR} e^{-y}\,\ln y\,dy - \ln t\int_{t\varepsilon}^{tR} e^{-y}\,dy\right)\,.
\end{aligned} \tag{6.53}$$

Let us consider the four summands embraced by the parentheses:

$$e^{-Rt}\,\ln R \,\to\, 0 \qquad \text{se } R\to\infty\,, \tag{6.54}$$

$$\int_{t\varepsilon}^{tR} e^{-y}\,\ln y\,dy \,\to\, \gamma \qquad \text{se } R\to\infty\,,\ \varepsilon\to 0\,, \tag{6.55}$$

$$\ln t\int_{t\varepsilon}^{tR} e^{-y}\,dy \,\to\, \ln t \qquad \text{se } R\to\infty\,,\ \varepsilon\to 0\,, \tag{6.56}$$

whereas the term $e^{-\varepsilon t}\,\ln\varepsilon$ diverges as $\varepsilon\to 0$. On EFG we have: $s = \varepsilon\,e^{i\theta}$, $ds = i\varepsilon e^{i\theta}\,d\theta$

$$\int_{EFG} e^{st}\,\frac{\gamma - \ln s}{s}\,ds = \int_{\pi}^{-\pi} e^{\varepsilon t e^{i\theta}}\,\frac{\gamma - \ln\varepsilon - i\theta}{\varepsilon e^{i\theta}}\,i\varepsilon e^{i\theta}\,d\theta = -i\int_{-\pi}^{\pi} e^{\varepsilon t e^{i\theta}}(\gamma - \ln\varepsilon - i\theta)\,d\theta\,. \tag{6.57}$$

Since $e^{\varepsilon t e^{i\theta}}$ converges to 1 uniformly in θ as ε goes to zero, we get

$$\lim_{\varepsilon\to 0}\left[-i\int_{-\pi}^{\pi} e^{\varepsilon t e^{i\theta}}\,(\gamma - i\theta)\,d\theta\right] \,=\, -2\pi i\gamma\,. \tag{6.58}$$

Applying de L'Hôpital's rule, we obtain

$$\lim_{\varepsilon\to 0}\left[\ln\varepsilon\left(i\int_{-\pi}^{\pi} e^{\varepsilon t e^{i\theta}}\,d\theta \,-\, 2\pi i e^{-\varepsilon t}\right)\right] \,=\, 0\,. \tag{6.59}$$

Taking into account (6.49), (6.50), (6.53) - (6.59), we conclude

$$f(t) \,=\, H(t)\,\ln t\,.$$

□

Exercise 6.20. By using the Riemann-Fourier inversion formula, compute

$$\mathcal{L}^{-1}\left\{\frac{s}{s^2+\omega^2}\,e^{-a\sqrt{s}}\right\}$$

where $a,\,\omega\in\mathbb{R}$, $a>0$ and $\sqrt{s}$ denotes the arithmetic root if s is a real positive.
The inverse transform can be expressed in terms of an integral that cannot be elementarily computed.
Verify that the solution *at regime* has an oscillatory behavior, by proving that this integral is an infinitesimal as t goes to infinity.

Solution - We consider the complex plane with a cut along the half-line of negative real numbers and choose this analytic branch of $\sqrt{s}$:

$$s=\varrho\,e^{i\vartheta}\,,\qquad |\vartheta|<\pi\,,\qquad \sqrt{s}=\varrho^{1/2}e^{i\vartheta/2}\,.$$

For any fixed $x>0$, the Riemann-Fourier formula gives:

$$\mathcal{L}^{-1}\left\{\frac{s\,e^{-a\sqrt{s}}}{s^2+\omega^2}\right\}=\frac{1}{2\pi i}\int_{x-i\infty}^{x+i\infty}\frac{s\,e^{st-a\sqrt{s}}}{s^2+\omega^2}\,ds\,.$$

The function $\varphi(s)=\dfrac{s\,e^{st-a\sqrt{s}}}{s^2+\omega^2}$ is analytic in the complex plane except the half-line of negative real numbers and the two simple poles $\pm i\omega$.

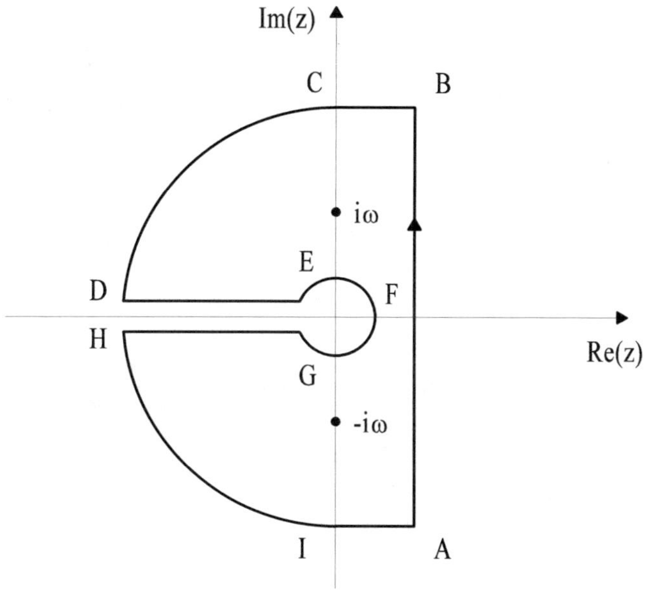

Figure 6.5 - Path $ABCDEFGHI$ in Exercise 6.20.

We refer to the loop $C_{R,\varepsilon}$ in Figure 6.5, with $0<\varepsilon<1$, $R>2$.

$$|\varphi(s)|\le\frac{C}{|s|}\,e^{xt}\qquad \forall s\in BC\bigcup IA \text{ se } R>2\,. \tag{6.60}$$

For any $s\in CD$ we have $\pi/2<\vartheta<\pi$, hence $\operatorname{Re}\sqrt{s}\ge 0$.

Then, by setting $f(s)=\dfrac{s\,e^{-a\sqrt{s}}}{s^2+\omega^2}$, we get

$$|f(s)| \le \frac{C}{|s|} \qquad \forall s \in CD\,.$$

Analogously we have

$$|f(s)| \le \frac{C}{|s|} \qquad \forall s \in HI\,.$$

If $t > 0$, the Jordan Lemma (Lemma 1.34(iii)) provides

$$\lim_{R\to+\infty} \int_{CD\bigcup HI} \varphi(s)\,ds = 0\,. \tag{6.61}$$

It is easy to check that

$$\lim_{\varepsilon\to 0} \int_{EFG} \varphi(s)\,ds = 0\,. \tag{6.62}$$

Let us consider now the two integrals over the two edges of the cut:

$$\int_{DE} \varphi(s)\,ds = \int_R^\varepsilon \frac{\varrho\, e^{-\varrho t - ia\sqrt{\varrho}}}{\varrho^2+\omega^2}\,d\varrho\,, \tag{6.63}$$

$$\int_{GH} \varphi(s)\,ds = \int_\varepsilon^R \frac{\varrho\, e^{-\varrho t + ia\sqrt{\varrho}}}{\varrho^2+\omega^2}\,d\varrho\,. \tag{6.64}$$

$$\operatorname{res}\big(\varphi(s), i\omega\big) = \frac{i\omega \exp(i\omega t - a\sqrt{\varrho}\,e^{i\pi/4})}{2\omega i} = \frac{1}{2}\exp\Big(-a\sqrt{\frac{\omega}{2}} + i\Big(\omega t - a\sqrt{\frac{\omega}{2}}\Big)\Big) \tag{6.65}$$

$$\operatorname{res}\big(\varphi(s), -i\omega\big) = \frac{-i\omega \exp(-i\omega t - a\sqrt{\varrho}\,e^{-i\pi/4})}{-\,2\omega i} = \frac{1}{2}\exp\Big(-a\sqrt{\frac{\omega}{2}} - i\Big(\omega t - a\sqrt{\frac{\omega}{2}}\Big)\Big). \tag{6.66}$$

By taking into account (6.60)-(6.66), we summarize

$$\begin{aligned}
&\lim_R \int_{AB} \varphi(s)\,ds = \\
&= 2\pi i\,\big(\operatorname{res}(\varphi, i\omega) + \operatorname{res}(\varphi, -i\omega)\big) - \lim_{R,\varepsilon}\Big(\int_{DE}\varphi(s)\,ds + \int_{GH}\varphi(s)\,ds\Big) \\
&= 2\pi i\, e^{-a\sqrt{\omega/2}}\cos(\omega t - a\sqrt{\omega/2}) - 2i\int_0^\infty \frac{\varrho\, e^{-\varrho t}}{\varrho^2+\omega^2}\sin(a\sqrt{\varrho})\,d\varrho\,.
\end{aligned}$$

$$\mathcal{L}^{-1}\left\{\frac{s\,e^{-a\sqrt{s}}}{s^2+\omega^2}\right\} = H(t)\left(e^{-a\sqrt{\omega/2}}\cos(\omega t - a\sqrt{\omega/2}) - \frac{1}{\pi}\int_0^\infty \frac{\varrho\, e^{-\varrho t}}{\varrho^2+\omega^2}\sin(a\sqrt{\varrho})\,d\varrho\right).$$

Now we examine the behavior of solutions at regime.

$$\left|\frac{\varrho\, e^{-\varrho t}}{\varrho^2+\omega^2}\sin(a\sqrt{\varrho})\right| \le \frac{\varrho\, e^{-\varrho t}}{\varrho^2+\omega^2} \in L^1(0,+\infty)\ \ \forall t \ge 1 \tag{6.67}$$

$$\lim_{t\to+\infty} \frac{\varrho\, e^{-\varrho t}}{\varrho^2+\omega^2}\sin(a\sqrt{\varrho}) = 0\ \ \forall \varrho > 0\,. \tag{6.68}$$

By the dominated convergence theorem (Theorem 2.5), (6.67) and (6.68) we get

$$\lim_{t\to+\infty} \int_0^\infty \frac{\varrho\, e^{-\varrho t}}{\varrho^2+\omega^2}\sin(a\sqrt{\varrho})\,d\varrho = 0\,,$$

therefore for t "large" the behavior of the solution is described (up to a numerically negligible correction) by

$$e^{-a\sqrt{\omega/2}}\cos(\omega t - a\sqrt{\omega/2})\,.$$

□

6.2.2 Differential equations, integro-differential equations and equations with delay

Exercise 6.21. Solve the initial boundary value problem for the harmonic oscillator ($\omega \in \mathbb{R},\ \omega \neq 0$), by applying the Laplace transform:

$$\begin{cases} y''(t) + \omega^2\, y(t) = f(t) & t > 0 \\ y(0) = A \\ y'(0) = B \end{cases} \tag{6.69}$$

Solution - Setting $Y = \mathcal{L}\{y\}$, we have

$$(s^2 + \omega^2)\, Y = \mathcal{L}\{f\} + sA + B$$

$$Y(s) = \frac{\mathcal{L}\{f\}}{s^2 + \omega^2} + \frac{sA + B}{s^2 + \omega^2}$$

therefore

$$y(t) = \frac{1}{\omega} \int_0^t f(\tau)\, \sin\Big(\omega(t - \tau)\Big)\, d\tau + A\, \cos(\omega t) + \frac{B}{\omega}\, \sin(\omega t).$$

Remark -

- The solution y of the homogeneous problem ($f \equiv 0$) is a simple harmonic oscillator of minimum period $2\pi/\omega$, if $(A, B) \neq (0, 0)$.
- The driven harmonic oscillator y, solution of the non homogeneous ($f \not\equiv 0$) problem (6.69), cannot be $2\pi/\omega$ periodic if f is not $2\pi/\omega$ periodic too; then the solution of the non homogeneous problem is periodic with period $2\pi/\omega$ if and only if f is $2\pi/\omega$ periodic and there is no resonance, say iff f is $2\pi/\omega$ periodic and

$$\int_0^{2\pi/\omega} f(\tau)\, \sin(\omega\tau)\, d\tau = 0\,, \qquad \int_0^{2\pi/\omega} f(\tau)\, \cos(\omega\tau)\, d\tau = 0\,.$$

- The driven harmonic oscillator y, solution of the non homogeneous ($f \not\equiv 0$) problem (6.69), is not periodic in case of resonance, that is when f is $2\pi/\omega$ periodic and at least one of this two integrals is different from zero:

$$\int_0^{2\pi/\omega} f(\tau)\, \sin(\omega\tau)\, d\tau \qquad \text{and} \qquad \int_0^{2\pi/\omega} f(\tau)\, \cos(\omega\tau)\, d\tau\ ;$$

  in the resonance case the frequency ω is present in the Fourier series expansion of f.

□

Exercise 6.22. Set

$$f(t) = \begin{cases} e^{-t} & 1 < t < 2, \\ 0 & \text{otherwise,} \end{cases}$$

and solve:

$$\begin{cases} y''(t) + 2\, y'(t) + y(t) = f(t) & t > 0\,, \\ y(0) = 0\,, \\ y'(0) = 0\,. \end{cases}$$

Solution - If $Y = \mathcal{L}\{y\}$, we have

$$(s^2 + 2s + 1)\, Y = \mathcal{L}\{f\}\,, \qquad Y(s) = \frac{\mathcal{L}\{f\}}{(s+1)^2}\,,$$

$$y(t) = f * \mathcal{L}^{-1}\left\{\frac{1}{(s+1)^2}\right\} = f * \{e^{-t}\, t\} = \int_0^t f(\tau)\,(t-\tau)\,e^{-(t-\tau)}\,d\tau\,.$$

Hence

$$\text{if } 0 < t < 1 \qquad y(t) = 0$$

$$\text{if } 1 < t < 2 \qquad y(t) = \int_1^t e^{-\tau}\,(t-\tau)\,e^{-(t-\tau)}\,d\tau = $$
$$= e^{-t}\int_1^t (t-\tau)\,d\tau = e^{-t}\left[t\tau - \frac{\tau^2}{2}\right]_1^t = \frac{1}{2}(t-1)^2\,e^{-t}$$

$$\text{if } t > 2 \qquad y(t) = e^{-t}\int_1^2 (t-\tau)\,d\tau = \frac{1}{2}(2t-3)\,e^{-t}\,.$$

We emphasize that computing the transform of f was not necessary at all. □

Exercise 6.23. By using the Laplace transform, find the solution $y \in C^1([0,+\infty))$ of the following Cauchy problem

$$\begin{cases} 2\,y''(x) - 3\,y'(x) - 2\,y(x) = e^{-x} & x > 0 \\ y(0) = 1 \\ y'(0) = 2\,. \end{cases}$$

Solution - The problem can be re-formulated as follows, by applying the Laplace transform:

$$2\,\mathcal{L}\{y''\} - 3\,\mathcal{L}\{y'\} - 2\,\mathcal{L}\{y\} = \mathcal{L}\{e^{-x}\}$$

$$2s^2\mathcal{L}\{y\} - 2s - 4 - 3s\,\mathcal{L}\{y\} + 3 - 2\mathcal{L}\{y\} = \frac{1}{s+1}$$

$$(2s^2 - 3s - 2)\mathcal{L}\{y\} = \frac{1}{s+1} + 2s + 1$$

$$\mathcal{L}\{y\} = \frac{2s^2+3s+2}{(s+1)(s-2)(2s+1)}\,. \tag{6.70}$$

Now we set:

$$Y(s) = \frac{2s^2+3s+2}{(s+1)(s-2)(2s+1)} = \frac{c_1}{s+1} + \frac{c_2}{s-2} + \frac{c_3}{2s+1}\,. \tag{6.71}$$

By equating the numerators in (6.71), we obtain the following algebraic equation:

$$2s^2 + 3s + 2 = c_1\,(s-2)(2s+1) + c_2\,(s+1)(2s+1) + c_3\,(s+1)(s-2)$$

that implies:

$$2s^2 + 3s + 2 = s^2\,(2c_1 + 2c_2 + c_3) + s\,(-3c_1 + 3c_2 - c_3) + (-2c_1 + c_2 - 2c_3)$$

hence we get the linear system of three equations in three unknowns (c_1, c_2, c_3):

$$\begin{cases} 2c_1 + 2c_2 + c_3 = 2 \\ -3c_1 + 3c_2 - c_3 = 3 \\ -2c_1 + c_2 - 2c_3 = 2\,, \end{cases}$$

whose solution is $\left(\frac{1}{3}, \frac{16}{15}, -\frac{4}{5}\right)$.

Then equation (6.71) becomes:

$$Y(s) = \frac{2s^2+3s+2}{(s+1)(s-2)(2s+1)} = \frac{1}{3(s+1)} + \frac{16}{15(s-2)} - \frac{2}{5\left(s+\frac{1}{2}\right)}.$$

By anti-transforming, we obtain the solution of the given Cauchy problem:

$$y(x) = \frac{1}{3}\,e^{-x} + \frac{16}{15}\,e^{2x} - \frac{2}{5}\,e^{-\frac{1}{2}x} \qquad x \geq 0\,.$$

An alternative and faster way to solve the Cauchy problem is as follows.
By (6.70), after setting $Y = \mathcal{L}\{y\} = N/D$, we get:

$$N = 2s^2+3s+2$$

$$D' = (s-2)(2s+1) + (s+1)(2s+1) + 2(s+1)(s-2).$$

Since D has only simple zeros, we can apply the Heaviside expansion formula (6.26):

$$y(x) = \frac{N(-1)}{D'(-1)}\,e^{-x} + \frac{N(2)}{D'(2)}\,e^{2x} + \frac{N(-1/2)}{D'(-1/2)}\,e^{-\frac{1}{2}x} = \frac{1}{3}\,e^{-x} + \frac{16}{15}\,e^{2x} - \frac{2}{5}\,e^{-\frac{1}{2}x} \qquad x \geq 0.$$

□

Exercise 6.24. Solve the following system in the unknowns $x = x(t)$, $y = y(t)$, by using the Laplace transform:

$$\begin{cases} x'' - x + y' - y = 0 & t > 0\,, \\ -2x' - 2x + y'' - y = e^{-t} & t > 0\,, \\ x(0) = 0, \quad x'(0) = -1, \quad y(0) = 1, \quad y'(0) = 1. \end{cases}$$

Solution - After setting $X(s) = \mathcal{L}\{x\}$, $Y(s) = \mathcal{L}\{y\}$, we transform the two differential equations, by taking into account the initial data and formulas (6.10),(6.11) for transforming derivatives:

$$\begin{cases} s^2X + 1 - X + sY - 1 - Y = 0 \\ -2sX - 2X + s^2Y - s - 1 - Y = (s+1)^{-1}\,. \end{cases}$$

From the first equation we get $Y = -(s+1)X$; by inserting this expression in the second one:

$$\begin{cases} X = -\dfrac{1}{s^2+1} - \dfrac{1}{(s^2+1)(s+1)^2} \\ Y = \dfrac{s+1}{s^2+1} + \dfrac{1}{(s^2+1)(s+1)} \end{cases}$$

$$\begin{cases} X = \left(\dfrac{1}{2}\,\dfrac{s}{s^2+1} - \dfrac{1}{2}\,\dfrac{1}{s+1} - \dfrac{1}{2}\,\dfrac{1}{(s+1)^2}\right) - \dfrac{1}{s^2+1} \\ Y = \dfrac{1}{2}\,\dfrac{s}{s^2+1} - \dfrac{3}{2}\,\dfrac{1}{s^2+1} + \dfrac{1}{2}\,\dfrac{1}{s+1} \end{cases}$$

$$\begin{cases} x(t) = \left(\dfrac{1}{2}\cos t - \sin t - \dfrac{1}{2}(1+t)e^{-t}\right) H(t) \\ y(t) = \left(\dfrac{1}{2}\cos t + \dfrac{3}{2}\sin t + \dfrac{1}{2}e^{-t}\right) H(t)\,. \end{cases}$$

□

Exercise 6.25. Solve the system

$$\begin{cases} x(t) = 1 + \displaystyle\int_0^t \Big(y(\tau) - x(\tau)\Big)\, d\tau & t > 0\,, \\ y(t) = 2\, H(t-1) + \displaystyle\int_0^t \big(y(\tau) - x(\tau)\big)\, d\tau & t > 0\,. \end{cases}$$

Solution - By setting $X = \mathcal{L}\{x\}$, $Y = \mathcal{L}\{y\}$, we obtain

$$\begin{cases} X = \dfrac{1}{s} + \dfrac{1}{s}\,(Y - X) \\ \\ Y = 2\,\dfrac{e^{-s}}{s} + \dfrac{1}{s}\,(Y - X) \end{cases}$$

$$\begin{cases} X = \dfrac{1}{s} - \dfrac{1}{s^2} + 2\,\dfrac{e^{-s}}{s^2} \\ \\ Y = -\dfrac{1}{s^2} + 2\,\dfrac{e^{-s}}{s^2} + 2\,\dfrac{e^{-s}}{s} \end{cases}$$

$$\begin{cases} x(t) = 1 - t + 2\,(t-1)\,H(t-1) = \begin{cases} 1-t & 0 < t < 1 \\ t-1 & 1 < t \end{cases} \\ \\ y(t) = -t + 2\,(t-1)\,H(t-1) + 2\,H(t-1) = \begin{cases} -t & \text{if } 0 < t < 1\,, \\ +t & \text{if } 1 < t\,. \end{cases} \end{cases}$$

□

Exercise 6.26. 1) Denoting J_1 the Bessel function of first kind of order 1, compute

$$\mathcal{L}\left\{ H(t)\,\frac{J_1(t)}{t} \right\}.$$

2) Determine the solutions that are $\mathcal{L}$-transformable functions of the following problem (hint: take into account item 1):

$$\begin{cases} 2\,y''(t) + \dfrac{1}{2}\,y(t) = \displaystyle\int_0^t y(\tau)\,y''(t-\tau)\,d\tau & t > 0 \\ y(0) = 0 \\ y'(0) = -\dfrac{1}{2} \end{cases}$$

Solution - 1) It is known (Table 10.7.1) that $\mathcal{L}\{J_1 H\} = 1 - \dfrac{s}{\sqrt{s^2+1}}$. Then

$$\mathcal{L}\left\{ \frac{J_1(t)}{t}\,H(t) \right\} = \int_s^{\infty} \left(1 - \frac{\sigma}{\sqrt{\sigma^2+1}}\right) d\sigma = \Big[\sigma - \sqrt{\sigma^2+1}\Big]_s^{\infty} = \sqrt{s^2+1} - s.$$

2) Set $Y = \mathcal{L}\{y\}$:

$$s^2\,Y^2 - 2\,s^2\,Y - 1 = 0\,,$$

$$Y = \frac{s^2 \pm \sqrt{s^4+s^2}}{s^2} = \frac{s \pm \sqrt{s^2+1}}{s}\,.$$

Since Y must be the transform of an $\mathcal{L}$ transformable function, Y must be analytic and infinitesimal at infinity. It is acceptable only

$$Y = \frac{s - \sqrt{s^2+1}}{s},$$

Taking into account item 1: $\quad y(t) = \mathcal{L}^{-1}\left\{\frac{s-\sqrt{s^2+1}}{s}\right\} = -\int_0^t \frac{J_1(\tau)}{\tau}\, d\tau \quad t \geq 0\,.$

□

Exercise 6.27. Solve the following problem (*differential equation with delay*):

$$\begin{cases} \text{find the } \mathcal{L} \text{ transformable solution } y \text{ of} \\ y'(t) = y(t-1) + H(t) \quad \text{in } \mathcal{D}'(\mathbb{R})\,. \end{cases}$$

Solution - After setting $Y = \mathcal{L}\{y\}$, we transform the differential equation and obtain an algebraic equation in the transformed unknown:

$$(s - e^{-s})\, Y(s) = \frac{1}{s}\,.$$

The (non algebraic) equation $s = e^{-s}$ in the unknown $s \in \mathbb{C}$ has infinitely many solutions and they cannot be explicitly computed, but only numerically approximated. Nevertheless the real parts of all solutions are uniformly bounded from above. Precisely

$$\operatorname{Re} s > 1 \quad \Longrightarrow \quad |e^{-s}| = e^{-\operatorname{Re} s} < 1 < |s|\,, \qquad \left|\frac{e^{-s}}{s}\right| < 1\,,$$

therefore $s - e^{-s} \neq 0$ if $\operatorname{Re} s > 1$: we can divide by $(s - e^{-s})$, find an explicit formula for Y and subsequently expand Y in power series (by exploiting the geometric series with common ratio of modulus less than 1):

$$Y(s) = \frac{1}{s(s-e^{-s})} = \frac{1}{s^2}\frac{1}{1-e^{-s}/s} = \frac{1}{s^2}\sum_{k=0}^{+\infty}\frac{e^{-ks}}{s^k} = \sum_{k=0}^{+\infty}\frac{e^{-ks}}{s^{k+2}} \qquad \operatorname{Re} s > 1\,.$$

Exploiting the identity

$$\mathcal{L}\left\{\frac{1}{(k+1)!}(t-k)^{k+1}H(t-k)\right\} = \frac{e^{-ks}}{s^{k+2}}$$

we "formally" compute the inverse transform term-wise:

$$y(t) \stackrel{?}{=} \sum_{k=0}^{+\infty}\frac{1}{(k+1)!}(t-k)^{k+1}H(t-k)\,.$$

Notice that the term-wise anti-transform is correct, because the series

$$\sum_{k=0}^{+\infty}\frac{e^{-\lambda t}}{(k+1)!}(t-k)^{k+1}H(t-k)$$

converges in $L^1(\mathbb{R})$ $\forall\,\lambda > 1$: indeed the terms in the expansion of y can be bounded from above by terms of the expansion of exponential function:

$$\frac{(t-k)^{k+1}}{(k+1)!}H(t-k) \leq \frac{t^{k+1}}{(k+1)!}H(t) \qquad \forall k \in \mathbb{N},\ \forall t \geq 0\,,$$

therefore (Theorem 6.12)

$$y(t) = \sum_{k=0}^{+\infty} \frac{1}{(k+1)!}\,(t-k)^{k+1} H(t-k)\,.$$

Remark - In general the Laplace transform method allows to solve differential or integro-differential equations when the unknown appears also with delayed argument (provided the delay does not appear in the highest order derivative). A "thick" initial condition is required, if delayed arguments appear: information at initial time only is not enough, exact knowledge of data is needed for a time interval of length equal to the amount of maximum delay. In the particular case of this exercise the thick initial condition is implicitly given by requiring that the support of the solution is contained in $[0, +\infty)$. □

Exercise 6.28. By using the $\mathcal{L}$ transform, find the solution $y = y(x,t)$ of the following transport problem

$$\begin{cases} \dfrac{\partial y}{\partial t} + \dfrac{\partial y}{\partial x} = 0\,, & x > 0,\ t > 0\,, \\[2ex] y(x,0) = 0\,, & x > 0\,, \\ y(0,t) = e^{-t}\,, & t > 0\,. \end{cases}$$

Solution - After setting $Y(x,s) = \mathcal{L}_t\{y(x,t)\}(s) := \displaystyle\int_0^{+\infty} e^{-st} y(x,t)\,dt$, we obtain

$$\mathcal{L}_t\left\{\frac{\partial y}{\partial t}\right\} = s\,Y - y(x,0) = s\,Y(x,s)$$

$$\mathcal{L}_t\left\{\frac{\partial y}{\partial x}\right\} = \frac{\partial Y}{\partial x}(x,s)\,.$$

Transforming the equation and the condition on $y(0,t)$ we get the family of ordinary differential equations (dependent on the parameter s)

$$\begin{cases} s\,Y + Y_x = 0\,, & x > 0\,, \\[2ex] Y(0,s) = \dfrac{1}{s+1}\,, \end{cases}$$

whose solution is

$$Y(x,s) = \frac{e^{-sx}}{s+1}\,, \qquad x > 0.$$

This solution is the transform of a transformable function.
Anti-transforming we obtain the solution of the given problem:

$$y(x,t) = H(t-x)\,e^{x-t}\,.$$

□

Exercise 6.29. By using the $\mathcal{L}$ transform, find the solution $y = y(x,t)$ of problem

$$\begin{cases} \dfrac{1}{x+1}\dfrac{\partial y}{\partial t} + \dfrac{\partial y}{\partial x} = 0\,, & x > 0,\ t > 0\,, \\[2ex] y(x,0) = 1\,, & x > 0\,, \\ y(0,t) = 0\,, & t > 0\,. \end{cases}$$

Solution - After setting $Y(x,s) = \mathcal{L}_t\{y(x,t)\}(s) := \displaystyle\int_0^{+\infty} e^{-st} y(x,t)\,dt$ and transforming the equation, we obtain

$$\frac{\partial Y}{\partial x} = -\frac{s}{x+1}Y + \frac{1}{x+1},$$

$$\begin{aligned} Y(x,s) &= e^{-s\ln(x+1)}\left(C(s) + \int_0^x \frac{1}{y+1}e^{s\ln(y+1)}\,dy\right) = \\ &= (x+1)^{-s}\left(C(s) + \int_0^x (y+1)^{s-1}\,dy\right) = \frac{C(s)}{(x+1)^s} + \frac{1}{s} - \frac{1}{s(x+1)^s}. \end{aligned}$$

By transforming the condition on $y(0,t)$ we obtain

$$Y(0,s) \equiv 0, \qquad C(s) \equiv 0,$$

$$Y(x,s) = \frac{1}{s} - \frac{1}{s\,(x+1)^s} = \frac{1}{s} - \frac{1}{s}e^{-s\ln(x+1)}.$$

If $x > 0$ then $\ln(x+1) > 0$. Therefore $s \to Y(x,s)$ is a transform for every $x > 0$. Anti-transforming we obtain the solution of the problem:

$$y(x,t) = 1 - H\big(t - \ln(x+1)\big), \qquad x > 0,\ t > 0.$$

Remark - This solution is discontinuous, in particular it is not of class C^1. Nevertheless it solves the given differential equation in the distributional sense in the whole open set $\Omega = \{(x,t)\colon x>0, t>0\}$. To prove this claim, we refer to Exercise 3.27. □

Exercise 6.30. By using the $\mathcal{L}$ transform, find the solution $y = y(x,t)$ of the problem

$$\begin{cases} \dfrac{\partial y}{\partial t} - \dfrac{\partial^2 y}{\partial x^2} = 3\,e^{-t}\sin(2x), & 0 < x < \frac{\pi}{2},\ t > 0, \\ y(x,0) = \sin(2x) - \sin(x), & 0 < x < \frac{\pi}{2}, \\ y(0,t) = 0, & t > 0, \\ y(\pi/2,t) = -e^{-t}, & t > 0. \end{cases}$$

Solution - We set $Y(x,s) = \mathcal{L}_t\{y(x,t)\}(s)$ and transform the problem:

$$\begin{cases} s\,Y(x,s) - Y_{xx}(x,s) = \dfrac{3\sin(2x)}{s+1} + \sin(2x) - \sin(x), & 0 < x < \frac{\pi}{2}, \\ Y(0,s) = 0 \qquad Y(\pi/2,s) = -\dfrac{1}{s+1}. \end{cases}$$

The general integral of the homogeneous equation is

$$Y_1(x,s) = C_1(s)\sinh(\sqrt{s}\,x) + C_2(s)\cosh(\sqrt{s}\,x).$$

A particular integral of the non-homogeneous equation is given by

$$Y_2(s,x) = -\frac{\sin(x)}{s+1} + \frac{\sin(2x)}{s+1}.$$

Therefore the general integral is $Y = Y_1 + Y_2$. By imposing the conditions at $x = 0$ and $x = \pi/2$, we get $C_1 \equiv 0$, $C_2 \equiv 0$. We conclude:

$$\begin{aligned} Y(x,s) &= \frac{1}{s+1}\big(\sin(2x) - \sin(x)\big), \\ y(x,t) &= \big(\sin(2x) - \sin(x)\big)\,e^{-t}. \end{aligned}$$

□

6.2.3 *Differential equations in the distributional framework*

Exercise 6.31. Prove that

The problem
"*Give an $\mathcal{L}$ transformable distribution f and a constant $a \in \mathbb{R}$, find an*

$$\text{$\mathcal{L}$ transformable distribution y that solves} \quad y' + ay = f \quad \text{in } \mathcal{D}'(\mathbb{R})\text{"} \tag{6.72}$$

has a unique solution given by

$$y = \left(H(t)\, e^{-at}\right) * f\,. \tag{6.73}$$

Notice that $\left(H(t)\, e^{-at}\right)$ is the fundamental solution in $\mathbb{R}$ with support contained in $[0, +\infty)$ (hence unique) of the first order differential operator $aI + d/dt$.

Solution - We transform the differential equation, by applying (6.10) and the relation $1/(s+a) = \mathcal{L}\{H(t)e^{-at}\}$; then we get the claim by Theorem 6.21. □

Exercise 6.32. Prove that

The problem
"*Given an $\mathcal{L}$ transformable distribution g and a constant $\omega > 0$, find an*

$$\text{$\mathcal{L}$ transformable distribution v that solves} \quad v'' + \omega^2 v = g \quad \text{in } \mathcal{D}'(\mathbb{R})\text{"} \tag{6.74}$$

has a unique solution given by

$$v = \left(H(t)\,\frac{\sin(\omega t)}{\omega}\right) * g\,. \tag{6.75}$$

Notice that $\left(H(t)\,\dfrac{\sin(\omega t)}{\omega}\right)$ is the (unique) fundamental solution in $\mathbb{R}$ with support contained in $[0, +\infty)$ of the second order differential operator $\omega^2 I + d^2/dt^2$.

Solution - We transform the differential equation and apply (6.11):

$$(s^2 + \omega^2)\, V(s) \;=\; G(s)\,.$$

The identity $1/(s^2 + \omega^2) = \mathcal{L}\{H(t)\sin(\omega t)/\omega\}$ and Theorem 6.21 prove the claim. □

Exercise 6.33. For each one of the following functions establish whether it is the Laplace transform of an $\mathcal{L}$ transformable distribution:

$$e^{1/s}\,, \qquad e^{-|s|}\,, \qquad e^{-s^2}\,, \qquad \frac{e^{-s}}{\sin s}\,, \qquad e^{-s}\,.$$

Remark - The reader is invited to compare the answers of the present exercise with those of Exercise 6.5.

Solution - $e^{-|s|}$ and $e^{-s}/\sin s$ are neither transform of functions nor transform of transformable distributions because they are not analytic functions in any right half-plane.
Though $e^{1/s}$ is not the transform of a function, $e^{1/s}$ is the transform of a distribution, because it is entire (therefore it verifies (6.42)) and it is bounded in a right half-plane (therefore it verifies (6.43)), namely:

$$\left|e^{1/s}\right| = \left|e^{\mathrm{Re}\, 1/s}\right| = \left|e^{\mathrm{Re}(\overline{s}/|s|^2)}\right| = \left|e^{\mathrm{Re}(s/|s|^2)}\right| \leq e \qquad \forall s: \ \mathrm{Re}\, s \geq 1\,.$$

$$e^{1/s} = 1 + \sum_{k=1}^{+\infty} \frac{1}{k!\, s^k} = \mathcal{L}\{\delta\} + \sum_{k=1}^{+\infty} \mathcal{L}\left\{\frac{t^{k-1} H(t)}{k!(k-1)!}\right\}\,.$$

Thus it is possible to compute the transform term-wise (Theorems 6.12 and 6.24):

$$e^{1/s} = \mathcal{L}\left\{\delta + \sum \frac{t^{k-1} H(t)}{k!(k-1)!}\right\}\,.$$

even if e^{-s^2} is an entire analytic function, e^{-s^2} is not the Laplace transform of any $\mathcal{L}$ transformable distribution, indeed along the half-line starting from origin tilted by $\pi/3$ radians (that is $s = x(1 + i\sqrt{3}),\ x > 0$) we get: $|e^{-s^2}| = |e^{-x^2(-2+2i\sqrt{3})}| = e^{2x^2}$; hence the necessary condition (6.43) is not met.
Though e^{-s} is not the transform of a function, e^{-s} is the transform of a distribution:

$$e^{-s} = \mathcal{L}\{\delta(t-1)\}\,.$$

□

Exercise 6.34. Find the $\mathcal{L}$ transformable functions y that solve

$$\begin{cases} (t\, y(t))' - t\, \cos t = \displaystyle\int_0^t J_0(\tau)\, J_0(t-\tau)\, d\tau & t > 0 \\ y(0) = 0 \end{cases}$$

where J_0 denotes the Bessel function of first kind, of order 0.
Solution - By setting $Y = \mathcal{L}\{y\}$, we get

$$\begin{aligned} \mathcal{L}\{(t\, y)'\} &= \mathcal{L}\{y + t\, y'\} = -s\, Y' \\ \mathcal{L}\{t\, \cos t\} &= -\frac{d}{ds}\frac{s}{s^2+1} \\ \mathcal{L}\{J_0 * J_0\} &= \frac{1}{s^2+1}\,. \end{aligned}$$

Hence we find the differential equation

$$-s\, Y' - \frac{2s^2}{(s^2+1)^2} = 0$$

to be solved in the unknown $Y = Y(s)$, for $s > 0$. We obtain

$$Y(s) = \frac{1}{s^2+1} + C\,.$$

If y is a transformable function then Y must be an infinitesimal at infinity, this is equivalent to require $C = 0$, hence we obtain the unique solution:

$$y(t) = \sin(t)\,, \qquad t > 0\,.$$

□

Exercise 6.35. Find the $\mathcal{L}$ transformable distributions y that solve

$$(t\,y(t))' - t\,\cos t = \big(J_0\,H(t)\big) * \big(J_0\,H(t)\big)(t) \qquad \mathcal{D}'(\mathbb{R})$$

where J_0 denotes the Bessel function of first kind, of order 0.

Solution - Transforming the equation we obtain $Y(s) = \frac{1}{s^2+1} + C$, that is the transform of an $\mathcal{L}$ transformable distribution for every value of the constant C. Therefore the general solution in the setting of $\mathcal{L}$ transformable distributions is $y(t) = \sin(t)\,H(t) + C\,\delta(t)$, where δ is the Dirac distribution centered at zero. □

Exercise 6.36. Denoting by J_1 the Bessel function of first kind of order 1 and using the results of Exercise 6.26, determine all solutions of the following problem:

$$\begin{cases} \text{find an } \mathcal{L}\text{ transformable distribution } y : \\ 2\,y''(t) + \dfrac{1}{2}\,y(t) = y * y'' \\ y(0) = 0 \\ y'(0) = -\dfrac{1}{2}\,. \end{cases}$$

Solution - The problem has 2 solutions y_1, y_2 that we compute with the $\mathcal{L}$ transform. Proceeding as in the solution of Exercise 6.26, we obtain

$$Y_1(s) = \frac{s - \sqrt{s^2+1}}{s} \qquad Y_2(s) = \frac{s + \sqrt{s^2+1}}{s}\,.$$

Y_1 is an $\mathcal{L}$ transformable function, and by the identity $\mathcal{L}\{J_1(t)H(t)/t\} = \sqrt{s^2+1} - s$, proved in the solution of Exercise 6.26, we found

$$y_1(t) = \mathcal{L}^{-1}\{Y_1\} = -\left(\int_0^t \frac{J_1(\tau)}{\tau}\,d\tau\right) H(t) \qquad t \geq 0$$

Y_2 is not an $\mathcal{L}$ transformable function, nevertheless Y_2 is an $\mathcal{L}$ transformable distribution (since it is analytic and bounded in $\{\mathrm{Re}\,s \geq 1\}$):

$$\begin{aligned} y_2(t) &= \mathcal{L}^{-1}\{Y_2\} = \mathcal{L}^{-1}\left\{\frac{s + \sqrt{s^2+1}}{s}\right\} = \mathcal{L}^{-1}\left\{1 + \sqrt{1 + 1/s^2}\right\} = \\ &= \delta(t) + \mathcal{L}^{-1}\left\{(1+s^{-2})^{-1/2}\right\}. \end{aligned}$$

By exploiting the binomial series (Table 10.2) we anti-transform term-wise formally:

$$\begin{aligned} \mathcal{L}^{-1}\left\{(1+s^{-2})^{-1/2}\right\} &= \mathcal{L}^{-1}\left\{\sum_{n=0}^{+\infty}\binom{-1/2}{n} s^{-2n}\right\} \stackrel{?}{=} \sum_{n=0}^{+\infty}\binom{-1/2}{n}\mathcal{L}^{-1}\left\{s^{-2n}\right\} = \\ &= \delta(t) + \sum_{n=1}^{+\infty}\frac{(-1)^n}{2^n}\,\frac{t^{2n-1}}{(2n-1)!}\,H(t) \end{aligned}$$

The resulting series is dominated by the exponential series of t, therefore if we multiply it by $e^{-\lambda t}$ with $\lambda > 1$ it converges in $L^1(\mathbb{R})$ (and also in $\mathcal{S}'$). Thus the anti-transformation (equality $\stackrel{?}{=}$) is (not only formally) correct, due to Theorem 6.12 and 6.24.

By summarizing $Y_2(t) = 2\delta(t) + \displaystyle\sum_{n=1}^{+\infty}\frac{(-1)^n}{2^n}\,\frac{t^{2n-1}}{(2n-1)!}\,H(t)\,.$ □

Exercise 6.37. Find all $\mathcal{L}$ transformable solutions of the system

$$\begin{cases} x'(t) = -2\,y(t) - \delta(t-\pi) & \mathcal{D}'(\mathbb{R}), \\ y'(t) = x(t) + \chi_{[0,\pi/2]}(t) & \mathcal{D}'(\mathbb{R}), \end{cases}$$

where $\chi_{[0,\pi/2]}$ is the indicator function of the interval $[0,\pi/2]$.

Solution - We set $X(s)=\mathcal{L}(x(t))$, $Y(s)=\mathcal{L}(y(t))$ and we transform the system obtaining

$$\begin{cases} s\,X(s) = -2\,Y(s) - e^{-\pi s} \\ s\,Y(s) = X(s) + \dfrac{1}{s} - \dfrac{e^{-\pi s/2}}{s}. \end{cases}$$

From the second equation we get

$$X(s) = s\,Y(s) - \frac{1}{s} + \frac{e^{-\pi s/2}}{s}$$

and, by inserting this in the first equation,

$$s\left(s\,Y(s) - \frac{1}{s} + \frac{e^{-\pi s/2}}{s}\right) = -2\,Y(s) - e^{-\pi s}.$$

Hence

$$Y(s) = \frac{1}{s^2+2} - \frac{e^{-\pi s/2}}{s^2+2} - \frac{e^{-\pi s}}{s^2+2},$$

$$X(s) = \frac{s}{s^2+2} - \frac{s\,e^{-\pi s/2}}{s^2+2} - \frac{s\,e^{-\pi s}}{s^2+2} - \frac{1}{s} + \frac{e^{-\pi s/2}}{s}.$$

Eventually, by anti-transforming:

$$x(t) = H(t)\cos(\sqrt{2}\,t) - H(t-\pi/2)\cos\left(\sqrt{2}\,(t-\pi/2)\right) - H(t-\pi)\cos\left(\sqrt{2}\,(t-\pi)\right) - \chi_{[0,\pi/2]}(t),$$

$$y(t) = \frac{1}{\sqrt{2}}\left(H(t)\sin(\sqrt{2}\,t) - H(t-\pi/2)\sin\left(\sqrt{2}\,(t-\pi/2)\right) - H(t-\pi)\sin\left(\sqrt{2}\,(t-\pi)\right)\right).$$

□

Exercise 6.38. Find the $\mathcal{L}$ transformable solutions of the system

$$\begin{cases} x'(t) = y(t) - \delta(t-\pi) + \delta'(t) & \mathcal{D}'(\mathbb{R}) \\ y'(t) = -x(t) + H(t) & \mathcal{D}'(\mathbb{R}). \end{cases}$$

Solution - We set $X(s)=\mathcal{L}(x(t))$, $Y(s)=\mathcal{L}(y(t))$ and transform the system:

$$\begin{cases} sX(s) = Y(s) - e^{-\pi s} + s \\ sY(s) = -X(s) + 1/s. \end{cases}$$

By inserting the second equation multiplied by s in the first one:

$$1 - s^2 Y(s) = Y(s) - e^{-\pi s} + s.$$

By solving the algebraic equation:

$$(s^2+1)\,Y(s) = e^{-\pi s} - s + 1,$$

$$Y(s) = \frac{e^{-\pi s}}{s^2+1} - \frac{s}{s^2+1} + \frac{1}{s^2+1}.$$

From

$$X(s) = -s\,y(s) + \frac{1}{s}$$

we obtain

$$X(s) = -s\Big(\frac{e^{-\pi s}}{s^2+1} - \frac{s}{s^2+1} + \frac{1}{s^2+1}\Big) + \frac{1}{s}$$

$$X(s) = -\frac{se^{-\pi s}}{s^2+1} + 1 - \frac{1}{s^2+1} - \frac{s}{s^2+1} + \frac{1}{s}\,,$$

now anti-transforming we find

$$\begin{cases} x(t) &= -H(t-\pi)\cos(t-\pi) + \delta(t) - H(t)\sin t - H(t)\cos t + H(t) \\ y(t) &= H(t-\pi)\sin(t-\pi) + H(t)\sin t - H(t)\cos t\,. \end{cases}$$

□

Exercise 6.39. Set $t^+ = t\,H(t)$ and face the problem:

determine a Laplace transformable function u *s.t.* $\quad u'' * u = t^+ \quad$ in $\mathbb{R}$.

Say whether there exist solutions and in the affirmative case compute them, by motivating the answers.

Solution - We apply the Laplace transform to both sides of the equation, by taking into account $\mathcal{L}\{u''\} = s^2\mathcal{L}\{u\}$, $\mathcal{L}\{u * v\} = \mathcal{L}\{u\}\,\mathcal{L}\{v\}$ and $\mathcal{L}\{t^+\} = 1/s^2$.
After setting $U(s) = \mathcal{L}\{u(t)\}$ we obtain

$$s^2\big(U(s)\big)^2 = \frac{1}{s^2}\,, \qquad \text{hence} \qquad \big(U(s)\big)^2 = \frac{1}{s^4}\,.$$

Therefore, for every $s \in \mathbf{C}$ s.t. $\mathrm{Re}\, s > 0$, we have either $u(s) = 1/s^2$ or $u(s) = -1/s^2$. This condition is compatible with infinitely many possible choices of the function $U(s)$. Nevertheless U must be an analytic function of s in the half-plane of convergence, since it is a Laplace transform; therefore, if a solution u exists and U is the transform, then there are only two admissible choices:

$$U_1(s) = 1/s^2 \ \ \forall s \in \mathbf{C} \text{ s.t. } \mathrm{Re}\, s > 0\,, \qquad U_2(s) = -1/s^2 \ \ \forall s \in \mathbf{C} \text{ s.t. } \mathrm{Re}\, s > 0\,.$$

By injectivity of the transform, it is possible to determine all solutions of the proposed problem by anti-transforming U_1 and U_2.
We can conclude that the problem has exactly two solutions: the transformable functions

$$u_1(t) = t^+ \ \ t \in \mathbb{R}\,, \qquad u_2(t) = -t^+ \ \ t \in \mathbb{R}\,.$$

Since $u_1'(t) = H(t)$, $u_1''(t) = \delta(t)$ and $u_2'(t) = -H(t)$, $u_2''(t) = -\delta(t)$, we obtain

$$u_1'' * u = \delta * u_1 = t^+\,, \qquad u_2'' * u = -\delta * u_2 = t^+\,,$$

therefore both u_1 and u_2 solve the given problem. □

Exercise 6.40. By motivating the answer, establish whether there are solutions of the following problem and in the affirmative case compute them.

Determine the $\mathcal{L}$ transformable distributions u s.t. $\quad u' * u' = \dfrac{1}{6}\,t^3\,H(t) \quad$ in $\mathbb{R}$.

Solution - Let us apply the Laplace transform to both sides of the equation:

$$\begin{aligned} s^2\left(U(s)\right)^2 &= \frac{1}{s^4} & \forall s:\ \operatorname{Re} s > 0 \\ \left(U(s)\right)^2 &= \frac{1}{s^6} & \forall s:\ \operatorname{Re} s > 0 \\ U(s) &= \pm\frac{1}{s^3} & \forall s:\ \operatorname{Re} s > 0\,. \end{aligned}$$

The last relation is compatible with infinitely many choices of the function U, nevertheless the function $s \to U(s)$ must be analytic in the half-plane of convergence because it is an $\mathcal{L}$ transform. Therefore only two choices are acceptable:

$$U_1(s) = \frac{1}{s^3}\,, \qquad U_2(s) = -\frac{1}{s^3}\,,$$

accordingly, we find exactly two solutions of the given problem:

$$u_1(s) = \frac{1}{2}\,t^2\,H(t)\,, \qquad u_2(s) = -\frac{1}{2}\,t^2\,H(t)\,.$$

□

Exercise 6.41. Solve the following problems.

1) Given a sequence $w_n(t) = \sum_{k=0}^{n} \delta(t-k)$, compute $W_n = \mathcal{L}\{w_n\}$, prove that w_n converges in the space of $\mathcal{L}$ transformable distributions and determine its limit w.

2) Given the sequence $f_n(t) = n\chi_{[0,1/n]}$, compute $F_n = \mathcal{L}\{f_n\}$, prove that f_n converges in the space of $\mathcal{L}$ transformable distributions and determine its limit f.

3) Given the sequence $g_n(t) = n\big(\delta(t-1) - \delta(t-1-1/n)\big)$, compute $G_n = \mathcal{L}\{g_n\}$, prove that g_n converges in the space of $\mathcal{L}$ transformable distributions and determine its limit g.

Solution - 1) After setting $w(t) = \sum_{k=0}^{+\infty} \delta(t-k)$, we have

$$\int w_n\psi = \sum_{k=0}^{n} \psi(k) \to \sum_{k=0}^{+\infty} \psi(k) = \int w\psi \qquad \forall\psi \in \mathcal{S}'(\mathbb{R})\,,$$

that is $w_n \to w$ in $\mathcal{S}'(\mathbb{R})$ and hence in the space of $\mathcal{L}$ transformable distributions too.

$$W_n(s) = \mathcal{L}\{w_n\} = \sum_{k=0}^{n} e^{-sk} \;\longrightarrow\; \frac{1}{1-e^{-s}} = \mathcal{L}\{w\}$$

where the convergence is point-wise in $\{\operatorname{Re} s > 1\}$ for the properties of the geometric series. Furthermore the convergence is uniform in compact subsets of $\{\operatorname{Re} s > 1\}$, according to Theorem 6.22.

2) By $f_n(t) = n\chi_{[0,1/n]} = n\big(H(t) - H(t-1/n)\big)$ we get

$$F_n = \mathcal{L}\{f_n\} = n\,\mathcal{L}\{H(t) - H(t-1/n)\} = n\left(\frac{1}{s} - \frac{e^{-s/n}}{s}\right) = \frac{n}{s}\left(1 - e^{-s/n}\right).$$

f_n converges to $f = \delta$ in $\mathcal{E}'(\mathbb{R})$ and hence in the space of $\mathcal{L}$ transformable distributions. Precisely $e^{-\lambda t} f_n \to \delta$ in $\mathcal{S}'(\mathbb{R})$ if $n \to +\infty$, for every $\lambda \in \mathbb{R}$: F_n and F are entire analytic functions.

Notice that $F_n(s) \to 1 = \mathcal{L}\{\delta\}$ point-wise (it can be proved by using de L'Hôpital's rule), moreover the convergence is uniform on compact subsets of $\mathbb{C}$ according to Theorem 6.22.

3) We compute $G_n = \mathcal{L}\{g_n\}$:

$$\mathcal{L}\{g_n\} = n\,\mathcal{L}\{\delta(t-1)-\delta(t-1-1/n)\} = n\left(e^{-s}-e^{-s(1+1/n)}\right) = se^{-s}\,\frac{(1-e^{-s/n})}{s/n}\,.$$

$g_n(t)=f_n'(t-1)$ converges to $g(t)=f'(t-1)=\delta'(t-1)$ in $\mathcal{E}'(\mathbb{R})$ and hence in the space of $\mathcal{L}$ transformable distributions. Precisely $e^{-\lambda t}g_n \to e^{-\lambda t}\delta'(t-1)$ in $\mathcal{S}'(\mathbb{R})$ if $n\to+\infty$, for every $\lambda\in\mathbb{R}$: G_n and G are entire analytic functions.
Notice that $G_n(s)\to G(s)$ point-wise (it can be proved by using de L'Hôpital's rule), moreover the convergence is uniform on compact subsets of $\mathbb{C}$ according to Theorem 6.22.
□

Exercise 6.42. Solve the following problems.
1) Set $u(t)=J_0(t)H(t)$, where J_0 and H denotes the Bessel function of first kind of order 0 and the Heaviside function respectively, and prove

$$t\,u''(t)+u'(t)+t\,u(t) = 0 \qquad \text{in } \mathcal{D}'(\mathbb{R})\,.$$

2) Deduce from the previous item

$$\mathcal{L}\{J_0H\} = \frac{1}{\sqrt{s^2+1}}\,,$$

where we consider the analytic branch of the complex root that coincides with the arithmetic square root on half-line of positive real numbers.
3) Compute the transform of J_0H, starting directly from the differential equation solved by the Bessel function of first kind of order 0 in $\mathbb{R}^+$ and taking into account the values of J_0 and J_0' in 0.
4) Prove the identity:

$$\big(J_0\,H\big)*\big(J_0\,H\big) = \sin(t)\,H(t)\,.$$

Solution - 1) J_0 is an entire function that solves the differential equation $z\,J_0''(z)+J_0'(z)+z\,J_0(z) = 0\ \forall z\in\mathbb{C}$. In particular

$$t\,J_0''(t)+J_0'(t)+t\,J_0(t) = 0 \qquad t\in\mathbb{R}\,. \tag{6.76}$$

By using the Leibniz rule we compute the distributional derivative of $u=J_0H$ in $\mathbb{R}$:

$$\begin{aligned} u' &= J_0'\,H+J_0\,H' = J_0'\,H+J_0\,\delta = \boxed{J_0(0)=1} \\ &= J_0'\,H+\delta \end{aligned}$$

$$\begin{aligned} u'' &= J_0''\,H+J_0'\,H'+\delta' = J_0''\,H+J_0'\,\delta+\delta' = \boxed{J_0'(0)=0,\ J_0'H'=J_0'(0)\delta=0} \\ &= J_0''\,H+\delta'\,. \end{aligned}$$

By inserting these derivatives and taking into account (6.76) and $t\delta'(t)=-\delta(t)$, we obtain:

$$t\,u''+u'+t\,u = \Big(t\,J_0''(t)+J_0'(t)+t\,J_0(t)\Big)H(t)\ +\ t\,\delta'\ +\ \delta = 0\,.$$

2) Transforming the differential equation $tu''+u'+tu=0$ in $\mathcal{D}'(\mathbb{R})$ we obtain

$$-\frac{d}{ds}(s^2U)+s\,U-\frac{d}{ds}U = 0$$

hence $(s^2+1)U'+sU=0$ and, through separation of variables, $\dfrac{U'}{U} = -\dfrac{s}{s^2+1}$, whose solution is

$$\ln U(s)-\ln U(s_0) = -\frac{1}{2}\,\ln(s^2+1)+\frac{1}{2}\,\ln(s_0^2+1)\,,\ \text{ that is}$$

$$U(s) = \frac{C}{\sqrt{s^2+1}}.$$

We are left to to compute the value of the integration constant C. Since $J_0 \in AC_{loc}$ and $J_0 H$, $J_0' H$ are $\mathcal{L}$ transformable functions, we can apply the initial value Theorem:

$$C = \lim_{\mathrm{Re}\, s \to +\infty} \frac{s\,C}{\sqrt{s^2+1}} = \lim_{\mathrm{Re}\, s \to +\infty} s\,U(s) = u(0_+) = J_0(0) = 1$$

and conclude
$$U(s) = \frac{1}{\sqrt{s^2+1}}.$$

3) In this item it is suggested an alternative procedure (with respect to the one of items 1 and 2) to compute the transform of $u = J_0 H$.
We consider the following Cauchy problem in the right half-line, which is solved by the Bessel function J_0:

$$\begin{cases} t\,J_0''(t) + J_0'(t) + t\,J_0(t) = 0 & t > 0 \\ J_0(0) = 1 \\ J_0'(0) = 0\,. \end{cases}$$

By $\mathcal{L}$ transform we obtain

$$-\frac{d}{ds}(s^2 U) + s\,U - \frac{d}{ds}U = \frac{d}{ds}\big(s\,u(0)\big) - u(0) = 0$$

hence the O.D.E. $(s^2+1)U' + sU = 0$, that can be solved in U and anti-transformed as in item 2.

4) By item 2 we obtain

$$\mathcal{L}\big\{\big(J_0\,H(t)\big) * \big(J_0\,H\big)\big\} = \frac{1}{\sqrt{s^2+1}}\,\frac{1}{\sqrt{s^2+1}} = \frac{1}{s^2+1} = \mathcal{L}\{\sin(t)\,H(t)\}\,,$$

hence the identity to be proved follows by the injectivity of the $\mathcal{L}$ transform. □

Exercise 6.43. By using $\mathcal{L}$ transform, find the solution $y = y(x,t)$ of

$$\begin{cases} \dfrac{\partial y}{\partial t} + \dfrac{\partial y}{\partial x} = 0\,, & x > 0,\ t > 0\,, \\ y(x,0) = e^x\,, & x > 0\,, \\ y(0,t) = e^t\,, & t > 0\,. \end{cases}$$

Solution - After setting $Y(x,s) = \mathcal{L}_t\{y(x,t)\}(s) := \int_0^{+\infty} e^{-st} y(x,t)\,dt$, we obtain

$$\mathcal{L}_t\left\{\frac{\partial y}{\partial t}\right\} = s\,Y - y(x,0) = s\,Y(x,s)\,, \qquad \mathcal{L}_t\left\{\frac{\partial y}{\partial x}\right\} = \frac{\partial Y}{\partial x}(x,s)\,.$$

Transforming the partial differential equation and the condition on $y(0,t)$ we obtain the (one parameter) family of ordinary differential equations

$$\begin{cases} s\,Y + Y_x = e^x\,, & x > 0\,, \\ Y(0,s) = \dfrac{1}{s-1}\,, \end{cases}$$

whose solution is $Y(x,s) = \dfrac{e^x}{s+1} + C(s)e^{-sx}$, $x > 0$, with $C(s) = \dfrac{2}{s^2-1} = \mathcal{L}\{2\sinh(t)H(t)\}$.
This solution is the transform of an $\mathcal{L}$ transformable function.
By anti-transforming we obtain the solution of the given problem:

$$y(x,t) = e^{x-t}H(t) + \left(e^{t-x} - e^{x-t}\right)H(t-x) = e^{|x-t|}\,.$$ □

Remark- This solution does not belong to C^1. Nevertheless it solves the differential equation in the space of distributions in the one set $\Omega=\{(x,t): x>0, t>0\}$. To prove it we refer to Exercise 3.27.

Exercise 6.44. The problem of finding solutions with support in $\mathbb{R}^+$ for a non homogeneous linear differential equation with constant coefficients is solvable with the Laplace transform method even if the forcing term is not an $\mathcal{L}$ transformable function.

Let us define $L = \sum_{j=0}^{k} c_j\, d^j/dt^j$ (constant coefficients linear differential operator of order k), where $c_k \neq 0$, and the derivatives d^j/dt^j are understood in the sense of distributions of $\mathcal{D}'(\mathbb{R})$ and the polynomial $\mathcal{P}(s) = \sum_{j=0}^{k} c_j\, s^j$ (associated to the operator in the symbolic calculus).

1) Prove the following statement.

Given $g \in L^1_{loc}(\mathbb{R})$ with $\operatorname{spt} g \subset [0,+\infty)$, the problem

$$(*) \quad w \in L^1_{loc}(\mathbb{R}) : w^{(j)} \in AC_{loc}(\mathbb{R}),\ j \leq k-1, \quad Lw = g \text{ in } \mathcal{D}'(\mathbb{R}),\ \operatorname{spt} w \subset [0,\infty)$$

has a unique solution w and this solution verifies for every $T > 0$

$$w(t) = w_T(t) \quad \text{q.o. in } (-\infty, T),$$

where w_T is the unique $\mathcal{L}$ transformable solution of

$$(**) \qquad L\, w_T = g\, \chi_{(-\infty,T]} \quad \text{in } \mathcal{D}'(\mathbb{R}).$$

By summarizing, problem $(*)$ has a unique solution w that verifies

$$w(t) = w_T(t) = \mathcal{L}^{-1}\left\{ \frac{\mathcal{L}\{g\,\chi_{(-\infty,T]}\}}{\mathcal{P}(s)} \right\} \qquad \forall t, T > 0 \,:\, 0 < t < T. \tag{6.77}$$

Referring to Sections 6.1.4 and 6.1.8, formulate statements similar to the previous one for

2) Cauchy problem in $(0,\infty)$ with non homogeneous initial data at 0 and $g \in L^1_{loc}([0,\infty))$;
3) problem of finding the solution with support in $\mathbb{R}^+$ for a linear differential equation in $\mathbb{R}$ with constant coefficients whose forcing term is a distribution $g \in \mathcal{D}'(\mathbb{R})$ with $\operatorname{spt} g \subset [0,+\infty)$ but not $\mathcal{L}$ transformable.

Solution - 1) Any solution of the problem $(**)$, if restricted to the interval $(0,T)$, solves

$$(***) \quad L\, w_T = g \ (0,T), \quad w^{(j)} \in AC(0,T),\ 0 \leq j \leq k-1, \quad w^{(j)}(0) = 0,\ 0 \leq j \leq k-1.$$

By the Cauchy-Lipschitz Theorem, $(***)$ admits a unique solution $\forall T > 0$, furthermore

$$w_T = w_L \qquad \text{q.o. } (0,T), \qquad \forall T, L :\ 0 < L < T.$$

Therefore we can define the function w by

$$w(t) := w_T(t) \qquad \forall t :\ 0 < t < T;$$

clearly such w solves $(*)$ since, $\forall T > 0$, it solves $Lw = g$ in $(-\infty, T)$. The uniqueness of the solution of $(*)$ follows from the fact that every solution must coincide with w_T in $(-\infty, T)$.

Formula (6.77) can be proved by noticing that $g_T := g\,\chi_{(-\infty,T]} \in L^1(\mathbb{R})\ \forall\, T$, so g_T is an $\mathcal{L}$ transformable function with abscissa of convergence $\lambda_{g_T} \leq 0\ \forall\, T$.

2) With arguments similar to those of item 1, we obtain the following statement.

Given $g \in L^1_{loc}\big([0,\infty)\big)$ and $w_j \in \mathbb{C}$, the problem

$$\begin{cases} w \in L^1_{loc}\big([0,\infty)\big), \text{ s.t. } w^{(j)} \in AC_{loc}\big([0,\infty)\big),\ j \leq k-1, \\ Lw = g \quad \text{in } \mathcal{D}'(0,\infty), \quad w^{(j)} = w_j\,,\ j \leq k-1, \end{cases} \tag{6.78}$$

admits a unique solution w and this solution verifies for every $T > 0$

$$w(t) = w_T(t) \text{ q.o. in } (-\infty, T)\,,$$

where w_T is the unique $\mathcal{L}$ transformable solution of

$$L\,w_T \;=\; g\,\chi_{(0,T]} \quad \text{in } \mathcal{D}'(0,\infty).$$

By summarizing, the problem (6.78) has a unique solution w verifying

$$w(t) = w_T(t) = \mathcal{L}^{-1}\left\{ \frac{\mathcal{L}\left\{ g\,\chi_{(-\infty,T]} + \sum_{h=0}^{k}\sum_{j=0}^{h-1} c_h w_j s^{h-j-1}\right\}}{\mathcal{P}(s)} \right\} \tag{6.79}$$

$\forall t, T > 0\ :\ 0 < t < T\,.$

3) About the problem where a generic distribution g with support in $[0,+\infty)$ is the datum, the uniqueness follows by the support condition. The solution can be obtained exactly as in item 1, but one point: since we cannot use "the rectangular window" $\chi_{(-\infty,T]}$, we replace it by a "smooth" window (multiplier of a distribution, see Definition 3.13), for instance:

$$\psi_T(t) \in C^\infty(\mathbb{R}), \quad 0 \leq \psi_T \leq 1, \quad \psi_T(t) = 1, \text{ if } t \leq T\,, \quad \psi_T(t) = 0, \text{ if } t \geq T+1\,.$$

In this way, by using the $\mathcal{L}$ transformable distributions $\widetilde{g}_T := \psi_T g$ (all of them with abscissa of convergence $\lambda_{\widetilde{g}_T} \leq 0$), we obtain the following statement.

Given $g \in \mathcal{D}'(\mathbb{R})$ and $\operatorname{spt} g \subset [0,+\infty)$ the problem

$$w \in \mathcal{D}'(\mathbb{R}) \text{ s.t. } \quad Lw = g \quad \text{in } \mathcal{D}'(\mathbb{R}) \quad \text{and} \quad \operatorname{spt} w \subset [0,\infty) \tag{6.80}$$

admits a unique solution w and this solution verifies for every $T > 0$

$$w(t) = w_T(t) \quad \text{q.o. in } (-\infty, T)\,,$$

where w_T is the unique solution in the space of $\mathcal{L}$ transformable distributions of

$$L\,\widetilde{w}_T \;=\; \psi_T\, g \quad \text{in } \mathcal{D}'(\mathbb{R}).$$

By summarizing, the problem (6.80) has a unique solution w that verifies

$$w = \widetilde{w}_T = \mathcal{L}^{-1}\left\{ \frac{\mathcal{L}\{\psi_T\, g\}}{\mathcal{P}(s)} \right\} \qquad \text{in } \mathcal{D}'(0,T)\,,\ \forall T > 0\,. \tag{6.81}$$

□

Chapter 7
Transport and propagation models

Transport models describe phenomena connected with matter shift. For instance, the transport of a pollutant substance in a river. The transport can be described by a first order differential equation. Wave propagation depends substantially on the space dimension n. In dimension $n = 1$ (vibrating string) it can be seen as the superposition of two transports in opposite directions. In dimension $n = 2$ we may think to the configuration of surface waves in the water: if a stone is thrown in a pond, then the initial solicitation is described by a Dirac delta centered at the stone falling point and circular waves are generated, drawing circles whose largest radius is increasing in time. In dimension $n = 3$ we may think to a sound produced somewhere: the resulting acoustic wave propagates in the air on spheres of increasing radii (not on balls, in contrast to $n = 2$): after we heard the noise, the signal is gone forever.

In this chapter we study the transport in the most elementary case. This enables us to describe the wave propagation in the vibrating string by superposing two transports. With the method of the spherical means, we then solve the problem of wave propagation in dimension $n = 3$ by using the solution in the 1-dimensional case. Finally, by projecting the 3-dimensional wave on a plane, the wave propagation problem is solved also in the planar case $n = 2$.

7.1 Transport equation

Let $\mathbf{x} = (x_1, ..., x_n) \in \mathbb{R}^n$ and $t \in \mathbb{R}_+$. For each differentiable function $u = u(\mathbf{x}, t)$, we denote

$$\nabla_{\mathbf{x}} u = (u_{x_1}, ..., u_{x_n}) \qquad \nabla_{\mathbf{x},t} u = (u_{x_1}, ..., u_{x_n}, u_t) \ .$$

We solve the Cauchy problem for the transport equation assuming that the propagation speed is constant and equal to $\mathbf{b} \in \mathbb{R}^n$:

$$\begin{cases} u_t(\mathbf{x},t) + \mathbf{b} \cdot \nabla_{\mathbf{x}} u(\mathbf{x},t) = f(\mathbf{x},t) & \text{in } \mathbb{R}^n \times \mathbb{R}_+ \\ u(\mathbf{x},0) = g(\mathbf{x}) & \text{in } \mathbb{R}^n. \end{cases} \tag{7.1}$$

If $f = 0$ (absence of source term) the problem becomes:

$$\begin{cases} u_t(\mathbf{x},t) + \mathbf{b} \cdot \nabla_{\mathbf{x}} u(\mathbf{x},t) = 0 & \text{in } \mathbb{R}^n \times \mathbb{R}_+ \\ u(\mathbf{x},0) = g(\mathbf{x}) & \text{in } \mathbb{R}^n \ . \end{cases} \tag{7.2}$$

Setting $\mathbf{v} = (\mathbf{b}, 1) \in \mathbb{R}^{n+1}$, the equation (7.2) can be written as

$$\frac{\partial u}{\partial \mathbf{v}} = (\nabla_{\mathbf{x},t} u) \cdot \mathbf{v} = 0$$

which states that u is constant along the lines with directing vector $\mathbf{v}$. These lines are called characteristic lines and they have parametric equations of the form:

$$\begin{cases} \mathbf{x}(s) = \mathbf{x}_0 + s\mathbf{b} \\ t(s) = t_0 + s \end{cases} \qquad (s \in \mathbb{R}) \qquad \forall (\mathbf{x}_0, t_0) \in \mathbb{R}^{n+1} \ .$$

Therefore, the solution of (7.2) is

$$u(\mathbf{x},t) = u(\mathbf{x} - t\mathbf{b}, 0) = g(\mathbf{x} - t\mathbf{b}) \tag{7.3}$$

showing that the initial datum is "transported" along the characteristic lines. Let us now solve the full equation (7.1) with the delay method of Duhamel. We define the function $z : \mathbb{R} \to \mathbb{R}$ by

$$z(s) = u(\mathbf{x} + s\mathbf{b}, t + s), \quad s \in \mathbb{R} \ .$$

By differentiating, we obtain $z'(s) = (\nabla_{\mathbf{x}} u) \cdot \mathbf{b} + u_t = f(\mathbf{x} + s\mathbf{b}, t + s)$. Thus, by integrating between $-t$ and 0 we get

$$u(\mathbf{x},t) - g(\mathbf{x} - t\mathbf{b}) = z(0) - z(-t) = \int_{-t}^{0} f(\mathbf{x} + s\mathbf{b}, t + s) ds$$

$$\stackrel{\sigma = t+s}{=} \int_0^t f(\mathbf{x} + (\sigma - t)\mathbf{b}, \sigma) d\sigma \ .$$

Summarizing, the solution of (7.1) is given by

$$u(\mathbf{x},t) = g(\mathbf{x} - t\mathbf{b}) + \int_0^t f(\mathbf{x} + (\sigma - t)\mathbf{b}, \sigma) d\sigma \ . \tag{7.4}$$

Formula (7.4) shows that the solution u depends not only on the initial datum g transported along the characteristic but also on the contribution of the

forcing term f on the part of the characteristic line that "comes before" $(\mathbf{x}, t)$.
Let us now consider the case where the source term f is proportional to the mass u, $f(\mathbf{x}, t) = -cu(\mathbf{x}, t)$ with $c \in \mathbb{R}$. Therefore (7.2) becomes:

$$\begin{cases} u_t(\mathbf{x}, t) + \mathbf{b} \cdot \nabla_{\mathbf{x}} u(\mathbf{x}, t) + c\,u(\mathbf{x}, t) = 0 & \text{in } \mathbb{R}^n \times \mathbb{R}_+ \\ u(\mathbf{x}, 0) = g(\mathbf{x}) & \text{in } \mathbb{R}^n \,. \end{cases} \tag{7.5}$$

We set $v(\mathbf{x}, t) = e^{ct} u(\mathbf{x}, t)$; hence $u_t = -ce^{-ct} v + e^{-ct} v_t$ and v solves (7.2). By (7.3), we know that the solution of problem (7.5) is given by

$$u(\mathbf{x}, t) = e^{-ct} g(\mathbf{x} - t\mathbf{b}).$$

This form expresses the fact that the initial datum propagates along the characteristic line with a damping coefficient (if $c > 0$) or dilation coefficient (if $c < 0$) that increases in the time.
Finally, if in (7.5) we add a source term source f, the problem becomes

$$\begin{cases} u_t(\mathbf{x}, t) + \mathbf{b} \cdot \nabla_{\mathbf{x}} u(\mathbf{x}, t) + cu(\mathbf{x}, t) = f(\mathbf{x}, t) & \text{in } \mathbb{R}^n \times \mathbb{R}_+ \\ u(\mathbf{x}, 0) = g(\mathbf{x}) & \text{in } \mathbb{R}^n \,; \end{cases} \tag{7.6}$$

whose solution is given by

$$u(\mathbf{x}, t) = e^{-ct} g(\mathbf{x} - t\mathbf{b}) + \int_0^t e^{c(\sigma - t)} f(\mathbf{x} + (\sigma - t)\mathbf{b}, \sigma) d\sigma.$$

7.2 Equation of the vibrating string

Let $c > 0$ and consider the equation of a vibrating string of infinite length

$$\begin{cases} u_{tt} - c^2 u_{xx} = f(x, t) & (x, t) \in \mathbb{R} \times \mathbb{R}_+ \\ u(x, 0) = g(x) & x \in \mathbb{R} \\ u_t(x, 0) = h(x) & x \in \mathbb{R} \,, \end{cases} \tag{7.7}$$

where c represents the propagation speed of the wave, f is an external force, g and h are respectively the position and the transversal velocity of the string at time $t = 0$. Let us split (7.7) in the two following problems

$$\begin{cases} u_{tt} - c^2 u_{xx} = 0 & (x, t) \in \mathbb{R} \times \mathbb{R}_+ \\ u(x, 0) = g(x) & x \in \mathbb{R} \\ u_t(x, 0) = h(x) & x \in \mathbb{R} \end{cases} \tag{7.8}$$

$$\begin{cases} u_{tt} - c^2 u_{xx} = f(x, t) & (x, t) \in \mathbb{R} \times \mathbb{R}_+ \\ u(x, 0) = 0 & x \in \mathbb{R} \\ u_t(x, 0) = 0 & x \in \mathbb{R} \,. \end{cases} \tag{7.9}$$

If $u_1 = u_1(x,t)$ and $u_2 = u_2(x,t)$ solve respectively (7.8) and (7.9), the function $u = u_1 + u_2$ solves (7.7). As we will see in Section 7.2.4, the solution of problem (7.9) can be deduced from that of problem (7.8). To solve (7.8) we can proceed in three different ways.

7.2.1 Decomposition in two transport equations

We reduce the equation in problem (7.8) to a system by setting:

$$v_t(x,t) + cv_x(x,t) = 0 , \quad u_t(x,t) - c\,u_x(x,t) = v(x,t) . \tag{7.10}$$

We observe that if u and v solve (7.10), then the initial conditions in (7.8) become

$$u(x,0) = g(x) , \quad v(x,0) = u_t(x,0) - cu_x(x,0) = h(x) - cg'(x) \quad (x \in \mathbb{R}) . \tag{7.11}$$

Indeed, u solves (7.8) if and only if the pair (u,v) solves (7.10)-(7.11). The first equation in (7.10) is a homogeneous transport equation. Therefore, taking into account the initial condition, by (7.3) we find

$$v(x,t) = v(x-ct,0) = h(x-ct) - cg'(x-ct).$$

It then follows that the second equation in (7.10) becomes the non-homogeneous transport equation $u_t(x,t) - cu_x(x,t) = h(x-ct) - cg'(x-ct)$. By (7.4), and performing the change of variable $y = x + ct - 2c\sigma$, its solution with $u(x,0) = g(x)$ is given by

$$u(x,t) = g(x+ct) + \frac{1}{2c}\int_{x-ct}^{x+ct} [h(y) - cg'(y)]\,dy ,$$

from which, we derive the d'Alembert formula for the homogeneous equation

$$u(x,t) = \frac{1}{2}[g(x+ct) + g(x-ct)] + \frac{1}{2c}\int_{x-ct}^{x+ct} h(y)dy , \tag{7.12}$$

that yields the solution of problem (7.8).

7.2.2 Solving the problem through Fourier transform

Suppose that $g \in C^2(\mathbb{R})$ and $h \in C^1(\mathbb{R})$ are Fourier transformable; this means that they also belong to one of the spaces considered in Chapter 4. Let us solve (7.8) formally by applying to both sides of equation (7.8) the Fourier transform with respect to the x variable only: by (4.80) we know that the transformation with respect x and the differentiation with respect to t switch and, therefore the calculation is justified. By (7.8) and by (4.28) we obtain

$$\begin{aligned}\partial_t^2\widehat{u}(\xi,t) &= \int_{\mathbb{R}} e^{-i\xi x}\partial_t^2 u(x,t)\,dx = c^2\int_{\mathbb{R}} e^{-i\xi x}\partial_x^2 u(x,t)\,dx\\ &= c^2\mathcal{F}(\partial_x^2 u(x,t),\xi) = -c^2\xi^2\widehat{u}(\xi,t).\end{aligned}$$

So, we obtain the second order Cauchy problem

$$\begin{cases}\partial_t^2\widehat{u}(\xi,t)+c^2\xi^2\widehat{u}(\xi,t)=0\\ \widehat{u}(\xi,0)=\widehat{g}(\xi)\\ \partial_t\widehat{u}(\xi,0)=\widehat{h}(\xi)\end{cases}$$

that has the solution

$$\widehat{u}(\xi,t)=\widehat{g}(\xi)\cos(c\xi t)+\widehat{h}(\xi)\frac{\sin(c\xi t)}{c\xi}.$$

By anti-transforming, using (4.6), (4.50) and $\widehat{p(x/ct)} = 2\sin(c\xi t)/\xi$, we get

$$u(x,t)=\frac{g(x+ct)+g(x-ct)}{2}+\frac{1}{2c}h(x)*p\left(\frac{x}{ct}\right)\qquad\forall x\in\mathbb{R}\,,\ t>0$$

and we obtain the d'Alembert formula (7.12) again.

7.2.3 Solving the problem with the fundamental solution

If $g=0$, then we may use the fundamental solution to solve (7.8). Let H be the Heaviside function and put

$$K(x,z,t)=\frac{1}{2c}[H(x-z+ct)-H(x-z-ct)]. \tag{7.13}$$

The function K is called fundamental solution of the wave equation in dimension $n=1$ with pole at $z\in\mathbb{R}$. By (7.12), the solution of (7.8) with $g=0$ can be written as

$$u(x,t)=\frac{1}{2c}\int_{x-ct}^{x+ct}h(y)\,dy=\int_{\mathbb{R}}K(x,y,t)h(y)\,dy\,.$$

Therefore, for every given $z\in\mathbb{R}$, it is possible to interpret the function K defined in (7.13) as the solution of the Cauchy problem

$$\begin{cases} u_{tt} - c^2 u_{xx} = 0 & \text{in } \mathbb{R} \times \mathbb{R}_+ \\ u(x,0) = 0 & \text{in } \mathbb{R} \\ u_t(x,0) = \delta_z(x) & \text{in } \mathbb{R} \end{cases} \tag{7.14}$$

where $\delta_z(x) = \delta(x-z)$ is the Dirac delta centered at $z \in \mathbb{R}$.

7.2.4 The Duhamel delay method

To solve the non-homogeneous problem (7.9) we could proceed using the Fourier transform, see the solution of the problem (4.83). Nevertheless, we will use here the Duhamel delay method. Let us introduce the delay parameter $s \geq 0$ and let $w = w(x,t;s)$ be the solution of the problem

$$\begin{cases} w_{tt}(x,t;s) - c^2 w_{xx}(x,t;s) = 0 & x \in \mathbb{R},\ t > s \\ w(x,s;s) = 0 & x \in \mathbb{R} \\ w_t(x,s;s) = f(x,s) & x \in \mathbb{R}\ . \end{cases}$$

By d'Alembert formula (7.12), its solution takes the form

$$w(x,t;s) = \frac{1}{2c} \int_{x-c(t-s)}^{x+c(t-s)} f(y,s)dy\ . \tag{7.15}$$

Define the function

$$u(x,t) = \int_0^t w(x,t;s)ds\ . \tag{7.16}$$

By differentiating we find

$$u_t(x,t) = w(x,t;t) + \int_0^t w_t(x,t;s)ds = \int_0^t w_t(x,t;s)ds$$

$$u_{tt}(x,t) = w_t(x,t;t) + \int_0^t w_{tt}(x,t;s)ds = f(x,t) + \int_0^t w_{tt}(x,t;s)ds$$

$$u_{tt}(x,t) - c^2 u_{xx}(x,t) = f(x,t) + \int_0^t [w_{tt}(x,t;s) - c^2 w_{xx}(x,t;s)]ds = f(x,t)\ .$$

Moreover, $u(x,0) = u_t(x,0) = 0$. Therefore, by (7.15) and (7.16), the solution of (7.9) is given by

$$u(x,t) = \frac{1}{2c} \int_0^t \int_{x-cs}^{x+cs} f(y,t-s)dy\,ds\ . \tag{7.17}$$

Summarizing, by adding (7.12) and (7.17), we infer that the unique solution of the problem (7.7) is given by the d'Alembert formula:

$$u(x,t) = \frac{g(x+ct)+g(x-ct)}{2} + \frac{1}{2c}\int_{x-ct}^{x+ct} h(y)dy$$
$$+\frac{1}{2c}\int_0^t \int_{x-cs}^{x+cs} f(y,t-s)dyds.$$

We observe that the domain of integration of the external force f corresponds to a triangle, see Figure 4.9.

7.2.5 Separation of variables in bounded intervals

Assuming $c > 0$, let us solve the Cauchy-Dirichlet problem for a vibrating string of finite length $\ell > 0$ with fixed endpoints, that is,

$$\begin{cases} u_{tt} - c^2 u_{xx} = 0 & x \in (0,\ell)\,,\ t > 0 \\ u(x,0) = g(x) & x \in (0,\ell) \\ u_t(x,0) = h(x) & x \in (0,\ell) \\ u(0,t) = u(\ell,t) = 0 & t \in \mathbb{R}_+\,. \end{cases} \tag{7.18}$$

Let us solve (7.18) by separating variables. Seeking solutions of $(7.18)_1$ in the form $u(x,t) = X(x)T(t)$, we obtain

$$\frac{T''(t)}{c^2 T(t)} = \frac{X''(x)}{X(x)}$$

from which we infer that there exists $\lambda \in \mathbb{C}$ such that

$$T''(t) + c^2\lambda^2 T(t) = 0, \qquad X''(x) + \lambda^2 X(x) = 0. \tag{7.19}$$

Imposing the boundary conditions $X(0) = X(\ell) = 0$ (obtained from $(7.18)_4$), we find nontrivial solutions X if and only if $\lambda = m\pi/\ell$ with $m \in \mathbb{N}\setminus\{0\}$ and $X(x) = X_m(x) = \sin\left(\frac{m\pi}{\ell}x\right)$. For given $\lambda = m\pi/\ell$, by (7.19) we see that X_m is coupled with $T_m(t) = C\cos(cm\pi t/\ell) + D\sin(cm\pi t/\ell)$ for $C, D \in \mathbb{R}$. Let us look for solutions of $(7.18)_1$ and $(7.18)_4$ that are the superposition of separated variable solutions:

$$u(x,t) = \sum_{m=1}^{\infty}\left[\alpha_m \cos\left(\frac{cm\pi}{\ell}t\right) + \beta_m \sin\left(\frac{cm\pi}{\ell}t\right)\right]\sin\left(\frac{m\pi}{\ell}x\right). \tag{7.20}$$

By formally term-wise differentiating of the series we get

$$u_t(x,t) = \sum_{m=1}^{\infty}\left[-\alpha_m \frac{cm\pi}{\ell}\sin\left(\frac{cm\pi}{\ell}t\right) + \beta_m \frac{cm\pi}{\ell}\cos\left(\frac{cm\pi}{\ell}t\right)\right]\sin\left(\frac{m\pi}{\ell}x\right)$$

and imposing the condition $(7.18)_3$ we find

$$\sum_{m=1}^{\infty} \beta_m \frac{cm\pi}{\ell} \sin\left(\frac{m\pi}{\ell}x\right) = h(x) \qquad \forall x \in (0,\ell)$$

which implies that $\frac{cm\beta_m\pi}{\ell}$ are the Fourier coefficients of h, seen as an extended odd function in the interval $(-\ell,\ell)$. From (5.10) with $T = 2\ell$ we obtain

$$\beta_m = \frac{2}{cm\pi}\int_0^{\ell} h(x)\sin\left(\frac{m\pi}{\ell}x\right)dx.$$

To determine α_m, we make use of (7.20) and the initial condition $(7.18)_2$:

$$\sum_{m=1}^{\infty} \alpha_m \sin\left(\frac{m\pi}{\ell}x\right) = g(x) \qquad \forall x \in (0,\ell)$$

and therefore α_m are the Fourier coefficients of g, seen as an extended odd function in the interval $(-\ell,\ell)$. From (5.10) with $T = 2\ell$ we obtain

$$\alpha_m = \frac{2}{\ell}\int_0^{\ell} g(x)\sin\left(\frac{m\pi}{\ell}x\right)dx.$$

Remark 7.1. In formula (7.20), we can use Werner's formulas (Table 10.1.6) to obtain

$$\cos\left(\frac{cm\pi}{\ell}t\right)\sin\left(\frac{m\pi}{\ell}x\right) = \frac{1}{2}\left[\sin\left(\frac{m\pi}{\ell}(x+ct)\right) + \sin\left(\frac{m\pi}{\ell}(x-ct)\right)\right]$$

and similarly for the product $\sin\left(\frac{cm\pi}{\ell}t\right)\sin\left(\frac{m\pi}{\ell}x\right)$: so (7.20) is the restriction to $(0,l)\times\mathbb{R}_+$ of the solution of the same partial differential equation in $\mathbb{R}\times\mathbb{R}_+$ with odd and $2l$ periodic initial data. This different way to write the solution u highlights the phenomenon of traveling waves with speed propagation $\pm c$ already observed in d'Alembert formula (7.12). The only difference is that for the string of finite length this phenomenon is "free" if the signal comes from the initial condition (corresponding to $x \in (0,\ell)$ and $t = 0$), whereas it is "constrained" when the signal comes form some boundary conditions (corresponding to $x \in \{0,\ell\}$ and $t > 0$).

Similar computations can be carried on for the Cauchy-Neumann problem for a string that can move freely at the endpoints and whose initial profile is horizontal:

$$\begin{cases} u_{tt} - c^2 u_{xx} = 0 & (x,t) \in (0,\ell)\times\mathbb{R}_+ \\ u(x,0) = g(x) & x \in (0,\ell) \\ u_t(x,0) = h(x) & x \in (0,\ell) \\ u_x(0,t) = u_x(\ell,t) = 0 & t \in \mathbb{R}_+ \,. \end{cases} \tag{7.21}$$

Separating variables, we get again (7.19). Imposing the boundary conditions $X'(0) = X'(\ell) = 0$ (derived from $(7.21)_4$), we obtain nontrivial solutions X

if and only if $\lambda = m\pi/\ell$ with $m \in \mathbb{N}$ (0 included!) and $X_0(x) = 1$, $X_m(x) = \cos\left(\frac{m\pi}{\ell}x\right)$ for $m \geq 1$. For a given $\lambda = m\pi/\ell$, by (7.19) we infer that X_m is coupled with $T_0(t) = Ct + D$ and $T_m(t) = C\cos(cm\pi t/\ell) + D\sin(cm\pi t/\ell)$ for $m \geq 1$ ($C, D \in \mathbb{R}$). Therefore, instead of (7.20) we obtain

$$u(x,t) = \beta_0 t + \alpha_0 + \sum_{m=1}^{\infty}\left[\alpha_m \cos\left(\frac{cm\pi}{\ell}t\right) + \beta_m \sin\left(\frac{cm\pi}{\ell}t\right)\right]\cos\left(\frac{m\pi}{\ell}x\right).$$

By differentiating by series we obtain

$$u_t(x,t) = \beta_0 + \sum_{m=1}^{\infty}\left[\beta_m \frac{cm\pi}{\ell}\cos\left(\frac{cm\pi}{\ell}t\right) - \alpha_m \frac{cm\pi}{\ell}\sin\left(\frac{cm\pi}{\ell}t\right)\right]\cos\left(\frac{m\pi}{\ell}x\right)$$

and imposing the condition $(7.21)_3$ we have

$$\beta_0 + \sum_{m=1}^{\infty}\beta_m \frac{cm\pi}{\ell}\cos\left(\frac{m\pi}{\ell}x\right) = h(x) \qquad \forall x \in (0,\ell)$$

that implies β_0 and all $\frac{cm\beta_m\pi}{\ell}$ are the Fourier coefficients of h, seen as function extended as even function in the interval $(-\ell,\ell)$. By (5.10) with $T = 2\ell$ we obtain (be careful to the factor 2 over β_0!)

$$\beta_0 = \frac{1}{\ell}\int_0^\ell h(x)dx, \qquad \beta_m = \frac{2}{cm\pi}\int_0^\ell h(x)\cos\left(\frac{m\pi}{\ell}x\right)dx.$$

By $(7.21)_2$ the α_m are instead the Fourier coefficients of g extended even in the interval $(-\ell,\ell)$. By (5.10) with $T = 2\ell$ we obtain

$$\alpha_0 = \frac{1}{\ell}\int_0^\ell g(x)dx, \qquad \alpha_m = \frac{2}{\ell}\int_0^\ell g(x)\cos\left(\frac{m\pi}{\ell}x\right)dx.$$

We conclude by emphasizing that it is possible to solve (7.18) and (7.21) also when an external force is present. We apply again the Duhamel method. For instance, in the case of a Dirichlet condition, the problem

$$\begin{cases} u_{tt} - c^2 u_{xx} = f(x,t) & (x,t) \in (0,\ell)\times\mathbb{R}_+ \\ u(x,0) = 0 & x \in (0,\ell) \\ u_t(x,0) = 0 & x \in (0,\ell) \\ u(0,t) = u(\ell,t) = 0 & t \in \mathbb{R}_+ \end{cases}$$

is delayed with a parameter $s \geq 0$ and transformed into the problem

$$\begin{cases} w_{tt}(x,t;s) - c^2 w_{xx}(x,t;s) = 0 & (x,t) \in (0,\ell)\times(s,+\infty) \\ w(x,s;s) = 0 & x \in (0,\ell) \\ w_t(x,s;s) = f(x,s) & x \in (0,\ell) \\ w(0,t;s) = w(\ell,t;s) = 0 & t \in (s,+\infty). \end{cases}$$

After that the solution w is determined by separating variables, we set $u(x,t) = \int_0^t w(x,t;s)\,ds.$

7.3 Wave equation in dimensions 2 and 3

The wave equation (with propagation speed $c > 0$) in the whole space $\mathbb{R}^n$ can be written as

$$\begin{cases} u_{tt} - c^2 \Delta u = f(\mathbf{x},t) & (\mathbf{x},t) \in \mathbb{R}^n \times \mathbb{R}_+ \\ u(\mathbf{x},0) = g(\mathbf{x}) & \mathbf{x} \in \mathbb{R}^n \\ u_t(\mathbf{x},0) = h(\mathbf{x}) & \mathbf{x} \in \mathbb{R}^n \ . \end{cases} \tag{7.22}$$

As for (7.7), we decompose (7.22) into the two following problems

$$\begin{cases} u_{tt} - c^2 \Delta u = 0 & (\mathbf{x},t) \in \mathbb{R}^n \times \mathbb{R}_+ \\ u(\mathbf{x},0) = g(\mathbf{x}) & \mathbf{x} \in \mathbb{R}^n \\ u_t(\mathbf{x},0) = h(\mathbf{x}) & \mathbf{x} \in \mathbb{R}^n \end{cases} \tag{7.23}$$

$$\begin{cases} u_{tt} - c^2 \Delta u = f(\mathbf{x},t) & (\mathbf{x},t) \in \mathbb{R}^n \times \mathbb{R}_+ \\ u(\mathbf{x},0) = 0 & \mathbf{x} \in \mathbb{R}^n \\ u_t(\mathbf{x},0) = 0 & \mathbf{x} \in \mathbb{R}^n \ . \end{cases} \tag{7.24}$$

If $u_1 = u_1(\mathbf{x},t)$ and $u_2 = u_2(\mathbf{x},t)$ solve (7.23) and (7.24), the function $u = u_1 + u_2$ solves (7.22). Arguing as in Section 7.2.4, the solution of problem (7.24) can be deduced from that of problem (7.23).

7.3.1 Solution in dimension $n = 3$

If $n = 3$, we solve (7.23) by using the spherical mean method. For a given function $w = w(\mathbf{x},t)$ of class C^2, we define its spherical mean by:

$$W(\mathbf{x};r,t) = \fint_{\partial B_r(\mathbf{x})} w(\mathbf{y},t)\, d\sigma(\mathbf{y}) = \frac{1}{4\pi} \int_{|\mathbf{y}|=1} w(\mathbf{x}+r\mathbf{y},t)\, d\sigma(\mathbf{y})$$

for each $\mathbf{x} \in \mathbb{R}^3$, $t > 0$, $r > 0$. Then we set $W(\mathbf{x};0,t) = w(\mathbf{x},t)$ and $W(\mathbf{x};r,t) = W(\mathbf{x};-r,t)$ for $r < 0$ in such a way that $r \mapsto W(\mathbf{x};r,t)$ is an even function. It is possible to prove ([8, Chapter 5]) that if $u \in C^2(\mathbb{R}^3 \times [0,+\infty))$ is the solution to (7.23), then, letting $U = U(\mathbf{x};r,t)$ be the spherical mean of u and $G = G(r)$ and $H = H(r)$ the spherical mean of g and h respectively (independent of time!), for every $\mathbf{x} \in \mathbb{R}^3$ we have $U(\mathbf{x};\cdot,\cdot) \in C^2(\mathbb{R} \times [0,+\infty))$ and

$$\begin{cases} \frac{1}{c^2}U_{tt} = U_{rr} + \frac{2}{r}U_r & (r,t) \in \mathbb{R} \times \mathbb{R}_+ \\ U(r,0) = G(r), \quad U_t(r,0) = H(r) & r \in \mathbb{R}. \end{cases} \tag{7.25}$$

We are so led to solve (7.25). Setting $V = rU$, we obtain $V_r = U + rU_r$, $V_{rr} = U_r + U_r + rU_{rr} = 2U_r + rU_{rr}$, $V_{tt} = rU_{tt}$; after substitution, problem (7.25) is transformed into the following 1-dimensional wave equation

$$\begin{cases} V_{tt} - c^2 V_{rr} = 0 & (r,t) \in \mathbb{R} \times \mathbb{R}_+ \\ V(r,0) = rG(r) & r \in \mathbb{R} \\ V_t(r,0) = rH(r) & r \in \mathbb{R}. \end{cases}$$

From d'Alembert formula (7.12) we get:

$$V(r,t) = \frac{(r+ct)G(r+ct) + (r-ct)G(r-ct)}{2} + \frac{1}{2c}\int_{r-ct}^{r+ct} sH(s)\, ds$$

and, since the spherical means are extended as even functions, we deduce

$$U(\mathbf{x}; r,t) = \frac{(ct+r)G(ct+r) - (ct-r)G(ct-r)}{2r} + \frac{1}{2cr}\int_{ct-r}^{ct+r} sH(s)\, ds.$$

Taking the limit as $r \to 0$ we find

$$u(\mathbf{x},t) = t\ H(ct) + \frac{d}{dt}[t\ G(ct)]$$

which yields

$$u(\mathbf{x},t) = \frac{1}{4\pi c^2 t}\int_{|\mathbf{y}-\mathbf{x}|=ct} h(\mathbf{y})\, d\sigma(\mathbf{y}) + \frac{\partial}{\partial t}\frac{1}{4\pi c^2 t}\int_{|\mathbf{y}-\mathbf{x}|=ct} g(\mathbf{y})\, d\sigma(\mathbf{y})\ ; \tag{7.26}$$

this is called Kirchhoff formula for the homogeneous wave equation.
We can write (7.26) as

$$u(\mathbf{x},t) = \fint_{\partial B_{ct}(\mathbf{x})} \{t\ h(\mathbf{y}) + g(\mathbf{y}) + \nabla g(\mathbf{y}) \cdot (\mathbf{y}-\mathbf{x})\}\, d\sigma(\mathbf{y}), \tag{7.27}$$

see [5, Section 3.4.2] for the details. This is the solution of (7.23).
Using the Duhamel method introduced in Section 7.2.4, we infer that the solution of (7.24) is given by

$$u(\mathbf{x},t) = \frac{1}{4\pi c^2}\int_{B_{ct}(\mathbf{x})} \frac{f(\mathbf{y}, t - c^{-1}|\mathbf{x}-\mathbf{y}|)}{|\mathbf{x}-\mathbf{y}|}\, d\mathbf{y}. \tag{7.28}$$

Observe that the integration domain in (7.28) is a 3-dimensional ball for any fixed time; the radius of the ball increases with t, therefore the dependence domain of the solution u is contained in a solid cone in the variables $(\mathbf{y}, \tau) \in \mathbb{R}^4$.

Actually (7.28) coincides with the first summand in complete Kirchhoff formula found in Exercise 4.57: so the dependence domain of the solution u is exactly the bounded one-fold conical surface $\{(\mathbf{y},\tau) \in \mathbb{R}^3 \times \mathbb{R}_+ : \ |\mathbf{x}-\mathbf{y}| = c(t-\tau) > 0\}$ in $\mathbb{R}^4$ (see Figure 4.13).
We call fundamental solution (focussed at $\mathbf{z}$) of the wave equation in dimension $n = 3$ the distribution $K(\mathbf{z},t) \in \mathcal{D}'(\mathbb{R}^3)$, depending on the two parameters $\mathbf{z} \in \mathbb{R}^3$ and $t \geq 0$, implicitly defined by

$$\langle K(\mathbf{z},t), \varphi \rangle = \int_{\partial B_{ct}(\mathbf{z})} \frac{\varphi(\mathbf{y})}{4\pi c|\mathbf{z}-\mathbf{y}|} d\sigma(\mathbf{y}) \qquad \forall \varphi \in \mathcal{D}(\mathbb{R}^3). \tag{7.29}$$

Note that the right hand side of (7.29) is meaningful for more general test functions, for instance we can compute $\langle K(\mathbf{z},t), \varphi \rangle$ for all $\varphi \in C^0(\mathbb{R}^3)$. If $g = 0$, by (7.27) the solution u of (7.23) is given by

$$u(\mathbf{x},t) = \fint_{\partial B_{ct}(\mathbf{x})} t\, h(\mathbf{y})\, d\sigma(\mathbf{y}).$$

From the equivalence $\mathbf{y} \in \partial B_{ct}(\mathbf{x}) \Leftrightarrow ct = |\mathbf{y}-\mathbf{x}|$, we deduce that

$$u(\mathbf{x},t) = \int_{\partial B_{ct}(\mathbf{x})} \frac{h(\mathbf{y})}{4\pi c|\mathbf{x}-\mathbf{y}|} d\sigma(\mathbf{y}) = \langle K(\mathbf{x},t), h \rangle.$$

Therefore the solution u can be expressed through a suitable duality between the fundamental solution and the initial datum. Moreover, the distribution $K(\mathbf{z},t)$ can be seen as the solution of the Cauchy problem (see [5]):

$$\begin{cases} u_{tt} - c^2 \Delta u = 0 & \text{in } \mathbb{R}^3 \times \mathbb{R}_+ \\ u(\mathbf{x},0) = 0 & \text{in } \mathbb{R}^3 \\ u_t(\mathbf{x},0) = \delta_{\mathbf{z}} & \text{in } \mathbb{R}^3. \end{cases} \tag{7.30}$$

7.3.2 Solution in dimension $n = 2$

We consider the Cauchy problem

$$\begin{cases} u_{tt} - c^2 \Delta u = 0 & (\mathbf{x},t) \in \mathbb{R}^2 \times \mathbb{R}_+ \\ u(\mathbf{x},0) = g(\mathbf{x}) & \mathbf{x} \in \mathbb{R}^2 \\ u_t(\mathbf{x},0) = h(\mathbf{x}) & \mathbf{x} \in \mathbb{R}^2. \end{cases} \tag{7.31}$$

We view the solution $u(x_1,x_2,t)$ of (7.31) as a solution of the same equation in dimension $n = 3$, with 2-dimensional data

$$g = g(x_1,x_2) \qquad h = h(x_1,x_2).$$

We introduce the following notations: with $\mathbf{x} = (x_1, x_2, x_3)$ we denote the generic point of $\mathbb{R}^3$, with $\mathbf{x}' = (x_1, x_2)$ the generic point of $\mathbb{R}^2$, with ∇ the gradient in $\mathbb{R}^3$ and with $\nabla_{\mathbf{x}'}$ the gradient in $\mathbb{R}^2$. Set $w(x_1, x_2, x_3, t) := u(x_1, x_2, t)$, then by Kirchhoff formula (7.27) and since w does not depend on x_3, we have for $\mathbf{x} = (x_1, x_2, 0)$:

$$u(\mathbf{x}', t) = w(\mathbf{x}, t) = \fint_{\partial B_{ct}(\mathbf{x})} \{t\ h(\mathbf{y}) + g(\mathbf{y}) + \nabla g(\mathbf{y}) \cdot (\mathbf{y} - \mathbf{x})\}\ d\sigma(\mathbf{y})$$

$$= \frac{1}{4\pi c^2 t^2} \int_{\partial B_{ct}(\mathbf{x})} [t\ h(\mathbf{y}) + g(\mathbf{y}) + \nabla_{\mathbf{x}'} g(\mathbf{y}) \cdot (\mathbf{y}' - \mathbf{x}')]\ d\sigma(\mathbf{y}).$$

Moreover, due to the independence of x_3 of the integrated functions and by the symmetry of the sphere $B_{ct}(\mathbf{x})$ with respect to the plane $x_3 = 0$, we have

$$u(\mathbf{x}', t) = \frac{2}{4\pi c^2 t^2} \int_{\substack{|\mathbf{y}-\mathbf{x}|=ct \\ \mathbf{y}_3>0}} \{th(\mathbf{y}) + g(\mathbf{y}) + \nabla_{\mathbf{x}'} g(\mathbf{y}) \cdot (\mathbf{y}' - \mathbf{x}')\} d\sigma(\mathbf{y})$$

$$= \frac{1}{2\pi c^2 t^2} \int_{|\mathbf{y}'-\mathbf{x}'|<ct} \frac{t\,h(\mathbf{y}') + g(\mathbf{y}') + \nabla_{\mathbf{x}'} g(\mathbf{y}') \cdot (\mathbf{y}' - \mathbf{x}')}{\sqrt{c^2t^2 - |\mathbf{y}' - \mathbf{x}'|^2}}\, ct\, d\mathbf{y}'$$

$$= \frac{1}{2} \fint_{B_{ct}(\mathbf{x}')} \frac{ct\{t\ h(\mathbf{y}') + g(\mathbf{y}') + \nabla_{\mathbf{x}'} g(\mathbf{y}') \cdot (\mathbf{y}' - \mathbf{x}')\}}{\sqrt{c^2t^2 - |\mathbf{y}' - \mathbf{x}'|^2}}\, d\mathbf{y}'. \tag{7.32}$$

Let us emphasize that $B_{ct}(\mathbf{x}')$ represents the 2-dimensional disk centered at $\mathbf{x}' \in \mathbb{R}^2$ of radius ct. Formula (7.32) represents the solution of (7.31).
By using the Heaviside function H, we define the fundamental solution of the wave equation in dimension $n = 2$ with pole in $\mathbf{z} \in \mathbb{R}^2$ the function

$$K(\mathbf{x}, \mathbf{z}, t) = \frac{1}{2\pi c} \frac{H(c^2t^2 - |\mathbf{x} - \mathbf{z}|^2)}{\sqrt{c^2t^2 - |\mathbf{x} - \mathbf{z}|^2}}.$$

Note that $K(\cdot, \mathbf{z}, t) \in L^1(\mathbb{R}^2)$. If $g = 0$, by (7.32), the solution of (7.31) can be written as

$$u(\mathbf{x}, t) = \int_{\mathbb{R}^2} K(\mathbf{x}, \mathbf{y}, t) h(\mathbf{y}) d\mathbf{y}. \tag{7.33}$$

Therefore, for every fixed $\mathbf{z} \in \mathbb{R}^2$, we can interpret the function $K(\mathbf{x}, \mathbf{z}, t)$ as the solution of the Cauchy problem

$$\begin{cases} u_{tt} - c^2 \Delta u = 0 & (\mathbf{x}, t) \in \mathbb{R}^2 \times \mathbb{R}_+ \\ u(\mathbf{x}, 0) = 0 & \mathbf{x} \in \mathbb{R}^2 \\ u_t(\mathbf{x}, 0) = \delta_{\mathbf{z}}(\mathbf{x}) & \mathbf{x} \in \mathbb{R}^2. \end{cases} \tag{7.34}$$

This formal interpretation has a rigorous justification too, by taking a sequence of regular data $h_k(\mathbf{x})$ that converges to $\delta(\mathbf{x} - \mathbf{z})$ and passing to the limit; we omit the details of the proof. In Chapter 4 (Exercise 4.57) we gave a precise meaning to singular initial data (in the present case $u_t(\mathbf{x},0)=\delta_{\mathbf{z}}(\mathbf{x})$) for wave equation in any space dimension n, by extending the problem to $\mathbb{R}^{n+1}$.

7.4 Exercises

Exercise 7.1. Determine the characteristics and the solution of the transport problem

$$\begin{cases} u_t - u_x - u_y = 2y & \text{if } (x, y, t) \in \mathbb{R}^2 \times \mathbb{R}_+ \\ u(x, y, 0) = e^x - y^2 & \text{if } (x, y) \in \mathbb{R}^2 \ . \end{cases}$$

Solution - For $s > 0$, the characteristics are given by

$$\begin{cases} x(s) = x_0 - s \\ y(s) = y_0 - s \\ t(s) = t_0 + s \ . \end{cases}$$

We put $z(s) = u(x - s, y - s, t + s)$; then, setting $\dot{z}(s) = dz/ds$,

$$\dot{z}(s) = -u_x(x - s, y - s, t + s) - u_y(x - s, y - s, t + s) + u_t(x - s, y - s, t + s) = 2(y - s)$$

and integrating over $[-t, 0]$ we get

$$u(x, y, t) - e^{x+t} + (y + t)^2 = u(x, y, t) - u(x + t, y + t, 0) = z(0) - z(-t) =$$

$$= 2 \int_{-t}^{0} (y - s)\, ds = 2yt + t^2$$

that provides $u(x, y, t) = e^{x+t} - y^2$. □

Exercise 7.2. Find the characteristics and the solution of the transport equation:

$$\begin{cases} u_t + u_x - 2u_y = 4t^3 & (x, y, t) \in \mathbb{R} \times \mathbb{R} \times \mathbb{R}^+ \\ u(x, y, 0) = \arctan(x + y) & (x, y) \in \mathbb{R} \times \mathbb{R} \ . \end{cases}$$

Solution - The parametric equations of the characteristics are ($s > 0$)

$$\begin{cases} x(s) = x_0 + s \\ y(s) = y_0 - 2s \\ t(s) = t_0 + s \ . \end{cases}$$

We set $z(s) = u(x + s, y - 2s, t + s)$; then

$$\dot{z}(s) = u_x(x + s, y - 2s, t + s) - 2u_y(x + s, y - 2s, t + s) + u_t(x + s, y - 2s, t + s) = 4(t + s)^3$$

and integrating over $[-t, 0]$ we have

$$u(x, y, t) - \arctan(x + y + t) = z(0) - z(-t) = 4 \int_{-t}^{0} (t + s)^3 ds = t^4$$

that provides $u(x, y, t) = \arctan(x + y + t) + t^4$. □

Exercise 7.3. Determine the characteristics and the solution of the transport problem with damping term

$$\begin{cases} u_t - u_x - u_y + 3u = 0 & \text{if } (x, y, t) \in \mathbb{R}^2 \times \mathbb{R}_+ \\ u(x, y, 0) = e^x - y^2 & \text{if } (x, y) \in \mathbb{R}^2 \ . \end{cases}$$

Solution - Putting $v = e^{3t}u$ we find that v solves the problem

$$\begin{cases} v_t - v_x - v_y = 0 & \text{if } (x,y,t) \in \mathbb{R}^2 \times \mathbb{R}_+ \\ v(x,y,0) = e^x - y^2 & \text{if } (x,y) \in \mathbb{R}^2 \ . \end{cases}$$

The characteristics are

$$\begin{cases} x(s) = x_0 - s \\ y(s) = y_0 - s \\ t(s) = t_0 + s \end{cases}$$

and therefore

$$v(x,y,t) = v(x+t, y+t, 0) = e^{x+t} - (y+t)^2.$$

Summarizing, we have $u(x,y,t) = e^{-3t} v(x+t,y+t,0) = e^{x-2t} - e^{-3t}(y+t)^2$. □

Exercise 7.4. Determine the characteristics and the solution of the problem

$$\begin{cases} u_t - 2u_x + 3u_y = e^{x+y} & (x,y,t) \in \mathbb{R}^2 \times (0,\infty) \\ u(x,y,0) = (\ln|xy|)^4 & (x,y) \in \mathbb{R}^2 \ . \end{cases}$$

Observe that the initial datum is not regular, hence the phenomenon of singularities propagation along characteristics has to be expected.

Solution - The parametric equations of the characteristics are

$$\begin{cases} x(s) = x_0 - 2s \\ y(s) = y_0 + 3s \\ t(s) = t_0 + s \end{cases} \qquad (s \in \mathbb{R}).$$

Setting $z(s) = u(x-2s, y+3s, t+s)$ and $\dot{z}(s) = dz/ds$, we have

$$\dot{z}(s) = -2u_x(x-2s,y+3s,t+s) + 3u_y(x-2s,y+3s,t+s) + u_t(x-2s,y+3s,t+s) = e^{x+y+s}$$

and after integration we obtain

$$u(x,y,t) - \big[\ln|(x+2t)(y-3t)|\big]^4 = z(0) - z(-t) = \int_{-t}^{0} e^{x+y+s}\, ds = e^{x+y}(1-e^{-t}),$$

that is, $u(x,y,t) = [\ln|(x+2t)(y-3t)|]^4 + e^{x+y}(1-e^{-t})$. Such a solution is singular if and only if $x = -2t$ or $y = 3t$ which are the planes containing all the characteristic lines containing the points of the kind $(x,0,0)$ or $(0,y,0)$ where the initial datum is singular. □

Exercise 7.5. Using the Laplace transform, solve the transport problem

$$\begin{cases} u_t + u_x = 0 & \text{for } x > 0 \text{ e } t > 0 \\ u(x,0) = e^x & \text{for } x > 0 \\ u(0,t) = e^{-t} & \text{for } t > 0 \ . \end{cases}$$

Solution - We apply the Laplace transform with respect to the variable t and set $U(x,s) = \mathcal{L}_t\{u(x,t)\}$. By (6.10), the equation becomes $U_x(x,s) + sU(x,s) = u(x,0) = e^x$. Considering s as a parameter, we can solve this linear ordinary differential equation and obtain

$$U(x,s) = \frac{e^x}{s+1} + C(s)\, e^{-sx}.$$

Transforming the boundary condition at $x = 0$, we obtain $U(0,s) = \dfrac{1}{s+1}$, from which we infer that $C(s) \equiv 0$ and, by anti-transforming, $u(x,t) = e^{x-t}$. □

Exercise 7.6. Solve the following Cauchy problem for the vibrating string equation:

$$\begin{cases} u_{tt} - u_{xx} = 2 & x \in \mathbb{R},\ t > 0 \\ u(x,0) = x^2 + e^x & x \in \mathbb{R} \\ u_t(x,0) = e^x & x \in \mathbb{R} \ . \end{cases}$$

Solution - By d'Alembert formula (7.12), the problem

$$\begin{cases} v_{tt} - v_{xx} = 0 & x \in \mathbb{R},\ t > 0 \\ v(x,0) = x^2 + e^x & x \in \mathbb{R} \\ v_t(x,0) = e^x & x \in \mathbb{R} \end{cases}$$

has the solution

$$v(x,t) = \frac{(x+t)^2 + e^{x+t} + (x-t)^2 + e^{x-t}}{2} + \frac{1}{2}\int_{x-t}^{x+t} e^y \, dy = x^2 + t^2 + e^{x+t} \ .$$

The problem

$$\begin{cases} w_{tt} - w_{xx} = 2 & x \in \mathbb{R},\ t > 0 \\ w(x,0) = 0 & x \in \mathbb{R} \\ w_t(x,0) = 0 & x \in \mathbb{R} \end{cases}$$

has (trivially!) the solution $w(x,t) = t^2$ (it is not necessary to apply a formula). Hence, the solution of the original problem is $u(x,t) = v(x,t) + w(x,t) = x^2 + 2t^2 + e^{x+t}$. □

Exercise 7.7. Consider the equation of the vibrating string:

$$(*) \qquad \begin{cases} u_{tt} - 4u_{xx} = f(t) & (x,t) \in \mathbb{R} \times \mathbb{R}_+ \\ u(x,0) = x & x \in \mathbb{R} \\ u_t(x,0) = x & x \in \mathbb{R} \ . \end{cases}$$

Find the solution of $(*)$ in the cases where $f(t) \equiv 0$ and $f(t) = e^t$.

Solution - We solve

$$\begin{cases} u_{tt} - 4u_{xx} = 0 & (x,t) \in \mathbb{R} \times \mathbb{R}_+ \\ u(x,0) = x & x \in \mathbb{R} \\ u_t(x,0) = x & x \in \mathbb{R} \end{cases}$$

by using d'Alembert formula (7.12); the solution is given by

$$u_1(x,t) = \frac{1}{2}[(x+2t) + (x-2t)] + \frac{1}{4}\int_{x-2t}^{x+2t} y \, dy = x + xt \ .$$

Using the superposition principle, to find the second solution it suffices to add to u_1 the solution of the problem

$$\begin{cases} u_{tt} - 4u_{xx} = e^t & (x,t) \in \mathbb{R} \times \mathbb{R}_+ \\ u(x,0) = 0 & x \in \mathbb{R} \\ u_t(x,0) = 0 & x \in \mathbb{R} \ ; \end{cases}$$

by Duhamel's method (7.17) the solution of the second problem is given by

$$u_2(x,t) = \frac{1}{4}\int_0^t \int_{x-2s}^{x+2s} e^{t-s} \, dy \, ds = e^t - 1 - t \ .$$

Finally, by adding u_1 and u_2 we obtain the solution of $(*)$ with $f(t) = e^t$:

$$u(x,t) = x + xt + e^t - 1 - t \ .$$

□

Exercise 7.8. Find the solution u of the problem

$$\begin{cases} u_{tt} - u_{xx} + 6u_t + 9u = 2e^{-3t} & (x,t) \in \mathbb{R} \times \mathbb{R}_+ \\ u(x,0) = x + \sin x & x \in \mathbb{R} \\ u_t(x,0) = \cos x - 3x - 3\sin x & x \in \mathbb{R} \,. \end{cases}$$

Hint: put $u(x,t) = e^{at}U(x,t)$ for a suitable $a \in \mathbb{R}$.

Solution - Following the suggestion, we set $u(x,t) = e^{-3t}U(x,t)$. We find

$$u_{xx} = e^{-3t}U_{xx}, \; u_t = -3e^{-3t}U + e^{-3t}U_t, \; u_{tt} = 9e^{-3t}U - 6e^{-3t}U_t + e^{-3t}U_{tt} \,.$$

Therefore, U solves the problem

$$\begin{cases} U_{tt} - U_{xx} = 2 & (x,t) \in \mathbb{R} \times \mathbb{R}_+ \\ U(x,0) = x + \sin x & x \in \mathbb{R} \\ U_t(x,0) = \cos x & x \in \mathbb{R} \,. \end{cases} \tag{7.35}$$

We first solve the problem (7.35) with zero forcing term:

$$\begin{cases} v_{tt} - v_{xx} = 0 & (x,t) \in \mathbb{R} \times \mathbb{R}_+ \\ v(x,0) = x + \sin x & x \in \mathbb{R} \\ v_t(x,0) = \cos x & x \in \mathbb{R} \,. \end{cases}$$

From the d'Alembert formula (7.12), we infer

$$v(x,t) = \frac{1}{2}[x + t + \sin(x+t) + x - t + \sin(x-t)] + \frac{1}{2}\int_{x-t}^{x+t} \cos y \, dy = x + \sin(x+t) \,.$$

Considering $s \geq 0$ as parameter, let $w = w(x,t;s)$ be the solution of the problem

$$\begin{cases} w_{tt}(x,t;s) - w_{xx}(x,t;s) = 0 & x \in \mathbb{R}, \; t > s \\ w(x,s;s) = 0 & x \in \mathbb{R} \\ w_t(x,s;s) = 2 & x \in \mathbb{R} \,. \end{cases}$$

Again by d'Alembert formula (7.12), its solution has the form

$$w(x,t;s) = \frac{1}{2}\int_{x-t+s}^{x+t-s} 2 \, dy = 2(t-s) \,.$$

Then we define

$$z(x,t) = \int_0^t w(x,t;s) ds = t^2 \,,$$

that solves the problem with zero initial data:

$$\begin{cases} z_{tt} - z_{xx} = 2 & (x,t) \in \mathbb{R} \times \mathbb{R}_+ \\ z(x,0) = 0 & x \in \mathbb{R} \\ z_t(x,0) = 0 & x \in \mathbb{R} \,. \end{cases}$$

Summarizing, the solution U of (7.35) can be obtained by adding v and z

$$U(x,t) = v(x,t) + z(x,t) = x + t^2 + \sin(x+t)$$

and the solution of the initial problem is given by $u(x,t) = [x + t^2 + \sin(x+t)]e^{-3t}$. □

Exercise 7.9. Solve the problem

$$\begin{cases} u_{tt} - 4u_{xx} = 0 & \text{for } x \in (0,\pi) \text{ e } t > 0 \\ u(x,0) = \pi x - x^2 & \text{for } x \in (0,\pi) \\ u_t(x,0) = 0 & \text{for } x \in (0,\pi) \\ u(0,t) = u(\pi,t) = 0 & \text{for } t > 0 \end{cases}$$

and discuss the regularity of the solution.

Solution - By separating variables, we look for a solution of the equation in the form $u(x,t) = X(x)T(t)$. By arguing as in Section 7.2.5, see formula (7.19), we deduce that there exists $\lambda \in \mathbb{C}$ such that

$$T''(t) + 4\lambda^2 T(t) = 0, \qquad X''(x) + \lambda^2 X(x) = 0\,,$$

therefore there exist $A, B, C, D \in \mathbb{R}$ such that

$$X(x) = A\cos(\lambda x) + B\sin(\lambda x), \qquad T(t) = C\cos(2\lambda t) + D\sin(2\lambda t).$$

Imposing the boundary conditions $X(0) = X(\pi) = 0$ we obtain $A = 0$, $\lambda = m$ with $m \in \mathbb{N} \setminus \{0\}$ and $X(x) = k\sin(mx)$. We construct the series, by summing solutions with separated variables,

$$u(x,t) = \sum_{m=1}^{\infty} [\alpha_m \cos(2mt) + \beta_m \sin(2mt)] \sin(mx).$$

After term-wise differentiating the series and imposing the initial condition $u_t(x,0) = 0$, we find $\beta_m = 0$ for each $m \geq 1$. Therefore,

$$u(x,t) = \sum_{m=1}^{\infty} \alpha_m \cos(2mt) \sin(mx)\,.$$

To determine the coefficients α_m we use the initial condition $u(x,0) = \pi x - x^2$. In order to expand in sine series, the function $x \mapsto \pi x - x^2$ must be extended oddly to the whole interval $[-\pi, \pi]$. From (5.3) we deduce

$$\alpha_m = \frac{2}{\pi} \int_0^{\pi} [\pi x - x^2] \sin(mx) dx = \frac{4}{\pi m^3} [1 - (-1)^m].$$

Therefore the solution is

$$u(x,t) = \frac{4}{\pi} \sum_{m=1}^{\infty} \frac{1 - (-1)^m}{m^3} \cos(2mt) \sin(mx).$$

Concerning its regularity, by setting $g(x) = \pi x - x^2$ and $h(x) = 0$, we notice that the compatibility conditions $g(0) = g(\pi) = h(0) = h(\pi) = 0$ are satisfied and they guarantee that $u \in C^1$. On the other hand $g''(0) = g''(\pi) = 2 \neq 0$, therefore $u \notin C^2$, with propagation of the discontinuity of the second derivatives along the characteristics $x - 2t = 0$ and $x + 2t = \pi$ and along their reflections across the lines $x = 0$ and $x = \pi$. □

Exercise 7.10. Solve the problem

$$\begin{cases} u_{tt} - 4u_{xx} = 8 & \text{for } x \in (0,\pi) \text{ and } t > 0 \\ u(x,0) = 0 & \text{for } x \in (0,\pi) \\ u_t(x,0) = 0 & \text{for } x \in (0,\pi) \\ u(0,t) = 0 & \text{for } t > 0 \\ u(\pi,t) = 0 & \text{for } t > 0. \end{cases}$$

This problem represents a string fixed at both endpoints, initially at rest in the straight configuration and undergoing a force of uniform intensity $f = 8$, constant with respect to time.

Solution - We leave to the reader the solution of this exercise using Duhamel's method. Alternatively, we can look for a function $v(x,t)$ such that $v_{tt} - v_{xx} = 8$. We pick $v(x,t) = \pi x - x^2$ that satisfies also $v(0,t) = v(\pi,t) = 0$. But then, $w = v - u$ satisfies

$$\begin{cases} w_{tt} - 4w_{xx} = 0 & \text{for } x \in (0,\pi) \text{ and } t > 0 \\ w(x,0) = \pi x - x^2 & \text{for } x \in (0,\pi) \\ w_t(x,0) = 0 & \text{for } x \in (0,\pi) \\ w(0,t) = 0 & \text{for } t > 0 \\ w(\pi,t) = 0 & \text{for } t > 0. \end{cases}$$

Referring to Exercise 7.9, we deduce that

$$w(x,t) = \frac{4}{\pi} \sum_{m=1}^{\infty} \frac{1-(-1)^m}{m^3} \cos(2mt)\sin(mx).$$

Therefore, $u(x,t) = v(x,t) - w(x,t) = \pi x - x^2 - \frac{4}{\pi} \sum_{m=1}^{\infty} \frac{1-(-1)^m}{m^3} \cos(2mt)\sin(mx)$. □

Exercise 7.11. Solve the problem

$$\begin{cases} u_{tt} - 4u_{xx} = 0 & \text{for } x \in (0,\pi) \text{ and } t > 0 \\ u(x,0) = 0 & \text{for } x \in (0,\pi) \\ u_t(x,0) = \sin x & \text{for } x \in (0,\pi) \\ u(0,t) = 0 & \text{for } t > 0 \\ u(\pi,t) = 0 & \text{for } t > 0. \end{cases}$$

Solution - We proceed by separation of variables. We look for solutions of the equation in the form $u(x,t) = X(x)T(t)$ and, arguing as for (7.19), we deduce that there exists $\lambda \in \mathbb{C}$ such that

$$T''(t) + 4\lambda^2 T(t) = 0, \qquad X''(x) + \lambda^2 X(x) = 0$$

and therefore there exist $A, B, C, D \in \mathbb{R}$ such that

$$X(x) = A\cos(\lambda x) + B\sin(\lambda x), \qquad T(t) = C\cos(2\lambda t) + D\sin(2\lambda t).$$

Imposing the boundary conditions $X(0) = X(\pi) = 0$ we obtain $A = 0$, $\lambda = m$ with $m \in \mathbb{N} \setminus \{0\}$ and $X(x) = c\sin(mx)$. Then we construct a series by adding separated variables solutions,

$$u(x,t) = \sum_{m=1}^{\infty} [\alpha_m \cos(2mt) + \beta_m \sin(2mt)] \sin(mx).$$

The initial condition $u(x,0) = 0$ implies that $\alpha_m = 0$ for every $m \geq 1$. By imposing the other initial condition we obtain

$$\sum_{m=1}^{\infty} \beta_m \frac{cm\pi}{\ell} \sin\left(\frac{m\pi}{\ell} x\right) = \sin x \qquad \forall x \in (0,\pi)$$

that implies $\beta_1 = 1/2$ and $\beta_m = 0 \; \forall m \geq 2$.
Therefore the solution is $u(x,t) = \frac{1}{2}\sin(2t)\sin(x)$. □

Exercise 7.12. The motion of the points of a string initially at rest, with one end moving in a prescribed way, can be modeled through the problem

$$\begin{cases} u_{tt} - \frac{1}{4}u_{xx} = 0 & \text{for } x > 0 \text{ and } t > 0 \\ u(x,0) = 0 & \text{for } x > 0 \\ u_t(x,0) = 0 & \text{for } x > 0 \\ u(0,t) = f(t) & \text{for } t > 0\ . \end{cases}$$

To have compatible data, we assume $f(0) = 0$ and f continuous. Assuming that the solution is bounded, find it by using the Laplace transform and discuss its regularity.

Solution - We perform the partial Laplace transform with respect to the variable t and we set $U(x,s) = \mathcal{L}_t\{u(x,t)\}$. By (6.11) the equation becomes $s^2U(x,s) - \frac{1}{4}U_{xx}(x,s) = su(x,0) + u_t(x,0) = 0$. Considering now s as a parameter, we can solve this linear ordinary differential equation of the second order and obtain

$$U(x,s) = c_1(s)e^{2sx} + c_2(s)e^{-2sx}.$$

To determine $c_1(s)$ and $c_2(s)$ we impose the transformed boundary conditions: $U(0,s) = \mathcal{L}_t\{u(0,t)\}(s) = \mathcal{L}\{f\}(s)$ and $\lim_{x\to\infty} \mathcal{L}_t\{u(x,t)\} = 0$. The second one provides $c_1(s) = 0$ whereas the first one yields $c_2(s) = \mathcal{L}\{f\}(s)$. Therefore, $U(x,s) = \mathcal{L}\{f\}(s)\cdot e^{-2sx}$ and, by anti-transforming, $u(x,t) = f(t-2x)H(t-2x)$. This solution is 0 for $t < 2x$ and also the limit as $t - 2x \to 0$ with $t > 2x$ is 0 (here we use $f(0) = 0$); therefore u is a continuous function in $[0,\infty)\times[0,\infty)$. The regularity of u is the same as that of f. □

Exercise 7.13. Consider the problem (7.8) and set $v(y,z) = u(\frac{y-z}{2}, \frac{y+z}{2c})$.
(i) Determine the set where v is defined and deduce from (7.8) a differential equation of the second order solved by v.
(ii) Determine the general solution of the equation found.
(iii) Transform the initial conditions for u in (7.8) in initial conditions for the function v and derive the d'Alembert formula (7.12) using the result of item (i).

Solution - (i) From $x = \frac{y-z}{2}$ and $t = \frac{y+z}{2c}$ we deduce $y = ct + x$, $z = ct - x$; threfore $(y,z) \in \mathbb{R}^2$. Moreover, $u(x,t) = v(ct+x, ct-x)$, from which we infer $u_x = v_y - v_z$, $u_t = cv_y + cv_z$, $u_{xx} = v_{yy} - 2v_{yz} + v_{zz}$, $u_{tt} = c^2v_{yy} + 2c^2v_{yz} + c^2v_{zz}$. Hence, $0 = u_{tt} - c^2u_{xx} = 4c^2v_{yz}$ and the sought equation is $v_{yz} = 0$.
(ii) $v(y,z) = F(y) + G(z)$ with F and G being any differentiable functions.
(iii) The line $t = 0$ in the plane (x,t) becomes the line $z = -y$ in the plane (y,z). Using the initial conditions in (7.8) and taking into account item (i), we obtain $g(y) = u(y,0) = v(y,-y)$ and $h(y) = u_t(y,0) = cv_y(y,-y) + cv_z(y,-y)$. □

Exercise 7.14. Let $R > 0$ and suppose that $g \in C^2(\mathbb{R})$ and $h \in C^1(\mathbb{R})$ have support contained in $[-R,R]$. Let u be the solution of (7.8) with such g and h. For every $t > 0$, determine a bounded interval that contains the support of $x \mapsto u(x,t)$. Defining the energies

$$\text{kinetic:}\quad K(t) = \frac{1}{2}\int_{\mathbb{R}} u_t^2(x,t)\,dx, \qquad \text{potential:}\quad P(t) = \frac{c^2}{2}\int_{\mathbb{R}} u_x^2(x,t)\,dx,$$

prove that there is energy conservation: $K(t) + P(t)$=constant. Prove that $K(t) = P(t)$ for every $t \geq T := R/c$ (energy equipartition), even if $K(0) \neq P(0)$.

Solution - By d'Alembert formula (7.12) we infer that $\operatorname{spt} u(\cdot,t) \subset [-R-ct, R+ct]$. Since $u \in C^2$ and it has compact support with respect to x for each t, we can differentiate the two energies (by switching integration and differentiation with respect to t) and obtain

$$P'(t) = c^2\int_{\mathbb{R}} u_x(x,t)u_{xt}(x,t)\,dx$$

$$K'(t) = \int_{\mathbb{R}} u_t(x,t)u_{tt}(x,t)\,dx = c^2\int_{\mathbb{R}} u_t(x,t)u_{xx}(x,t)\,dx = -c^2\int_{\mathbb{R}} u_{xt}(x,t)u_x(x,t)\,dx$$

where we have used the differential equation in (7.8) and an integration by parts; the boundary terms are zero because $\operatorname{spt} u(\cdot,t)$ is compact. Therefore, $P'(t)+K'(t)=0$ and $P(t)+K(t)$ is constant.
By d'Alembert formula (7.12) we deduce

$$u_x(x,t)=\frac{1}{2}[g'(x+ct)+g'(x-ct)]+\frac{1}{2c}[h(x+ct)-h(x-ct)],$$

$$u_t(x,t)=\frac{c}{2}[g'(x+ct)-g'(x-ct)]+\frac{1}{2}[h(x+ct)+h(x-ct)].$$

By squaring, we obtain

$$\begin{aligned}c^2u_x^2(x,t) &= \frac{c^2}{4}[g'(x+ct)^2+2g'(x+ct)g'(x-ct)+g'(x-ct)^2]\\ &\quad+\frac{1}{4}[h(x+ct)^2-2h(x+ct)h(x-ct)+h(x-ct)^2]\\ &\quad+\frac{c}{2}[g'(x+ct)h(x+ct)-g'(x+ct)h(x-ct)+g'(x-ct)h(x+ct)-g'(x-ct)h(x-ct)],\\ u_t^2(x,t) &= \frac{c^2}{4}[g'(x+ct)^2-2g'(x+ct)g'(x-ct)+g'(x-ct)^2]\\ &\quad+\frac{1}{4}[h(x+ct)^2+2h(x+ct)h(x-ct)+h(x-ct)^2]\\ &\quad+\frac{c}{2}[g'(x+ct)h(x+ct)+g'(x+ct)h(x-ct)-g'(x-ct)h(x+ct)-g'(x-ct)h(x-ct)].\end{aligned}$$

Hence, the energy equipartition holds true under the following conditions:

$$g'(x+ct)g'(x-ct)\equiv 0 \quad\text{and}\quad h(x+ct)h(x-ct)\equiv 0 \qquad \forall x\in\mathbb{R},\ t\geq T.$$

Observing that, by the assumption on the support, we have

$$g'(x+ct)g'(x-ct)\neq 0 \Rightarrow |x|<R-ct\,,$$

$$|x|<R-ct \Leftrightarrow ct-R<x<R-ct\,,$$

$$ct-R<x<R-ct \Rightarrow t<\frac{R}{c}\,.$$

Therefore, if we contradict the last statement (that is, $t\geq R/c$), then the initial statement is also contradicted (that is, $g'(x+ct)g'(x-ct)=0\ \forall x$). Similar observations hold for the other terms that are not shared by K and P, that is $t\geq R/c \Rightarrow h(x+ct)h(x-ct)=g'(x+ct)h(x-ct)=g'(x-ct)h(x+ct)=0\ \forall x$. Therefore, $c^2u_x^2(x,t)=u_t^2(x,t)$ for each $t\geq T=R/c$ and each $x\in\mathbb{R}$. Integrating with respect to x over $\mathbb{R}$, we obtain that $K(t)=P(t)$ for every $t\geq T$. □

Exercise 7.15. Let $c>0$. Determine the solution of the problem

$$\begin{cases} u_{tt}-c^2\Delta u=0 & (\mathbf{x},t)\in\mathbb{R}^2\times\mathbb{R}_+\\ u(\mathbf{x},0)=0 & \mathbf{x}\in\mathbb{R}^2\\ u_t(\mathbf{x},0)=x_1^2 & \mathbf{x}\in\mathbb{R}^2.\end{cases}$$

Solution - By (7.33) we get

$$u(\mathbf{x},t)=\frac{1}{2\pi c}\int_{|\mathbf{y}-\mathbf{x}|<ct}\frac{y_1^2}{\sqrt{c^2t^2-|\mathbf{y}-\mathbf{x}|^2}}\,d\mathbf{y}.$$

We evaluate the previous integral by introducing polar coordinates, setting $y_1=x_1+\rho\cos\theta$ and $y_2=x_2+\rho\sin\theta$:

$$u(\mathbf{x},t) = \frac{1}{2\pi c}\int_0^{2\pi}\int_0^{ct} \frac{(x_1+\rho\cos\theta)^2\,\rho}{\sqrt{c^2t^2-\rho^2}}\,d\rho\,d\theta.$$

Expanding the numerator and computing the three integrals separately we have

$$\frac{1}{2\pi c}\int_0^{2\pi}\int_0^{ct} \frac{x_1^2\,\rho}{\sqrt{c^2t^2-\rho^2}}\,d\rho\,d\theta = \frac{x_1^2}{c}\int_0^{ct}\frac{\rho}{\sqrt{c^2t^2-\rho^2}}\,d\rho = x_1^2 t.$$

$$\frac{1}{2\pi c}\int_0^{2\pi}\int_0^{ct} \frac{2x_1\,\rho^2\cos\theta}{\sqrt{c^2t^2-\rho^2}}\,d\rho\,d\theta = \frac{1}{2\pi c}\int_0^{2\pi}\cos\theta\,d\theta\int_0^{ct}\frac{2x_1\,\rho^2}{\sqrt{c^2t^2-\rho^2}}\,d\rho = 0.$$

$$\frac{1}{2\pi c}\int_0^{2\pi}\int_0^{ct} \frac{\rho^3\cos^2\theta}{\sqrt{c^2t^2-\rho^2}}\,d\rho\,d\theta = \frac{1}{2\pi c}\int_0^{2\pi}\cos^2\theta\,d\theta\int_0^{ct}\frac{\rho^3}{\sqrt{c^2t^2-\rho^2}}\,d\rho =$$

$$= \frac{1}{2c}\left(\int_0^{ct} 2\rho\sqrt{c^2t^2-\rho^2}\,d\rho - \left[\rho^2\sqrt{c^2t^2-\rho^2}\right]_0^{ct}\right) = \frac{c^2t^3}{3}.$$

Adding these terms we obtain $u(\mathbf{x},t) = x_1^2 t + \dfrac{c^2t^3}{3}$. □

Exercise 7.16. Let $c > 0$. Determine the solution of the problem

$$\begin{cases} u_{tt} - c^2\Delta u = 0 & (\mathbf{x},t)\in\mathbb{R}^3\times\mathbb{R}_+ \\ u(\mathbf{x},0) = 1 & \mathbf{x}\in\mathbb{R}^3 \\ u_t(\mathbf{x},0) = 2 & \mathbf{x}\in\mathbb{R}^3. \end{cases}$$

Solution - Using (7.27) we obtain

$$u(\mathbf{x},t) = \fint_{\partial B_{ct}(\mathbf{x})} (2t+1)\,d\sigma(\mathbf{y}) = 2t+1.$$

It was possible to get this (trivial!) solution also by just looking carefully at the problem, without using (7.27). □

Exercise 7.17. Let u be the solution of the problem

$$\begin{cases} u_{tt} - 4\Delta u = 0 & (\mathbf{x},t)\in\mathbb{R}^3\times\mathbb{R}_+ \\ u(\mathbf{x},0) = 1 & \mathbf{x}\in\mathbb{R}^3 \\ u_t(\mathbf{x},0) = |\mathbf{x}|^2 + 3x_1 & \mathbf{x}\in\mathbb{R}^3. \end{cases}$$

Determine the value of $u(\mathbf{0},t)$ for each $t > 0$.

Solution - Using (7.27) we obtain

$$u(\mathbf{0},t) = \frac{1}{16\pi t^2}\int_{\partial B_{2t}(\mathbf{0})} \{t(|\mathbf{y}|^2 + 3y_1) + 1\}\,d\sigma(\mathbf{y}).$$

The integral of $3y_1$ is zero because its argument is odd with respect to y_1 and the sphere $\partial B_{2t}(\mathbf{0})$ is symmetric with respect to the origin. The integral of 1 is equal to 1 (it is a mean value!). Eventually, $|\mathbf{y}|^2 = 4t^2$ on $\partial B_{2t}(\mathbf{0})$. Therefore, $u(\mathbf{0},t) = 1 + 4t^3$. □

Exercise 7.18. Solve the problem

$$\begin{cases} u_{tt} - 9\Delta u = 0 & (\mathbf{x},t)\in\mathbb{R}^3\times\mathbb{R}_+ \\ u(\mathbf{x},0) = x_1^3 + 2x_2^3 + 3x_3^3 & \mathbf{x}\in\mathbb{R}^3 \\ u_t(\mathbf{x},0) = 1 & \mathbf{x}\in\mathbb{R}^3. \end{cases}$$

Solution - We observe that the solution of the problem with $u(\mathbf{x},0) = 0$ and $u_t(\mathbf{x},0) = 1$ is given by $u(\mathbf{x},t) = t$. By the superposition principle we can restrict our attention to the problem with $u(\mathbf{x},0) = x_1^3 + 2x_2^3 + 3x_3^3$ and $u_t(\mathbf{x},0) = 0$. Again by the superposition principle it suffices to solve the problem

$$\begin{cases} u_{tt} - 9\Delta u = 0 & (\mathbf{x},t) \in \mathbb{R}^3 \times \mathbb{R}_+ \\ u(\mathbf{x},0) = x_1^3 & \mathbf{x} \in \mathbb{R}^3 \\ u_t(\mathbf{x},0) = 0 & \mathbf{x} \in \mathbb{R}^3. \end{cases} \tag{7.36}$$

Once (7.36) is solved, we can replace x_1 with x_2 or x_3 and multiply by the corresponding coefficients. To solve (7.36) it is convenient to use (7.26):

$$u(\mathbf{x},t) = \frac{\partial}{\partial t}\frac{1}{36\pi t}\int_{|\mathbf{y}-\mathbf{x}|=3t} y_1^3 \, d\sigma(\mathbf{y}) \; .$$

We can write

$$y_1^3 = (y_1 - x_1 + x_1)^3 = (y_1 - x_1)^3 + 3(y_1 - x_1)^2 x_1 + 3(y_1 - x_1)x_1^2 + x_1^3$$

and observe that the terms $(y_1 - x_1)^3$ and $(y_1 - x_1)$ have zero integral by a symmetry argument. Therefore,

$$\begin{aligned} u(\mathbf{x},t) &= \frac{\partial}{\partial t}\frac{1}{36\pi t}\int_{|\mathbf{y}-\mathbf{x}|=3t} \left[3(y_1 - x_1)^2 x_1 + x_1^3\right] d\sigma(\mathbf{y}) \\ &= \frac{\partial}{\partial t}\left[t x_1^3 + \frac{x_1}{6\pi t}\int_{\substack{|\mathbf{y}-\mathbf{x}|=3t \\ y_1 > x_1}} (y_1 - x_1)^2 \, d\sigma(\mathbf{y})\right] \end{aligned}$$

(we set $\mathbf{y}' = (y_2, y_3)$ and $\mathbf{x}' = (x_2, x_3)$)

$$= x_1^3 + \frac{\partial}{\partial t}\left[\frac{x_1}{6\pi t}\int_{|\mathbf{y}'-\mathbf{x}'|<3t} \frac{3t\left[9t^2 - (y_2 - x_2)^2 - (y_3 - x_3)^2\right]}{\sqrt{9t^2 - (y_2 - x_2)^2 - (y_3 - x_3)^2}} \, d\sigma(\mathbf{y}')\right]$$

$$= x_1^3 + \frac{x_1}{2\pi}\frac{\partial}{\partial t}\int_{|\mathbf{y}'-\mathbf{x}'|<3t} \sqrt{9t^2 - (y_2 - x_2)^2 - (y_3 - x_3)^2} \, d\sigma(\mathbf{y}')$$

$$= x_1^3 + x_1 \frac{\partial}{\partial t}\int_0^{3t} \rho\sqrt{9t^2 - \rho^2} \, d\rho$$

where we have switched to polar coordinates: $y_2 = x_2 + \rho\cos\theta$ and $y_3 = x_3 + \rho\sin\theta$. Summarizing,

$$u(\mathbf{x},t) = x_1^3 + 27x_1 t^2$$

solves (7.36). Applying the superposition method we find that the solution of the original problem is $u(\mathbf{x},t) = x_1^3 + 27x_1t^2 + 2(x_2^3 + 27x_2t^2) + 3(x_3^3 + 27x_3t^2) + t$. □

Exercise 7.19. Solve the problem

$$\begin{cases} u_{tt} - \Delta u = 0 & (\mathbf{x},t) \in \mathbb{R}^3 \times \mathbb{R}_+ \\ u(\mathbf{x},0) = |\mathbf{x}|^2 & \mathbf{x} \in \mathbb{R}^3 \\ u_t(\mathbf{x},0) = |\mathbf{x}|^2 & \mathbf{x} \in \mathbb{R}^3. \end{cases}$$

Solution - The data are radially symmetric. Rewriting the problem with $r = |\mathbf{x}|$ and exploiting Table 10.6.4, we obtain (7.25):

$$\begin{cases} u_{tt}(r,t) - u_{rr}(r,t) - \frac{2}{r}u_r(r,t) = 0 & (r,t) \in \mathbb{R} \times \mathbb{R}_+ \\ u(r,0) = r^2, \quad u_t(r,0) = r^2 & r \in \mathbb{R}. \end{cases}$$

Setting $v(r,t) = ru(r,t)$, we obtain the 1-dimensional wave problem

$$\begin{cases} v_{tt}(r,t) - v_{rr}(r,t) = 0 & (r,t) \in \mathbb{R} \times \mathbb{R}_+ \\ v(r,0) = r^3, \quad v_t(r,0) = r^3 & r \in \mathbb{R}. \end{cases}$$

By d'Alembert formula (7.12) we get:

$$v(r,t) = \frac{(r+t)^3 + (r-t)^3}{2} + \frac{1}{2}\int_{r-t}^{r+t} s^3\, ds = r^3 + 3rt^2 + r^3 t + rt^3.$$

Back to the original variables, we find $u(\mathbf{x},t) = |\mathbf{x}|^2 + 3t^2 + |\mathbf{x}|^2 t + t^3$. □

Exercise 7.20. Assume that $c > 0$, $\Omega \subset \mathbb{R}^n$ is an open bounded set, both g and h belong to $C^2(\overline{\Omega})$ and u is the solution of the Cauchy-Neumann problem

$$\begin{cases} u_{tt} - c^2 \Delta u = 0 & (\mathbf{x},t) \in \Omega \times \mathbb{R}_+ \\ u(\mathbf{x},0) = g(\mathbf{x}) & \mathbf{x} \in \Omega \\ u_t(\mathbf{x},0) = h(\mathbf{x}) & \mathbf{x} \in \Omega \\ \frac{\partial u}{\partial \nu} = 0 & (\mathbf{x},t) \in \partial\Omega \times \mathbb{R}_+ \ . \end{cases}$$

Prove that $\displaystyle\int_\Omega u_t(\mathbf{x},t)\, d\mathbf{x} = \int_\Omega h(\mathbf{x})\, d\mathbf{x}$, $\forall t \geq 0$.

Solution - By differentiating, switching the derivative with the integral and using the equation we obtain

$$\frac{d}{dt}\int_\Omega u_t(\mathbf{x},t)\, d\mathbf{x} = \int_\Omega u_{tt}(\mathbf{x},t)\, d\mathbf{x} = c^2 \int_\Omega \Delta(\mathbf{x},t)\, d\mathbf{x} = c^2 \int_{\partial\Omega} \frac{\partial u}{\partial \nu}(\mathbf{y},t)\, d\sigma(\mathbf{y}) = 0$$

where we have used the Divergence Theorem (Formula 2, Section 10.5) and the Neumann condition on $\partial\Omega$. We deduce that

$$t \mapsto \int_\Omega u_t(\mathbf{x},t)\, d\mathbf{x}$$

is constant and it is therefore equal to its value for $t = 0$. □

7.5 Exercises without detailed solution

Exercise 7.21. Find the characteristic lines and the solutions of the following transport equations:

$$(1) \begin{cases} u_t - 2u_x = 2e^x & (x,t) \in \mathbb{R} \times \mathbb{R}^+ \\ u(x,0) = e^x + \sin x & x \in \mathbb{R}\ ; \end{cases}$$

$$(2) \begin{cases} u_t - u_x - u_y = 2e^{x+y} & (x,y,t) \in \mathbb{R} \times \mathbb{R} \times \mathbb{R}^+ \\ u(x,y,0) = \arctan(x+y) - e^{x+y} & (x,y) \in \mathbb{R} \times \mathbb{R}\ ; \end{cases}$$

$$(3) \begin{cases} u_t - u_x + 2u_y = e^x & (x,y,t) \in \mathbb{R} \times \mathbb{R} \times \mathbb{R}^+ \\ u(x,y,0) = y & (x,y) \in \mathbb{R} \times \mathbb{R}\ ; \end{cases}$$

$$(4) \begin{cases} u_t - u_x + 2u_y = 1 + \sin x & (x,y,t) \in \mathbb{R} \times \mathbb{R} \times \mathbb{R}^+ \\ u(x,y,0) = \ln(1+e^y) & (x,y) \in \mathbb{R} \times \mathbb{R}\ ; \end{cases}$$

$$(5) \begin{cases} u_t + 2u_x + u_y = x^2 & (x,y,t) \in \mathbb{R} \times \mathbb{R} \times \mathbb{R}^+ \\ u(x,y,0) = e^y & (x,y) \in \mathbb{R} \times \mathbb{R}\ ; \end{cases}$$

$$(6) \begin{cases} u_t + u_x - 2u_y = x + 1 & (x,y,t) \in \mathbb{R} \times \mathbb{R} \times \mathbb{R}^+ \\ u(x,y,0) = \sin(x+y) & (x,y) \in \mathbb{R} \times \mathbb{R}\ ; \end{cases}$$

$$(7)\ \begin{cases} u_t + 2u_x - u_y = x & (x,y,t) \in \mathbb{R}\times\mathbb{R}\times\mathbb{R}^+ \\ u(x,y,0) = ye^x & (x,y) \in \mathbb{R}\times\mathbb{R}\ ; \end{cases}$$

$$(8)\ \begin{cases} u_t + u_x - u_y + u = 0 & (x,y,t) \in \mathbb{R}\times\mathbb{R}\times\mathbb{R}^+ \\ u(x,y,0) = x^2 + y & (x,y) \in \mathbb{R}\times\mathbb{R}\ ; \end{cases}$$

$$(9)\ \begin{cases} u_t(x,t) - 2\,u_x(x,t) = \sin(t) & (x,t) \in \mathbb{R}\times(0,\infty) \\ u(x,0) = \cos(x^3) & x \in \mathbb{R}\ ; \end{cases}$$

$$(10)\ \begin{cases} u_t - u_x + 2u_y = (x+t)\cos(y) & (x,y,t) \in \mathbb{R}\times\mathbb{R}\times(0,\infty) \\ u(x,y,0) = \sin(xy) & (x,y) \in \mathbb{R}\times\mathbb{R}\ . \end{cases}$$

Solution - (1) $x(s) = x_0 - 2s$, $t(s) = t_0 + s$, $u(x,t) = -e^x + 2e^{x+2t} + \sin(x+2t)$.
(2) $x(s) = x_0 - s$, $y(s) = y_0 - s$, $t(s) = t_0 + s$, $u(x,y,t) = \arctan(x+y+2t) - e^{x+y}$.
(3) $x(s) = x_0 - s$, $y(s) = y_0 + 2s$, $t(s) = t_0 + s$, $u(x,y,t) = e^{x+t} - e^x + y - 2t$.
(4) $x(s) = x_0 - s$, $y(s) = y_0 + 2s$, $t(s) = t_0 + s$, $u(x,y,t) = \ln(1+e^{y-2t}) + t + \cos x - \cos(x+t)$.
(5) $x(s) = x_0 + 2s$, $y(s) = y_0 + s$, $t(s) = t_0 + s$, $u(x,y,t) = e^{y-t} + \frac{x^3-(x-2t)^3}{6}$.
(6) $x(s) = x_0 + s$, $y(s) = y_0 - 2s$, $t(s) = t_0 + s$, $u(x,y,t) = \sin(x+y+t) + (x+1)t - \frac{t^2}{2}$.
(7) $x(s) = x_0 + s$, $y(s) = y_0 - s$, $t(s) = t_0 + s$, $u(x,y,t) = (y+t)e^{x-2t} + tx - t^2$.
(8) $x(s) = x_0 + s$, $y(s) = y_0 - s$, $t(s) = t_0 + s$, $u(x,t) = e^{-t}[(x-t)^2 + y + t]$.
(9) $x(s) = x_0 - 2s$, $t(s) = t_0 + s$, $u(x,t) = \cos((x+2t)^3) + 1 - \cos(t)$.
(10) $x(s) = x_0 - s$, $y(s) = y_0 + 2s$, $t(s) = t_0 + s$, $u(x,y,t) = \sin[(x+t)(y-2t)] + \frac{x+t}{2}[\sin(y) - \sin(y-2t)]$. □

Exercise 7.22. Solve the following wave equations

$$(1)\ \begin{cases} u_{tt} - u_{xx} = e^t - e^x & (x,t) \in \mathbb{R}\times\mathbb{R}_+ \\ u(x,0) = 2e^x + 1 & x \in \mathbb{R} \\ u_t(x,0) = e^x + 1 & x \in \mathbb{R}\ ; \end{cases}$$

$$(2)\ \begin{cases} u_{tt} - 4u_{xx} = 2x & (x,t) \in \mathbb{R}\times\mathbb{R}_+ \\ u(x,0) = x^3 & x \in \mathbb{R} \\ u_t(x,0) = 1 & x \in \mathbb{R}\ ; \end{cases}$$

$$(3)\ \begin{cases} u_{tt} - 4u_{xx} = 2x & (x,t) \in \mathbb{R}\times\mathbb{R}_+ \\ u(x,0) = x^2 & x \in \mathbb{R} \\ u_t(x,0) = x & x \in \mathbb{R}\ ; \end{cases}$$

$$(4)\ \begin{cases} u_{tt} - u_{xx} = e^{x+t} & (x,t) \in \mathbb{R}\times\mathbb{R}_+ \\ u(x,0) = 0 & x \in \mathbb{R} \\ u_t(x,0) = x & x \in \mathbb{R}\ ; \end{cases}$$

$$(5)\ \begin{cases} u_{tt} - 4u_{xx} = xt & (x,t) \in \mathbb{R}\times\mathbb{R}_+ \\ u(x,0) = 0 & x \in \mathbb{R} \\ u_t(x,0) = 0 & x \in \mathbb{R}\ ; \end{cases}$$

$$(6)\ \begin{cases} u_{tt} - 9u_{xx} = e^{x+t} & (x,t) \in \mathbb{R}\times\mathbb{R}_+ \\ u(x,0) = 0 & x \in \mathbb{R} \\ u_t(x,0) = 0 & x \in \mathbb{R}\ ; \end{cases}$$

$$(7)\ \begin{cases} u_{tt}(x,t) - 4u_{xx}(x,t) = 2x & (x,t) \in \mathbb{R}\times\mathbb{R}_+ \\ u(x,0) = \sinh(x) & x \in \mathbb{R} \\ u_t(x,0) = 2\cosh(x) & x \in \mathbb{R}\ ; \end{cases}$$

Solution - (1) $u(x,t) = e^x + e^t + e^{x+t}$. (2) $u(x,t) = x^3 + 13xt^2 + t$. (3) $u(x,t) = x^2 + 4t^2 + xt + xt^2$. (4) $u(x,t) = xt + \frac{te^{x+t}}{2} + \frac{e^{x-t}}{4} - \frac{e^{x+t}}{4}$. (5) $u(x,t) = \frac{xt^3}{6}$. (6) $u(x,t) = \frac{e^{x+3t}}{12} - \frac{e^{x+t}}{8} + \frac{e^{x-3t}}{24}$. (7) $u(x,t) = v(x,t) + z(x,t) = \sinh(x+2t) + xt^2$. □

Exercise 7.23. Solve the problem

$$\begin{cases} u_{tt} - c^2 u_{xx} = 0 & (x,t) \in (0,\ell) \times \mathbb{R}_+ \\ u(x,0) = g(x) & x \in (0,\ell) \\ u_t(x,0) = 0 & x \in (0,\ell) \\ u(0,t) = u(\ell,t) = 0 & t \in \mathbb{R}_+ \end{cases}$$

in the following cases:
(1) $c = 2$, $\ell = \pi$, $g(x) = \sin x$; (2) $c = 1$, $\ell = \pi$, $g(x) = e^x$; (3) $c = 3$, $\ell = 1$, $g(x) = x$.
Solution - (1) $u(x,t) = \cos(2t)\sin(x)$;
(2) $u(x,t) = \dfrac{2}{\pi}\displaystyle\sum_{m=1}^{\infty} \frac{m[1 + (-1)^{m+1}e^{\pi}]}{m^2+1}\cos(mt)\sin(mx)$;
(3) $u(x,t) = \dfrac{2}{\pi}\displaystyle\sum_{m=1}^{\infty} \frac{(-1)^{m+1}}{m}\cos(3m\pi t)\sin(m\pi x)$. □

Exercise 7.24. Solve the problem

$$\begin{cases} u_{tt} - c^2 u_{xx} = 0 & (x,t) \in (0,\ell) \times \mathbb{R}_+ \\ u(x,0) = g(x) & x \in (0,\ell) \\ u_t(x,0) = h(x) & x \in (0,\ell) \\ u_x(0,t) = u_x(\ell,t) = 0 & t \in \mathbb{R}_+ \end{cases}$$

in the following cases:
(1) $c = 1$, $\ell = \pi$, $g(x) = 0$, $h(x) = 1 + \cos x$;
(2) $c = 2$, $\ell = \pi$, $g(x) = \cos(3x) + \cos(4x)$, $h(x) = 0$;
(3) $c = 2$, $\ell = 1$, $g(x) = h(x) = x$.
Solution - (1) $u(x,t) = t + \cos(x)\sin(t)$; (2) $u(x,t) = \cos(3x)\cos(6t) + \cos(4x)\cos(8t)$;
(3) $u(x,t) = \dfrac{t+1}{2} + \displaystyle\sum_{m=1}^{\infty}\left[\frac{2[(-1)^m - 1]}{m^2\pi^2}\cos(2m\pi t) + \frac{(-1)^m - 1}{m^3\pi^3}\sin(2m\pi t)\right]\cos(m\pi x)$. □

Exercise 7.25. Solve the problem

$$\begin{cases} u_{tt} - 4u_{xx} = 6t + x & (x,t) \in (0,1) \times \mathbb{R}_+ \\ u(x,0) = u_t(x,0) = 0 & x \in (0,1) \\ u_x(0,t) = u_x(1,t) = 0 & t \in \mathbb{R}_+ \ . \end{cases}$$

Solution - $u(x,t) = t^3 + \dfrac{t^2}{4} + \displaystyle\sum_{m=1}^{\infty} \frac{(-1)^m - 1}{2m^4\pi^4}[1 - \cos(2m\pi t)]\cos(m\pi x)$. □

Exercise 7.26. Solve the problem

$$\begin{cases} u_{tt} - \frac{1}{\pi^2}u_{xx} = xe^{-t} & (x,t) \in (0,1) \times \mathbb{R}_+ \\ u(x,0) = u_t(x,0) = 0 & x \in (0,1) \\ u(0,t) = u(1,t) = 0 & t \in \mathbb{R}_+ \ . \end{cases}$$

Solution - $u(x,t) = \dfrac{2}{\pi}\displaystyle\sum_{m=1}^{\infty} \frac{(-1)^{m-1}}{m^2(1+m^2)}[me^{-t} - m\cos(mt) + \sin(mt)]\cos(m\pi x)$. □

Exercise 7.27. Solve the problem

$$\begin{cases} u_{tt} - 4u_{xx} = 1 & \text{if } x \in (0,\pi) \text{ and } t > 0 \\ u(x,0) = \sin x & \text{if } x \in (0,\pi) \\ u_t(x,0) = 2\sin x & \text{if } x \in (0,\pi) \\ u(0,t) = u(\pi,t) = 0 & \text{if } t > 0. \end{cases}$$

Solution - By exploiting the superposition principle and the solutions of Exercises 7.10, 7.11 and 7.23 (1), we obtain

$$u(x,t) = [\cos(2t) + \sin(2t)]\sin(x) + \frac{\pi x - x^2}{8} - \frac{1}{2\pi}\sum_{m=1}^{\infty} \frac{1-(-1)^m}{m^3}\cos(2mt)\sin(mx). \qquad \square$$

Exercise 7.28. Using the Laplace transform solve the problem

$$\begin{cases} u_{tt} - u_{xx} = 0 & (x,t) \in (0,\infty) \times (0,\infty) \\ u(x,0) = 0, \quad u_t(x,0) = 1 & x \in [0,\infty) \\ u(0,t) = 0 & t \in [0,\infty). \end{cases}$$

Discuss the propagation of the singularities along the characteristics.

Solution - The function

$$U(x,s) = \frac{1}{s^2} - \frac{1}{s^2}\, e^{-xs}$$

solves the transformed problem. By anti-transforming we find $u(x,t) = (t)^+ - (t-x)^+ = \min\{x,t\}$ for $x, t > 0$. The characteristic half-line $t = x$ ($x > 0$) is the place where the function u fails to be of class C^1; namely

$$\lim_{t \to x^-} u_t(x,t) = 1 \neq 0 = \lim_{t \to x^+} u_t(x,t)\ , \qquad \lim_{x \to t^-} u_x(x,t) = 1 \neq 0 = \lim_{x \to t^+} u_t(x,t).$$

This can be explained by the fact that $\lim_{x \to 0^+} u_t(x,0) = 1 \neq 0 = \lim_{t \to 0^+} u_t(0,t)$ and this singularity at the origin propagates along the characteristic line $t = x$. $\square$

Exercise 7.29. Let g be an harmonic function in $\mathbb{R}^n$. Determine the solution of

$$\begin{cases} u_{tt} - \Delta u = 0 & (x,t) \in \mathbb{R}^n \times (0,\infty) \\ u(x,0) = g(x), \quad u_t(x,0) = 0 & x \in \mathbb{R}^n. \end{cases}$$

Solution - $u(x,t) = g(x)$. $\square$

Exercise 7.30. Determine the solution of the problem

$$\begin{cases} u_{tt} - 3\Delta u = 0 & (\mathbf{x},t) \in \mathbb{R}^2 \times \mathbb{R}_+ \\ u(\mathbf{x},0) = 1 & \mathbf{x} \in \mathbb{R}^2 \\ u_t(\mathbf{x},0) = x_2^2 & \mathbf{x} \in \mathbb{R}^2. \end{cases}$$

Solution - By Exercise 7.15 and superposition principle we obtain $u(\mathbf{x},t) = x_2^2 t + t^3 + 1$. $\square$

Exercise 7.31. Let $c > 0$. Solve the problem

$$\begin{cases} u_{tt} - c^2\Delta u = 0 & (\mathbf{x},t) \in \mathbb{R}^3 \times \mathbb{R}_+ \\ u(\mathbf{x},0) = x_1^2 & \mathbf{x} \in \mathbb{R}^3 \\ u_t(\mathbf{x},0) = 0 & \mathbf{x} \in \mathbb{R}^3. \end{cases}$$

Solution - $u(\mathbf{x},t) = x_1^2 + c^2t^2$. $\square$

Exercise 7.32. Solve the problem

$$\begin{cases} u_{tt} - \Delta u = |\mathbf{x}|^2 & (\mathbf{x},t) \in \mathbb{R}^3 \times \mathbb{R}_+ \\ u(\mathbf{x},0) = 0 & \mathbf{x} \in \mathbb{R}^3 \\ u_t(\mathbf{x},0) = 0 & \mathbf{x} \in \mathbb{R}^3. \end{cases}$$

Solution - Using the Duhamel principle and arguing as in Exercise 7.19 we find $u(\mathbf{x},t) = \dfrac{|\mathbf{x}|^2t^2}{2} + \dfrac{t^4}{4}$. Alternatively, one can use (7.27). $\square$

Chapter 8
Diffusion models

Diffusion is a process governed by partial differential equations, in which the variable considered (be it a temperature, a flow or a probability density) varies in time, and in each point it tends to coincide with its average in a local neighborhood of the same point. Diffusion is therefore a phenomenon which does not entail "motion" of molecules, but instead a transfer of "signal" among them. To better understand, we think of a closed room that has a scent inside: if we open the window in the middle of winter, immediately we feel the cold invade the room while the scent (molecules) takes more time to disappear, because molecules have to be transported outside the room. For this reason it is customary to say that diffusion has an infinite propagation speed. In this phenomenon also comes into play convection, which has a finite propagation velocity, instead (see Chapter 7).
In this chapter we consider first the Cauchy problem for the heat equation in the space $\mathbb{R}^n$ (given the initial temperature, compute the temperature in the whole domain at later instants), and thereafter, the Cauchy-Dirichlet problem (or Cauchy-Neumann) in domains bounded in space, where it is also necessary to have information on the temperature distribution on the boundary of the domain. We continue then with the study of the stationary equation, with particular regard to harmonic functions and their important properties. Finally, we describe the method of the separation of variables, which is each time adjusted to the different problem considered.

8.1 Heat Equation

8.1.1 The Cauchy problem in $\mathbb{R}^n$

Consider the Cauchy problem

$$\begin{cases} u_t - \Delta u = f(\mathbf{x}, t) & (\mathbf{x}, t) \in \mathbb{R}^n \times \mathbb{R}_+ \\ u(\mathbf{x}, 0) = g(\mathbf{x}) & \mathbf{x} \in \mathbb{R}^n. \end{cases} \tag{8.1}$$

The solution u can be obtained by adding the solutions of the problems

$$\begin{cases} u_t - \Delta u = 0 & (\mathbf{x}, t) \in \mathbb{R}^n \times \mathbb{R}_+ \\ u(\mathbf{x}, 0) = g(\mathbf{x}) & \mathbf{x} \in \mathbb{R}^n, \end{cases} \tag{8.2}$$

$$\begin{cases} u_t - \Delta u = f(\mathbf{x}, t) & (\mathbf{x}, t) \in \mathbb{R}^n \times \mathbb{R}_+ \\ u(\mathbf{x}, 0) = 0 & \mathbf{x} \in \mathbb{R}^n. \end{cases} \tag{8.3}$$

To solve (8.2) we proceed via Fourier transform, assuming that g belongs to one of the spaces considered in Chapter 4. Applying to both members of equation (8.2) a Fourier transform, with respect to the space variable alone, we get

$$\begin{aligned} \partial_t \widehat{u}(\boldsymbol{\xi}, t) &= \int_{\mathbb{R}^n} e^{-i\boldsymbol{\xi}\cdot\mathbf{x}} \partial_t u(\mathbf{x}, t)\, d\mathbf{x} = \int_{\mathbb{R}^n} e^{-i\boldsymbol{\xi}\cdot\mathbf{x}} \Delta u(\mathbf{x}, t)\, d\mathbf{x} \\ &= \mathcal{F}(\Delta u(\mathbf{x}, t), \boldsymbol{\xi}) = -|\boldsymbol{\xi}|^2 \widehat{u}(\boldsymbol{\xi}, t). \end{aligned}$$

By transforming also the initial data, we infer that $\widehat{u}(\boldsymbol{\xi}, t)$ solves the Cauchy problem

$$\begin{cases} \partial_t \widehat{u}(\boldsymbol{\xi}, t) + |\boldsymbol{\xi}|^2 \widehat{u}(\boldsymbol{\xi}, t) = 0 \\ \widehat{u}(\boldsymbol{\xi}, 0) = \widehat{g}(\boldsymbol{\xi}) \end{cases}$$

and therefore $\widehat{u}(\boldsymbol{\xi}, t) = \widehat{g}(\boldsymbol{\xi}) e^{-|\boldsymbol{\xi}|^2 t}$. Back-transforming, from (4.50) we get

$$u(\mathbf{x}, t) = \mathcal{F}^{-1}(\widehat{u}(\boldsymbol{\xi}, t), \mathbf{x}) = \mathcal{F}^{-1}(e^{-|\boldsymbol{\xi}|^2 t}, \mathbf{x}) * g(\mathbf{x}).$$

From the Table 10.8.6 we know that

$$\mathcal{F}^{-1}(e^{-|\boldsymbol{\xi}|^2 t}, \mathbf{x}) = \frac{1}{(4\pi t)^{n/2}}\, e^{-\frac{|\mathbf{x}|^2}{4t}}.$$

From this, we infer that the solution of (8.2) is given by

$$u(\mathbf{x}, t) = \frac{1}{(4\pi t)^{n/2}} \int_{\mathbb{R}^n} e^{-\frac{|\mathbf{x}-\mathbf{y}|^2}{4t}} g(\mathbf{y})\, d\mathbf{y}. \tag{8.4}$$

Remark 8.1. The adopted procedure is surely valid if $g \in L^1(\mathbb{R}^n) \cup L^2(\mathbb{R}^n)$ or $g \in \mathcal{S}'(\mathbb{R}^n)$. However, looking carefully to the various steps, we notice that there are no formal obstacles also with more general initial data g, provided that the convolution (8.4) may be computed.

For any $n \geq 1$ we define as fundamental solution of the heat equation, with a pole in $\mathbf{z} \in \mathbb{R}^n$, the function

$$K(\mathbf{x}, \mathbf{z}, t) = \frac{1}{(4\pi t)^{n/2}} e^{-\frac{|\mathbf{x}-\mathbf{z}|^2}{4t}}. \tag{8.5}$$

Using K we can rewrite (8.4) as

$$u(\mathbf{x},t) = \int_{\mathbb{R}^n} K(\mathbf{x},\mathbf{y},t)g(\mathbf{y})\, d\mathbf{y}. \tag{8.6}$$

Hence, $K(\mathbf{x},\mathbf{z},t)$ can be seen as solution of the problem

$$\begin{cases} u_t - \Delta u = 0 & (\mathbf{x},t) \in \mathbb{R}^n \times \mathbb{R}_+ \\ u(\mathbf{x},0) = \delta_{\mathbf{z}} & \mathbf{x} \in \mathbb{R}^n \end{cases}$$

and this accounts for the name of fundamental solution for the function K. To solve (8.3) we use again the Duhamel method, already employed in Section 7.2.4 for the solution of the vibrating string equation. Considering $s \geq 0$ as a parameter, let $w = w(\mathbf{x},t;s)$ be the solution of the problem

$$\begin{cases} w_t(\mathbf{x},t;s) - \Delta w(\mathbf{x},t;s) = 0 & \mathbf{x} \in \mathbb{R}^n,\ t > s \\ w(\mathbf{x},s;s) = f(\mathbf{x},s) & \mathbf{x} \in \mathbb{R}^n\ . \end{cases} \tag{8.7}$$

By (8.4), its solution is

$$w(\mathbf{x},t;s) = \frac{1}{[4\pi(t-s)]^{n/2}} \int_{\mathbb{R}^n} e^{-\frac{|\mathbf{x}-\mathbf{y}|^2}{4(t-s)}} f(\mathbf{y},s)\, d\mathbf{y}\ .$$

Now put

$$u(\mathbf{x},t) = \int_0^t w(\mathbf{x},t;s) ds\ .$$

With some computation (see [5]) one finds that this u solves (8.3). Using the fundamental solution K and the expression (8.6), we get

$$u(\mathbf{x},t) = \int_0^t \left(\int_{\mathbb{R}^n} K(\mathbf{x},\mathbf{y},s) f(\mathbf{y},t-s)\, d\mathbf{y} \right) ds.$$

Then, the unique solution of the problem (8.1) is given by

$$u(\mathbf{x},t) = \int_{\mathbb{R}^n} K(\mathbf{x},\mathbf{y},t)g(\mathbf{y})\, d\mathbf{y} + \int_0^t \left(\int_{\mathbb{R}^n} K(\mathbf{x},\mathbf{y},s) f(\mathbf{y},t-s)\, d\mathbf{y} \right) ds.$$

Remark 8.2. There are different kinds of heat equation. For example, the equation $u_t - \Delta u + cu = f$ is a heat equation with a damping term, and can be solved with the same trick adopted for (7.5). Diffusion and transfer phenomena can be combined via the equation $u_t - \Delta u + \mathbf{b} \cdot \nabla u = f$ ($\mathbf{b} \in \mathbb{R}^n$), adding, in case, a damping term. All these (linear) equations exhibit quite similar solution methods, and therefore we do not describe them here.

8.1.2 The Cauchy-Dirichlet problem in bounded domains

For any bounded open domain $\Omega \subset \mathbb{R}^n$ and for any $T > 0$ we set $\Omega_T = \Omega \times (0, T)$. The set defined by

$$\partial_p \Omega_T = (\partial\Omega \times [0, T]) \cup (\Omega \times \{t = 0\}) \tag{8.8}$$

is called *parabolic boundary* of Ω_T.

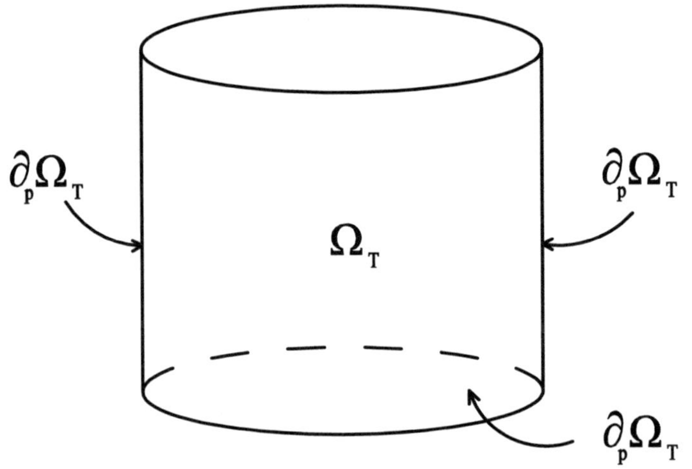

Figure 8.1 - Parabolic boundary.

We state the fundamental

Theorem 8.3. (Maximum Principle). *Let $\Omega \subset \mathbb{R}^n$ be an open bounded domain and let $u \in C^2(\overline{\Omega}_T)$ obey*

$$u_t - \Delta u \leq 0 \qquad \text{in } \Omega_T. \tag{8.9}$$

Then $\max\limits_{\overline{\Omega}_T} u = \max\limits_{\partial_p \Omega_T} u$. Moreover, if Ω is connected and there exists $(x_0, t_0) \in \overline{\Omega}_T \setminus \partial_p \Omega_T$ such that $\max\limits_{\overline{\Omega}_T} u = u(x_0, t_0)$ then u is constant in $\overline{\Omega}_{t_0}$.

If in (8.9) the inequality is reverted, the statement involves the minimum.

Definition 8.4. *Let $\Omega \subset \mathbb{R}^n$ be an open bounded domain. A function $u \in C^2(\overline{\Omega}_T)$ satisfying (8.9) is called subcaloric. A function satisfying the opposite inequality is called supercaloric. A function satisfying both inequalities, i.e. $u_t - \Delta u = 0$ in Ω_T is called caloric.*

A subcaloric (supercaloric) function describes therefore a non positive (non negative) heat source.

Let us now consider in some detail the one-dimensional case, i.e. a finite rod with length $\ell > 0$. If the temperature is kept equal to zero at the endpoints, we get the *Cauchy-Dirichlet problem*:

$$\begin{cases} u_t - u_{xx} = 0 & (x,t) \in (0,\ell) \times \mathbb{R}_+ \\ u(0,t) = u(\ell,t) = 0 & t \in \mathbb{R}_+ \\ u(x,0) = g(x) & x \in (0,\ell) \ . \end{cases} \tag{8.10}$$

We proceed by separation of variables. First, we look for solutions of $(8.10)_1$ of the kind $u(x,t) = X(x)T(t)$; we get

$$\frac{T'(t)}{T(t)} = \frac{X''(x)}{X(x)} \ ,$$

from which,

$$\exists \lambda \in \mathbb{C} \ , \quad T'(t) + \lambda T(t) = 0 \ , \quad X''(x) + \lambda X(x) = 0 \ . \tag{8.11}$$

To the equation for X we associate the boundary conditions $(8.10)_2$, i.e. $X(0) = X(\ell) = 0$; the eigenvalues and eigenfunctions of this problem are $\lambda_m = m^2\pi^2/\ell^2$ and $X_m(x) = \sin(m\pi x/\ell)$ for every $m \geq 1$. For $\lambda = \lambda_m$, the equation in T has solutions proportional to $T_m(t) = e^{-m^2\pi^2 t/\ell^2}$. Hence, we look for solutions of (8.10) of the kind

$$u(x,t) = \sum_{m=1}^{\infty} c_m e^{-m^2\pi^2 t/\ell^2} \sin\left(\frac{m\pi x}{\ell}\right) \ . \tag{8.12}$$

to compute the coefficients c_m we require the condition $u(x,0) = g(x)$; we obtain

$$g(x) = \sum_{m=1}^{\infty} c_m \sin\left(\frac{m\pi x}{\ell}\right) \ .$$

The coefficients c_m are therefore the Fourier coefficients of g prolonged as an odd function to $(-\ell,\ell)$:

$$c_m = \frac{2}{\ell} \int_0^\ell g(x) \sin\left(\frac{m\pi x}{\ell}\right) dx \, .$$

Remind that, in order that a Fourier series converges in L^2, it suffices for g^2 to be integrable over $(0, \ell)$, see (5.6).

Remark 8.5. If in (8.10) we suppose $g \in L^2(0, \ell)$ then by the Parseval Theorem 5.1 the series $\sum c_m^2$ is convergent. Hence, $\lim_{m\to\infty} c_m = 0$. By differentiating (8.12) (any number of times!) with respect to x, we obtain that the differentiated series converges for any $t > 0$ because of the negative exponential. As a consequence, Theorem 5.2 guarantees that $u(\cdot, t) \in C^\infty(0, \ell)$ for any $t > 0$; this means that, even if $u(\cdot, 0) = u_0 = g \in L^2(0, \ell)$, the solution regularizes instantaneously in the space variable for $t > 0$. This is the regularizing effect of the heat operator, already seen in Exercise 4.54 of Chapter 4.

If the bar of length $\ell > 0$ is isolated at the extremes, then there is no heat flux, and we get the *Cauchy-Neumann problem*:

$$\begin{cases} u_t - u_{xx} = 0 & (x, t) \in (0, \ell) \times \mathbb{R}_+ \\ u_x(0, t) = u_x(\ell, t) = 0 & t \in \mathbb{R}_+ \\ u(x, 0) = g(x) & x \in (0, \ell) \, . \end{cases}$$

By separating the variables, we reach again (8.11). To the equation for X we associate the boundary conditions $X'(0) = X'(\ell) = 0$. The eigenvalues of this problem are $\lambda_0 = 0$ and $\lambda_m = m^2\pi^2/\ell^2$, and the relevant eigenfunctions are $X_0(x) = 1$ and $X_m(x) = \cos(m\pi x/\ell)$. For $\lambda = \lambda_m$, the equation in T has solutions proportional to $T_0(t) = 1$ and $T_m(t) = e^{-m^2\pi^2 t/\ell^2}$. We look for solutions of the kind

$$u(x, t) = c_0 + \sum_{m=1}^{\infty} c_m e^{-m^2\pi^2 t/\ell^2} \cos\left(\frac{m\pi x}{\ell}\right) \, .$$

Imposing the condition $u(x, 0) = g(x)$, we obtain

$$g(x) = c_0 + \sum_{m=1}^{\infty} c_m \cos\left(\frac{m\pi x}{\ell}\right) \, .$$

The coefficients c_m are therefore the Fourier coefficients of g extended as an even function to $(-\ell, \ell)$:

$$c_0 = \frac{1}{\ell} \int_0^\ell g(x)\, dx, \qquad c_m = \frac{2}{\ell} \int_0^\ell g(x) \cos\left(\frac{m\pi x}{\ell}\right) dx \, .$$

One can solve the above problems even in the presence of a forcing term, by applying the method of Duhamel. For example, in the case of Dirichlet boundary conditions, the problem

$$\begin{cases} u_t - u_{xx} = f(x,t) & (x,t) \in (0,\ell) \times \mathbb{R}_+ \\ u(0,t) = u(\ell,t) = 0 & t \in \mathbb{R}_+ \\ u(x,0) = 0 & x \in (0,\ell) \end{cases}$$

can be delayed with the parameter $s \geq 0$ and transformed into the problem

$$\begin{cases} w_t(x,t;s) - w_{xx}(x,t;s) = 0 & (x,t) \in (0,\ell) \times (s,+\infty) \\ w(0,t;s) = w(\ell,t;s) = 0 & t \in (s,+\infty) \\ w(x,s;s) = f(x,s) & x \in (0,\ell). \end{cases}$$

Once we have found the solution w by separation of variables, we set $u(x,t) = \int_0^t w(x,t;s)\,ds$.

8.2 The stationary case

8.2.1 The fundamental solution of the Laplace equation

If in (8.1) we suppose $f = f(\mathbf{x})$ to be time independent, then it becomes interesting to look for stationary solutions, they are also time independent. It is therefore useless to impose an initial condition. If $f \equiv 0$, then the equation is named after Laplace. We look for solutions u with radial symmetry to the Laplace equation $\Delta u = 0$ in $\mathbb{R}^n \setminus \{0\}$

$$u(\mathbf{x}) = v(r) \qquad r = |\mathbf{x}| = \Big(\sum_{j=1}^{n} x_j^2\Big)^{\frac{1}{2}}.$$

Since it results $\qquad \dfrac{\partial r}{\partial x_i} = \dfrac{x_i}{r}, \qquad \dfrac{\partial^2 r}{\partial x_i^2} = \dfrac{1}{r} - \dfrac{x_i^2}{r^3},$

$$u_{x_i} = v'(r)\frac{x_i}{r}, \qquad u_{x_i x_i} = v''(r)\frac{x_i^2}{r^2} + v'(r)\left(\frac{1}{r} - \frac{x_i^2}{r^3}\right),$$

we get the ordinary differential equation

$$\Delta u = v''(r) + \frac{n-1}{r} v'(r) = 0 \qquad (r > 0). \tag{8.13}$$

This can also be rewritten as $(r^{n-1}v'(r))' = 0$ and by integration we obtain

$$v(r) = \begin{cases} b\,\ln r + c & n = 2 \\ b\,r^{2-n} + c & n \geq 3. \end{cases} \tag{8.14}$$

With a suitable choice of the constants b and c we obtain the function

$$\Phi(\mathbf{x},\mathbf{y}) = \begin{cases} -\frac{1}{2\pi} \ln|\mathbf{x}-\mathbf{y}| & n = 2 \\ \frac{1}{(n-2)\sigma_{n-1}} |\mathbf{x}-\mathbf{y}|^{2-n} & n \geq 3, \end{cases} \tag{8.15}$$

and we call it fundamental solution of the Laplace equation, focalized in the point $\mathbf{y}$. We recall that $\sigma_{n-1} = |\partial B_1|$, see (2.13), and that the choice of the constants b and c in (8.14) has been done in such a way that

$$-\Delta_{\mathbf{x}}\Phi(\mathbf{x}, 0) = \delta_0 \qquad \text{in } \mathcal{D}'(\mathbb{R}^n)$$

with δ_0 Dirac mass at $\mathbf{x} = 0$. That is, it holds

$$-\int_{\mathbb{R}^n} \Phi(\mathbf{x}, 0)\Delta\varphi \, d\mathbf{x} = \varphi(0) \qquad \forall\varphi \in \mathcal{D}(\mathbb{R}^n). \tag{8.16}$$

We note that we can reach this conclusion also via the Fourier transform, see (4.94). Analogously, we have that

$$-\Delta_{\mathbf{x}}\Phi(\mathbf{x}, \mathbf{y}) = \delta_{\mathbf{y}}$$

with $\delta_{\mathbf{y}}$ Dirac delta function centered at $\mathbf{y}$; with the notations of Chapter 4, see for example (4.73) for $n = 3$, we have $\Phi(\mathbf{x}, \mathbf{y}) = U(\mathbf{x} - \mathbf{y})$. But then, if $f \in C_c^2(\mathbb{R}^n)$, the function

$$u(\mathbf{x}) = \int_{\mathbb{R}^n} \Phi(\mathbf{x}, \mathbf{y}) \; f(\mathbf{y}) \; d\mathbf{y} \tag{8.17}$$

solves $-\Delta u = f$ on $\mathbb{R}^n$. This can be rigorously verified by deriving under integration, see [5].

8.2.2 Harmonic functions

Definition 8.6. *Let $\Omega \subset \mathbb{R}^n$ be an open set. A function $u \in C^2(\Omega)$ satisfying $\Delta u = 0$ in Ω is called harmonic in Ω.*

Harmonic functions have remarkable properties. Let us have a look to the most important.

Theorem 8.7. (mean value formula). *Let $\Omega \subset \mathbb{R}^n$ be an open set and let u be harmonic in Ω. Then, for any $\mathbf{x} \in \Omega$ and for any $R > 0$ such that $\overline{B_R(\mathbf{x})} \subset \Omega$, we have*

$$u(\mathbf{x}) = \fint_{\partial B_R(\mathbf{x})} u(\mathbf{y}) d\sigma(\mathbf{y}) = \fint_{B_R(\mathbf{x})} u(\mathbf{y}) d\mathbf{y} \, . \tag{8.18}$$

A kind of mean value property also holds for superharmonic and subharmonic functions, that are C^2 functions satisfying, respectively, $\Delta u \leq 0$ and $\Delta u \geq 0$.

Theorem 8.8. *Let $\Omega \subset \mathbb{R}^n$ be an open set and let $u \in C^2(\Omega)$.*
(i) If $\Delta u \leq 0$ (in such a case, u is called superharmonic) then for any $\mathbf{x} \in \Omega$ and for any $R > 0$ such that $\overline{B_R(\mathbf{x})} \subset \Omega$, we have

$$u(\mathbf{x}) \geq \fint_{\partial B_R(\mathbf{x})} u(\mathbf{y})d\sigma(\mathbf{y}) \qquad \text{and} \qquad u(\mathbf{x}) \geq \fint_{B_R(\mathbf{x})} u(\mathbf{y})d\mathbf{y}.$$

(ii) If $\Delta u \geq 0$ (in such a case, u is called subharmonic) then for any $\mathbf{x} \in \Omega$ and for any $R > 0$ such that $\overline{B_R(\mathbf{x})} \subset \Omega$, we have

$$u(\mathbf{x}) \leq \fint_{\partial B_R(\mathbf{x})} u(\mathbf{y})d\sigma(\mathbf{y}) \qquad \text{and} \qquad u(\mathbf{x}) \leq \fint_{B_R(\mathbf{x})} u(\mathbf{y})d\mathbf{y}.$$

The mean value formulae have as a consequence the following

Theorem 8.9. (maximum principle). *Let $\Omega \subset \mathbb{R}^n$ be an open bounded set and let $u \in C^2(\Omega) \cap C^0(\overline{\Omega})$.*
If u is subharmonic in Ω then

$$\max_{\overline{\Omega}} u = \max_{\partial\Omega} u .$$

If u is superharmonic in Ω then

$$\min_{\overline{\Omega}} u = \min_{\partial\Omega} u .$$

If u is harmonic in Ω then

$$\min_{\partial\Omega} u \leq u(x) \leq \max_{\partial\Omega} u \qquad \forall x \in \overline{\Omega}.$$

Moreover, if Ω is a connected set and u is not constant, then u attains a maximum (if it is subharmonic) or a minimum (if it is superharmonic) only on $\partial\Omega$.

8.2.3 Separation of variables

In case of simple planar geometries, that is when $\Omega = (0,1)^2$ (the unit square of $\mathbb{R}^2$) or $\Omega = B_1(\mathbf{0})$ (the unit disc of $\mathbb{R}^2$), we can solve the Laplace equation via separation of variables.

8.2.3.1 Laplace equation in a square

First, we study the *Dirichlet problem* on the square $(0,1)^2 = (0,1) \times (0,1)$:

$$\begin{cases} u_{xx} + u_{yy} = 0 & \text{in } (0,1)^2 \\ u(x,0) = f(x) & \text{for } x \in (0,1) \\ u(x,1) = g(x) & \text{for } x \in (0,1) \\ u(0,y) = \phi(y) & \text{for } y \in (0,1) \\ u(1,y) = \psi(y) & \text{for } y \in (0,1) \ . \end{cases} \tag{8.19}$$

For sake of simplicity we suppose that $f, g, \phi, \psi \in L^2(0,1)$. Since the problem is linear, we can treat separately the 4 conditions on the sides of the square and add the corresponding solutions. Whence, suppose that

$$g \equiv \phi \equiv \psi \equiv 0 \tag{8.20}$$

and let us seek solutions of (8.19) of the kind

$$u(x,y) = X(x)Y(y) \ . \tag{8.21}$$

Then, $u_{xx}(x,y) + u_{yy}(x,y) = X''(x)Y(y) + X(x)Y''(y) = 0$; dividing this equation by $X(x)Y(y)$, we find

$$\frac{X''(x)}{X(x)} = -\frac{Y''(y)}{Y(y)} \ .$$

But a function depending only on x can be equal to a function depending only on y only if both functions are equal to the same constant. Therefore,

$$\exists \lambda \in \mathbb{R} \ , \qquad X''(x) + \lambda X(x) = 0 \ , \quad Y''(y) - \lambda Y(y) = 0 \ ; \tag{8.22}$$

to these equation we should associate the boundary conditions

$$X(0) = X(1) = Y(1) = 0 \ . \tag{8.23}$$

By imposing that X is not identically zero, by (8.22) and (8.23) it follows that it must necessarily be $\lambda = \lambda_m = m^2\pi^2$ ($m \in \mathbb{N} \setminus \{0\}$) and also $X_m(x) = A\sin(m\pi x)$; but then $Y_m(y) = Be^{m\pi y} + Ce^{-m\pi y}$ and, imposing $Y_m(1) = 0$, we deduce $C = -Be^{2m\pi}$ and then $Y_m(y) = D_m \sinh[m\pi(1-y)]$. Looking for a solution u of the form (8.21), with X and Y satisfying (8.22)-(8.23), up to a multiplicative constant, we find infinitely many solutions:

$$u_m(x, y) = \sin(m\pi x)\sinh[m\pi(1-y)] \qquad (m \in \mathbb{N}\setminus\{0\}) . \tag{8.24}$$

Since problem (8.22)-(8.23) is linear, any linear combination (even infinite, provided it converges in a suitable sense!) of solutions of type (8.24) is also a solution; we are therefore led to seek solutions of (8.19)-(8.20) of the type

$$u(x, y) = \sum_{m=1}^{\infty} c_m \sin(m\pi x)\sinh[m\pi(1-y)] , \tag{8.25}$$

where the constants c_m should be determined by imposing the last condition $u(x, 0) = f(x)$. But then,

$$f(x) = u(x, 0) = \sum_{m=1}^{\infty} c_m \sinh(m\pi)\sin(m\pi x) , \qquad x \in (0, 1) ;$$

therefore, $c_m \sinh(m\pi)$ are the Fourier coefficients of the odd extension to $(-1, 1)$ of f, see (5.10):

$$c_m \sinh(m\pi) = b_m = 2\int_0^1 f(x)\sin(m\pi x)\, dx .$$

Summarizing, the solution of (8.19)-(8.20) is given by

$$u(x, y) = \sum_{m=1}^{\infty} \frac{b_m}{\sinh(m\pi)} \sin(m\pi x)\sinh[m\pi(1-y)] .$$

Having assumed that $f \in L^2(0, 1)$, this series converges in L^2.
Let us now consider the *Neumann problem*

$$\begin{cases} u_{xx} + u_{yy} = 0 & \text{in } (0,1)^2 \\ -u_y(x, 0) = f(x) & \text{for } x \in (0, 1) \\ u_y(x, 1) = g(x) & \text{for } x \in (0, 1) \\ -u_x(0, y) = \phi(y) & \text{for } y \in (0, 1) \\ u_x(1, y) = \psi(y) & \text{for } y \in (0, 1) , \end{cases} \tag{8.26}$$

where we observe that the signs $\pm$ on the normal derivatives are used to give the outgoing direction to the same derivatives. From the divergence Theorem (Formula 2, Table 10.5) the problem (8.26) is not solvable unless the following compatibility condition holds:

$$\int_0^1 [f(s) + g(s) + \varphi(s) + \psi(s)]\, ds = 0 . \tag{8.27}$$

The problem (8.26) is linear, so we can proceed as for the Dirichlet problem (8.19) assuming that (8.20) is valid, and that $f \in L^2(0,1)$; in such a case, the condition (8.27) becomes

$$\int_0^1 f(s)\,ds = 0 \ . \tag{8.28}$$

Proceeding as for the problem (8.19), we can separate the variables and lseek solutions u of the form (8.21). We get again the ordinary differential equations (8.22), but now with the boundary conditions

$$X'(0) = X'(1) = Y'(1) = 0 \ . \tag{8.29}$$

As for the Dirichlet problem, imposing that X is a nontrivial function, we get $\lambda_m = m^2\pi^2$ ($m \in \mathbb{N}$) and $X_m(x) = A\cos(m\pi x)$. For $\lambda = \lambda_m$ we then obtain $Y_m(y) = Be^{m\pi y} + Ce^{-m\pi y}$; by requiring $Y_m'(1) = 0$, we infer $C = Be^{2m\pi}$ and hence $Y_m(y) = D_m \cosh[m\pi(1-y)]$. Seeking a solution u of the kind (8.21) we get, up to a multiplicative constant, infinitely many solutions:

$$u_m(x,y) = \cos(m\pi x)\cosh[m\pi(1-y)] \qquad (m \in \mathbb{N}) \ .$$

Recalling assumptions (8.20) and (8.28), we are finally led to seek solutions of (8.26) of the type

$$u(x,y) = \sum_{m=0}^{\infty} c_m \cos(m\pi x)\cosh[m\pi(1-y)] \ , \tag{8.30}$$

where the constants c_m can be obtained from the condition $u_y(x,0) = -f(x)$. Whence,

$$f(x) = -u_y(x,0) = \pi \sum_{m=1}^{\infty} m\, c_m \sinh(m\pi)\cos(m\pi x) \ , \qquad x \in (0,1) \ ;$$

therefore, $m\pi c_m \sinh(m\pi)$ are the Fourier coefficients of the even extension to $(-1,1)$ of f:

$$m\pi c_m \sinh(m\pi) = b_m = 2\int_0^1 f(x)\cos(m\pi x)\,dx \qquad (m \in \mathbb{N}) \ .$$

We observe that the previous equality would be false for $m = 0$ if (8.28) were not valid, in agreement with the fact that if the compatibility condition is not valid, then there are no solutions to the Neumann problem associated to the Poisson equation. In conclusion, all the solutions of (8.26) (with the assumptions (8.20) and (8.28)) are given by

$$u(x,y) = C + \frac{1}{\pi}\sum_{m=1}^{\infty} \frac{b_m}{m\,\sinh(m\pi)} \cos(m\pi x)\cosh[m\pi(1-y)] \ ,$$

where $C \in \mathbb{R}$ is an arbitrary constant.

8.2.3.2 Laplace equation in a disc

Consider the *Dirichlet problem* for the Laplace equation in the unit disc of $\mathbb{R}^2$:

$$\begin{cases} u_{xx} + u_{yy} = 0 & \text{if } x^2 + y^2 < 1 \\ u(x,y) = g(x,y) & \text{if } x^2 + y^2 = 1 \,. \end{cases} \tag{8.31}$$

We rewrite the problem (8.31) in polar coordinates $x = \rho\cos\theta$, $y = \rho\sin\theta$. By differentiating with respect to ρ and θ the function $u(\rho\cos\theta, \rho\sin\theta)$ we get

$$\frac{\partial u}{\partial \rho} = u_x \cos\theta + u_y \sin\theta \,, \qquad \frac{\partial^2 u}{\partial \rho^2} = u_{xx}\cos^2\theta + 2u_{xy}\cos\theta\sin\theta + u_{yy}\sin^2\theta \,,$$

$$\frac{\partial^2 u}{\partial \theta^2} = u_{xx}\,\rho^2\sin^2\theta - 2u_{xy}\,\rho^2\cos\theta\sin\theta + u_{yy}\,\rho^2\cos^2\theta - u_x\,\rho\cos\theta - u_y\,\rho\sin\theta \,.$$

Hence,

$$u_{xx} + u_{yy} = \frac{\partial^2 u}{\partial \rho^2} + \frac{1}{\rho}\frac{\partial u}{\partial \rho} + \frac{1}{\rho^2}\frac{\partial^2 u}{\partial \theta^2}$$

and the problem (8.31) becomes

$$\begin{cases} \dfrac{\partial^2 u}{\partial \rho^2} + \dfrac{1}{\rho}\dfrac{\partial u}{\partial \rho} + \dfrac{1}{\rho^2}\dfrac{\partial^2 u}{\partial \theta^2} = 0 & \text{if } \rho \in [0,1) \\ u(1,\theta) = g(\theta) \end{cases} \tag{8.32}$$

where $u = u(\rho,\theta)$ and g has also been written in polar coordinates: since it is defined for $\rho = 1$, it depends only on θ. We seek solutions of equation (8.32) of the kind $u(\rho,\theta) = R(\rho)\Theta(\theta)$. By substitution and dividing, we arrive to

$$\frac{\Theta''(\theta)}{\Theta(\theta)} = -\left(\rho^2\frac{R''(\rho)}{R(\rho)} + \rho\frac{R'(\rho)}{R(\rho)}\right) .$$

Both members are constant; we look for (nontrivial) solutions of the equations

$$\Theta''(\theta) = \lambda\Theta(\theta) \,, \qquad \rho^2 R''(\rho) + \rho R'(\rho) + \lambda R(\rho) = 0 \,, \qquad (\lambda \in \mathbb{R}) \,.$$

Clearly, Θ must be 2π-periodic; we get (see Remark 8.10): $\lambda_m = -m^2$ $(m \in \mathbb{N})$ and $\Theta(\theta) = a\cos(m\theta) + b\sin(m\theta)$.
If $\lambda = \lambda_0 = 0$ the equation in R can be easily integrated and it yields the general solution $R(\rho) = a\ln\rho + b$. If $\lambda = \lambda_m$ $(m \geq 1)$ we seek power-like solutions, $R(\rho) = \rho^a$: we find $a^2 = m^2$ and therefore $R(\rho) = a\rho^m + b\rho^{-m}$ is the general solution. Since we seek harmonic solutions in $B_1(\mathbf{0})$ and hence regular at the origin, we must exclude logarithms and negative powers. Whence, the only possible solutions with separate variables are

$$u_0(\rho,\theta) = \frac{a_0}{2}, \quad u_m(\rho,\theta) = \rho^m \big(a_m \cos(m\theta) + b_m \sin(m\theta)\big) \quad (m \in \mathbb{N} \setminus \{0\}).$$

Because of the linearity of the problem, we seek solutions of the type

$$u(\rho,\theta) = \sum_{m=0}^{\infty} u_m(\rho,\theta) .$$

Imposing the Dirichlet condition $u(1,\theta) = g(\theta)$ we find that a_m and b_m are Fourier coefficients of g, see (5.3):

$$a_m = \frac{1}{\pi}\int_0^{2\pi} g(\theta)\cos(m\theta)\,d\theta , \qquad b_m = \frac{1}{\pi}\int_0^{2\pi} g(\theta)\sin(m\theta)\,d\theta$$

and the solution of (8.32) is given by

$$u(\rho,\theta) = \frac{a_0}{2} + \sum_{m=1}^{\infty} \rho^m \big(a_m \cos(m\theta) + b_m \sin(m\theta)\big) .$$

Consider now the *Neumann problem* for the Laplace equation on the unit disc of $\mathbb{R}^2$:

$$\begin{cases} u_{xx} + u_{yy} = 0 & \text{if } x^2+y^2 < 1 \\ \dfrac{\partial u}{\partial \nu}(x,y) = g(x,y) & \text{if } x^2+y^2 = 1 . \end{cases} \tag{8.33}$$

In polar coordinates the problem (8.33) becomes

$$\begin{cases} \dfrac{\partial^2 u}{\partial \rho^2} + \dfrac{1}{\rho}\dfrac{\partial u}{\partial \rho} + \dfrac{1}{\rho^2}\dfrac{\partial^2 u}{\partial \theta^2} = 0 & \text{if } \rho \in [0,1) \\ \\ \dfrac{\partial u}{\partial \rho}(1,\theta) = g(\theta) , \end{cases} \tag{8.34}$$

where we assume fulfilled the compatibility condition obtained from the divergence Theorem:

$$\int_0^{2\pi} g(\theta)\,d\theta = 0 . \tag{8.35}$$

By separating the variables, and proceeding as in the case of (8.32), we obtain

$$u(\rho,\theta) = C + \sum_{m=1}^{\infty} \rho^m \big(a_m \cos(m\theta) + b_m \sin(m\theta)\big) ,$$

where $C \in \mathbb{R}$ is an arbitrary constant. Imposing the Neumann condition we find that

$$g(\theta) = \frac{\partial u}{\partial \rho}(1,\theta) = \sum_{m=1}^{\infty} m\big(a_m \cos(m\theta) + b_m \sin(m\theta)\big)$$

and therefore a_m and b_m are proportional to the Fourier coefficients of g:

$$a_m = \frac{1}{m\pi}\int_0^{2\pi} g(\theta)\,\cos(m\theta)\,d\theta\ , \qquad b_m = \frac{1}{m\pi}\int_0^{2\pi} g(\theta)\,\sin(m\theta)\,d\theta\ .$$

Remark 8.10. For completeness we give the details of the steps which allow to determine all the 2π periodic solutions of the differential equation $\Theta''(\theta) = \lambda\,\Theta(\theta)$, by studying the following eigenvalue problem.
Find a constant $\mu \in \mathbb{C}$ and a nontrivial, 2π periodic function v which solves

$$-\,v''(x) \;=\; \mu\, v(x) \qquad x \in \mathbb{R}\,.$$

All the solutions of this eigenvalue problem are given by the pairs (μ_m, V_M), where the eigenvalues are $\mu_m = m^2,\ m = 0, 1, 2, \ldots$ and the corresponding eigenfunctions are $v_m = A\cos(mx) + B\sin(mx)$.
To prove this statement it is enough to observe that the general solution of the equation is given by $A\,\exp(\sqrt{-\mu}\,x) +\ B\,\exp(-\sqrt{-\mu}\,x)$. The periodicity implies $0 \equiv v(x) - v(x+2\pi) = \big(1 - \exp(2\pi\sqrt{-\mu})\big)\,g(x)$, where $g(x) = A\exp(\sqrt{-\mu}\,x) + B\exp(-\sqrt{-\mu}\,x)$ is a nontrivial analytic function of x, for any μ. Hence g is identically zero if and only if $A = B = 0$, but these choices are not acceptable because we are interested in solutions that are not identically zero. So the factor $\big(1 - \exp(2\pi\sqrt{-\mu})\big)$ must cancel, and then $\mu = m^2,\ m = 0, 1, \ldots$.
The numbers μ_m are called eigenvalues of the operator $-d^2/dx^2$ in $L^2(0, 2\pi)$ with periodic boundary conditions, whereas the functions v_m are called eigenfunctions: these names are motivated by the analogy with the algebraic problem of determining the eigenvalues of a square matrix (and the corresponding linear transformation from $\mathbb{R}^n$ to $\mathbb{R}^n$).

8.2.3.3 Radially symmetric solutions

We can also solve the Poisson problem in the case where the source has radial symmetry, $f = f(|\mathbf{x}|)$, and the boundary datum is constant. For example, consider the following *Dirichlet problem* for the Poisson equation in the unit ball of $\mathbb{R}^n$:

$$\begin{cases} \Delta u = f(|\mathbf{x}|) & \text{if } |\mathbf{x}| < 1 \\ u(\mathbf{x}) = \alpha \in \mathbb{R} & \text{if } |\mathbf{x}| = 1\ . \end{cases} \tag{8.36}$$

Assuming that $f \in C^1(B_1(\mathbf{0}))$, we look for radial solutions and proceed as for (8.13); we are then led to seek solutions of the problem

$$\begin{cases} u''(r) + \frac{n-1}{r}u'(r) = f(r) & r \in (0,1) \\ u(1) = \alpha,\ u'(0) = 0\ . \end{cases}$$

The condition $u(1) = \alpha$ is the Dirichlet condition, while $u'(0) = 0$ is a request for regularity of the solution: if u is regular and with radial symmetry, necessarily its gradient vanishes at the origin. Solving this problem we obtain the unique solution of problem (8.36):

$$u(r) = \alpha - \int_r^1 \left(\int_0^s t^{n-1} f(t)\, dt \right) \frac{ds}{s^{n-1}} \,. \tag{8.37}$$

Consider also the *Neumann problem* for the Poisson equation in $B_1(\mathbf{0}) \subset \mathbb{R}^n$, always assuming that $f \in C^1(B_1(\mathbf{0}))$:

$$\begin{cases} \Delta u = f(|\mathbf{x}|) & \text{if } |\mathbf{x}| < 1 \\ \frac{\partial u}{\partial \nu}(\mathbf{x}) = \alpha \in \mathbb{R} & \text{if } |\mathbf{x}| = 1 \,. \end{cases} \tag{8.38}$$

First, we should verify the compatibility condition given by the divergence Theorem:

$$\int_{|\mathbf{x}|<1} f(|\mathbf{x}|)\, d\mathbf{x} = \alpha \int_{|\mathbf{x}|=1} d\sigma(\mathbf{y}) = \sigma_{n-1}\, \alpha \tag{8.39}$$

where σ_{n-1} is the surface measure of $\partial B_1(\mathbf{0})$, see (2.14). If (8.39) is not satisfied, then (8.38) admits no solution.
If (8.39) is satisfied, then (8.38) admits infinitely many solutions, equal up to an additive constant. In this case, proceeding just as for (8.36), we must seek solutions to the problem

$$\begin{cases} u''(r) + \frac{n-1}{r} u'(r) = f(r) & r \in (0,1) \\ u'(1) = \alpha, \;\; u'(0) = 0 \,. \end{cases}$$

Solving this problem, we get infinitely many solutions:

$$u(r) = c - \int_r^1 \left(\int_0^s t^{n-1} f(t)\, dt \right) \frac{ds}{s^{n-1}} \qquad (c \in \mathbb{R}). \tag{8.40}$$

8.3 Exercises

Exercise 8.1. By separation of variables, solve the following heat equation:

$$\begin{cases} u_t - u_{xx} = 0 & (x,t) \in (0,\pi) \times \mathbb{R}^+ \\ u(0,t) = u(\pi,t) = 0 & t \in \mathbb{R}^+ \\ u(x,0) = \sin x + \sin 2x & x \in (0,\pi) \ . \end{cases}$$

Solution - Let us seek solutions of the kind $u(x,t) = X(x)T(t)$; we get

$$\frac{T'(t)}{T(t)} = \frac{X''(x)}{X(x)} \ ,$$

and hence there exists $\lambda \in \mathbb{R}$ such that

$$T'(t) + \lambda T(t) = 0 \text{ for } t \in (0,\infty) \ , \quad X''(x) + \lambda X(x) = 0 \text{ for } x \in (0,\pi) \ .$$

To the equation for X we associate the condition $X(0) = X(\pi) = 0$; eigenvalues and eigenfunctions of this problem are $\lambda_m = m^2$ and $\varphi_m(x) = \sin(mx)$. For $\lambda = \lambda_m$, the equation in T has solutions proportional to $T_m(t) = e^{-m^2 t}$. Therefore we look for solutions of the type

$$u(x,t) = \sum_{m=1}^{\infty} c_m e^{-m^2 t} \sin(mx) \ . \tag{8.41}$$

To determine the coefficients c_m we impose the condition $u(x,0) = \sin x + \sin 2x$; we obtain

$$\sin x + \sin 2x = \sum_{m=1}^{\infty} c_m \sin(mx) \ ,$$

that is $c_1 = c_2 = 1$ while $c_m = 0$ for any $m \geq 3$. Substituting in (8.41) we find the solution $u(x,t) = e^{-t} \sin x + e^{-4t} \sin(2x)$. □

Exercise 8.2. Consider the following Cauchy-Dirichlet problem:

$$\begin{cases} u_t - u_{xx} = 0 & (x,t) \in (0,\pi) \times \mathbb{R}^+ \\ u(0,t) = u(\pi,t) = 0 & t \in \mathbb{R}^+ \\ u(x,0) = \pi x - x^2 & x \in (0,\pi) \ . \end{cases}$$

Without solving it, show that the solution u satisfies $0 < u(x,t) < \frac{\pi^2}{4}$ $\forall x \in (0,\pi)$ $\forall t > 0$. Separating the variables, then determine the solution.

Solution - For every $T > 0$ we can apply Theorem 8.3: on the parabolic boundary of $(0,\pi) \times (0,T)$ we have $\max u = u(\frac{\pi}{2}, 0) = \frac{\pi^2}{4}$ and $\min u = u(0,t) = 0$. The required inequalities follow, and they are strict since u is not constant.
From Exercise 5.7 we know that the Fourier coefficients of $g(x) = \pi x - x^2$ are

$$\frac{2}{\pi} \int_0^{\pi} (\pi x - x^2) \, \sin(mx) \, dx = \frac{4}{\pi \, m^3} [1 + (-1)^{m+1}] \, .$$

Therefore, the solution of the problem is

$$u(x,t) = \sum_{m=1}^{\infty} \frac{4}{\pi \, m^3} [1 + (-1)^{m+1}] \, e^{-m^2 t} \, \sin(mx).$$

□

Exercise 8.3. Consider the problem

$$\begin{cases} u_t - u_{xx} = 12x^2 & (x,t) \in (0,1) \times \mathbb{R}^+ \\ u(x,0) = 0 & x \in (0,1) \\ u(0,t) = u(1,t) = 0 & t \in \mathbb{R}^+ . \end{cases}$$

(a) Determine the stationary solution that satisfies the boundary conditions at $x \in \{0,1\}$.
(b) Without computing it, prove that the solution u satisfies $0 < u(x,t) < x - x^4$ $\forall x \in (0,1)$ $\forall t > 0$.

Solution - (a) We have to solve the problem $-v''(x) = 12x^2$ with the conditions $v(0) = v(1) = 0$; we find $v(x) = x - x^4$.
(b) The inequality $u(x,0) > 0$ follows from Theorem 8.3: u is supercaloric and on the parabolic boundary of $(0,1) \times (0,T)$ (for any $T > 0$) results $u = 0$. For the other inequality, we observe that $w = u - v$ satisfies

$$\begin{cases} w_t - w_{xx} = 0 & (x,t) \in (0,1) \times \mathbb{R}^+ \\ w(x,0) = x^4 - x < 0 & x \in (0,1) \\ w(0,t) = w(1,t) = 0 & t \in \mathbb{R}^+ . \end{cases}$$

By applying again Theorem 8.3, we infer that $w < 0$ and therefore $u(x,t) < v(x,t) = x - x^4$ $\forall x \in (0,1)$ $\forall t > 0$. □

Exercise 8.4. Let $v(x,y,t) = e^x \sin y + y^3 + 6yt$. Compute $v_t - \Delta v$ and $v(x,y,0)$. Then find the solution of the problem

$$\begin{cases} u_t - \Delta u = e^x \sin y + y^3 & (x,y,t) \in \mathbb{R}^2 \times \mathbb{R}_+ \\ u(x,y,0) = e^x \sin y + y^3 & (x,y) \in \mathbb{R}^2 . \end{cases}$$

Solution - We have $v_t - \Delta v = 0$ and $v(x,y,0) = e^x \sin y + y^3$. The solution u is obtained as the sum of the solutions v and w of the problems

$$\begin{cases} v_t - \Delta v = 0 & (x,y,t) \in \mathbb{R}^2 \times \mathbb{R}_+ \\ v(x,y,0) = e^x \sin y + y^3 & (x,y) \in \mathbb{R}^2 \end{cases}$$

$$\begin{cases} w_t - \Delta w = e^x \sin y + y^3 & (x,y,t) \in \mathbb{R}^2 \times \mathbb{R}_+ \\ w(x,y,0) = 0 & (x,y) \in \mathbb{R}^2 . \end{cases}$$

Clearly, the solution of the first problem is precisely the given function $v(x,y,t) = e^x \sin y + y^3 + 6yt$. For the solution of the second equation, we consider $s \geq 0$ as a parameter, and we take as $z = z(x,y,t;s)$ the solution of the problem

$$\begin{cases} z_t(x,y,t;s) - \Delta z(x,y,t;s) = 0 & (x,y) \in \mathbb{R}^2, \ t > s \\ z(x,y,s;s) = e^x \sin y + y^3 & (x,y) \in \mathbb{R}^2 . \end{cases}$$

By what we have seen above, we have $z(x,y,t;s) = e^x \sin y + y^3 + 6y(t-s)$. The solution to the second problem will then be $w(x,y,t) = \int_0^t z(x,y,t;s)ds = t(e^x \sin y + y^3) + 3yt^2$. In conclusion, $u(x,y,t) = v(x,y,t) + w(x,y,t) = (t+1)(e^x \sin y + y^3) + 6yt + 3yt^2$. □

Exercise 8.5. Find the solution of the problem

$$\begin{cases} u_t - u_{xx} = \sin t + x \cos t & (x,t) \in \mathbb{R} \times \mathbb{R}_+ \\ u(x,0) = 0 & x \in \mathbb{R} . \end{cases}$$

Hint: consider the function $v(x,t) = u(x,t) - x \sin t + \cos t$.

Solution - Put $v(x,t) = u(x,t) - x \sin t + \cos t$, then v is a solution of the problem

$$\begin{cases} v_t - v_{xx} = 0 \ (x,t), \in \mathbb{R} \times \mathbb{R}_+ \\ v(x,0) = 1 \quad x \in \mathbb{R}, \end{cases}$$

whose solution is given by the formula

$$v(x,t) = \frac{1}{2\sqrt{\pi t}} \int_{\mathbb{R}} e^{-\frac{(x-y)^2}{4t}}\, dy = 1.$$

Hence the solution of the original problem is $u(x,t) = 1 + x \sin t - \cos t$. □

Exercise 8.6. Find the solution u of the problem

$$\begin{cases} u_t - u_{xx} = 1 & (x,t) \in \mathbb{R} \times \mathbb{R}_+ \\ u(x,0) = 1 & x \in \mathbb{R}\ . \end{cases}$$

Solution - The solution u is the sum of the solutions v and w to the problems

$$\begin{cases} v_t - v_{xx} = 0 & (x,t) \in \mathbb{R} \times \mathbb{R}_+ \\ v(x,0) = 1 & x \in \mathbb{R} \end{cases} \qquad \begin{cases} w_t - w_{xx} = 1 & (x,t) \in \mathbb{R} \times \mathbb{R}_+ \\ u(x,0) = 0 & x \in \mathbb{R}\ . \end{cases}$$

The solution of the first problem is $v(x,t) = 1$; this can be seen directly, or also (by using heavy weapons!) by using the formula

$$v(x,t) = \frac{1}{2\sqrt{\pi t}} \int_{\mathbb{R}} e^{-(x-y)^2/4t}\, dy = \frac{1}{2\sqrt{\pi t}} \int_{\mathbb{R}} e^{-y^2/4t}\, dy = \frac{1}{\sqrt{\pi}} \int_{\mathbb{R}} e^{-y^2}\, dy = 1\ .$$

For the solution of the second equation we consider $s \geq 0$ as parameter, and we define $z = z(x,t;s)$ to be the solution of the problem

$$\begin{cases} z_t(x,t;s) - z_{xx}(x,t;s) = 0 & x \in \mathbb{R},\ t > s \\ z(x,s;s) = 1 & x \in \mathbb{R}\ . \end{cases}$$

By what we have just seen, we have $z(x,t;s) \equiv 1$. The solution of the second problem will then be $w(x,t) = \int_0^t z(x,t;s)ds = t$. Summarizing, $u(x,t) = t + 1$. □

Exercise 8.7. In the point $(1, \frac{1}{4})$, find the value of the solution of the problem

$$\begin{cases} u_t - u_{xx} = 0 & \text{if } (x,t) \in \mathbb{R} \times \mathbb{R}_+ \\ u(x,0) = e^{-2x} & \text{if } x \in \mathbb{R}\ . \end{cases}$$

Solution - Making use of Remark 8.1, we can apply (8.4) and get

$$u\left(1, \frac{1}{4}\right) = \frac{1}{\sqrt{\pi}} \int_{-\infty}^{+\infty} e^{-(1-y)^2} e^{-2y}\, dy = \frac{e^{-1}}{\sqrt{\pi}} \int_{-\infty}^{+\infty} e^{-y^2}\, dy = e^{-1}.$$

Observe that it was not necessary to find explicitly the solution, which is however given by $u(x,t) = e^{-2x+4t}$. □

Exercise 8.8. Consider the problem

$$\begin{cases} w_t - w_{xx} = 0 & (x,t) \in (0,\pi) \times (0,\infty) \\ w(x,0) = \sin x & x \in (0,\pi) \\ w(0,t) = 0,\ w(\pi,t) = 1 & t \in [0,\infty). \end{cases} \tag{8.42}$$

Solve (8.42) proceeding both by separation of variables and by Laplace transformation; then, discuss both methods.

Solution - We note that the data on the parabolic boundary is not continuous in $(\pi, 0)$; as we shall see, this does not create problems because of the regularizing effect of the heat equation.
We decompose the problem (8.42) in the two subproblems

$$(P_1) \quad \begin{cases} u_t - u_{xx} = 0 & x \in (0,\pi) \times (0,\infty) \\ u(x,0) = 0 & x \in (0,\pi) \\ u(0,t) = 0, \; u(\pi,t) = 1 & t \in [0,\infty) \end{cases}$$

$$(P_2) \quad \begin{cases} v_t - v_{xx} = 0 & (x,t) \in (0,\pi) \times (0,\infty) \\ v(x,0) = \sin x & x \in (0,\pi) \\ v(0,t) = 0, \; v(\pi,t) = 0 & t \in [0,\infty). \end{cases}$$

The solution of (8.42) is the sum of the solutions of (P_1) and (P_2).
To solve (P_1), proceed first by separating the variables. Let $z(x,t) = u(x,t) - \frac{x}{\pi}$ so that z solves

$$(P_3) \quad \begin{cases} z_t - z_{xx} = 0 & x \in (0,\pi) \times (0,\infty) \\ z(x,0) = -\frac{x}{\pi} & x \in (0,\pi) \\ u(0,t) = u(\pi,t) = 0 & t \in [0,\infty). \end{cases}$$

Seeking separate variables solutions of (P_3), $z(x,t) = X(x)T(t)$, we still get (8.11). To the equation for X we associate the boundary conditions $X(0) = X(\pi) = 0$; the eigenvalues and the eigenfunctions of this problem are $\lambda_m = m^2$ and $X_m(x) = \sin(mx)$ for every $m \geq 1$. For $\lambda = \lambda_m$, the equation in T has solutions proportional to $T_m(t) = e^{-m^2 t}$. So let us seek solutions of (P_3) of the type

$$z(x,t) = \sum_{m=1}^{\infty} c_m e^{-m^2 t} \sin(mx) \, .$$

To fix the coefficients c_m we impose the condition $z(x,0) = -\frac{x}{\pi}$; we get

$$-\frac{x}{\pi} = \sum_{m=1}^{\infty} c_m \sin\left(\frac{m\pi x}{\ell}\right) \, .$$

Hence:

$$c_m = -\frac{2}{\pi^2} \int_0^{\pi} x \sin(mx)\, dx = \frac{(-1)^m\, 2}{m\,\pi} \, .$$

Finally,

$$z(x,t) = \sum_{m=1}^{\infty} (-1)^m \, \frac{2}{m\pi} \, e^{-m^2 t} \, \sin(mx) \, ,$$

$$u(x,t) = \frac{x}{\pi} + \sum_{m=1}^{\infty} (-1)^m \, \frac{2}{m\pi} \, e^{-m^2 t} \, \sin(mx) \, . \tag{8.43}$$

Let us now solve (P_1) by using the Laplace transform. Denoting by U the partial Laplace transform (with respect to t) of the solution u, from (P_1) we obtain the following transformed problem:

$$\begin{cases} s\, U(x,s) - U_{xx}(x,s) = 0 & 0 < x < \pi \\ U(0,s) = 0 \\ U(\pi,s) = \frac{1}{s}. \end{cases}$$

Note that 1 can be transformed and $Re(s) > 0$. The general integral of this (ordinary) differential equation is given by $U(x,s) = a(s)e^{x\sqrt{s}} + b(s)e^{-x\sqrt{s}}$. Imposing the boundary conditions, we find

$$U(x,s) = \frac{1}{s} \, \frac{\sinh(x\sqrt{s})}{\sinh(\pi\sqrt{s})}.$$

By antitransforming (and this is not trivial at all!) we still find (8.43) as a solution of the problem (P_1). This dual approach (separation of variables and Laplace transforms) may be used to prove that the antitransform of U coincides with (8.43).
Let us now deal with (P_2). Denoting by V the partial Laplace transform with respect to t of the solution v, from (P_2) we obtain the following transformed problem:

$$\begin{cases} s\,V(x,s) - V_{xx}(x,s) = \sin x & 0 < x < \pi \\ V(0,s) = 0 \\ V(\pi,s) = 0. \end{cases}$$

By solving the ordinary equation with the boundary value problem we obtain $V(x,s) = \frac{\sin x}{s+1}$ where $v(x,t) = e^{-t}\sin x$, a solution that we could also find by separating the variables as in Exercise 8.1. Finally, the solution of (8.42) is given by

$$w(x,t) = u(x,t) + v(x,t) = e^{-t}\sin x + \frac{x}{\pi} + \sum_{m=1}^{\infty} (-1)^m \,\frac{2}{m\pi}\, e^{-m^2 t}\, \sin(mx).$$

□

Exercise 8.9. Let $g \in L^1(\mathbb{R}^n)$ and consider the problem (8.2). Denoting by $u = u(\mathbf{x},t)$ its solution, show that

$$\int_{\mathbb{R}^n} u(\mathbf{x},t)\,d\mathbf{x}$$

is independent from t (*conservation of the total heat quantity* in an isolated system).

Solution - Notice that for (8.4), not only we have $u(\cdot,t) \in L^1(\mathbb{R}^n)$ for any $t > 0$, but also $u(\mathbf{x},t) = o(1)$ of exponential type for $|\mathbf{x}| \to \infty$. This eliminates the terms at "infinity" in the following application of the divergence Theorem:

$$\frac{d}{dt}\int_{\mathbb{R}^n} u(\mathbf{x},t)\,d\mathbf{x} = \int_{\mathbb{R}^n} u_t(\mathbf{x},t)\,d\mathbf{x} = \int_{\mathbb{R}^n} \Delta u(\mathbf{x},t)\,d\mathbf{x}$$

$$= \lim_{R\to\infty}\int_{B_R} \Delta u(\mathbf{x},t)\,d\mathbf{x} = \lim_{R\to\infty}\int_{\partial B_R} \frac{\partial u}{\partial \nu}(\mathbf{y},t)\,d\sigma(\mathbf{y}) = 0$$

where B_R is the ball of radius R centered into the origin. Since its derivative vanishes, the considered integral is constant. □

Exercise 8.10. Find the solution to the problem

$$\begin{cases} u_t - u_{xx} = x^3 & x \in \mathbb{R},\ t \in (0,\infty) \\ u(x,0) = x^3 & x \in \mathbb{R}. \end{cases}$$

Solution - First, we solve the problem

$$\begin{cases} u_t - u_{xx} = 0 & x \in \mathbb{R},\ t \in (0,\infty) \\ u(x,0) = x^3 & x \in \mathbb{R}. \end{cases} \tag{8.44}$$

For (8.4) we have

$$u(x,t) = \frac{1}{\sqrt{4\pi t}}\int_{\mathbb{R}} e^{-\frac{(y-x)^2}{4t}}\, y^3\,dy.$$

We can rewrite $y^3 = (y - x + x)^3 = (y-x)^3 + 3(y-x)^2 x + 3(y-x)x^2 + x^3$ and obtain

$$u(x,t) = \frac{1}{\sqrt{4\pi t}}\int_{\mathbb{R}} e^{-\frac{(y-x)^2}{4t}}\,[(y-x)^3 + 3(y-x)^2 x + 3(y-x)x^2 + x^3]\,dy.$$

Notice that with the change of variable $z = y - x$ we have

$$\int_{\mathbb{R}} e^{-\frac{(y-x)^2}{4t}} (y-x)^3 \, dy = \int_{\mathbb{R}} e^{-\frac{z^2}{4t}} z^3 \, dz = 0$$

since the integral is odd. Analogously,

$$\int_{\mathbb{R}} e^{-\frac{(y-x)^2}{4t}} (y-x)x^2 \, dy = x^2 \int_{\mathbb{R}} e^{-\frac{z^2}{4t}} z \, dz = 0.$$

Let us now deal with the quadratic term; with an integration by parts we get

$$\frac{3x}{\sqrt{4\pi t}} \int_{\mathbb{R}} e^{-\frac{(y-x)^2}{4t}} (y-x)^2 \, dy = \frac{6xt}{\sqrt{4\pi t}} \int_{\mathbb{R}} e^{-\frac{(y-x)^2}{4t}} \, dy = 6xt$$

where we used the formula for the Gauss integral:

$$\frac{1}{\sqrt{4\pi t}} \int_{\mathbb{R}} e^{-\frac{(y-x)^2}{4t}} \, dy = \frac{1}{\sqrt{\pi}} \int_{\mathbb{R}} e^{-\tau^2} \, d\tau = 1.$$

Using again the Gauss integral, we infer that

$$\frac{1}{\sqrt{4\pi t}} \int_{\mathbb{R}} e^{-\frac{(y-x)^2}{4t}} x^3 \, dy = x^3.$$

By adding together the four integrals obtained, we deduce $u(x,t) = x^3 + 6xt$, which is the solution of (8.44). Next, by using the Duhamel method, we solve

$$\begin{cases} u_t - u_{xx} = x^3 & x \in \mathbb{R}, \ t \in (0,\infty) \\ u(x,0) = 0 & x \in \mathbb{R}. \end{cases} \tag{8.45}$$

The auxiliary problem (8.7) becomes

$$\begin{cases} w_t(x,t;s) - w_{xx}(x,t;s) = 0 & x \in \mathbb{R}, \ t > s \\ w(x,s;s) = x^3 & x \in \mathbb{R} \end{cases}$$

and, by what we have just seen, its solution is given by $w(x,t;s) = x^3 + 6x(t-s)$. The solution of (8.45) is therefore

$$u(x,t) = \int_0^t [x^3 + 6x(t-s)] ds = x^3 t + 3xt^2.$$

By the superposition principle, the solution of the original problem is $u(x,t) = x^3 + 6xt + x^3 t + 3xt^2$. □

Exercise 8.11. Solve the problem

$$\begin{cases} u_{xx} + u_{yy} = 0 & \text{in } (0,1)^2 \\ u(x,0) = x & \text{for } x \in (0,1) \\ u(x,1) = 0 & \text{for } x \in (0,1) \\ u(0,y) = y & \text{for } y \in (0,1) \\ u(1,y) = 0 & \text{for } y \in (0,1) \end{cases}$$

and determine whether its solution is continuous in $[0,1]^2$.

Solution - It is sufficient to solve the problem

$$\begin{cases} \Delta v = 0 & \text{in } (0,1)^2 \\ v(x,0) = x & x \in (0,1) \\ v = 0 & \text{on the remaining part of } \partial(0,1)^2 \end{cases} \tag{8.46}$$

Setting $v(x, y) = X(x)\ Y(y)$, we get:

$$\begin{cases} X'' + \lambda\ X(x) = 0 \\ Y''(y) - \lambda\ Y(y) = 0 \\ X(0) = X(1) = Y(1) = 0 \end{cases}$$

The eigenvalues are $\lambda_m = m^2\ \pi^2$ and therefore

$$X_m(x) = \sin(m\pi x) \quad Y_m(y) = C_m\ \sinh[m\pi(1-y)]\,.$$

After imposing

$$x = \sum_{m=1}^{\infty} C_m\ \sinh(m\pi)\sin(m\pi x)$$

we obtain

$$C_m\ \sinh(m\pi) = 2\int_0^1 x\ \sin(m\pi x)\ dx = \frac{2(-1)^{m+1}}{m\pi}$$

and therefore

$$v(x, y) = \sum_{m=1}^{\infty} \frac{2(-1)^{m+1}}{m\pi}\frac{1}{\sinh(m\pi)}\sin(m\pi x)\sinh[m\pi(1-y)]$$

solves (8.46). But then $u(x, y) = v(x, y) + v(y, x)$ solves the initial problem. Moreover, $u \notin C^0([0, 1]^2)$ because $\lim\limits_{y\to 1} u(0, y) \neq \lim\limits_{x\to 0} u(x, 1)$. □

Exercise 8.12. Find the solution of the problem

$$\begin{cases} \Delta u = 0 & \text{in } (0, 1)^2 \\ u(x, 0) = 0 & 0 < x < 1 \\ u(x, 1) = x & 0 < x < 1 \\ u(0, y) = 0 & 0 < y < 1 \\ u_x(1, y) = 0 & 0 < y < 1. \end{cases}$$

Solution - By separating variables we write $u(x, y) = v(x)w(y)$ and we get

$$\frac{v''(x)}{v(x)} = -\frac{w''(y)}{w(y)} = \lambda,$$

where λ is a constant. By solving the boundary value problem

$$\begin{cases} v''(x) - \lambda v(x) = 0 \text{ in } (0, 1) \\ v(0) = v'(1) = 0 \end{cases}$$

we get $\lambda_k = -\frac{(2k+1)^2\pi^2}{4}$, $k \geq 0$, and $v_k = \sin\left(\frac{(2k+1)\pi x}{2}\right)$. The equation for w becomes $w''(y) + \lambda_k w(y) = 0$, whose general solution is given by

$$w(y) = C_1 \sinh\left(\frac{(2k+1)\pi y}{2}\right) + C_2 \cosh\left(\frac{(2k+1)\pi y}{2}\right).$$

From the condition $w(0) = 0$ we obtain $C_2 = 0$. Hence we find the family of solutions $u_k(x, y) = \sin\left(\frac{(2k+1)\pi x}{2}\right)\sinh\left(\frac{(2k+1)\pi y}{2}\right)$. We look for a solution of the form

$$u(x, y) = \sum_{k=0}^{+\infty} c_k \sin\left(\frac{(2k+1)\pi x}{2}\right)\sinh\left(\frac{(2k+1)\pi y}{2}\right),$$

satisfying the condition $u(x,1) = x$. By expanding x in Fourier series, which is odd in the interval $(-1,1)$, we obtain

$$x = \sum_{k=0}^{+\infty} b_k \sin\left(\frac{(2k+1)\pi x}{2}\right),$$

where

$$b_k = 2\int_0^1 x \sin\left(\frac{(2k+1)\pi x}{2}\right) dx = \frac{(-1)^k\, 8}{\pi^2(2k+1)^2}.$$

By choosing

$$c_k = b_k \left[\sinh\left(\frac{(2k+1)\pi}{2}\right)\right]^{-1}$$

we obtain the desired solution. □

Exercise 8.13. Determine the solution of the problem

$$\begin{cases} \Delta u = 2 & \text{in } (0,1)^2 \\ u(0,y) = 0 & \text{for } y \in (0,1) \\ u(1,y) = 1 & \text{for } y \in (0,1) \\ u(x,0) = 0 & \text{for } x \in (0,1) \\ u(x,1) = 0 & \text{for } x \in (0,1). \end{cases}$$

Solution - Let us first seek a function v (the simplest possible!) satisfying $\Delta v = 2$, without worrying about the boundary conditions. Take $v(x,y) = x^2$; then

$$\begin{cases} \Delta v = 2 & \text{in } (0,1)^2 \\ v(0,y) = 0 & \text{for } y \in (0,1) \\ v(1,y) = 1 & \text{for } y \in (0,1) \\ v(x,0) = x^2 & \text{for } x \in (0,1) \\ v(x,1) = x^2 & \text{for } x \in (0,1). \end{cases}$$

Then we set $w = v - u$ so that w satisfies

$$\begin{cases} \Delta w = 0 & \text{in } (0,1)^2 \\ w(0,y) = 0 & \text{for } y \in (0,1) \\ w(1,y) = 0 & \text{for } y \in (0,1) \\ w(x,0) = x^2 & \text{for } x \in (0,1) \\ w(x,1) = x^2 & \text{for } x \in (0,1). \end{cases} \tag{8.47}$$

We can now proceed by separation of variables, by setting $w(x,y) = X(x)Y(y)$. Let us start from (8.22)

$$\exists\lambda \in \mathbb{R}\,, \qquad X''(x) + \lambda X(x) = 0\,, \quad Y''(y) - \lambda Y(y) = 0\,,$$

to which we associate the boundary conditions $X(0) = X(1) = 0$. Requiring that X does not vanish identically, we deduce that $\lambda = \lambda_m = m^2\pi^2$ ($m \in \mathbb{N} \setminus \{0\}$) and $X_m(x) = A\sin(m\pi x)$. For such λ we have $Y_m(y) = Be^{m\pi y} + Ce^{-m\pi y}$, and a function satisfying the first three equations in (8.47) is given by

$$w(x,y) = \sum_{m=1}^{\infty} \left(\alpha_m e^{m\pi y} + \beta_m e^{-m\pi y}\right) \sin(m\pi x)$$

with the coefficients α_m and β_m to be determined. To this aim, we impose the last two conditions in (8.47):

$$x^2 = w(x,0) = \sum_{m=1}^{\infty} (\alpha_m + \beta_m) \sin(m\pi x)$$

$$x^2 = w(x,1) = \sum_{m=1}^{\infty} \left(\alpha_m e^{m\pi} + \beta_m e^{-m\pi}\right) \sin(m\pi x).$$

We find $\alpha_m + \beta_m = \alpha_m e^{m\pi} + \beta_m e^{-m\pi} = \gamma_m$ where γ_m are the Fourier coefficients of x^2 over $[0,1]$, extended as odd function to the whole $[-1,1]$ and also 2-periodic. Therefore, by (5.10),

$$\gamma_m = 2\int_0^1 x^2 \sin(m\pi x)\, dx = 4\,\frac{(-1)^m - 1}{(\pi m)^3} + 2\,\frac{(-1)^{m+1}}{\pi m}$$

and, by solving the system,

$$\alpha_m = \frac{\gamma_m}{e^{m\pi}+1}\,, \qquad \beta_m = \frac{\gamma_m\, e^{m\pi}}{e^{m\pi}+1}.$$

With this choice of the coefficients, the solution of the original problem is

$$u(x,y) = v(x,y) - w(x,y) = x^2 - \sum_{m=1}^{\infty} \left(\alpha_m e^{m\pi y} + \beta_m e^{-m\pi y}\right) \sin(m\pi x).$$

□

Exercise 8.14. Solve

$$\begin{cases} \Delta u = 0 & \text{if } x^2+y^2<1 \\ u(x,y) = xy(1-2y^2) & \text{if } x^2+y^2=1\,. \end{cases}$$

Solution - Writing the operator Δ in polar coordinates, with the notation of (8.32) we get

$$g(\theta) = \cos\theta\sin\theta(1-2\sin^2\theta) = \frac{\sin 2\theta}{2}\cos 2\theta = \frac{\sin 4\theta}{4}\,;$$

hence all the Fourier coefficients of g vanish, apart for $b_4 = 1/4$. Consequently,

$$u(\rho,\theta) = \frac{\rho^4}{4}\sin 4\theta = \frac{1}{4}\,\mathrm{Im}\left[\rho(cos\theta + i\sin\theta)\right]^4 = \frac{1}{4}\,\mathrm{Im}(x+iy)^4$$

and then $u(x,y) = x^3y - xy^3$. □

Exercise 8.15. Let u be a solution of the problem

$$\begin{cases} \Delta u = e^x & \text{if } x^2+y^2<1 \\ u(x,y) = 7+y-x^2 & \text{if } x^2+y^2=1\,. \end{cases}$$

Determine the maximum value of $u(x,y)$ for $x^2+y^2 \le 1$.

Solution - The function u is subharmonic. By Theorem 8.9, u attains its maximum when $x^2+y^2=1$ where $u(x,y) = 7+y-x^2 = y^2+y+6$. The maximum value is achieved at $y=1$ and therefore $\max u = 8$. □

Exercise 8.16. Show that the problem

$$\begin{cases} \Delta u = 10(x^2+y^2)^{3/2} & \text{if } x^2+y^2<1 \\ \frac{\partial u}{\partial \nu}(x,y) = 2 & \text{if } x^2+y^2=1 \end{cases}$$

admits infinitely many solutions, and determine them all.

Solution - Set $f(\rho) = 10\rho^3$ and write the integral in polar coordinates, then the compatibility condition (8.39) for the given problem becomes

$$2\pi \int_0^1 \rho\, f(\rho)\, d\rho = 20\pi \int_0^1 \rho^4\, d\rho = 4\pi = 2 \int_{x^2+y^2=1} d\omega = 2\sigma_1$$

and it is therefore satisfied. The problem admits infinitely many solutions given by (8.40): $u(r) = c + \frac{2}{5}r^5$ or $u(x,y) = c + \frac{2}{5}(x^2+y^2)^{5/2}$ with $c \in \mathbb{R}$. □

Exercise 8.17. Show that the solution v of the problem

$$\begin{cases} \Delta v = 1 + \cos^4 x & \text{if } x^2 + y^2 < 1 \\ v(x,y) = 1 & \text{if } x^2 + y^2 = 1 \end{cases}$$

satisfies $v(x,y) < \frac{x^2+y^2+3}{4}$ for $x^2 + y^2 < 1$.

Solution - Let us first solve the problem

$$\begin{cases} \Delta u = 1 & \text{if } x^2 + y^2 < 1 \\ u(x,y) = 1 & \text{if } x^2 + y^2 = 1\ . \end{cases}$$

By (8.37), we have

$$u(r) = \frac{r^2+3}{4} \implies u(x,y) = \frac{x^2+y^2+3}{4}.$$

Then note that $\Delta(v-u) = \cos^4 x \geq 0$ for $x^2+y^2 < 1$ and $v - u = 0$ for $x^2 + y^2 = 1$. By Theorem 8.9 we finally have $v - u < 0$ for $x^2 + y^2 < 1$. □

Exercise 8.18. Let $B_1(\mathbf{0})$ be the unit disc of $\mathbb{R}^2$. Let $u \in C^2(B_1(\mathbf{0})) \cap C^0(\overline{B_1(\mathbf{0})})$ be the solution of the problem

$$\begin{cases} \Delta u = 0 & \text{if } x^2 + y^2 < 1 \\ u(x,y) = x^2 + y^3 & \text{if } x^2 + y^2 = 1\ . \end{cases}$$

Compute $u(0,0)$.

Solution - By applying Theorem 8.7 we get

$$u(0,0) = \frac{1}{2\pi} \int_{x^2+y^2=1} u(\mathbf{z}) d\sigma(\mathbf{z}).$$

Since the function $y \mapsto y^3$ is odd, we can rewrite the previous integral as

$$u(0,0) = \frac{1}{2\pi} \int_0^{2\pi} \cos^2\theta\, d\theta = \frac{1}{2},$$

where $\partial B_1(\mathbf{0})$ has been parametrized by $x = \cos\theta$ and $y = \sin\theta$ ($\theta \in [0, 2\pi]$), therefore $\dot{x}^2 + \dot{y}^2 \equiv 1$. □

Exercise 8.19. Let $B = \{\mathbf{x} \in \mathbb{R}^n;\ |\mathbf{x}| < 1\}$. Find for which values of $\alpha \in \mathbb{R}$ the problem

$$\begin{cases} \Delta u = |\mathbf{x}|^4 & \mathbf{x} \in B \\ \frac{\partial u}{\partial \nu}(\mathbf{x}) = \alpha & \mathbf{x} \in \partial B \end{cases}$$

admits solutions and determine them.

Solution - Going to spherical coordinates, we have

$$\int_B \Delta u = \int_B |\mathbf{x}|^4 = \sigma_{n-1} \int_0^1 r^{n+3}\, dr = \frac{\sigma_{n-1}}{n+4}.$$

On the other hand, $\int_{\partial B} \alpha = \alpha\, \sigma_{n-1}$. Hence, the compatibility condition (8.39) is satisfied for $\alpha = \frac{1}{n+4}$. Only in this case solutions exist and, by substituting the data in (8.40), they are given by

$$u(\mathbf{x}) = \frac{|\mathbf{x}|^6}{6(n+4)} + c$$

for any $c \in \mathbb{R}$. □

Exercise 8.20. Let $\Omega \subset \mathbb{R}^n$ be a bounded open domain. Determine all the solutions of the problem

$$\begin{cases} \Delta u = ue^u & \text{in } \Omega \\ u = 0 & \text{over } \partial\Omega \,. \end{cases}$$

Solution - By multiplying the equation for u and by an integration by parts we get

$$-\int_\Omega |\nabla u|^2 = \int_\Omega u^2 e^u.$$

The LHS member is ≤ 0 while the RHS member is ≥ 0; both vanish if and only if $u \equiv 0$. the only solution is therefore $u = 0$.
We remark that the same conclusion holds if in place of ue^u there is a nonlinearity $f(u)$ satisfying $uf(u) \geq 0$. □

Exercise 8.21. Let $\Omega \subset \mathbb{R}^2$ be an open bounded domain. Determine all the solutions of the problem

$$\begin{cases} \Delta u = 0 & \text{in } \Omega \\ u(x,y) = e^{2x}\cos(2y) & \text{if } (x,y) \in \partial\Omega \,. \end{cases}$$

Solution - $u(x,y) = e^{2x}\cos(2y)$; this follows from the uniqueness of the solution and from the fact that the Dirichlet datum is a harmonic function on $\mathbb{R}^2$. □

Exercise 8.22. Let $\Omega = \{(x,y,z) \in \mathbb{R}^3;\ x^4 + y^4 + z^4 < 1\}$ and let u be the solution of the problem

$$\begin{cases} \Delta u = 0 & \text{in } \Omega \\ u = x^2 + y^2 & \text{over } \partial\Omega \,. \end{cases}$$

Determine $\max_{\overline{\Omega}} u$ and $\min_{\overline{\Omega}} u$ and the points where these extremes are reached.

Solution - By Theorem 8.9, u attains absolute maximum and minimum in $\overline{\Omega}$ only on $\partial\Omega$. Hence, we must find the absolute extremes of x^2+y^2 under the constraint $x^4+y^4+z^4 = 1$. The minimum is clearly 0 and is reached in the two points $(0,0,\pm 1)$. The maximum is reached on $\partial\Omega$ when $z = 0$; we must therefore maximize $x^2 + y^2$ under the constraint $x^4 + y^4 = 1$; we find the 4 points $(\pm 1/\sqrt[4]{2}, \pm 1/\sqrt[4]{2}, 0)$ where the function has the value $\sqrt{2} = \max_{\overline{\Omega}} u$. □

Exercise 8.23. Let $\Omega \subset \mathbb{R}^n$ be an open bounded domain, with regular boundary. Consider functions $\alpha, \beta, \phi \in C^0(\partial\Omega)$ such that $\alpha \geq 0$ and $\beta > 0$. Show that the problem

$$\begin{cases} \Delta u = 0 & \text{in } \Omega \\ \alpha \frac{\partial u}{\partial \nu} + \beta u = \phi & \text{on } \partial\Omega \end{cases}$$

admits at most one solution.

Solution - The present problem is named after Robin. Notice that it coincides with a Dirichlet problem if $\alpha \equiv 0$, and with a Neumann problem if $\beta \equiv 0$ (this case is excluded by the hypothesis). Suppose by contradiction that the problem admits two solutions u_1 and u_2. If we put $v = u_1 - u_2$, then v satisfies

$$\begin{cases} \Delta v = 0 & \text{in } \Omega \\ \alpha \frac{\partial v}{\partial \nu} + \beta v = 0 & \text{on } \partial\Omega. \end{cases}$$

By multiplying the equation by v and integrating by parts we obtain

$$0 = \int_\Omega v\Delta v = -\int_\Omega |\nabla v|^2 + \int_{\partial\Omega} v\frac{\partial v}{\partial \nu}.$$

The boundary condition can be rewritten as $v = -\frac{\alpha}{\beta}\frac{\partial v}{\partial \nu}$ and therefore,

$$\int_\Omega |\nabla v|^2 = -\int_{\partial\Omega} \frac{\alpha}{\beta}\left(\frac{\partial v}{\partial \nu}\right)^2 \leq 0.$$

This implies that $\nabla v \equiv 0$, which means that v is constant in Ω. But then $\frac{\partial v}{\partial \nu} = 0$ on $\partial\Omega$ and therefore also $v = 0$ on $\partial\Omega$. This shows that $v \equiv 0$ in Ω. □

8.4 Exercises without detailed solution

Exercise 8.24. By separating the variables, solve the Cauchy-Dirichlet problem for the heat equation

$$\begin{cases} u_t - u_{xx} = 0 & (x,t) \in (0,\pi) \times \mathbb{R}^+ \\ u(0,t) = u(\pi,t) = 0 & t \in \mathbb{R}^+ \\ u(x,0) = g(x) & x \in (0,\pi) \end{cases}$$

in the following cases: (1) $g(x) = \sin 3x - 3\sin x$, (2) $g(x) = 1$.

Solution - (1) $u(x,t) = -3e^{-t}\sin x + e^{-9t}\sin 3x$.

(2) $u(x,t) = \dfrac{2}{\pi}\displaystyle\sum_{m=1}^{\infty} \frac{1-(-1)^m}{m} e^{-m^2 t}\sin(mx).$ □

Exercise 8.25. By separating the variables, solve the problem:

$$\begin{cases} u_t - u_{xx} = 0 & (x,t) \in (0,1) \times \mathbb{R}^+ \\ u(0,t) = 0, \ u(1,t) = e^{-t} & t \in \mathbb{R}^+ \\ u(x,0) = x & x \in (0,1) \ . \end{cases}$$

Solution - $u(x,t) = xe^{-t} + \dfrac{2}{\pi}\displaystyle\sum_{m=1}^{\infty} \frac{(-1)^m}{m(m^2\pi^2-1)}\left(e^{-m^2\pi^2 t} - e^{-t}\right)\sin(m\pi x).$ □

Exercise 8.26. Solve the problem

$$\begin{cases} u_t - u_{xx} = f(x,t) & (x,t) \in (0,\ell) \times \mathbb{R}^+ \\ u(0,t) = u(\ell,t) = 0 & t \in \mathbb{R}^+ \\ u(x,0) = 0 & x \in (0,\ell) \end{cases}$$

in the following cases: (1) $\ell = \pi$, $f(x,t) = \sin x + 1$, (2) $\ell = 1$, $f(x,t) = x - x^2 + 2t$, (3) $\ell = \pi$, $f(x,t) = e^{-t}\sin x$.

Solution - (1) $u(x,t) = (1-e^{-t})\sin x + \dfrac{2}{\pi}\displaystyle\sum_{m=1}^{\infty} \frac{1-(-1)^m}{m^3}(1-e^{-m^2 t})\sin(mx).$

(2) $u(x,t) = tx(1-x)$. (3) $u(x,t) = te^{-t}\sin x$. □

Exercise 8.27. Show that $u(x,t) = e^{x+t}$ solves the problem

$$\begin{cases} u_t - u_{xx} = 0 & (x,t) \in \mathbb{R} \times \mathbb{R}_+ \\ u(x,0) = e^x & x \in \mathbb{R} \ . \end{cases}$$

Then infer the solutions of the problem

$$\begin{cases} v_t - v_{xx} = e^x & (x,t) \in \mathbb{R} \times \mathbb{R}_+ \\ v(x,0) = 0 & x \in \mathbb{R} \ . \end{cases}$$

Solution - $v(x,t) = e^{x+t} - e^x$. □

Exercise 8.28. Let $v(x,t) = x^2 + 2t$. Compute $v_t - v_{xx}$ and $v(x,0)$. Then deduce the solution of the problem

$$\begin{cases} u_t - u_{xx} = x^2 & (x,t) \in \mathbb{R} \times \mathbb{R}_+ \\ u(x,0) = x^2 & x \in \mathbb{R} \ . \end{cases}$$

Solution - $u(x,t) = x^2 t + t^2 + x^2 + 2t$. □

Exercise 8.29. Given $v(x,t) = e^{x+t} + x^2 + 2t$, compute $v_t - v_{xx}$ and $v(x,0)$. From that, infer the solutions of the problem

$$\begin{cases} u_t - u_{xx} = x^2 + e^x & (x,t) \in \mathbb{R} \times \mathbb{R}_+ \\ u(x,0) = x^2 + e^x & x \in \mathbb{R} \ . \end{cases}$$

Solution - $u(x,t) = x^2 t + t^2 + x^2 + 2t + 2e^{x+t} - e^x$. □

Exercise 8.30. Assuming that u satisfies $u_t - \Delta u = 0$ in $\mathbb{R}^n \times (0,\infty)$, show that:
(i) if $u_\lambda(x,t) = u(\lambda x, \lambda^2 t)$, then $(u_\lambda)_t - \Delta(u_\lambda) = 0$ in $\mathbb{R}^n \times (0,\infty)$ for any $\lambda \in \mathbb{R}$.
(ii) if $v(x,t) = x \cdot \nabla u + 2t u_t$, then $v_t - \Delta v = 0$ in $\mathbb{R}^n \times (0,\infty)$.

Exercise 8.31. Using the Laplace transform, determine the transformable solution $\mathcal{L}$ of the following Cauchy-Dirichlet problem for the heat equation on the half line:

$$\begin{cases} u_t - u_{xx} = 0 & (x,t) \in (0,\infty) \times (0,\infty) \\ u(x,0) = 1 & x \in [0,\infty) \\ u(0,t) = 0 & t \in [0,\infty). \end{cases} \tag{8.48}$$

Solution - Denoting by U the partial Laplace transform with respect to t of the solution u, we get the transformed problem:

$$\begin{cases} s\,U(x,s) - U_{xx}(x,s) = 1 & x > 0 \\ U(0,s) = 0, \ \lim_{x\to\infty} U(x,s) = 0 \end{cases} \tag{8.49}$$

and its solution is given by

$$U(x,s) = \frac{1}{s} - \frac{1}{s}\, e^{-x\sqrt{s}}.$$

To compute the Laplace antitransform of $s \mapsto e^{-x\sqrt{s}}/s$, for any $s = \rho e^{i\vartheta} \in \mathbb{C}$ with $\rho > 0$ and $\vartheta \in (-\pi,\pi)$ we set

$$\sqrt{s} = \sqrt{\rho} e^{i\vartheta/2}.$$

The the function $s \mapsto \sqrt{s}$ is a holomorphic function (single valued) in $\mathbb{C}$ minus the set of non positive real numbers. Chosen $\alpha > 0$ and antitransforming:

$$\mathcal{L}^{-1}\left(\frac{e^{-x\sqrt{s}}}{s}, t\right) = \frac{1}{2\pi i}\int_{\alpha - i\infty}^{\alpha + i\infty} e^{st}\frac{e^{-x\sqrt{s}}}{s} ds\,.$$

For $0 < \varepsilon < R$, consider the close path C as in the following figure

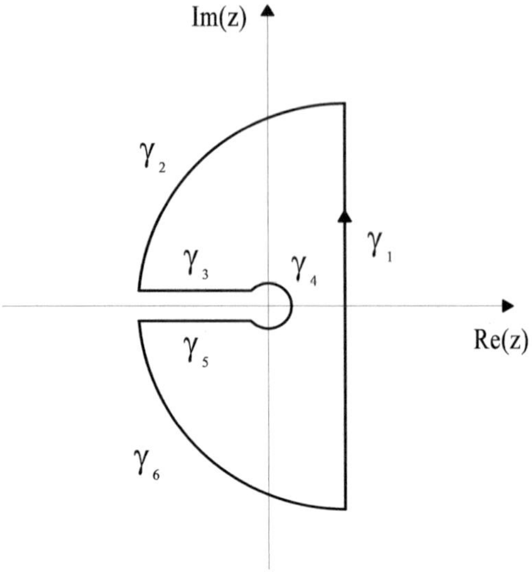

Figure 8.2 - Path needed for the solution of Exercise 8.31 .

where ε is the radius of the arc of circle γ_4, R is the the radius of the two arcs of circle γ_2 and γ_6, and α is the abscissa of the vertical segment. Then let $\varepsilon \to 0^+$ and $R \to +\infty$: the integral over γ_4 goes to 0, the integral over $\gamma_2 \cup \gamma_6$ also goes to 0 in view of Lemma 1.34 (iii). Taking into account the multi-valued behavior which gives different values on the two sides of the cut on $\mathbb{R}_-$, by Theorem 1.20 and the inversion formula (6.29) we obtain:

$$u(x,t) = -\frac{1}{2\pi i} \lim_{\substack{\varepsilon\to 0+ \\ R\to+\infty}} \int_{\gamma_3\cup\gamma_5} e^{st}\, \frac{e^{-x\sqrt{s}}}{s}\, ds = \frac{1}{\pi}\int_0^{\infty} e^{-\rho t}\, \frac{\sin(x\sqrt{\rho})}{\rho}\, d\rho$$

which is therefore the solution of (8.48). □

Exercise 8.32. Show that the functions

$$\text{(i) } u(x,y) = xy \qquad \text{(ii) } u(x,y) = x^3 - 3xy^2 \qquad \text{(iii) } u(x,y) = e^x \cos y$$

are harmonic on $\mathbb{R}^2$ (actually, they are so in any dimension $n \geq 2$, if we set $x_1 = x$ and $x_2 = y$).

Exercise 8.33. Let $B = \{\mathbf{x} \in \mathbb{R}^3;\ |\mathbf{x}| < 1\}$. Solve

$$\begin{cases} \Delta u = |\mathbf{x}|^2 + 1 & \mathbf{x} \in B \\ u(\mathbf{x}) = 1 & \mathbf{x} \in \partial B\ . \end{cases}$$

Solution - For (8.37) we have $u(\mathbf{x}) = \dfrac{|\mathbf{x}|^4}{20} + \dfrac{|\mathbf{x}|^2}{6} + \dfrac{47}{60}$. □

Exercise 8.34. Solve

$$\begin{cases} \Delta u = 0 & \text{if } x^2 + y^2 < 1 \\ u(x,y) = g(x,y) & \text{if } x^2 + y^2 = 1 \end{cases}$$

with (1) $g(x,y) = 2x^2 - 1$, (2) $g(x,y) = 2x^2y - y^3$, (3) $g(x,y) = xy^2 - 2x^3$, (4) $g(x,y) = 4x^2$, (5) $g(x,y) = 8x^4 - 8x^2 + 1$.

Solution - (1) $u(x,y) = x^2 - y^2$, (2) $u(x,y) = -\frac{3}{4}y^3 + \frac{9}{4}x^2y - \frac{1}{4}y$, (3) $u(x,y) = -\frac{3}{4}x^3 + \frac{9}{4}xy^2 - \frac{5}{4}x$, (4) $u(x,y) = 2(1 + x^2 - y^2)$, (5) $u(x,y) = x^4 + y^4 - 6x^2y^2$. □

Exercise 8.35. Let B be the unit ball in $\mathbb{R}^2$. Solve

$$\begin{cases} \Delta u = g(|\mathbf{x}|) & \text{if } \mathbf{x} \in B \\ u = \gamma & \text{if } \mathbf{x} \in \partial B \end{cases}$$

with (1) $g(r) = 16r^2 - 4$ and $\gamma = 0$, (2) $g(r) = 4(1 + r^2)e^{r^2}$ and $\gamma = e$, (3) $g(r) = 4\cos(r^2) - 4r^2 \sin(r^2)$ and $\gamma = 0$.

Solution - (1) $u(\mathbf{x}) = |\mathbf{x}|^4 - |\mathbf{x}|^2$, (2) $u(\mathbf{x}) = e^{|\mathbf{x}|^2}$, (3) $u(\mathbf{x}) = \sin(|\mathbf{x}|^2) - \sin 1$. □

Exercise 8.36. Find the solutions of the Neumann problem

$$\begin{cases} u_{xx} + u_{yy} = 0 & \text{in } (0,1)^2 \\ -u_y(x,0) = -1 + 3x^2 & \text{for } x \in (0,1) \\ u_y(x,1) = 0 & \text{for } x \in (0,1) \\ -u_x(0,y) = 0 & \text{for } y \in (0,1) \\ u_x(1,y) = 1 - 2y & \text{for } y \in (0,1). \end{cases}$$

Solution - A solution is the sum of the solutions of the problems

$$\begin{cases} v_{xx} + v_{yy} = 0 & \text{in } (0,1)^2 \\ -v_y(x,0) = 0 & \text{for } x \in (0,1) \\ v_y(x,1) = 0 & \text{for } x \in (0,1) \\ -v_x(0,y) = 0 & \text{for } y \in (0,1) \\ v_x(1,y) = 1 - 2y & \text{for } y \in (0,1), \end{cases}$$

$$\begin{cases} w_{xx} + w_{yy} = 0 & \text{in } (0,1)^2 \\ -w_y(x,0) = -1 + 3x^2 & \text{for } x \in (0,1) \\ w_y(x,1) = 0 & \text{for } x \in (0,1) \\ -w_x(0,y) = 0 & \text{for } y \in (0,1) \\ w_x(1,y) = 0 & \text{for } y \in (0,1). \end{cases}$$

We observe that the compatibility condition (8.27) is satisfied for both problems. A solution of the first problem is given by

$$v(x,y) = \frac{4}{\pi^3} \sum_{n=1}^{\infty} \frac{(1 + (-1)^{n+1}) \cos(n\pi y) \cosh(n\pi x)}{n^3 \sinh(n\pi)}.$$

A solution of the second problem is given by

$$w(x,y) = \frac{12}{\pi^3} \sum_{n=1}^{\infty} \frac{(-1)^n \cos(n\pi x) \cosh(n\pi(1-y))}{n^3 \sinh(n\pi)}.$$

Any other solution is obtained by adding an arbitrary constant to the sum of these two solutions. □

Exercise 8.37. Show that if $\Delta u(\mathbf{x}) = 0$ in $\mathbb{R}^n$ and $\mathbb{M}$ ia any orthogonal matrix of order n, then $\Delta u(\mathbb{M}\mathbf{x}) = 0$.
More generally, we have that $\Delta u(\mathbb{M}\mathbf{x}) = \Delta u(\mathbf{x})$, $\forall u \in C^2(\mathbb{R}^n)$, $\forall \mathbb{M}$ orthogonal: this means that the Laplacian is invariant under rotations and reflections.

Solution - Hint: to understand the problem, consider first the case $n = 2$ with the matrix

$$\mathbb{M} = \begin{pmatrix} 0 & 1 \\ 1 & 0 \end{pmatrix}.$$

□

Exercise 8.38. Let B be the unit disc in $\mathbb{R}^2$ and let $u(x,y) = (x^2+y^2-1)x$. Verify that $u = 0$ over ∂B and compute Δu. Then find a solution of the problem

$$\begin{cases} \Delta v = 16x - 24y & \text{if } (x,y) \in B \\ v = 0 & \text{if } (x,y) \in \partial B. \end{cases}$$

Solution - We have $\Delta u(x,y) = 8x$ and therefore $\Delta u(y,x) = 8y$; by superposition principle we have $v(x,y) = 2u(x,y) - 3u(y,x) = (x^2+y^2-1)(2x-3y)$. □

Exercise 8.39. Let $\Omega \subset \mathbb{R}^n$ be an open bounded connected set, and let $u \in C^2(\Omega)$ be such that

$$\Delta u + u^4 = 0 \quad \text{in } \Omega.$$

Show that if $u \not\equiv 0$ then u cannot have a minimum in Ω.

Solution - Hint: use the fact that the function u is superharmonic in Ω. □

Chapter 9
Recap exercises and final examinations

This chapter shows some self-evaluation trials and survey problems, together with their solutions.
The chapter is divided in two parts.
The first part includes a collection of trials built with some multiple-choice questions, covering all the topics of the book; the complete solution are given in separate pages.
The second part proposes problems to be solved in detail, explaining all the answers and computations of the solution.
Every trial and problem included in the present chapter is related to several subjects contained in various chapters of the book.
Some of the following exercises have been proposed as finals: therefore, they can be considered as summarizing exercises or models of final examinations.

9.1 Self-evaluation trials with multiple-choice answer

In the following pages some self-evaluation trials are proposed.
Each one of the subsequent trials is composed by several questions.
Each question is followed by four statements: one and only one of the four statements is correct.
The solutions of all the trials are grouped together in the Section 9.1.2 .

9.1.1 Tests

Trial 1

1. Compute $I = \int_\gamma \left(\frac{iz}{z^2+1} + \exp \frac{1}{z^2} \right) dz$, where γ is the circumference with radius 2 and center at the origin, swept once with the counter-clockwise orientation.

a $I = 2\pi i$ b $I = 1$ c $I = -2\pi$ d $I = 0$.

2. Compute the Laplace transform of the function $f(t) = 2e^{-t}\cos^2(t)$.

a $F(s) = \frac{1}{s+1} + \frac{s}{s^2+4}$ b $F(s) = \frac{1}{s+1} + \frac{s+1}{(s+1)^2+4}$

c $F(s) = \frac{1}{s+1} + \frac{s+1}{(s+1)^2+2}$ d $F(s) = \frac{1}{s+1} + \frac{s}{s^2+2}$.

3. Set $f(x) = \frac{\sin(x^2)}{1+x^2}$ with $x \in \mathbb{R}$ and denote its Fourier transform by $\widehat{f}(\xi)$.

a $\widehat{f}(\xi) \in \mathcal{S}$ b $\widehat{f}(\xi)$ has compact support in $\mathbb{R}$

c $\widehat{f}(\xi) \in L^2(\mathbb{R})$ d $\lim_{\xi\to\pm\infty} \widehat{f}(\xi) = 1$.

4. Evaluate the distributional derivative of the function $f(x) = x^2H(x) + x^3H(1-x)$.

a $f'(x) = 2xH(x) + 3x^2H(1-x)$

b $f'(x) = 2x\delta(x) + 3x^2H(1-x)$

c $f'(x) = 2xH(x) + 3x^2\delta(1-x)$

d $f'(x) = 2xH(x) + 3x^2H(1-x) - \delta(x-1)$.

5. Consider the differential equation $u'' + 9u = \sin 3x$.

a The equation has infinitely many periodic solutions

b Among solutions of the equation, only one belongs to $C^2(\mathbb{R})$ and is not periodic

c The equation has one and only one periodic solution

d The equation has no periodic solution.

Trial 2

1. Compute $I = \int_\gamma \left(\cos \frac{1}{z^2} + \frac{z+2}{z^2-1} \right) dz$, where γ is the circumference with radius 2 and center at the origin, swept once with the counter-clockwise orientation.

a	$I = 2\pi i$	b	$I = 1$	c	$I = 2\pi$	d	$I = 0$.

2. Set $f(x) = e^{-x^2}\chi_{[-1,1]}$ and denote its Fourier transform by $\widehat{f}(\xi)$.

a	$\widehat{f}(\xi) \notin \mathcal{S}$	b	$\widehat{f}(\xi)$ is continuous, but not diffrentiable
c	$\widehat{f}(\xi) \in L^2(\mathbb{R}) \cap L^1(\mathbb{R})$	d	$\widehat{f}(\xi)$ is odd.

3. Let $u'' - 9u = \cos(3x)$ be a differential equation.

a The equation has infinitely many periodic solutions

b Among the solutions of the equation, there is only one solution in $\mathcal{C}^2(\mathbb{R})$ which is not periodic

c The equation has exactly one periodic solution

d The equation has no periodic solution.

4. Compute the Laplace transform of the function $f(t) = 2e^t \sin^2(2t)$.

a	$F(s) = \dfrac{1}{s-1} - \dfrac{s}{s^2+4}$	b	$F(s) = \dfrac{1}{s-1} - \dfrac{s-1}{(s-1)^2+4}$
c	$F(s) = \dfrac{1}{s-1} - \dfrac{s-1}{(s-1)^2+16}$	d	$F(s) = \dfrac{1}{s-1} - \dfrac{s}{s^2+16}$.

5. Compute the distributional derivative of the function
$f(x) = H(x - x^2) + \ln|x+3|$.

a $f'(x) = \delta(x) + \delta(x-1) + \frac{1}{x+3}$

b $f'(x) = \delta(x - x^2) - \frac{1}{x+3}$

c $f'(x) = \delta(x) - \delta(x-1) + \text{v.p.} \frac{1}{x+3}$

d $f'(x) = \delta(1-2x) + \text{v.p.} \frac{1}{x+3}$.

Trial 3

1. Given the differential equation $u'' + 9u = \sin(3x)$.

- a The equation has no 2π periodic solutions
- b The equation has a unique 2π periodic solution
- c The equation has infinitely many 2π periodic solutions
- d The equation has only one solution which is not 2π periodic.

2. Compute the Laplace transform of the function $f(t) = \displaystyle\int_0^t \tau \sin\tau \, d\tau$.

- a $F(s) = -\dfrac{2s}{(s^2+1)^2}$
- b $F(s) = \dfrac{2s}{(s^2+1)^2}$
- c $F(s) = \dfrac{s}{s^2+1}$
- d $F(s) = \dfrac{2}{(s^2+1)^2}$.

3. Set $f(x) = x^2 e^{-x^2}$ and denote its Fourier transform by $\widehat{f}(\xi)$.

- a $\widehat{f}(\xi)$ has compact support in $\mathbb{R}$
- b $\widehat{f}(\xi) \in \mathcal{S}$
- c $\widehat{f}(\xi) \in L^2(\mathbb{R})$, but $\widehat{f}(\xi) \notin L^1(\mathbb{R})$
- d $\lim_{\xi\to\pm\infty} \widehat{f}(\xi) = 1$.

4. Given the function

$$f(x) = \begin{cases} 0 & x < 0 \\ \pi & 0 \le x < 1 \\ -2\pi & 1 \le x < 10 \\ 0 & x \ge 10 \end{cases}$$

compute its distributional derivative.

- a $f'(x) = \delta(x-\pi) + \delta(x-3\pi) + \delta(x-2\pi)$
- b $f'(x) = \delta(x-\pi) - 3\delta(x-1+\pi) + \delta(x-10)$
- c $f'(x) = \delta(x) + \delta(x-1) + \delta(x-10)$
- d $f'(x) = \pi\delta(x) - 3\pi\delta(x-1) + 2\pi\delta(x-10)$.

5. Given $f(z) = \dfrac{2}{z^2(z-2)}$, compute $I = \displaystyle\int_\gamma f(z)\, dz$, where γ is the circumference with radius 1 and center at the origin, swept once with the counterclockwise orientation.

- a $I = 0$
- b $I = 2\pi i$
- c $I = -\pi i$
- d $I = \pi i$.

Trial 4

1. Let $u'' + 4u = \cos x$ be a differential equation in $\mathbb{R}$.

- a The equation has infinitely many 2π periodic solutions
- b The equation has no 2π periodic solutions
- c The equation has one and only one 2π periodic solution
- d The equation has at least one solution $u \in L^2(\mathbb{R})$.

2. Set $f(x) = (\cos^2 x)\chi_{[-1,1]}(x)$ and let $\widehat{f}(\xi)$ be its Fourier transform.

- a $\widehat{f} \in L^1(\mathbb{R})$
- b $\widehat{f}$ is periodic
- c $\widehat{f}$ is continue, and null at infinity
- d $\widehat{f}\,' \notin L^2(\mathbb{R})$.

3. Compute the Laplace transform of the function $f(t) = \displaystyle\int_0^t \frac{1}{\tau}(1 - e^{-\tau})\, d\tau$.

- a $F(s) = \dfrac{1}{s} \ln\left(1 - \dfrac{1}{s}\right)$
- b $F(s) = \ln\left(s + \dfrac{1}{s}\right)$
- c $F(s) = \dfrac{1}{s} \ln\left(1 + \dfrac{1}{s}\right)$
- d $F(s) = \dfrac{2}{s^2} \ln\left(1 - \dfrac{1}{s}\right)$.

4. Given $f(z) = \dfrac{\sin 3z}{z^4(z+3)}$, compute the integral $I = \displaystyle\int_\gamma f(z)\, dz$ where γ is the circumference with radius 1 and center at the origin, swept once with the counter-clockwise orientation.

- a $I = \dfrac{2}{3}\pi i$
- b $I = -\dfrac{25}{9}\pi i$
- c $I = 0$
- d $I = 2\pi i$.

5. Compute the distributional derivative of $g(x) = H(x-2) + \ln|x|$.

- a $g'(x) = \delta(x-2) - \dfrac{1}{x}$
- b $g'(x) = \delta(x-2) + \text{v.p.}\, \dfrac{1}{x}$
- c $g'(x) = \delta(x) + \dfrac{1}{x}$
- d $g'(x) = \delta(x) + \text{v.p.}\, \dfrac{1}{x}$.

Trial 5

1. $\mathcal{L}\left\{\int_0^t \cos\tau \, d\tau\right\} =$

[a] $\dfrac{1}{1+s^2}$ [b] $-\dfrac{1}{1+s^2}$ [c] $\arctan s$ [d] $\dfrac{s}{1+s^2}$.

2. The differential equation $-v'' + v = 14\delta$

[a] has a unique solution v in $L^2(\mathbb{R})$

[b] has no solution in $L^2(\mathbb{R})$

[c] has infinitely many solutions in $L^2(\mathbb{R})$

[d] has two solutions in $L^2(\mathbb{R})$.

3. $x^2\delta'' =$

[a] 2δ [b] 0 [c] $-2\delta'$ [d] $2\delta'$.

4. Given $f \in \mathcal{S}'(\mathbb{R}^N)$, the partial differential equation $u - \Delta u = f$ in the unknown $u = u(\mathbf{x})$ with $\mathbf{x} \in \mathbb{R}^N$

[a] has no solution in $\mathcal{S}'(\mathbb{R}^N)$

[b] has one solution $u \in H^{15}(\mathbb{R}^N) \iff f \in H^{13}(\mathbb{R}^N)$

[c] has infinitely many solutions in $\mathcal{S}'(\mathbb{R}^N)$

[d] ha one solution $u \in H^{15}(\mathbb{R}^N) \iff f \in H^{14}(\mathbb{R}^N)$.

5. The series $\displaystyle\sum_{k=-\infty}^{+\infty} \delta'''(t-k)$

[a] converges in $\mathcal{S}'$ to an odd periodic distribution

[b] converges in $\mathcal{S}'$ to a distribution which is not periodic

[c] is the Fourier transform of a periodic distribution

[d] is the Fourier series of a $C^1(\mathbb{R})$ periodic function.

Trial 6

1. The integral $I = \int_\gamma \left(\frac{e^{3z}}{z^2(z-2)} \right) dz$, where γ is the circumference with radius 1 and center at the origin, swept once with the counter-clockwise orientation, is equal to:

[a] $5\pi i$ [b] $7\pi i$ [c] $-\frac{7}{2}\pi$ [d] $-\frac{7}{2}\pi i$.

2. Given $f(x) = x^2 + H(1-x^2)$, then the distributional derivative of f is

[a] $f'(x) = 2x + \delta(x+1) - \delta(x-1)$ [b] $f'(x) = 2x + \delta(1-x^2)$

[c] $f'(x) = 2x + 2H(x)$ [d] $f'(x) = 2x - 2H(x)$.

3. If $\alpha \in \mathbb{R}$, then the differential equation $u'' + \alpha u = \sin(2x)$

[a] has no periodic solution $\forall \alpha > 0$

[b] has a unique periodic solution $\forall \alpha > 0$

[c] has infinitely many periodic solutions $\forall \alpha > 0$

[d] has infinitely many periodic solution if $\alpha = 1$.

4. Given the function $g(\xi) = e^{-9|\xi|}\,\mathrm{sign}(\xi)$, then:

[a] g is the Fourier transform of an even function u

[b] g is the Fourier transform of a function $u \in L^2(\mathbb{R})$

[c] g is not the Fourier transform of any function or distribution

[d] g is the Fourier transform of a function $u \in L^1(\mathbb{R})$.

5. The Laplace transform of $\int_0^t e^\tau \cos\tau \, d\tau$ is equal to

[a] $F(s) = \frac{s-1}{s^2-2s+1}$ [b] $F(s) = \frac{s-1}{s^3-2s^2+s}$

[c] $F(s) = \frac{s-1}{s^2-2s+2}$ [d] $F(s) = \frac{s-1}{s^3-2s^2+2s}$.

Trial 7

1. The functions $f_n(x) = n\chi_{[1,1+(1/n)]}(x)$, $n \geq 1$

- a converge uniformly to 0, in $\mathbb{R}$ as $n \to +\infty$
- b converge to $\delta(x-1)$ in $\mathcal{D}'(\mathbb{R})$
- c converge to 0 in $L^1(\mathbb{R})$
- d converge to 0 in $\mathcal{D}'(\mathbb{R})$.

2. $\mathcal{L}^{-1}\left(\dfrac{s}{s^2+4}\right)$ is equal to:

- a $2\cos(t)H(t)$ b $4\cos(t)H(t)$
- c $\cos(4t)H(t)$ d $\cos(2t)H(t)$.

3. $\mathcal{F}\left\{\text{v.p. } \dfrac{1}{x}\right\}(\xi) =$

- a $-i\pi\,\text{sign}(\xi)$ b $i\pi\,\text{sign}(\xi)$ c $-i\pi\delta(\xi)$ d $\pi\delta(\xi)$.

4. $\text{res}\left(\dfrac{z^3 e^{1/z^2}}{z^5+4}, z = \infty\right) =$

- a 0 b 1/4 c 1 d −1/4.

5. The Fourier series $\displaystyle\sum_{n=1}^{+\infty} \frac{1}{n^3}\cos nx$

- a is the Fourier series of a function $u \in \mathcal{C}^\infty(\mathbb{R})$
- b is the Fourier series of a continuous, differentiable and 2π periodic function
- c converges point-wise, but not uniformly in $\mathbb{R}$
- d isn't the Fourier series of any function $f \in L^2(\mathbb{R})$.

6. The integral $\displaystyle\int_{\mathbb{R}} \frac{xe^{ix}}{(1+x^2)^2}\,dx$, is equal to:

- a $-\dfrac{i}{2}\pi e^{-1}$ b 0 c $\dfrac{i}{2}\pi e^{-1}$ d $-\dfrac{i}{2}\pi e$.

7. The distribution $\sin(3x)(\delta(x) - \delta'(x-\pi))$ is equivalent to:

a	$3\delta(x-\pi)$	b	$\delta(x) - \delta(x-\pi)$
c	$-3\delta(x-\pi)$	d	$-\delta(x-\pi)$.

8. The differential equation $u''(x) + u(x) = \sin(x) - 2\cos(x), \ x \in \mathbb{R}$

- a has no solution in $C^2(\mathbb{R})$
- b has infinitely many 2π periodic solutions
- c has exactly one π periodic solution
- d hasn't any 2π periodic solutions .

9. Which one of the following statements is true:

- a if $f^{(n)}(x) \in L^1(\mathbb{R})$, then $\widehat{f}(\xi)$ is an infinitesimal of order n at least as $|\xi| \to \infty$
- b the function $f(x) = \text{sign}(x)$, $x \in \mathbb{R}$, is an absolutely continuous function
- c $C^0([0,1])$, endowed with the norm $\|u\|_{L^1(0,1)} := \int_0^1 |u(x)|\, dx$, is a Banach space
- d $\mathbb{R}^n$ can be endowed with different and non equivalent norms .

10. Denote the Heaviside function by $H(x)$ and fix $u \in \mathcal{S}'(\mathbb{R})$ such that $(xe^{-x}H(x)) * u = e^{-x}H(x)$. Then u is equal to

a	$-H(x)$	b	$-\delta(x) - \delta'(x)$	c	$\delta(x) + \delta'(x)$	d	$-\delta(x) - i\delta'(x)$.

Trial 8

1. $\mathcal{L}\left\{\delta^{(2)}(t)\right\}$ is equal to:

a $s\,H(s-2)$ b $s^2\,H(s)$ c s^2 d $s^2\,H(s-2)$.

2. The differential equation $u''+2u=\sin 2x$

a has a unique periodic solution

b has infinitely many periodic solutions

c has no periodic solution

d has no $L^\infty(\mathbb{R})$ solution.

3. The Laurent series of the function $f(z)=\dfrac{\cos z^2}{z}$ is

a $\displaystyle\sum_{n=0}^{+\infty}(-1)^n\frac{z^{4n-1}}{(4n)!}$ b $\displaystyle\sum_{n=0}^{+\infty}(-1)^n\frac{z^{4n-1}}{(2n)!}$

c $\displaystyle\sum_{n=-\infty}^{0}(-1)^n\frac{z^{4n-2}}{(2n)!}$ d $\displaystyle\sum_{n=0}^{+\infty}(-1)^n\frac{z^{4n-2}}{(4n)!}$.

4. Set $u(x)=\dfrac{x}{(4+x^2)^2}$, then:

a $\widehat{u}(\xi)=-i\dfrac{\pi}{4}\xi\exp\left(-|\xi|\right)$ b $\widehat{u}(\xi)=i\dfrac{\pi}{4}\xi\exp\left(-2|\xi|\right)$

c $\widehat{u}(\xi)=-i\dfrac{\pi}{4}\xi\exp\left(-2|\xi|\right)$ d $\widehat{u}(\xi)=i\dfrac{\pi}{4}\xi\exp\left(-|\xi|\right)$.

5. The integral $\displaystyle\int_\gamma\frac{z+1}{(z-1)^2(z-2)^2}\,dz$, where γ is the circumference $\{|z-2|=7\}$, swept once with the counter-clockwise orientation, is equal to:

a $-10\pi i$ b $10\pi i$ c πi d 0.

6. Which one of the following functions of the variable s is the Laplace transform of an $\mathcal{L}$ transformable function?

a $|\cos s|$ b $\dfrac{1}{s^{3/2}}$ c $\operatorname{Re}\left(\exp(s)\right)$ d $\dfrac{1}{1+|s|^2}$.

7. Set $v_n(x) = nx\exp(-n|x|)$ and $v(x) \equiv 0$. Then

a $v_n \xrightarrow{L^1} v$, but $v_n \not\xrightarrow{L^2} v$

b $v_n \xrightarrow{L^2} v$, but $v_n \not\xrightarrow{L^1} v$

c $v_n \xrightarrow{L^\infty} v$

d $v_n \xrightarrow{L^1} v$ and $v_n \xrightarrow{L^2} v$.

8. The distribution $x^2\delta'$ is equivalent to:

a $-\delta$

b $\text{p.v.}\,\dfrac{1}{x}$

c δ

d 0.

9. Set $u(x) = \sum_{n=1}^{+\infty} \dfrac{1}{2n}\sin(n\pi x)$,

a u is 2π periodic

b $u \in \mathcal{C}^1(\mathbb{R})$

c $u \notin L^2(0,2)$

d u is odd.

10. Select $u(x) \in L^1(\mathbb{R})$ such that $xu(x) \in L^1(\mathbb{R})$. Then

a $\mathcal{F}\{u\} \notin L^\infty(\mathbb{R})$

b $\lim_{\xi\to\infty} \mathcal{F}\{u\}(\xi) = 1$

c $\mathcal{F}\{u\} \in \mathcal{C}^1(\mathbb{R})$

d $\mathcal{F}\{u\} \notin \mathcal{C}^0(\mathbb{R})$.

Trial 9

1. The sequence of functions $f_n(x) = \dfrac{5}{n}\chi_{[0,n](x)}$, $x \in \mathbb{R}$ and $n \geq 1$,

a	converges to $f(x) \equiv 0$ in $L^1(\mathbb{R})$
b	doesn't converge to $f(x) \equiv 0$ in $L^2(\mathbb{R})$
c	converges to $f(x) \equiv 0$ in $L^\infty(\mathbb{R})$
d	doesn't belong to $L^1(\mathbb{R})$.

2. The integral $\displaystyle\oint_\gamma e^{3/z^2}\,dz$ where γ is the circumference with radius 5 and center at the origin, swept once with the counter-clockwise orientation, is equal to:

a $4\pi i$ b 0 c $-2\pi i$ d $-\pi i$.

3. The distribution $\sin(x)\,\delta''(x)$ is equal to

a 0 b $-2\delta'$ c δ' d δ .

4. $\mathcal{F}\left\{\cos(2x)\right\}(\xi) \;=$

a	$\pi\;(\delta(\xi+2)+\delta(\xi-2))$	b	$2\,\pi i\,\mathrm{sign}(\xi)$
c	$2\,\text{p.v.}\,\frac{1}{\xi}$	d	$2\,\pi\,\delta(\xi)$.

5. Set $u \in L^2(\mathbb{R})$ s.t. $u - u'' = 2\delta$. Then u is equal to:

a $\frac{1}{2}\,e^{-|x|}$ b $e^{-|x|}$ c e^{-x^2} d $\sqrt{\pi}\,e^{x^2}$.

6. The equation $-u'' + 7\,u \;=\; \dfrac{1}{1+x^2}$, in the unknown $u = u(x),\; x \in \mathbb{R}$

a	has no solutions in $L^2(\mathbb{R})$
b	has two distinct solutions in $L^2(\mathbb{R})$
c	has infinitely many solutions in $L^2(\mathbb{R})$
d	has exactly one solution in $L^2(\mathbb{R})$.

7. Which one of the following functions is analytic in $\mathbb{C}$?

a	$f(z) = \operatorname{Im} z$	b	$f(z) = \cos(z^9)$
c	$f(z) = z^2 + \|z\|^2$	d	$f(z) = z + \overline{z}\,.$

8. The differential equation $-u''(x) + u(x) = \sin(x),\ x \in \mathbb{R}$

a	has no $C^2(\mathbb{R})$ solutions
b	has exactly one 2π periodic solution
c	has exactly two π periodic solutions
d	has no 2π periodic solutions.

9. $\mathcal{F}\left\{\dfrac{d}{dx}\left(e^{-3x^2}\right)\right\}(\xi) \; =$

a	$i\xi\,\sqrt{\pi/3}\,e^{-\xi^2/12}$	b	$-i\xi\,\sqrt{\pi/3}\,e^{\xi^2/3}$
c	$-i\xi\,\sqrt{3\pi}\,e^{-\xi^2/3}$	d	$i\xi\,\sqrt{3\pi}\,e^{\xi^2/12}\,.$

10. The integral p.v. $\displaystyle\int_{\mathbb{R}} \frac{e^{ix}}{x}\,dx$ is equal to

a	$\pi\, i$	b	0	c	$2\pi\, i$	d	$-\pi\, i\,.$

Trial 10

1. The functions $f_n(x) = nx\, e^{-n|x|}$, $n \geq 1$

- [a] converge uniformly to $f(x) = 0$, in $\mathbb{R}$ as $n \to +\infty$
- [b] converge to $f(x) = 0$, in $\mathcal{D}'(\mathbb{R})$
- [c] don't converge to $f(x) = 0$, in $L^1(\mathbb{R})$
- [d] don't converge to $f(x) = 0$, in $L^2(\mathbb{R})$.

2. The integral $\displaystyle\oint_\gamma \frac{1}{(1+z^5)}\, dz$ where γ is the circumference with radius 2 and center at the origin, swept once with the counter-clockwise orientation, is equal to:

[a] 0 [b] $2\pi i$ [c] $-2\pi i$ [d] $4\pi i$.

3. The distribution $e^{-x}\,\delta(x)$ is equal to

[a] 0 [b] $-\delta$ [c] δ' [d] δ .

4. Let $H(x)$ be the Heaviside function. Then $\mathcal{F}\{H(x)\}(\xi) =$

[a] $\pi\,\delta(\xi) - i\,\text{p.v.}\,\dfrac{1}{\xi}$ [b] $\text{sign}(\xi)$

[c] $\pi\,\text{sign}(\xi)$ [d] $\pi\,\text{sign}(\xi) + \text{p.v.}\,\dfrac{1}{\xi}$.

5. The equation $e^{-x^2} * u(x) = \dfrac{d^2}{dx^2}\left(e^{-x^2}\right)$ in the unknown $u = u(x)$, $x \in \mathbb{R}$

- [a] has one solution in $S'(\mathbb{R})$
- [b] has two different solutions in $L^1(\mathbb{R})$
- [c] has infinitely many solutions in $L^1(\mathbb{R})$
- [d] has exactly one solution in $L^1(\mathbb{R})$.

6. The differential equation $u'' + 3u = e^{i4x}$, $x \in \mathbb{R}$, $u \in \mathcal{S}'(\mathbb{R})$

- [a] has no 2π periodic solutions
- [b] has exactly one 2π periodic solution
- [c] has infinitely many 2π periodic solutions
- [d] has exactly two distinct 2π periodic solutions .

7. Which one of the following statements is true:

- [a] $C^0([0,1])$, endowed with the norm $\|u\|_{C^0([0,1])} := \max_{x\in[0,1]} |u(x)|$, is a Hilbert space
- [b] $C^0([0,1])$, endowed with the norm $\|u\|_{C^0([0,1])} := \max_{x\in[0,1]} |u(x)|$, is a Banach space
- [c] the Heaviside function is an absolutely continuous function in $\mathbb{R}$
- [d] $\|u\| := \max_{x\in[0,1]} |u'(x)|$ is a norm in $C^1([0,1])$.

8. The Fourier series $\displaystyle\sum_{n=1}^{\infty} \frac{\pi}{1+n^2}\sin(nx)$, $x \in \mathbb{R}$

- [a] converges to a continuous function uniformly in $\mathbb{R}$
- [b] converges point-wise for any $x \in \mathbb{R}$, but not uniformly in $\mathbb{R}$
- [c] converges point-wise for no $x \in \mathbb{R}$
- [d] is the Fourier series of an even function.

9. If $g(x) = \arctan \dfrac{3x}{1+x^4}$, $x \in \mathbb{R}$, then its Fourier transform $\widehat{g}(\xi)$

- [a] is an imaginary and odd function
- [b] belongs to the Schwartz space $\mathcal{S}(\mathbb{R})$
- [c] is an even function
- [d] does not belong to $L^2(\mathbb{R})$.

10. The function $\dfrac{1-\cosh(z)}{2z^2}$ has (at $z = 0$)

- [a] a double pole
- [b] a removable singularity
- [c] an essential singularity
- [d] a simple zero.

Trial 11

1. The functions $f_n(x) = n\chi_{[1,1+(1/n)]}(x)$, $n \geq 1$

- **a** converge uniformly to 0, in $\mathbb{R}$ as $n \to +\infty$
- **b** converge to $\delta(x-1)$ in $\mathcal{D}'(\mathbb{R})$
- **c** converge to 0 in $L^1(\mathbb{R})$
- **d** converge to 0 in $\mathcal{D}'(\mathbb{R})$.

2. The function $z \mapsto \sin z$, $z \in \mathbf{C}$:

- **a** fulfils $\cos^2(z) + \sin^2(z) = 1$
- **b** fulfils $|\sin z| \leq 1$
- **c** is not an analytic function
- **d** is not a periodic function .

3. Every solution $u \in \mathcal{D}'(\mathbb{R})$ of $(x^2+x-2)^2\, u(x) = 0$ belongs to this kind:

- **a** $u(x) = c_1\delta(x-1) + c_2\delta'(x-1) + c_3\delta(x+2) + c_4\delta'(x+2)$
- **b** $u(x) = c_1\delta(x+1) + c_2\delta'(x+1) + c_3\delta(x-2) + c_4\delta'(x-2)$
- **c** $u(x) = c_1\delta(x-1) + c_2\delta(x+2)$, $c_i \in \mathbb{R}$
- **d** $u(x) = c_1\delta(x+1) + c_2\delta(x-2)$, $c_i \in \mathbb{R}$.

4. $\mathcal{F}\left\{\text{p.v.}\,\dfrac{5}{x}\right\}(\xi) \; =$

- **a** $-5\pi\, i\,\text{sign}(\xi)$
- **b** $5\pi\, i\,\text{sign}(\xi)$
- **c** $-5\pi\, i\delta(\xi)$
- **d** $5\pi\,\delta(\xi)$.

5. $\text{res}\left(\dfrac{z^4 e^{1/z^2}}{z^6+4}\,,\, z=0\right) \; =$

- **a** 0
- **b** 1/4
- **c** -1
- **d** $-1/4$.

6. Let $H(x)$ be Heaviside function and fix $u \in \mathcal{S}'(\mathbb{R})$ such that $(x\,e^{-x}\,H(x)) * u = e^{-x}\,H(x)$. Then u is equal to

- **a** $-\delta(x) - \delta'(x)$
- **b** $\delta(x) + \delta'(x)$
- **c** $-\delta(x) - i\delta'(x)$
- **d** $-H(x)$.

7. Which one of the following statements is true:

a	every sequence in the set $\{a \in l^2 : \|a\|_{l^2} = 1\}$ has a subsequence that is convergent in the l^2 norm
b	the function $f(x) = \text{sign}(x)$, $x \in \mathbb{R}$, is an absolutely continuous function
c	$C^0([0,1])$ endowed with the norm $\|u\|_{L^1(0,1)} := \int_0^1 \|u(x)\|\, dx$ is not a Banach space
d	$\mathbb{R}^n$ can be endowed with different and non equivalent norms.

8. The differential equation $u''(x) + u(x) = 6\cos(x),\ x \in \mathbb{R}$

a	has no solution in $C^2(\mathbb{R})$
b	has exactly one 2π periodic solution
c	has exactly one π periodic solution
d	has no 2π periodic solutions.

9. The differential equation $u - \Delta u = f$,
where $u \in \mathcal{S}'(\mathbb{R}^N)$ is the unknown and $f \in \mathcal{S}'(\mathbb{R}^N)$ is the datum,

a	has no solutions in $\mathcal{S}'(\mathbb{R}^N)$
b	$\{$ has a unique solution $u \in H^9(\mathbb{R}^N)\} \iff f \in H^8(\mathbb{R}^N)$
c	has infinitely many solutions in $\mathcal{S}'(\mathbb{R}^N)$
d	$\{$ has a unique solution $u \in H^{10}(\mathbb{R}^N)\} \iff f \in H^8(\mathbb{R}^N)$.

10. The integral $\displaystyle\int_{\mathbb{R}} \frac{xe^{ix}}{(1+x^2)^2}\, dx$, is equal to:

a	$-\frac{i}{2}\pi e^{-1}$	b	0	c	$\frac{i}{2}\pi e^{-1}$	d	$-\frac{i}{2}\pi e$.

Trial 12

1. $\mathcal{L}^{-1}\left\{\dfrac{2}{s^2+4}\right\}$ is equal to:

a	$\frac{1}{2}\sin(4t)\,H(t)$	b	$2\sin(2t)\,H(t)$
c	$\sin(4t)\,H(t)$	d	$\sin(2t)\,H(t)$.

2. The differential equation $u'' + 4u = \sin 2x$

- a has a unique periodic solution
- b has infinitely many periodic solutions
- c has no periodic solution
- d has no $\mathcal{C}^2(\mathbb{R})$ solutions .

3. The Laurent series of the function $f(z) = \dfrac{\sin z^2}{z^2}$

a	$\displaystyle\sum_{n=0}^{+\infty}(-1)^n\frac{z^{4n-2}}{(2n+1)!}$	b	$\displaystyle\sum_{n=0}^{+\infty}(-1)^n\frac{z^{4n}}{(2n+1)!}$
c	$\displaystyle\sum_{n=-\infty}^{0}(-1)^n\frac{z^{4n}}{(2n+1)!}$	d	$\displaystyle\sum_{n=0}^{+\infty}\frac{z^{4n}}{(2n+1)!}$.

4. Set $u(x) = \exp(-10x^2)$, then:

a	$\widehat{u}(\xi) = \sqrt{\frac{\pi}{10}}\exp\left(-\frac{\xi^2}{40}\right)$	b	$\widehat{u}(\xi) = \exp\left(-\frac{\xi^2}{40}\right)$
c	$\widehat{u}(\xi) = \sqrt{\pi}\exp\left(-\frac{\xi}{40}\right)$	d	$\widehat{u}(\xi) = \sqrt{\frac{\pi}{10}}\exp\left(-\frac{\xi^2}{10}\right)$.

5. The integral $\displaystyle\int_\gamma \frac{z}{(z-1)(z-2)}\,dz$, where γ is the circumference $\left\{|z-2| = \frac{1}{5}\right\}$, swept once with the counter-clockwise orientation, is equal to:

a	πi	b	$-4\pi i$	c	$4\pi i$	d	$-\pi i$.

6. Which one of the following functions of the variable s is the Laplace transform of an $\mathcal{L}$ transformable function?

a	$\exp(-\lvert s\rvert)$	b	$\frac{1}{\lvert s\rvert}$	c	$\frac{1}{\operatorname{Im} s}$	d	$\frac{2}{1+s^2}$.

7. Set $v_n(x) = \sqrt{n}\exp(-10n|x|)$ and $v(x) \equiv 0$. Then

a $v_n \xrightarrow{L^2} v$ **b** $v_n \xrightarrow{L^1} v$

c $v_n \xrightarrow{L^\infty} v$ **d** $v_n \longrightarrow v$ uniformly.

8. The distribution $x\delta'$ is equivalent to:

a $\text{p.v.}\,\dfrac{1}{x}$ **b** $-\delta$ **c** δ **d** 0.

9. The Fourier series of $u(x) = |\sin x|$

a is $\displaystyle\sum_{n=0}^{+\infty} c_n \cos(2nx)$ (suitable c_n) **b** is $\displaystyle\sum_{n=0}^{+\infty} \frac{1}{n^2+1}\cos(nx)$

c is $\displaystyle\sum_{n=0}^{+\infty} d_n \sin(nx)$ (suitable d_n) **d** cannot be defined.

10. Given the problem: find $u \in \mathcal{S}'(\mathbb{R})$ s.t. $-u'' + u = f$. We can say that

a $\{$it has a solution $u \in H^{k+2}(\mathbb{R})\} \Leftrightarrow \{f \in H^k(\mathbb{R})\}$

b it has no solutions if f is odd

c it has infinitely many solutions $\forall f \in L^2(\mathbb{R})$

d it has no solution if $f(x) = \dfrac{1}{1+x^2}$.

Trial 13

1. $\delta * \delta'$ is equal to:

| a | 1 | b | δ | c | δ' | d | δ''. |

2. The $\mathcal{L}$ transformable solution of the equation $y''(t) + y(t) = \delta(t - 2\pi)$, $t \in (-\infty, \infty)$, is:

| a | $H(t-2\pi)\cos t$ | b | $H(t)\sin t$ |
| c | $H(t-2\pi)\sin t$ | d | $H(t)\cos t$. |

3. The function $\dfrac{\exp(-x^2)}{x^{n+1}+1}$ is Lebesgue integrable in $\mathbb{R}$

| a | for no $n \in \mathbb{N}$ | b | for every $n \in \mathbb{N}$ |
| c | if n is even | d | if n is odd. |

4. The function $\exp(-|s|)$

a	is an entire analytic function
b	is the $\mathcal{L}$ transform of an $\mathcal{L}$ transformable function
c	is the $\mathcal{L}$ transform of an $\mathcal{L}$ transformable distribution
d	is not an $\mathcal{L}$ transform.

5. Set $u(x) = x^2$, $-\pi \le x \le \pi$.
Then the Fourier series of the 2π periodic extension of u in $\mathbb{R}$,

| a | contains sines and cosines | b | doesn't exist |
| c | contains sines only | d | contains cosines only. |

6. Set $f(z) = \exp\left(\dfrac{2}{z}\right)$. The residue of f at the origin is equal to:

| a | 3 | b | 0 | c | 1 | d | 2. |

7. The Fourier transform of the function $f(t) = \exp(-|t|)$ is equal to:

| a | $\dfrac{\pi}{1+x^2}$ | b | $\dfrac{2}{1+x^2}$ | c | $\dfrac{1}{1+x^2}$ | d | $\dfrac{2\pi}{1+x^2}$. |

8. The integral $\displaystyle\oint_{\Gamma} \frac{1}{z-3}\,dz$, with $\Gamma = \{z : |z| = 2\}$ swept once with the counter-clockwise orientation, is equal to:

[a] 0 [b] $2\pi i$ [c] $-\dfrac{\pi i}{2}$ [d] $\dfrac{\pi i}{2}$.

9. The Fourier transform of the function $f(x) = \dfrac{x \cos 2x}{1+x^2}$ is:

[a] purely imaginary and odd [b] real and even

[c] real and odd [d] purely imaginary and even.

10. If $z = x + iy$ then the function $f(z) = x^2 + 2ixy + y^2$

[a] is analytic for every $z \in \mathbb{C}$ such that $|z| > 1$

[b] is analytic in $\mathbb{C}$

[c] is not analytic in any open set of $\mathbb{C}$

[d] is analytic for every $z \in \mathbb{C}$ such that $|z| < 1$.

Trial 14

1. The function s^2

a is the $\mathcal{L}$ transform of an $\mathcal{L}$ transformable distribution, but it is not the $\mathcal{L}$ transform of an $\mathcal{L}$ transformable function

b is the $\mathcal{L}$ transform of neither an $\mathcal{L}$ transformable function nor an $\mathcal{L}$ transformable distribution

c is bounded if $s \in \mathbb{C} \setminus \{0\}$

d is the $\mathcal{L}$ transform of an $\mathcal{L}$ transformable function.

2. Set $u(x) = x$, with $x \in (-\pi, \pi)$.
Then the Fourier series of the 2π periodic extension of $u(x)$ in $\mathbb{R}$,

a has rapidly decreasing coefficients as $n \to \infty$

b contains cosines only

c contains only sines and the coefficients decrease with $O(1/n)$ order as $n \to \infty$

d has coefficients whose modulus is increasing as $n \to \infty$.

3. The Laplace transform of $f(t) = \displaystyle\int_0^t (t-x)\sin(2x)\,dx$ is equal to:

a $\dfrac{2}{\sin(2s)}$ b $\dfrac{2}{s^2(s^2+4)}$

c $\dfrac{(s-1)}{2s^2(s^2+4)}$ d $\dfrac{2s}{(s^2+4)}$.

4. The integral $\displaystyle\oint_\Gamma \frac{1}{z}\,dz$, with Γ the circumference with radius 1 and center at the origin, swept once with the counter-clockwise orientation, is equal to:

a $-\dfrac{\pi i}{2}$ b $4\pi i$ c 0 d $2\pi i$.

5. The Fourier transform of the function $f(x) = \dfrac{d^2}{dx^2}\left(\dfrac{1}{1+x^2}\right)$ is equal to:

a $-\pi e^{-\xi}$ b $\dfrac{\xi^2}{1+\xi^2}$

c $-\pi\xi^2 e^{-|\xi|}$ d $2\pi\xi e^{-|\xi|}$.

6. The Fourier transform of the function $f(x) = xe^{-|x|}$ is:

| a | a distribution in $\mathcal{D}'(\mathbb{R})$, but it is not a tempered distribution

| b | a continuous and even function

| c | a $C^\infty(\mathbb{R})$ odd function

| d | a real-valued and rapidly decreasing function.

7. Denote the open unit disk $\{(x,y) : x^2+y^2 < 1\} \subset \mathbb{R}^2$ by Ω and its boundary $\{(x,y) : x^2+y^2 = 1\}$ by $\partial\Omega$. Referring to the Neumann problem $\Delta u = 0$ in Ω with prescribed normal derivative $\dfrac{\partial u}{\partial n} = \sin\theta + \lambda$ if $(x,y) \in \partial\Omega$ (i.e. if $x = \cos\theta$ and $y = \sin\theta$ with $\theta \in [0, 2\pi]$).

| a | a solution u exists only if $\lambda = 0$

| b | a solution u exists only if λ is an integer number

| c | a solution u exists only if $\lambda = 2$

| d | a solution u exists for every real value of λ.

8. The distribution $x^2\,\delta'(x)$ is equal to:

| a | -1 | b | 0 | c | $-2!\,\delta$ | d | $-\delta$.

9. What can we say about the periodic solutions of the differential equation $u'' - 4u = g$ if the datum g is a periodic distribution, expressed by the Fourier series: $g(x) = \sum_{-\infty}^{+\infty} \gamma_k e^{ikx}$.

| a | there are infinitely many periodic solutions if $\gamma_2 = \gamma_{-2} = 0$

| b | a unique periodic solution exists for every 2π periodic data g

| c | the equation has no periodic solutions, for any g

| d | periodic solutions exist, if $\gamma_2 = 1$.

10. The function $f(x) = \dfrac{9x}{1+x^2}$

| a | belongs to $L^1(\mathbb{R})$

| b | belongs to $L^2(\mathbb{R})$

| c | has an entire analytic extension

| d | is not continuous in $\mathbb{R}$.

9.1.2 Trials answers

Trial 1 - Solutions

1. c 2. b 3. c 4. d 5. d

Trial 2 - Solutions

1. a 2. a 3. c 4. c 5. c

Trial 3 - Solutions

1. a 2. d 3. b 4. d 5. c

Trial 4 - Solutions

1. a 2. c 3. c 4. b 5. b

Trial 5 - Solutions

1. a 2. a 3. a 4. b 5. a

Trial 6 - Solutions

1. d 2. a 3. d 4. b 5. d

Trial 7 - Solutions

1. b	2. d	3. a	4. a	5. b	6. c
7. c	8. d	9. a	10. c		

Trial 8 - Solutions

1. c	2. a	3. b	4. c	5. d	6. b
7. d	8. d	9. d	10. c		

Trial 9 - Solutions

1. c	2. b	3. b	4. a	5. b	6. d
7. b	8. b	9. a	10. a		

Trial 10 - Solutions

1. b	2. a	3. d	4. a	5. a	6. b
7. b	8. a	9. a	10. b		

Trial 11 - Solutions

1. b	2. a	3. a	4. a	5. a	6. b
7. c	8. d	9. d	10. c		

Trial 12 - Solutions

1. d 2. c 3. b 4. a 5. c 6. d
7. b 8. b 9. a 10. a

Trial 13 - Solutions

1. c 2. c 3. d 4. d 5. d 6. d
7. b 8. a 9. a 10. c

Trial 14 - Solutions

1. a 2. c 3. b 4. d 5. c 6. c
7. a 8. b 9. b 10. b

9.2 Recap exercises

Solve the exercises listed below, by explaining all the answers and computations required by the solution.

Exercise 9.1. Given

$$f(t) = \sum_{k=0}^{+\infty} \delta(t-k)$$

1) Evaluate $\widehat{f}$.
2) Evaluate $\mathcal{L}\{f\}$.
3) Solve the problem:

$$\begin{cases} \text{find an } \mathcal{L} \text{ trasformable solution } y \text{ of} \\ y'(t) - y(t-1) = f(t) \quad \text{in } \mathcal{D}'(\mathbb{R}). \end{cases}$$

Solution - 1) We exploit these identities

$$\mathcal{F}\{\delta\} = 1, \qquad \mathcal{F}\{u(t-k)\}(x) = \widehat{u}(x)\, e^{-ikx}.$$

For any $\psi \in \mathcal{S}(\mathbb{R})$ the series $\sum_{k=0}^{+\infty} \psi(k)$ is absolutely convergent, hence

$$\langle f, \psi \rangle = \sum_{k=0}^{+\infty} \psi(k) \quad \forall \psi \in \mathcal{S}(\mathbb{R}).$$

Then the series $\sum_{k=0}^{+\infty} \delta(t-k)$ converges to f in $\mathcal{S}'(\mathbb{R})$ and can be transformed term-wise:

$$\widehat{f}(x) = \sum_{k=0}^{+\infty} e^{-ikx}.$$

2) We recall: $\mathcal{L}\{\delta\} = 1$ and $\mathcal{L}\{u(t-k)\} = e^{-ks}U(s)$.

The series $\sum_{k=0}^{+\infty} \delta(t-k)$ converges to f in $\mathcal{S}'(\mathbb{R})$, then it converges in the space of $\mathcal{L}$ transformable distributions (with $\lambda = 0$) and can be transformed term-wise:

$$\mathcal{L}\{f\}(s) = \sum_{k=0}^{+\infty} e^{-sk} = \frac{1}{1-e^{-s}} \qquad \operatorname{Re} s > 0.$$

3) After setting $Y(s) = \mathcal{L}\{y\}(s)$ we transform the equation:

$$sY(s) - e^{-s}Y(s) = \frac{1}{1-e^{-s}}$$

$$Y(s) = \frac{1}{(1-e^{-s})(s-e^{-s})}$$

$$\mathcal{L}^{-1}\left\{\frac{1}{1-e^{-s}}\right\} = \sum_{k=0}^{+\infty} \delta(t-k).$$

If $\operatorname{Re} s > 1$ then $|e^{-s}/s| < 1$, so we can exploit the explicit value of the sum of a geometric series of common ratio smaller than 1: after setting $s^+ = \max(s,0)\ \forall s \in \mathbb{R}$, we obtain

$$\mathcal{L}^{-1}\left\{\frac{1}{s-e^{-s}}\right\}=\mathcal{L}^{-1}\left\{\frac{1/s}{1-e^{-s}/s}\right\}=\mathcal{L}^{-1}\left\{\sum_{k=0}^{+\infty}\frac{1}{s}\left(\frac{e^{-s}}{s}\right)^k\right\}=\sum_{k=0}^{+\infty}\frac{((t-k)^+)^k}{k!}.$$

Term-wise anti-transformation is correct since, if $t>0$ then the achieved series has general term with modulus dominated by the corresponding general term in Taylor series expansion of $\exp(t)$ (convergent to $H(t)e^t$), therefore the achieved series is convergent in the space of $\mathcal{L}$ transformable distributions (after multiplication times $e^{-\lambda t}$ with $\lambda>1$, the series converges in L^1).

$$y(t)=\mathcal{L}^{-1}\left\{\frac{1}{1-e^{-s}}\right\}*\mathcal{L}^{-1}\left\{\frac{1}{s-e^{-s}}\right\}=\left(\sum_{h=0}^{+\infty}\delta(t-h)\right)*\left(\sum_{k=0}^{+\infty}\frac{((t-k)^+)^k}{k!}\right)=$$

$$=\sum_{h,k=0}^{+\infty}\frac{((t-(h+k))^+)^k}{k!}$$

Notice that, for any fixed $t\in\mathbb{R}$ the series has a finite number of terms different from 0. □

Exercise 9.2. Evaluate the distributional derivative of the function

$$f(x)=H(x)\cos(x)+\ln|x|.$$

Solution - Due to linearity of differentiation, we can differentiate separately the two terms, then add them together.
Both terms are L^1_{loc} function, so we obtain:

$$\langle (H(x)\cos x)',\varphi\rangle=-\langle H(x)\cos x,\varphi'\rangle=$$

$$=-\int_{\mathbb{R}}H(x)\cos(x)\varphi'(x)\,dx=-\int_{\mathbb{R}+}\cos(x)\varphi'(x)\,dx,\qquad\forall\varphi\in\mathcal{D}(\mathbb{R}).$$

Integrating by parts

$$\langle (H(x)\cos x)',\varphi\rangle=\varphi(0)-\int_{\mathbb{R}}H(x)\sin(x)\varphi(x)\,dx=\langle\delta(x)-H(x)\sin(x),\varphi\rangle,\quad\forall\varphi\in\mathcal{D}(\mathbb{R}).$$

Now we prove that

$$D_x\ln|x|=\text{v.p.}\,\frac{1}{x}:=\lim_{\varepsilon\to 0_+}\int_{|x|>\varepsilon}\frac{\varphi(x)}{x}\,dx\quad\text{ in }\mathcal{D}'(\mathbb{R}).\tag{9.1}$$

In fact

$$\langle(\ln|x|)',\varphi\rangle=-\langle\ln|x|,\varphi'\rangle=-\int_{\mathbb{R}}\ln|x|\varphi'(x)\,dx=-\lim_{\varepsilon\to 0_+}\int_{|x|>\varepsilon}\ln|x|\varphi'(x)\,dx.$$

Integrating by parts and dealing separately with the two intervals $(-\infty,-\varepsilon)$ and (ε,∞), we obtain

$$\langle(\ln|x|)',\varphi\rangle=\lim_{\varepsilon\to 0_+}\int_{|x|>\varepsilon}\frac{\varphi(x)}{x}\,dx+\lim_{\varepsilon\to 0_+}\left(\ln(\varepsilon)\varphi(\varepsilon)-\ln(\varepsilon)\varphi(-\varepsilon)\right).$$

The second limit is 0 since $\varphi\in\mathcal{D}(\mathbb{R})$ entails $(\varphi(-\varepsilon)-\varphi(\varepsilon)\sim 2\varphi'(0)\varepsilon)$ as $\varepsilon\to 0_+$ and, taking into account of the elementary limit $\varepsilon\ln\varepsilon\to 0_+$ as $\varepsilon\to 0_+$, we deduce(9.1).
Summarizing, we have proved the identity

$$f'(x)=\delta(x)-H(x)\sin(x)+\text{v.p.}\,\frac{1}{x}\quad\text{ in }\mathcal{D}'(\mathbb{R}).$$ □

Exercise 9.3. Given

$$f(t)=\sum_{k=-\infty}^{+\infty}\delta''(t-k)$$

1) State precisely in which sense the series in the definition of f is convergent.
2) Ascertain whether f is a periodic distribution.
3) Evaluate the Fourier transform $\widehat{f}$.
4) Evaluate the Fourier series of f.
5) Clarify whether $\widehat{f}$ is a periodic distribution and explain the answer.

Solution - 1) For any $\psi\in\mathcal{S}(\mathbb{R})$ the numerical series $\sum_{k=-\infty}^{+\infty}\psi''(k)$ is absolutely convergent, therefore

$$\langle f,\psi\rangle=\sum_{k=-\infty}^{+\infty}\psi''(k)=\lim_{N,M\to+\infty}\sum_{k=-N}^{+M}\psi''(k).$$

Then the series $\sum_{k=-\infty}^{+\infty}\delta''(t-k)$ converges in $\mathcal{S}'(\mathbb{R})$, the space of tempered distributions.

2) f is a periodic distribution whose period is $T=1$, indeed

$$\langle f(t),\psi(t-1)\rangle=\sum_{k=-\infty}^{+\infty}\psi''(k-1)=\sum_{k=-\infty}^{+\infty}\psi''(k)=\langle f(t),\psi(t)\rangle\qquad\forall\psi\in\mathcal{S}'(\mathbb{R}),$$

that is $f(t+1)=f(t)$ in $\mathcal{S}'(\mathbb{R})$.

3) We use the formulas

$$\mathcal{F}\{\delta\}=1,\qquad\mathcal{F}\{u(t-k)\}(x)=\widehat{u}(x)e^{-ikx},\qquad\mathcal{F}\{u''(t)\}(x)=-x^2\widehat{u}(x).$$

The series $\sum_{k=-\infty}^{+\infty}\delta''(t-k)$ converges in $\mathcal{S}'(\mathbb{R})$, then it can be transformed term-wise and its transform turns out to be a series converging in $\mathcal{S}'(\mathbb{R})$ too:

$$\widehat{f}(x)=-x^2\sum_{k=-\infty}^{+\infty}e^{-ikx}=-2\pi x^2\sum_{k=-\infty}^{+\infty}\delta(x-2k\pi)=-8\pi^3\sum_{k=-\infty}^{+\infty}k^2\delta(x-2k\pi).$$

4) In the Table 10.8.5, concerning the Fourier series expansion of comb_T, we choose $T=1$ and obtain

$$\sum_{k=-\infty}^{+\infty}\delta(t-k)=\sum_{k=-\infty}^{+\infty}e^{i2k\pi t}=1+2\sum_{n=1}^{+\infty}\cos(2n\pi t)$$

where both series are convergent in $\mathcal{S}'(\mathbb{R})$; therefore term-wise differentiation is correct on both sides; performing such differentiation we obtain series convergent in $\mathcal{S}'(\mathbb{R})$:

$$f(t)=\sum_{k=-\infty}^{+\infty}\delta''(t-k)=-4\pi^2\sum_{k=-\infty}^{+\infty}k^2e^{i2k\pi t}=-8\pi^2\sum_{n=1}^{+\infty}n^2\cos(2n\pi t).$$

5) The (real and even) distribution $\widehat{f}$ is not periodic because its Fourier transform $\mathcal{F}\{\widehat{f}\}(t)$ fulfils

$$\mathcal{F}\{\widehat{f}\}(t) = 2\pi\mathcal{F}^{-1}\{\widehat{f}\}(-t) = 2\pi\mathcal{F}^{-1}\{\widehat{f}\}(t) = 2\pi f(t) = 2\pi \sum_{k=-\infty}^{+\infty} \delta''(t-k)$$

that is this transform does not verify (for any $T > 0, A, B$) a formula of the kind

$$\sum_{k=-\infty}^{+\infty} c_k \delta(t - k2\pi/T) \qquad \text{con } |c_k| \leq A(1+|k|)^B \qquad \forall k \in \mathbb{Z}$$

(necessary and sufficient condition in order to be the Fourier transform of a T periodic signal).
Alternative solution: the distribution $\widehat{f}$ is not periodic since

$$\widehat{f}(x-T) = -8\pi^3 \sum_{k=-\infty}^{+\infty} k^2\delta(x - 2k\pi - T) \neq -8\pi^3 \sum_{k=-\infty}^{+\infty} k^2\delta(x-2k\pi) = \widehat{f}(x) \qquad \forall T \neq 0.$$

□

Exercise 9.4. Set

$$u_n(t) = n\, p(nt), \qquad t \in \mathbb{R},\ n = 1, 2, 3, ...$$

where $p(t) = 1$ if $|t| < 1$, $p(t) = 0$ else.

1) Establish whether the sequence u_n has a.e. point-wise limit as $n \to +\infty$.

2) Ascertain whether the sequence u_n has limit as $n \to +\infty$ with respect to the strong convergence in $L^1(\mathbb{R})$.

3) Prove that u_n has a distributional limit.

4) Study "a priori" the properties of the Fourier transform $\widehat{u}_n$.

5) Evaluate $\widehat{u}_n$.

6) Find the solutions $w_n \in \mathcal{S}'(\mathbb{R})$ of the equation $-w_n'' + w_n = u_n$ in $\mathcal{S}'(\mathbb{R})$.

7) Evaluate the strong limit in $L^2(\mathbb{R})$ of w_n, as $n \to +\infty$.

Solution - 1) Set $u(t) \equiv 0$, then $u_n \to u$ a.e. (precisely for any $x \neq 0$) as $n \to +\infty$, since, for any fixed $t \neq 0$, we have $u_n(t) = 0$ if $n > 1/|t|$, that is eventually with respect to n.

2)

$$\|u_n - u\|_{L^1(\mathbb{R})} = \|u_n\|_{L^1(\mathbb{R})} = 2, \tag{9.2}$$

then there is no convergence in $L^1(\mathbb{R})$ (and no convergence in any L^p with $1 \leq p < +\infty$), as $n \to +\infty$; indeed the convergence to a function z in L^p with $1 \leq p < +\infty$ would imply the a.e. convergence to z of (at least) a subsequence, that is $z \equiv 0$ due to item 1): a contradiction with (9.2).

3) For any $\varphi \in \mathcal{D}(\mathbb{R})$, the integral mean theorem entails the existence of $t_n \in (-1/n, 1/n)$:

$$\langle u_n, \varphi \rangle = \int_{-1/n}^{1/n} n\varphi(t)\, dt = 2\varphi(t_n) \to 2\varphi(0).$$

Therefore $u_n \to 2\delta$ in $\mathcal{D}'(\mathbb{R})$.

4) u_n real even $\Rightarrow \widehat{u}_n$ real even;

$u_n \in L^1 \cap L^2(\mathbb{R}) \Rightarrow \widehat{u}_n \in C^0 \cap L^\infty \cap L^2(\mathbb{R}),\ \widehat{u}_n(\xi) \to 0,\ |\xi| \to +\infty;$

$t^k u_n \in L^1\ \forall k \in \mathbb{N} \Rightarrow \widehat{u}_n \in C^\infty(\mathbb{R})$, moreover $(\widehat{u}_n)^h \in L^\infty$ and infinitesimal at $\infty\,\forall h$;

$t^k u_n \in L^2(\mathbb{R})\ \forall k \in \mathbb{N} \Rightarrow (\widehat{u}_n)^h \in L^2(\mathbb{R})\ \forall h.$

Due to the Paley Wiener Theorem, u_n compactly supported implies that $\widehat{u}_n$ is an entire function, that is there exists an analytic extension of $\widehat{u}_n$ whose domain is the whole complex plane.
Eventually $u_n \notin C^0(\mathbb{R})$ implies $\widehat{u}_n \notin L^1(\mathbb{R})$; indeed otherwise the continuity of u_n would be a consequence of Riemann-Lebesgue Lemma applied to the inverse Fourier transform.

5) It is known that $\widehat{p}(\xi) = 2\dfrac{\sin(\xi)}{\xi}$ and $\mathcal{F}\{u(at)\} = \dfrac{1}{a}\widehat{u}\left(\dfrac{\xi}{a}\right)$ for any $\mathcal{F}$ transformable u, we have

$$\widehat{u}_n(\xi) = 2n\frac{\sin(\xi/n)}{\xi}.$$

6) Applying $\mathcal{F}$ to both sides of the equation we obtain

$$(1+\xi^2)\widehat{w}_n(\xi) = \widehat{u}_n(\xi)$$

and, since $1+\xi^2 \neq 0$ for any $\xi \in \mathbb{R}$, we can divide:

$$\widehat{w}_n(\xi) = \frac{\widehat{u}_n(\xi)}{\xi^2+1} = \mathcal{F}\left\{\frac{e^{-|t|}}{2} * u_n\right\}(\xi).$$

Hence, anti-transforming

$$w_n(t) = \frac{e^{-|t|}}{2} * (np(nt)) = \frac{n}{2}\int_{t-1/n}^{t+1/n} e^{-|\tau|}\,d\tau =$$

$$\begin{cases} \dfrac{n}{2}\left(e^{t+1/n} - e^{t-1/n}\right) = n\sinh(1/n)e^t & \text{se } t \leq -1/n, \\ \dfrac{n}{2}\left(e^{t} - e^{t-1/n} + e^{1/n-t} - e^{-t}\right) = n\left(\sinh(t) - \sinh(t-1/n)\right) & \text{se } -1/n < t < 1/n, \\ \dfrac{n}{2}\left(e^{1/n-t} - e^{-1/n-t}\right) = n\sinh(1/n)e^{-t} & \text{se } t \geq 1/n. \end{cases}$$

7) The strong convergence in $L^2(\mathbb{R})$ holds true: $w_n \to e^{-|t|}$ in $L^2(\mathbb{R})$.
In fact, due to the Plancherel identity

$$2\pi\|w_n(t) - e^{-|t|}\|^2_{L^2(\mathbb{R})} = \|\mathcal{F}\{w_n(t) - e^{-|t|}\}(\xi)\|^2_{L^2(\mathbb{R})} =$$

$$= \left\|\frac{2}{\xi^2+1}\left(n\frac{\sin(\xi/n)}{\xi} - 1\right)\right\|^2_{L^2(\mathbb{R})} = \int_{\mathbb{R}} \frac{4}{(\xi^2+1)^2}\left(n\frac{\sin(\xi/n)}{\xi} - 1\right)^2 d\xi\,,$$

so, by the Lebesgue dominated convergence Theorem, the last integral tends to 0 (the integrand converges point-wise to 0 and is dominated by $16/(1+\xi^2)^2$), hence the claim. □

Exercise 9.5. Given

$$G(x) = \arctan x\,, \qquad x \in \mathbb{R},$$

answer the subsequent queries and justify all the replies.

a) "A priori" analyze the Fourier transform of $\widehat{G}$ of G.

b) Evaluate $\widehat{G}$.

c) Evaluate $\quad \dfrac{1}{2}\dfrac{1}{1+x^2} * \operatorname{sign}(x)\,.$

d) Given $\psi \in L^1(\mathbb{R})$, write the Fourier transform of $\widehat{\Psi}$ of

$$\Psi(x) = \int_{-\infty}^{x} \psi(t)\,dt$$

as a function of $\widehat{\psi}$.

e) Deduce the answer to query (b) from the identity proved in answering to query (d).

Solution - **a)** "A priori" analysis of G :

$G \in L^\infty(\mathbb{R}) \subset \mathcal{S}'(\mathbb{R}) \quad \Rightarrow \quad G$ is an $\mathcal{F}$ transformable distribution;

G is an odd real function $\quad \Rightarrow \quad \widehat{G}$ is an imaginary odd function ;

$G \notin L^1 \quad \Rightarrow \quad$ we have no information on the continuity of $\widehat{G}$, nevertheles we know $\widehat{G} \notin \mathcal{V}$, (nel calcolo in (b) si verifica che $\widehat{G} \notin C^0$);

$G \notin L^2 \quad \Rightarrow \quad \widehat{G} \notin L^2$;

$G' = 1/(1+x^2) \stackrel{\text{def}}{=} g \in L^1 \cap L^2(\mathbb{R}) \quad \Rightarrow \quad \xi\widehat{G}(\xi) \in C^0 \cap L^\infty \cap L^2(\mathbb{R})$, $\displaystyle\lim_{\xi\to\infty} \xi\widehat{G}(\xi) = 0$;

$G^{(k)} = g^{(k-1)} = \dfrac{d^k(1/(1+x^2))}{dx^{k-1}} \in L^1 \cap L^2(\mathbb{R})\ \forall k \geq 1 \quad \Rightarrow \quad \xi^k\widehat{G}(\xi) \in C^0 \cap L^\infty \cap L^2(\mathbb{R})\ \forall k \in \mathbb{N},\ k \geq 1$; da cui $\widehat{G}(\xi) = o(|\xi|^{-k})$ per $|\xi| \to +\infty$, $\forall k$;

$\forall k \in \mathbb{N}\ \exists h \in \mathbb{N}:\ x^k G^{(h)} \in L^1 \cap L^2(\mathbb{R}) \quad \Rightarrow \quad \left(\dfrac{d}{d\xi}\right)^k (\xi^h \widehat{G}) \in C^0 \Rightarrow \widehat{G} \in C^\infty(\mathbb{R} \setminus \{0\})$.

b) The identity

$$G'(x) = \frac{1}{1+x^2} \stackrel{def}{=} g(x)\,, \quad x \in \mathbb{R}\,, \qquad \widehat{g}(\xi) = \pi e^{-|\xi|}\,, \quad \xi \in \mathbb{R}\,,$$

leads to this non homogeneous division problem

$$i\,\xi\,\widehat{G}(\xi) = \widehat{g}(\xi) = \pi\,e^{-|\xi|}\,,$$

whose general solution is

$$\widehat{G}(\xi) = -i\pi\,e^{-|\xi|}\ \text{v.p.}\,\frac{1}{\xi} + C\,\delta(\xi)\,.$$

From the "a priori" analysis we know that $\widehat{G}$ is odd, then

$$\widehat{G}(\xi) \;=\; -i\,\pi\, e^{-|\xi|}\,\text{v.p.}\frac{1}{\xi}\,.$$

We remark that actually $\widehat{G}$ is neither a continuous nor an L^2 function.

c) We have to compute the convolution of $1/(1+x^2) \in L^1$ and $(1/2)\,\text{sign}(x) \in L^1_{loc} \cap L^\infty$, so we will obtain an $L^1_{loc} \cap L^\infty$ function.
In the computation we take into account that $1/(1+x^2)$ is an even function and $\arctan(x)$ is an odd function:

$$\frac{1}{2}\,\text{sign}(x) * \frac{1}{1+x^2} = \frac{1}{2}\int_{\mathbb{R}} \text{sign}(x-y)\frac{1}{1+y^2}\,dy = \frac{1}{2}\left(\int_{-\infty}^{-x} - \int_{x}^{+\infty} + \int_{-x}^{x}\right)\frac{1}{1+y^2}\,dy =$$

$$= \frac{1}{2}\int_{-x}^{x}\frac{1}{1+y^2}\,dy = \frac{1}{2}\left(\arctan(x) - \arctan(-x)\right) = \arctan(x)\,.$$

Alternatively, exploiting the item (b), we could evaluate

$$\frac{1}{2}\,\text{sign}(x) * \frac{1}{1+x^2} = \mathcal{F}^{-1}\,\mathcal{F}\left\{\frac{1}{1+x^2} * \frac{1}{2}\,\text{sign}\right\} = \mathcal{F}^{-1}\left\{\mathcal{F}\left\{\frac{1}{1+x^2}\right\}\,\mathcal{F}\left\{\frac{1}{2}\,\text{sign}\right\}\right\} =$$

$$= \mathcal{F}^{-1}\left\{-i\,\pi\, e^{-|\xi|}\;\text{v.p.}\,\frac{1}{\xi}\right\} = \arctan(x)\,.$$

d) $\psi \in L^1 \Rightarrow \widehat{\psi} \in C^0 \cap L^\infty$. By the identities

$$\Psi(x) = \int_{-\infty}^{x}\psi(t)\,dt = \int_{\mathbb{R}} H(x-t)\,\psi(t)\,dt \;=\; (H * \psi)(x)\,,$$

$$\widehat{H}(\xi) = \pi\,\delta(\xi) - i\;\text{v.p.}\,\frac{1}{\xi}\,,$$

where H denotes the Heaviside step function, we obtain

$$\widehat{\Psi}(\xi) \;=\; \widehat{\psi}(\xi)\,\widehat{H}(\xi) = \pi\widehat{\psi}(\xi)\delta(\xi) \;-\; i\,\widehat{\psi}(\xi)\,\text{v.p.}\,\frac{1}{\xi} \;= \pi\widehat{\psi}(0)\delta(\xi) \;-\; i\,\widehat{\psi}(\xi)\,\text{v.p.}\,\frac{1}{\xi} \;=$$

$$= \pi\left(\int_{\mathbb{R}}\psi\right)\delta(\xi) \;-\; i\,\widehat{\psi}(\xi)\,\text{v.p.}\,\frac{1}{\xi}\,.$$

e) By choosing $\psi(x) = g(x) = 1/(1+x^2)$ and $\Psi = G$ in (d), we transform the identity

$$G(x) \;=\; \arctan x \;=\; \int_{0}^{x}\frac{1}{1+t^2}\,dt \;=\; -\frac{\pi}{2} + \int_{-\infty}^{x}\frac{1}{1+t^2}\,dt$$

and we obtain

$$\widehat{G}(\xi) \;=\; -\frac{\pi}{2}(2\pi\delta(\xi)) \;+\; \pi\left(\int_{\mathbb{R}} g\right)\delta(\xi) \;-\; i\,\widehat{\psi}(\xi)\,\text{v.p.}\,\frac{1}{\xi} \;=\; -\,i\,\widehat{\psi}(\xi)\,\text{v.p.}\,\frac{1}{\xi}\,.$$

□

Exercise 9.6. Consider the equation

$$\varepsilon^4 v_\varepsilon'''' + v_\varepsilon = \delta' \qquad v_\varepsilon \in \mathcal{S}'(\mathbb{R}) \tag{9.3}$$

where ε is a parameter, $\varepsilon > 0$.
(1) prove that the problem (9.3) has a unique solution in $\mathcal{S}'(\mathbb{R})$;
(2) evaluate v_ε;
(3) study the regularity of v_ε;
(4) prove that $\|v_\varepsilon\|_{L^2(\mathbb{R})} \sim \varepsilon^{-3/2}$ per $\varepsilon \to 0_+$;
(5) evaluate $\lim\limits_{\varepsilon \to 0_+} v_\varepsilon$ in $\mathcal{S}'(\mathbb{R})$.

Solution - (1),(2) - Since $1 + \varepsilon^4\xi^4 \neq 0$, by applying the Fourier transform method, we obtain

$$\widehat{v_\varepsilon} = \frac{i\xi}{1 + \varepsilon^4\xi^4}\,.$$

the injectivity of $\mathcal{F}$ in $\mathcal{S}'(\mathbb{R})$ entails that the unique solution is

$$v_\varepsilon(x) = \mathcal{F}^{-1}\left(\frac{i\xi}{1 + \varepsilon^4\xi^4}\right) = \frac{1}{2\pi}\frac{d}{dx}\left(\int \frac{e^{i\xi x}}{1 + \varepsilon^4\xi^4}\, d\xi\right).$$

the analytic function of complex variable $z \mapsto \dfrac{e^{izx}}{1 + \varepsilon^4 z^4}$ has four simple poles:

$$z_j = \varepsilon^{-1} e^{i(-\pi/4 + j\pi/2)} \qquad j = 1, 2, 3, 4,$$

Using the Jordan Lemma with $x > 0$ (Lemma 1.34 (i)) and the de L'Hôpital rule

$$\int \frac{e^{i\xi x}}{1 + \varepsilon^4\xi^4}\, d\xi = 2\pi i \sum_{j=1,2} \operatorname{res}\left(\frac{e^{izx}}{1 + \varepsilon^4 z^4}, z_j\right) =$$

$$= 2\pi i \left[\frac{e^{ix\varepsilon^{-1}(\sqrt{2}/2 + i\sqrt{2}/2)}}{4\varepsilon^4\varepsilon^{-3}\left(-\frac{\sqrt{2}}{2} + i\frac{\sqrt{2}}{2}\right)} + \frac{e^{ix\varepsilon^{-1}(-\sqrt{2}/2 + i\sqrt{2}/2)}}{4\varepsilon^4\varepsilon^{-3}\left(\frac{\sqrt{2}}{2} + i\frac{\sqrt{2}}{2}\right)}\right],$$

where we have took into account

$$\begin{cases} z_1 = \varepsilon^{-1}\left(\dfrac{\sqrt{2}}{2} + i\dfrac{\sqrt{2}}{2}\right) = \varepsilon^{-1} e^{i\pi/4} \\[2ex] z_2 = \varepsilon^{-1}\left(-\dfrac{\sqrt{2}}{2} + i\dfrac{\sqrt{2}}{2}\right) = \varepsilon^{-1} e^{3\pi i/4}\,, \end{cases} \qquad \begin{cases} (\varepsilon z_1)^3 = \varepsilon z_2 \\[2ex] (\varepsilon z_2)^3 = \varepsilon z_1\,. \end{cases}$$

Then:

$$\int \frac{e^{i\xi x}}{1 + \varepsilon^4\xi^4}\, d\xi = \frac{\pi i}{\varepsilon\sqrt{2}} e^{-x/(\varepsilon\sqrt{2})}\left(\frac{e^{ix/(\varepsilon\sqrt{2})}}{i - 1} + \frac{e^{-ix/(\varepsilon\sqrt{2})}}{i + 1}\right) =$$

$$= \frac{\pi}{\varepsilon\sqrt{2}} e^{-x/(\varepsilon\sqrt{2})}\left(\cos\frac{x}{\varepsilon\sqrt{2}} + \sin\frac{x}{\varepsilon\sqrt{2}}\right) \qquad \text{se } x > 0.$$

Since $\dfrac{1}{1 + \varepsilon^4\xi^4}$ is an even function that belongs to $L^1(\mathbb{R})$ we deduce

$$\int \frac{e^{i\xi x}}{1+\varepsilon^4\xi^4}\,d\xi = \frac{\pi}{\varepsilon\sqrt{2}} e^{-|x|/(\varepsilon\sqrt{2})}\left(\cos\frac{x}{\varepsilon\sqrt{2}}+\sin\frac{|x|}{\varepsilon\sqrt{2}}\right) \qquad \forall x\in\mathbb{R}\,.$$

Therefore:

$$v_\varepsilon(x) = \frac{1}{2\sqrt{2}\varepsilon} e^{-|x|/(\varepsilon\sqrt{2})}\left[-\frac{\operatorname{sign}(x)}{\varepsilon\sqrt{2}}\left(\cos\frac{x}{\varepsilon\sqrt{2}}+\sin\frac{|x|}{\varepsilon\sqrt{2}}\right)+\right.$$

$$\left. + \frac{1}{\varepsilon\sqrt{2}}\left(-\sin\frac{x}{\varepsilon\sqrt{2}}+\operatorname{sign}(x)\cos\frac{x}{\varepsilon\sqrt{2}}\right)\right] = -\frac{e^{-|x|/(\varepsilon\sqrt{2})}}{2\,\varepsilon^2}\sin\frac{x}{\varepsilon\sqrt{2}}\,.$$

(3) v_ε is an odd function, continuous, vanishing at infinity, as one could foresee, due to the fact that $\widehat{v_\varepsilon}$ is odd and belongs to $L^1(\mathbb{R})$.
We have $v_\varepsilon, \widehat{v_\varepsilon} \in L^2(\mathbb{R})$ and

$$|\xi|^s|\widehat{v_\varepsilon}(\xi)| \in L^2 \Leftrightarrow 2s+2-8<-1 \Leftrightarrow s<\frac{5}{2}\,.$$

Hence

$$v_\varepsilon \in H^s(\mathbb{R}) \qquad \forall s<\frac{5}{2}.$$

Moreover v_ε has fast decay at infinity and belongs to $C^\infty(\mathbb{R}\setminus\{0\})$ as one could foresee without computing it, due to the prperties:

- $\widehat{v_\varepsilon} \in C^\infty\,,\ \widehat{v_\varepsilon}^{(h)} \in L^1\ \forall h\in\mathbb{N}\,,$
- $\forall n\ \exists\, h\ :\ \left(\xi^n\,(d^h/d\xi^h)\widehat{v_\varepsilon}\right) \in L^1(\mathbb{R}).$

The exponential decay of v_ε a $\pm\infty$ ($|v_\varepsilon| \le \frac{1}{\varepsilon^2}e^{-|x|/(\varepsilon\sqrt{2})}$) is coherent with the existence of an analytic extension $\widehat{v_\varepsilon}$ in the strip $\left\{\operatorname{Im} z<\frac{1}{\varepsilon\sqrt{2}}\right\}$.

(4)

$$2\pi\|v_\varepsilon\|^2_{L^2} = \|v_\varepsilon\|^2_{L^2} = \int \frac{\xi^2}{(1+\varepsilon^4\xi^4)^2}\,d\xi = \int \frac{t^2\varepsilon^{-2}}{(1+t^4)^2}\,\varepsilon^{-1}\,dt = C\,\varepsilon^{-3} \quad \forall\varepsilon>0$$

where $C = \displaystyle\int_{\mathbb{R}} \frac{t^2}{(1+t^4)^2}\,dt\,.$

So $\|v_\varepsilon\|_{L^2} \ = \sqrt{\dfrac{C}{2\pi}}\ \varepsilon^{-3/2}.$

(5) $\widehat{v_\varepsilon} \to i\xi$ in $\mathcal{S}'(\mathbb{R})$ as $\varepsilon\to 0_+$; indeed the Lebesgue Theorem (Theorem 2.5) gives:

$$\int \frac{i\xi}{1+\varepsilon^4\xi^4}\,\psi(\xi)\,d\xi \to \int i\xi\psi \qquad \forall\psi\in\mathcal{S}(\mathbb{R})$$

Then $v_\varepsilon \to \delta' = \mathcal{F}^{-1}(i\xi)$ in $\mathcal{S}'(\mathbb{R})$, since the inverse transform preserves the convergence in $\mathcal{S}'(\mathbb{R})$. □

Exercise 9.7. Given $n \in \mathbb{N}$ and this differential equation in the unknown u_n

$$-n^2 u_n'' - 2\,i\,n\,u_n' + 2\,u_n = n^2\,\delta' \qquad \text{in } \mathcal{S}'(\mathbb{R}),$$

a) evaluate $\widehat{u}_n$;
b) evaluate u_n ;
c) ascertain whether u_n is convergent in $L^1(\mathbb{R})$, $L^2(\mathbb{R})$ and $\mathcal{S}'(\mathbb{R})$, as $n \to +\infty$.

Solution - a) We set $\widehat{u}_n(\xi) = \mathcal{F}\{u_n(x)\}$ and use the identities $\widehat{v'} = i\,\xi\,\widehat{v}$ and $\widehat{\delta} = 1$ to transform the equation:

$$n^2\,\xi^2\,\widehat{u}_n + 2\,n\,\xi\,\widehat{u}_n + 2\,\widehat{u}_n = i\,n^2\,\xi,$$

since $n^2\,\xi^2 + 2\,n\,\xi + 2 \neq 0\ \forall \xi \in \mathbb{R}$, we can divide both sides of the equation:

$$\widehat{u}_n(\xi) = \frac{i\,n^2\,\xi}{n^2\,\xi^2 + 2\,n\,\xi + 2}\,.$$

b) After setting $w(x) = e^{-|x|}$, we find $\widehat{w}(\xi) = \frac{2}{1+\xi^2}$ and

$$\widehat{u}_n(\xi) = \frac{1}{2}\,n^2\,i\,\xi\,\widehat{w}(n\xi + 1) = \frac{n}{2}\,i\,\xi\,\Big(n\,\widehat{w}\big(n(\xi + 1/n)\big)\Big)$$

hence, using $\mathcal{F}\{v(x/n)\} = n\,\widehat{v}(n\xi)$ and $\mathcal{F}\left\{e^{-ix}v(x)\right\} = \widehat{v}(\xi + 1)$, we obtain:

$$\begin{aligned}
u_n(x) &= \frac{n}{2}\frac{d}{dx}\left(w(x/n)\,e^{-ix/n}\right) = \\
&= \frac{n}{2}\frac{d}{dx}\left(e^{-|x|/n}\,e^{-ix/n}\right) = \frac{n}{2}\frac{d}{dx}\exp(-|x|/n - ix/n) = \\
&= -\frac{n}{2}\left(\frac{\operatorname{sign}(x)}{n} + \frac{i}{n}\right)\exp(-|x|/n - ix/n) = -\frac{1}{2}\left(\operatorname{sign}(x) + i\right)e^{-|x|/n - ix/n}\,.
\end{aligned}$$

b') Alternatively we can anti-transform by exploiting complex variable techniques:
Setting

$$w_n(x) = \mathcal{F}^{-1}\left\{\frac{1}{n^2\,\xi^2 + 2\,n\,\xi + 2}\right\}$$

we obtain $\widehat{w}_n \in L^1(\mathbb{R}) \cap L^2(\mathbb{R})$ and $|\widehat{w}_n(\xi)| \leq \dfrac{2/n^2}{\xi^2}\ \forall \xi :\ |\xi| \geq 2/n\,.$

Moreover $\widehat{w}_n$ has only 2 isolated singularities (simple poles) at $\xi = z_{1,2} = (-1 \pm i)/n$.
Therefore we can apply the Jordan Lemma in the upper half-plane in order to evaluate $w_n(x)$ if $x > 0$, (in the lower half-plane in order to evaluate $w_n(x)$ if $x < 0$).

$$w_n(x) = \mathcal{F}^{-1}\left\{\frac{1}{n^2\,\xi^2 + 2\,n\,\xi + 2}\right\} = \frac{1}{2\pi}\int_{\mathbb{R}}\frac{e^{i\xi x}}{n^2\,\xi^2 + 2\,n\,\xi + 2}\,d\xi =$$

$$\underset{\boxed{\text{Jordan Lemma } (x>0)}}{=}\ \frac{2\pi i}{2\pi}\operatorname{res}\left(\frac{e^{i\xi x}}{n^2\,\xi^2 + 2\,n\,\xi + 2},\ \xi = z_1\right) =$$

$$\underset{\boxed{\text{Hôpital}}}{=}\ i\,\frac{e^{ixz_1}}{2n^2 z_1 + 2n} = i\,\frac{i\frac{x}{n}(-1+i)}{2n(-1+i+1)} = \frac{1}{2n}\exp(-(1+i)x/n)\,.$$

$$w_n(x) = \mathcal{F}^{-1}\left\{\frac{1}{n^2\,\xi^2 + 2\,n\,\xi + 2}\right\} = \frac{1}{2\pi}\int_{\mathbb{R}} \frac{e^{i\xi x}}{n^2\,\xi^2 + 2\,n\,\xi + 2}\,d\xi =$$

$$\overset{\boxed{\text{Jordan Lemma } (x<0)}}{=} \quad -\frac{2\pi i}{2\pi}\,\mathrm{res}\left(\frac{e^{i\xi x}}{n^2\,\xi^2 + 2\,n\,\xi + 2},\ \xi = z_2\right) =$$

$$\overset{\boxed{\text{Hôpital}}}{=} \quad -i\,\frac{e^{ixz_2}}{2n^2 z_2 + 2n} = -i\,\frac{i\frac{x}{n}(-1-i)}{2n(-1-i+1)} = \frac{1}{2n}\,\exp((1-i)x/n)\,.$$

The value $w_n(0) = \frac{1}{2n}$ is deduced by continuity ($\widehat{w}_n \in L^1(\mathbb{R})$).
Summarizing $w_n(x) = \dfrac{1}{2n}\,e^{-|x|/n}\,e^{-ix/n}$ and, taking into account of $\widehat{u}_n(\xi) = i\,n^2\,\xi\,\widehat{w}_n(\xi)$, we obtain:

$$u_n(x) = -\frac{1}{2}\,(\mathrm{sign}(x) + i)\,e^{-|x|/n}\,e^{-ix/n}\,.$$

c) u_n converges to $u(x) := -\dfrac{i}{2} - \dfrac{1}{2}\,\mathrm{sign}(x)$ point-wise (for any $x \neq 0$) and uniformly in any set of the kind $(-L, L)\setminus\{0\}$ with finite L.
Since $u \notin L^1(\mathbb{R})$ and $u \notin L^2(\mathbb{R})$, we conclude that u_n can converge to a limit w neither in $L^1(\mathbb{R})$ nor in $L^2(\mathbb{R})$ (since otherwise at least a subsequence would converge a.e., see Theorem 2.4).
The sequence u_n converges to $u(x) = -\dfrac{i}{2} - \dfrac{1}{2}\,\mathrm{sign}(x)$ in $\mathcal{S}'(\mathbb{R})$ (hence in $\mathcal{D}'(\mathbb{R})$ too), indeed the uniform convergence of u_n in any set $(-L, L)\setminus\{0\}$ with finite L together with the boundedness property ($|u_n(x)| \leq 1$, for any x) implies:

$$\lim_{n\to+\infty}\int_{\mathbb{R}} u_n\psi\,dx = \int_{\mathbb{R}} u\,\psi\,dx \quad \forall\psi \in \mathcal{S}'(\mathbb{R})\,.$$

□

Exercise 9.8. Given the square open set $\Omega = (0,\pi)\times(0,\pi) \subset \mathbb{R}^2$, solve this problem in the unknowns $\lambda \in \mathbb{C}$ and $u \in C^2(\Omega)\cap C^0(\overline{\Omega})$:

$$\begin{cases} -\Delta u = \lambda\,u & \Omega\,, \\ \quad\; u = 0 & \partial\Omega\,. \end{cases} \tag{9.4}$$

Solution - Obviously $u \equiv 0$ together with any complex number λ provides a trivial solution. The interesting case corresponds to nontrivial solutions u, if any; and this nontrivial solutions u can be found only for suitable values of λ. This values are called eigenvalues of the operator $-\Delta$, whereas the related solutions u are called eigenfunctions of the operator $-\Delta$ in the square Ω, with homogeneous Dirichlet condition. These names refer to the analogy with the problem of finding the eigenvalues and eigenvectors of a matrix, in Linear Algebra.

We state and solve the *eigenvalue problem* for the operator $-\Delta$ with homogeneous Dirichlet condition:

find $\lambda \in \mathbb{C}$ and $u \in C^2(\Omega)\cap C^0(\overline{\Omega})$ s.t. $u \not\equiv 0$ and (9.4) holds true.

The geometry of the domain suggests to exploit the separation of variables method.
Possible solutions of the kind $X(x)Y(y)$ must fulfil

$$-X''Y - XY'' = \lambda\,XY\,,$$

that is a constant $\mu \in \mathbb{C}$ ought to exist s.t.

$$-\frac{X''}{X} = \frac{Y''}{Y} + \lambda = \mu \,.$$

Here constant stands for a number independent of x and y. This entails two uncoupled problems:

$$\begin{cases} X'' + \mu X = 0 \\ X(0) = X(\pi) = 0\,, \end{cases} \quad \text{whose solutions are} \quad \begin{cases} \mu = n^2\,, & n = 1,2,\dots \\ X_n(x) = \sin(nx) & n = 1,2,\dots\,; \end{cases}$$

$$\begin{cases} Y'' + (\lambda - \mu)Y = 0 \\ Y(0) = Y(\pi) = 0\,, \end{cases} \quad \text{whose solutions are} \quad \begin{cases} \lambda - \mu = m^2\,, & m = 1,2,\dots \\ Y_m(x) = \sin(mx) & m = 1,2,\dots\,. \end{cases}$$

Of course any multiple of an eigenfunction is an eigenfunction too.
Moreover any eigenvalue of this particular eigenvalue problem is of the sum of two squared natural numbers

$$\lambda_{n,m} = n^2 + m^2\,, \qquad n, m = 1, 2, \dots\,,$$

and the eigenfunctions related to $\lambda_{n,m}$ are the multiples of

$$u_{n,m} = \sin(nx)\,\sin(my)\,.$$

We emphasize that the first eigenvalue $\lambda_{1,1}$ is simple, that is the space of its eigenfunctions has dimension 1. There are other eigenvalues that are simple too; but this is not true for all of them. For instance $\lambda = 50$ has multiplicity 3, indeed $50 = 1 + 49 = 25 + 25 = 49 + 1$, hence it corresponds to the three linearly independent eigenfunctions:

$$\sin(x)\sin(7y) \qquad \sin(5x)\sin(5y) \qquad \sin(7x)\sin(y)\,.$$

Any eigenvalue has finite multiplicity. □

Exercise 9.9. 1) Examine the issues of existence and uniqueness about the solution $u = u(x, y)$ of the subsequent problem, by formally computing the series expansion of the solution via separation of variables,

$$\begin{cases} -\Delta u = 0 & \Omega \\ \gamma u + \dfrac{\partial u}{\partial \nu} = g(\vartheta) & \partial\Omega\,, \end{cases} \tag{9.5}$$

where $\Omega = \{(x,y) \in \mathbb{R}^2 : \ x^2 + y^2 < 1\}$, $\vartheta \in [0, 2\pi]$, $x = \varrho\cos(\vartheta)$, $y = \varrho\sin(\vartheta)$, ν is the outward normal on the boundary $\partial\Omega$, $\gamma \neq 0$ is a real parameter and g is a given 2π periodic distribution.

2) Study the regularity of the solution u.

Solution - 1) This is a Robin problem, then the solution u (if it exists) is unique in the set $\{u \in C^2(\Omega) \cap C^1(\overline{\Omega})\}$ (it can be proved as in Exercise 8.23).
Since we do not know "a priori" the regularity of the solution (this regularity depends on the regularity of datum g), we start by formal computations.
we expand g in Fourier series: there exists a sequence of tempered coefficients α_n, β_n (that is, they fulfil: $\exists A,\, B :\ |\alpha_n| + |\beta_n| \leq A(1+n)^B \ \ \forall n \in \mathbb{N}$) s.t.

$$g(\vartheta) = \frac{\alpha_0}{2} + \sum_{n=1}^{+\infty} \big(\alpha_n \cos(n\vartheta) + \beta_n \sin(n\vartheta) \big)$$

and the series, defined for any $\vartheta \in \mathbb{R}$, converges to g in $\mathcal{S}'(\mathbb{R})$.
We express the 2-d Laplace operator in polar coordinates (Table 10.6.1): $\Delta = \partial_{\varrho\varrho} + (1/\varrho)\partial_\varrho + (1/\varrho^2)\partial_{\vartheta\vartheta}$. The separation of variables method provides the list of all harmonic functions of the kind $R(\varrho)\varphi(\vartheta)$, proceeding as in Section 8.2.3.2. Then starting from their linear combinations we find this series, formally representing the solution:

$$u(\varrho,\vartheta) \;=\; \frac{a_0}{2} \;+\; \sum_{n=1}^{+\infty} \varrho^n \big(a_n \cos(n\vartheta) + b_n \sin(n\vartheta) \big) \;.$$

We notice that, due to the condition that the coefficients are tempered, the formal series converges point-wise all over Ω and uniformly with respect to the variables x and y in every compact subset of Ω, that is $\rho \leq R$ for any fixed $R < 1$. The same properties hold true for any partial derivative. Then partial sums of the series converge point-wise in Ω and uniformly on compact subsets to a $C^2(\Omega)$ function; this sum of the series is an harmonic function in Ω, in particular it is real analytic in Ω.
However the sum u of the series could be highly irregular at the boundary. We postpone the regularity analysis up to the boundary, which is a relevant step in order to give a precise meaning to the boundary condition.
We differentiate formally (the evaluation is correct in the interior, but formal at the boundary!), we obtain

$$\left.\frac{\partial u}{\partial \nu}\right|_{\varrho=1} \;=\; \left.\frac{\partial u}{\partial \varrho}\right|_{\varrho=1} \;=\; \sum_{n=1}^{+\infty} \Big(n\, a_n \cos(n\vartheta) \;+\; n\, b_n \sin(n\vartheta) \Big)$$

and imposing the boundary condition ($\varrho = 1$):

$$\frac{\gamma a_0}{2} + \sum_{n=1}^{+\infty} \Big((\gamma+n) a_n \cos(n\vartheta) + (\gamma+n) b_n \sin(n\vartheta) \Big) = \frac{\alpha_0}{2} + \sum_{n=1}^{+\infty} \Big(\alpha_n \cos(n\vartheta) + \beta_n \sin(n\vartheta) \Big).$$

We deduce a system of infinitely many uncoupled equations in the unknowns a_n and b_n :

$$\begin{cases} \gamma\, a_0 \;=\; \alpha_0 \\ (\gamma+n)\, a_n \;=\; \alpha_n \;, \quad n = 1, 2, \ldots \\ (\gamma+n)\, b_n \;=\; \beta_n \;, \quad n = 1, 2, \ldots \end{cases}$$

- if $\gamma \neq -n \;\forall n = 0, 1, \ldots$ then there exists a unique solution: $\begin{cases} a_n = \dfrac{\alpha_n}{\gamma+n} & n = 0, 1, 2, \ldots \\ b_n = \dfrac{\beta_n}{\gamma+n} & n = 1, 2, \ldots \;; \end{cases}$
  and the formal solution that fulfils (formally) the boundary condition too is:

$$u(\varrho,\vartheta) \;=\; \frac{\alpha_0}{2\gamma} \;+\; \sum_{n=1}^{+\infty} \varrho^n \left(\frac{\alpha_n}{\gamma+n} \cos(n\vartheta) + \frac{\beta_n}{\gamma+n} \sin(n\vartheta) \right) \;.$$

- if $\exists\, \widetilde{n} \in \{0, 1, 2, \ldots\}$ s.t. $\gamma = -\widetilde{n}$, then there are two cases:
  if $\alpha_{\widetilde{n}} \neq 0$ or $\beta_{\widetilde{n}} \neq 0$, then there are no solutions;
  if $\alpha_{\widetilde{n}} = \beta_{\widetilde{n}} = 0$, then there are infinitely many solutions:

$$u(\varrho,\vartheta) \;=\; \frac{\alpha_0}{2\gamma} + C\, \cos(\widetilde{n}\vartheta) + D\, \sin(\widetilde{n}\vartheta) + \sum_{n \neq \widetilde{n}} \varrho^n \left(\frac{\alpha_n}{\gamma+n} \cos(n\vartheta) + \frac{\beta_n}{\gamma+n} \sin(n\vartheta) \right) \;.$$

2) From now on we assume that the conditions for the existence of solutions are fulfilled. The interior regularity has been clarified in item 1): u is real analytic in Ω, without additional assumption on g.
To analyze the regularity up to the boundary, which depends on the regularity of datum g, we refer to the Sobolev spaces of periodic functions (Definition 5.14 and Theorem 5.17).
If $g \in L^2_{2\pi} = \{ g \in L^2_{loc}(\mathbb{R}) : \; g \text{ is } 2\pi \text{ periodic } \}$, then

$$\sum_{n=1}^{+\infty}\left(|\alpha_n|^2+|\beta_n|^2\right) < +\infty \qquad \text{and hence} \qquad \sum_{n=1}^{+\infty}\left(|n\,a_n|^2+|n\,b_n|^2\right) < +\infty;$$

therefore, due to the inequalities

$$|a_n| \leq |n\,a_n|^2+\frac{1}{n^2}\,, \qquad |b_n| \leq |n\,b_n|^2+\frac{1}{n^2}\,,$$

the formal series of u converges uniformly in $\overline{\Omega}$ to the solution that turns out to be continuous in $\overline{\Omega}$.

If $g \in H^1_{2\pi} = \{g \in L^2_{loc}(\mathbb{R}) :\ g \text{ is } 2\pi \text{ periodic}\ ,\ g' \in L^2_{loc}(\mathbb{R})\}$, then

$$\sum_{n=1}^{+\infty}\left(|n\,\alpha_n|^2+|n\,\beta_n|^2\right) < +\infty \qquad \text{so} \qquad \sum_{n=1}^{+\infty}\left(|n^2\,a_n|^2+|n^2\,b_n|^2\right) < +\infty;$$

therefore, due to the inequalities

$$|n\,a_n|^2 \leq |n^2\,a_n|^2+\frac{1}{n^2}\,, \qquad |n\,b_n|^2 \leq |n^2\,b_n|^2+\frac{1}{n^2}\,,$$

the formal series of first partial derivatives $\partial u/\partial x$ and $\partial u/\partial y$ converge uniformly in $\overline{\Omega}$ to the correspondent derivatives that turns out to be continuous in $\overline{\Omega}$; in such case the solution belongs to $C^2(\Omega)\cap C^1(\overline{\Omega})$, then it is a classical solution of the given problem.

If $s > m+1/2$ and $g \in H^s_{2\pi} = \{g \in L^2_{loc}(\mathbb{R}) :\ g$ is 2π periodic and $\sum_n n^{2s}(|\alpha_n|^2+|\beta_n|^2) < +\infty\}$ (see Definition 5.16), then $u \in C^{m+1}(\overline{\Omega})$. □

Exercise 9.10. Analyze this differential equation in the unknown u_n

$$(*) \qquad u_n'''(x) + n^3\,u_n(x) \;=\; n^3\,\delta'(x)\,, \qquad u_n \in \mathcal{S}'(\mathbb{R})$$

where the parameter n denotes a nonnegative integer number.

(1) Solve $(*)$ in the case $n = 0$, discussing the issues of existence and uniqueness for the solutions.

(2) For any $n \geq 1$, prove that the equation $(*)$ has a unique solution u_n and evaluate its Fourier transform $\widehat{u}_n$.

(3) From the knowledge of the qualitative properties of $\widehat{u}_n$ perform an "a priori" analysis of u_n, for $n \geq 1$ (without the explicit evaluation of u_n).

(4) Ascertain whether there exists $u \in \mathcal{S}'(\mathbb{R})$ such that u_n converges to u in $\mathcal{S}'(\mathbb{R})$.

(5) Evaluate u_n, for $n \geq 1$.

(6) Prove that the problem

$$\text{find } w \in \mathcal{S}'(\mathbb{R}) : \qquad w''' + w = f \qquad \text{in } \mathcal{S}'(\mathbb{R})$$

has a unique solution for any $f \in \mathcal{S}'(\mathbb{R})$ and that such solution fulfils $w \in H^{k+3}(\mathbb{R})$ if and only if $f \in H^k(\mathbb{R})$, where $H^k(\mathbb{R})$ denotes the Sobolev space of functions that belong to $L^2(\mathbb{R})$ together with their derivatives up to the order k.

Solution - (1) If $n = 0$ then the equation $(*)$ reads as follows: $u_0''' = 0$ with $u_0 \in \mathcal{S}'(\mathbb{R})$. Applying the Fourier transform to both sides we find a homogeneous division problem in the distributional unknown $\widehat{u} \in \mathcal{S}'(\mathbb{R})$:

$$\xi^3\,\widehat{u}_0 \;=\; 0\,.$$

This problem has infinitely many solutions in $\mathcal{S}'(\mathbb{R})$, all of them can be expressed (with suitable choice of the constants k_0, k_1, k_2) by:

$$\widehat{u}_0(\xi) = k_0\,\delta(\xi) + k_1\,\delta'(\xi) + k_2\,\delta''(\xi)\,,$$

After setting $c_0 = k_0/(2\pi)$, $c_1 = k_1/(2\pi i)$, $c_2 = k_2/(-2\pi)$, by anti-transforming $\widehat{u}_n$ we get all the solutions of $(*)$ with $n = 0$: there are infinitely many solution and they can be expressed (with suitable choice of the constants c_0, c_1, c_2) by:

$$u_0(x) = c_0 + c_1\,x + c_2\,x^2\,.$$

The careful reader may notice that the solution above could be evaluated in a far easier (and faster!) way. However in order to prove that no different distributional solution exists in $\mathcal{S}'(\mathbb{R})$ a rigorous inspection is needed, as it is done in item (1); on this subject, we refer to Exercise 4.50 too.

(2) For any fixed integer n s.t. $n \geq 1$, setting $\widehat{u}_n = \mathcal{F}\{u_n\}$ where $u_n \in \mathcal{S}'(\mathbb{R})$ is any admissible solution of $(*)$, and taking into account of

$$i^3 = -i\,, \quad \widehat{\delta'} = i\xi\,, \quad \mathcal{F}\left\{\frac{d}{dx}u_n\right\} = i\,\xi\,\widehat{u}_n\,, \quad \mathcal{F}\left\{\frac{d^3}{dx^3}u_n\right\} = -\,i\,\xi^3\,\widehat{u}_n\,,$$

we transform the equation $(*)$ and obtain $-i\xi^3\widehat{u}_n + n^3\widehat{u}_n = in^3\xi$; by multiplying this last relation times i we get:

$$(\xi^3 + n^3 i)\,\widehat{u}_n = -\,n^3\,\xi\,.$$

The chain of inferences $\left\{\, \xi \in \mathbb{R} \Rightarrow \xi^3 \in \mathbb{R} \Rightarrow \xi^3 \neq n^3 i \ \forall n \geq 1 \,\right\}$ implies $\xi^3 + n^3 i \neq 0\ \forall \xi \in \mathbb{R}$. So we can divide times $(\xi^3 + n^3 i)$ and obtain:

$$(**) \qquad \widehat{u}_n(\xi) = -\,\frac{\xi}{(\xi/n)^3 + i} \in \mathcal{S}'(\mathbb{R})\,.$$

The injectivity of the Fourier transform gives existence and uniqueness of the solution for $(*)$ $\forall n \geq 1$: the solution is $u_n = \mathcal{F}^{-1}\{\widehat{u}_n\}$.

(3) "a priori" analysis of u_n based on the properties of $\widehat{u}_n$, as expressed by $(**)$.
$\widehat{u}_n$ is a rational function, whose denominator does not vanish in $\mathbb{R}$. Moreover

$$(***) \qquad \widehat{u}_n(\xi) \sim \frac{-n^3}{\xi^2} \quad \text{if } \xi \to \pm\infty\,.$$

In particular $\widehat{u}_n \in L^p(\mathbb{R})$, $\forall p \in [1, +\infty]$ and $\widehat{u}_n(\xi) = \big(-\xi/((\xi/n)^6 + 1)\big)\big((\xi/n)^3 - i\big)$. Then

$\widehat{u}_n$ = real even "minus" immaginary odd $\Rightarrow$ u_n real, neither even nor odd;

$\widehat{u}_n \in L^1 \cap L^2$ (Riemann-Lebesgue, Plancherel) $\Rightarrow$ $u_n \in C^0 \cap L^\infty \cap L^2$, $\lim_{|x|\to\infty} u_n(x) = 0$;

$(***) \Rightarrow \xi\,\widehat{u}_n \in L^2 \subset L^1_{loc}$ $\Rightarrow$ $u_n \in AC_{loc}(\mathbb{R})$, $u_n' \in L^2$, $u_n \in H^1(\mathbb{R})$;

$\widehat{u}_n \in C^\infty(\mathbb{R})$, $\widehat{u}_n^{(k)} \in L^1 \cap L^2$ $\Rightarrow$ $x^k u_n \in L^2 \cap L^\infty\ \forall k$ and $u_n(x) = o(|x|^{-k})$ at $\infty\ \forall k$, that together with $u \in L^\infty$ entails $u_n \in L^1(\mathbb{R})$;

$\forall k \in \mathbb{N}\ \exists h \in \mathbb{N} :\ \xi^k\,\widehat{u}_n^{(h)} \in L^1 \cap L^2$ $\Rightarrow$ $\forall k\ \exists h : \dfrac{d^k}{dx^k}(x^h u_n) \in C^0 \cap L^\infty \Rightarrow$ $\Rightarrow u_n \in C^\infty(\mathbb{R}\setminus\{0\})$;

$\widehat{u}_n$ is not entire (Paley-Wiener Theorem) $\Rightarrow$ u_n has non compact support.

(4) By $\widehat{u}_n(\xi) = -\xi/\big((\xi^3/n^3)+i\big)$ and

$$\left|\frac{\xi^3}{n^3}+i\right| = \sqrt{1+\xi^6/n^3} \geq 1 \qquad \forall \xi \in \mathbb{R},\ n \in \mathbb{N},$$

we deduce

$$|\widehat{u}_n(\xi)| \leq |\xi| \qquad \forall \xi \in \mathbb{R},\ n \in \mathbb{N}.$$

The previous estimate together with the fact that $\widehat{u}_n(\xi) \longrightarrow i\,\xi$ point-wise in $\mathbb{R}$ and uniformly on compact subsets of $\mathbb{R}$ as $n \to \infty$, implies $\widehat{u}_n(\xi) \longrightarrow i\,\xi$ in $\mathcal{S}'(\mathbb{R})$; hence $\forall \psi \in \mathcal{S}$ vale $\langle \widehat{u}_n, \psi \rangle \to \langle i\xi, \psi \rangle$.
Due to the continuity of $\mathcal{F}^{-1}$ in $\mathcal{S}'(\mathbb{R})$ we obtain:

$$u_n(x) = \mathcal{F}^{-1}\{\widehat{u}_n(\xi)\} \overset{\mathcal{S}'(\mathbb{R})}{\longrightarrow} \mathcal{F}^{-1}\{i\xi\} = \delta'\ .$$

(5) We consider the (unique!) analytic extension $z \mapsto \widehat{u}_n(z)$, whose domain is $\mathbb{C}\backslash\{z_1, z_2, z_3\}$

$$\widehat{u}_n(z) = -\frac{z}{(z/n)^3+i}\ .$$

There are 3 simple poles: the complex cubic roots z_j $(j=1,2,3)$ of $-n^3 i = n^3\, e^{\frac{3}{2}\pi i}$:

$$\begin{aligned} z_1 &= n\,e^{\frac{\pi}{2}i} = n\,i\,, \\ z_2 &= n\,\exp\left(\frac{\pi}{2}i+\frac{2}{3}\pi i\right) = n\,e^{\frac{7}{6}\pi i} = \left(-\frac{\sqrt{3}}{2}-\frac{i}{2}\right) n\,, \\ z_3 &= n\,\exp\left(\frac{\pi}{2}i+\frac{4}{3}\pi i\right) = n\,e^{\frac{11}{6}\pi i} = \left(+\frac{\sqrt{3}}{2}-\frac{i}{2}\right) n\,. \end{aligned}$$

By the de l'Hôpital rule (for evaluating the limit of indeterminate form of the kind 0/0) we get:

$$\begin{aligned} \mathrm{res}(e^{+izx}\widehat{u}_n(z), z=z_1) &= \lim_{z\to z_1} e^{+izx}\,(-z)\,\frac{(z-ni)}{(z/n)^3+i} = \\ &= e^{+iz_1x}\,(-z_1)\,\frac{1}{3\,z_1^2/n^3} = i\,\frac{n^2}{3}\,e^{-nx}\,, \end{aligned}$$

$$\begin{aligned} \mathrm{res}(e^{-izx}\widehat{u}_n(z), z=z_2) &= \lim_{z\to z_2} e^{izx}\,(-z)\,\frac{(z-z_2)}{(z/n)^3+i} = e^{iz_2x}\,(-z_2)\,\frac{1}{3\,z_2^2/n^3} = \\ &= \tfrac{n^2}{6}(\sqrt{3}-i)\,e^{(\frac{1}{2}-i\frac{\sqrt{3}}{2})nx}, \end{aligned}$$

$$\begin{aligned} \mathrm{res}(e^{izx}\widehat{u}_n(z), z=z_3) &= \lim_{z\to z_3} e^{-izx}\,(-z)\,\frac{(z-z_3)}{(z/n)^3+i} = e^{iz_3x}\,(-z_3)\,\frac{1}{3\,z_3^2/n^3} = \\ &= -\tfrac{n^2}{6}\,(\sqrt{3}+i)e^{(\frac{1}{2}+i\frac{\sqrt{3}}{2})nx}\,. \end{aligned}$$

Fix $x>0$, denote by S_R the half-circumference in the upper half-plane with center at 0 and radius $R>n$, and denote by γ the curve (swept once with the counter-clockwise orientation) made by joining S_r and the segment $[-R,R] \subset \mathbb{R}$: we apply the inverse Fourier transform formula (for functions in $L^1(\mathbb{R})$ with transform in $L^1(\mathbb{R})$), the Jordan Lemma to the half-circumference S_R (the multiplier $\widehat{u}_n(z)$ of e^{izx} in the integrand is an infinitesimal as $R \to \infty$) and the Residues Theorem

$$\begin{aligned} u_n(x) &= \frac{1}{2\pi}\int_{\mathbb{R}} e^{+i\xi x}\widehat{u}_n(\xi)\,d\xi = \frac{1}{2\pi}\,v.p.\int_{\mathbb{R}} e^{+i\xi x}\frac{-\xi}{(\xi/n)^3+i}\,d\xi = \\ &= \frac{1}{2\pi}\oint_{\gamma} e^{+izx}\frac{-z}{(z/n)^3+i}\,dz = \frac{2\pi i}{2\pi}\ \mathrm{res}(e^{+izx}\widehat{u}_n(z), z=z_1) = \\ &= i\,i\,\frac{n^2}{3}\,e^{-nx} = -\frac{n^2}{3}\,e^{-nx} \qquad \forall x>0\,. \end{aligned}$$

Fix $x<0$, denote by Σ_R the half-circumference in the lower half-plane, with center at 0 and radius $R>n$, and denote by η the curve (swept once with the clock-wise orientation) made by joining Σ_r and the segment $[-R,R]\subset\mathbb{R}$: we apply the inverse Fourier transform formula (for functions in $L^1(\mathbb{R})$ with transform in $L^1(\mathbb{R})$), the Jordan Lemma to the half-circumference Σ_R (the multiplier $\widehat{u}_n(z)$ of e^{izx} in the integrand is an infinitesimal as $R\to\infty$) and the Residues Theorem

$$\begin{aligned} u_n(x) &= \frac{1}{2\pi}\int_{\mathbb{R}} e^{i\xi x}\widehat{u}_n(\xi)\,d\xi = \frac{1}{2\pi}\,v.p.\int_{\mathbb{R}} e^{i\xi x}\,\widehat{u}_n(\xi)\,d\xi = -\frac{1}{2\pi}\oint_{\eta} e^{izx}\,\widehat{u}_n(z)\,dz = \\ &= -\frac{2\pi i}{2\pi}\left(\mathrm{res}(e^{izx}\widehat{u}_n(z), z=z_2) + \mathrm{res}(e^{izx}\widehat{u}_n(z), z=z_3)\right) = \\ &= -i\left(\frac{n^2}{6}(\sqrt{3}-i)\,e^{(\frac{1}{2}-i\frac{\sqrt{3}}{2})nx} - \frac{n^2}{6}(\sqrt{3}+i)\,e^{(\frac{1}{2}+i\frac{\sqrt{3}}{2})nx}\right) = \\ &= -i\frac{n^2}{6}e^{\frac{n}{2}x}\left\{(\sqrt{3}-i)\left(\cos\left(\frac{\sqrt{3}}{2}nx\right) - i\sin\left(\frac{\sqrt{3}}{2}nx\right)\right)\right. \\ &\qquad\qquad \left. -(\sqrt{3}+i)\left(\cos\left(\frac{\sqrt{3}}{2}nx\right) + i\sin\left(\frac{\sqrt{3}}{2}nx\right)\right)\right\} = \\ &= -\frac{n^2}{3}\left(e^{\frac{n}{2}x}\cos\left(\frac{\sqrt{3}}{2}nx\right) + \sqrt{3}\,\frac{n^2}{3}\,e^{\frac{n}{2}x}\sin\left(\frac{\sqrt{3}}{2}nx\right)\right) \qquad \forall x<0\,. \end{aligned}$$

Since u_n is continuous, we obtain $u_n(0) = \lim_{x\to 0} u_n(x) = -\dfrac{n^2}{3}$. Summarizing

$$u_n(x) = \begin{cases} -\dfrac{n^2}{3}\,e^{-nx} & \text{se } x\geq 0\,, \\ -\dfrac{n^2}{3}\,e^{\frac{n}{2}x}\cos\left(\dfrac{\sqrt{3}}{2}nx\right) - \dfrac{\sqrt{3}}{3}\,n^2\,e^{\frac{n}{2}x}\sin\left(\dfrac{\sqrt{3}}{2}nx\right) & \text{se } x<0\,. \end{cases} \tag{9.6}$$

This computed solution u_n verifies the properties that have been proved in (3).
(6) By Fourier transforming we find a relation that holds true for the Fourier transform of any admissible solution:

$$(1-i\,\xi^3)\,\widehat{w}(\xi) = \widehat{f}(\xi)\,.$$

The inferences $\left\{\xi\in\mathbb{R}\Rightarrow\xi^3\in\mathbb{R}\Rightarrow i\xi^3\neq 1\right\}$ entail that the multiplier of $\widehat{w}$ has no real zero. Then there exists exactly one solution w that fulfils:

$$\widehat{w} = \frac{\widehat{f}(\xi)}{1-i\,\xi^3}\,.$$

The Plancherel identity (4.36) and the transformation rules for derivatives (4.44) imply

$$w^{(j+3)}\in L^2 \iff \xi^{j+3}\widehat{w}\in L^2 \iff \frac{\xi^{j+3}}{1-i\,\xi^3}\,\widehat{f}\in L^2 \iff \xi^j\,\widehat{f}\in L^2 \iff f^{(j)}\in L^2;$$

this fact together with the inclusion

$$\{v\in L^2(\mathbb{R}) : v'\in L^2(\mathbb{R})\} \subset C^0\cap L^\infty(\mathbb{R})$$

leads to the conclusion:

- if $f\in L^2(\mathbb{R}) \Rightarrow \xi^h\widehat{w}\in L^2(\mathbb{R}),\ h=0,1,2 \Rightarrow$
$$\Rightarrow\ w,w',w'',w'''\in L^2(\mathbb{R}),\ \ w\in C^2 \ \text{ and } \ w,w',w''\in L^\infty;$$
- if $f,f',\dots,f^{(k)}\in L^2(\mathbb{R}) \quad\Rightarrow\quad \xi^h\widehat{w}\in L^2(\mathbb{R}),\ h=0,1,\dots,k+3 \quad\Rightarrow$
$$\Rightarrow w^{(h)}\in L^2(\mathbb{R}),\ h=0,1,\dots,k+3 \Rightarrow w\in C^{k+2},\ w^{(h)}\in L^\infty\ \ h=0,\dots,k+2\,.$$

Hence $w\in H^{k+3}(\mathbb{R})$ if and only if $f\in H^k(\mathbb{R})$ □

Chapter 10
Tables

10.1 Trigonometric functions

10.1.1 Addition and subtraction

$$\sin(\alpha+\beta) = \sin\alpha\ \cos\beta + \sin\beta\ \cos\alpha \qquad \sin(\alpha-\beta) = \sin\alpha\ \cos\beta - \sin\beta\ \cos\alpha$$

$$\cos(\alpha+\beta) = \cos\alpha\ \cos\beta - \sin\alpha\ \sin\beta \qquad \cos(\alpha-\beta) = \cos\alpha\ \cos\beta + \sin\alpha\ \sin\beta$$

$$\tan(\alpha+\beta) = \frac{\tan\alpha + \tan\beta}{1-\tan\alpha\ \tan\beta} \qquad \tan(\alpha-\beta) = \frac{\tan\alpha - \tan\beta}{1+\tan\alpha\ \tan\beta}$$

10.1.2 Double-angle formulae

$$\sin(2\alpha) = 2\sin\alpha\ \cos\alpha = \frac{2\tan\alpha}{1+\tan^2\alpha}$$

$$\cos(2\alpha) = \cos^2\alpha - \sin^2\alpha = 1-2\sin^2\alpha = 2\cos^2\alpha - 1 = \frac{1-\tan^2\alpha}{1+\tan^2\alpha}$$

$$\tan(2\alpha) = \frac{2\tan\alpha}{1-\tan^2\alpha}$$

10.1.3 Half-angle formulae

$$\sin\frac{\alpha}{2} = \pm\sqrt{\frac{1-\cos\alpha}{2}} \qquad \cos\frac{\alpha}{2} = \pm\sqrt{\frac{1+\cos\alpha}{2}}$$

$$\tan\frac{\alpha}{2} = \pm\sqrt{\frac{1-\cos\alpha}{1+\cos\alpha}} = \frac{1-\cos\alpha}{\sin\alpha} = \frac{\sin\alpha}{1+\cos\alpha}$$

10.1.4 Parametric change of variables

Set $t = \tan\dfrac{\alpha}{2}$, then : $\sin\alpha = \dfrac{2t}{1+t^2}$, $\cos\alpha = \dfrac{1-t^2}{1+t^2}$.

10.1.5 Sum-to-product formulae

$$\sin p + \sin q = 2\sin\frac{p+q}{2}\cos\frac{p-q}{2} \qquad \sin p - \sin q = 2\cos\frac{p+q}{2}\sin\frac{p-q}{2}$$

$$\cos p + \cos q = 2\cos\frac{p+q}{2}\cos\frac{p-q}{2} \qquad \cos p - \cos q = -2\sin\frac{p+q}{2}\sin\frac{p-q}{2}$$

10.1.6 Product-to-sum formulae (prostaphaeresis)

$$\sin\alpha\ \cos\beta = [\sin(\alpha+\beta) + \sin(\alpha-\beta)]\ /\ 2$$

$$\cos\alpha\ \cos\beta = [\cos(\alpha+\beta) + \cos(\alpha-\beta)]\ /\ 2$$

$$\sin\alpha\ \sin\beta = -\ [\cos(\alpha+\beta) - \cos(\alpha-\beta)]\ /\ 2$$

10.1.7 Useful identities

$$\int_{\mathbb{R}} e^{-x^2}\, dx = \sqrt{\pi}\,, \qquad \int_0^\infty x^n e^{-x}\, dx = n! \quad n \in \mathbb{N}\,,$$

$$\int (\sin x)^n\, dx = -\frac{1}{n}(\sin x)^{n-1}\cos x + \frac{n-1}{n}\int (\sin x)^{n-2}\, dx\,,$$

$$\int (\cos x)^n\, dx = \frac{1}{n}(\cos x)^{n-1}\sin x + \frac{n-1}{n}\int (\cos x)^{n-2}\, dx\,,$$

$$\int_0^{\pi/2} (\sin x)^n\, dx = \int_0^{\pi/2} (\cos x)^n\, dx = \begin{cases} \dfrac{1\cdot 3\cdot 5\cdots(n-1)}{2\cdot 4\cdot 6\cdots n}\,\dfrac{\pi}{2} & \text{if } n \text{ is even and } n \geq 2\,, \\[2ex] \dfrac{2\cdot 4\cdot 6\cdots(n-1)}{3\cdot 5\cdot 7\cdots n} & \text{if } n \text{ is odd and } n \geq 3\,, \end{cases}$$

$$\frac{\sin((n+1/2)x)}{2\,\sin(x/2)} = \frac{1}{2} + \cos(x) + \cos(2x) + \cdots + \cos(nx)\,.$$

10.1.8 Trigonometric functions of selected arcs

Radians	Degrees	sin	cos	tan
$\frac{\pi}{12}$	15°	$\frac{\sqrt{2}}{4}(\sqrt{3}-1)$	$\frac{\sqrt{2}}{4}(\sqrt{3}+1)$	$2-\sqrt{3}$
$\frac{\pi}{10}$	18°	$\frac{\sqrt{5}-1}{4}$	$\frac{\sqrt{10+2\sqrt{5}}}{4}$	$\frac{\sqrt{25-10\sqrt{5}}}{5}$
$\frac{\pi}{8}$	22° 30′	$\frac{\sqrt{2-\sqrt{2}}}{2}$	$\frac{\sqrt{2+\sqrt{2}}}{2}$	$\sqrt{2}-1$
$\frac{\pi}{6}$	30°	$\frac{1}{2}$	$\frac{\sqrt{3}}{2}$	$\frac{\sqrt{3}}{3}$
$\frac{\pi}{5}$	36°	$\frac{\sqrt{10-2\sqrt{5}}}{4}$	$\frac{1+\sqrt{5}}{4}$	$\sqrt{5-2\sqrt{5}}$
$\frac{\pi}{4}$	45°	$\frac{\sqrt{2}}{2}$	$\frac{\sqrt{2}}{2}$	1
$\frac{3\pi}{10}$	54°	$\frac{1+\sqrt{5}}{4}$	$\frac{\sqrt{10-2\sqrt{5}}}{4}$	$\frac{\sqrt{25+10\sqrt{5}}}{5}$
$\frac{\pi}{3}$	60°	$\frac{\sqrt{3}}{2}$	$\frac{1}{2}$	$\sqrt{3}$
$\frac{3\pi}{8}$	67° 30′	$\frac{\sqrt{2+\sqrt{2}}}{2}$	$\frac{\sqrt{2-\sqrt{2}}}{2}$	$\sqrt{2}+1$
$\frac{2\pi}{5}$	72°	$\frac{\sqrt{10+2\sqrt{5}}}{4}$	$\frac{\sqrt{5}-1}{4}$	$\sqrt{5+2\sqrt{5}}$
$\frac{5\pi}{12}$	75°	$\frac{\sqrt{2}}{4}(\sqrt{3}+1)$	$\frac{\sqrt{2}}{4}(\sqrt{3}-1)$	$2+\sqrt{3}$

10.2 Taylor series expansions

Function $f = f(z)$	Taylor series of $f(z)$	Radius
$\dfrac{1}{1-z}$	$\displaystyle\sum_{k=0}^{\infty} z^n$	$\lvert z\rvert < 1$
e^z	$\displaystyle\sum_{k=0}^{\infty} \frac{z^k}{k!}$	$z \in \mathbb{C}$
$\sin z$	$\displaystyle\sum_{k=0}^{\infty} (-1)^k \frac{z^{2k+1}}{(2k+1)!}$	$z \in \mathbb{C}$
$\cos z$	$\displaystyle\sum_{k=0}^{\infty} (-1)^k \frac{z^{2k}}{(2k)!}$	$z \in \mathbb{C}$
$\arctan z$	$\displaystyle\sum_{k=0}^{\infty} (-1)^k \frac{z^{2k+1}}{2k+1}$	$\lvert z\rvert < 1$
$\arcsin z$	$\displaystyle\sum_{k=0}^{\infty} \frac{(2k-1)!!}{(2k)!!} \frac{z^{2k+1}}{2k+1}$ $(*)$	$\lvert z\rvert < 1$
$\arccos z = \dfrac{\pi}{2} - \arcsin z$	$\displaystyle\frac{\pi}{2} - \sum_{k=0}^{\infty} \frac{(2k-1)!!}{(2k)!!} \frac{z^{2k+1}}{2k+1}$	$\lvert z\rvert < 1$
$\ln(1+z)$	$\displaystyle\sum_{k=1}^{\infty} (-1)^{k+1} \frac{z^k}{k}$	$\lvert z\rvert < 1$

$(*)$ The double factorial function $n!!$ denotes the product of integers with the same parity of n, where n is an integer ≥ -1. Examples: $6!! = 6 \cdot 4 \cdot 2$, $9!! = 9 \cdot 7 \cdot 5 \cdot 3 \cdot 1$.
For convenience the following conventions are adopted: $0!! = 1$, $(-1)!! = 1$.

$\sinh z$	$\displaystyle\sum_{k=0}^{\infty} \frac{z^{2k+1}}{(2k+1)!}$	$z \in \mathbb{C}$
$\cosh z$	$\displaystyle\sum_{k=0}^{\infty} \frac{z^{2k}}{(2k)!}$	$z \in \mathbb{C}$
$(1+z)^{\alpha}$	$\displaystyle\sum_{k=0}^{\infty} \binom{\alpha}{k} z^k$ (**)	$\lvert z\rvert < 1$ $\alpha \in \mathbb{R}$
$J_0(z)$	$\displaystyle\sum_{k=0}^{\infty} \frac{(-1)^k}{(k!)^2} \left(\frac{z}{2}\right)^{2k}$	$z \in \mathbb{C}$
$J_n(z)$	$\displaystyle\sum_{k=0}^{\infty} \frac{(-1)^k}{k!(k+n)!} \left(\frac{z}{2}\right)^{2k+n}$	$z \in \mathbb{C}$
$\mathrm{Sinc}(z) = \dfrac{\sin z}{z}$	$\displaystyle\sum_{k=0}^{\infty} \frac{(-1)^k}{(2k+1)!} z^{2k}$	$z \in \mathbb{C}$
$\mathrm{Si}(z) = \displaystyle\int_0^z \frac{\sin t}{t}\, dt = \int_0^z \mathrm{Sinc}(t)\, dt$	$\displaystyle\sum_{k=0}^{\infty} \frac{(-1)^k}{(2k)!(2k+1)^2} z^{2k+1}$	$z \in \mathbb{C}$
$\mathrm{erf}(z) = \dfrac{2}{\sqrt{\pi}} \displaystyle\int_0^z e^{-w^2}\, dw$	$\dfrac{2}{\sqrt{\pi}} \displaystyle\sum_{k=0}^{\infty} \frac{(-1)^k}{k!(2k+1)} z^{2k+1}$	$z \in \mathbb{C}$
$\mathrm{erfc}(z) = 1 - \mathrm{erf}(z)$	$1 + \dfrac{2}{\sqrt{\pi}} \displaystyle\sum_{k=0}^{\infty} \frac{(-1)^{k+1}}{k!(2k+1)} z^{2k+1}$	$z \in \mathbb{C}$

(**) The definition of binomial coefficient holds also in the case $\alpha \in \mathbb{R},\ k \in \mathbb{N}$, without any restriction on α, k, by setting: $\binom{\alpha}{k} = \frac{\alpha(\alpha-1)(\alpha-2)\cdot\dots\cdot(\alpha-k+1)}{k(k-1)(k-2)\cdot\dots\cdot 2\cdot 1}$.

10.3 Differential identities related to gradient, divergence, curl e Laplacian in $\mathbb{R}^3$

$\nabla = \mathbf{i}\frac{\partial}{\partial x} + \mathbf{j}\frac{\partial}{\partial y} + \mathbf{k}\frac{\partial}{\partial z}$ "nabla" operator

$\mathbf{F}(x,y,z) = F_1(x,y,z)\mathbf{i} + F_2(x,y,z)\mathbf{j} + F_3(x,y,z)\mathbf{k}$

$\nabla u(x,y,z) = \mathbf{grad}\, u(x,y,z) = \mathbf{i}\frac{\partial u}{\partial x} + \mathbf{j}\frac{\partial u}{\partial y} + \mathbf{k}\frac{\partial u}{\partial z}$

$\nabla \cdot \mathbf{F}(x,y,z) = \mathrm{div}\, \mathbf{F}(x,y,z) = \frac{\partial F_1}{\partial x} + \frac{\partial F_2}{\partial y} + \frac{\partial F_3}{\partial z}$

$$\nabla \times \mathbf{F}(x,y,z) = \mathbf{curl}\ \mathbf{F}(x,y,z) = \det \begin{bmatrix} \mathbf{i} & \mathbf{j} & \mathbf{k} \\ \frac{\partial}{\partial x} & \frac{\partial}{\partial y} & \frac{\partial}{\partial z} \\ F_1 & F_2 & F_3 \end{bmatrix} = \left(\frac{\partial F_3}{\partial y} - \frac{\partial F_2}{\partial z}\right)\mathbf{i} + \left(\frac{\partial F_1}{\partial z} - \frac{\partial F_3}{\partial x}\right)\mathbf{j} + \left(\frac{\partial F_2}{\partial x} - \frac{\partial F_1}{\partial y}\right)\mathbf{k}$$

$\nabla(uv) = u\nabla v + v\nabla u$

$\nabla \cdot (\mathbf{F} \times \mathbf{G}) = (\nabla \times \mathbf{F}) \cdot \mathbf{G} - \mathbf{F} \cdot (\nabla \times \mathbf{G})$

$\nabla \cdot (u\mathbf{F}) = (\nabla u) \cdot \mathbf{F} + u(\nabla \cdot \mathbf{F})$

$\nabla \times (\mathbf{F} \times \mathbf{G}) = \mathbf{F}(\nabla \cdot \mathbf{G}) - \mathbf{G}(\nabla \cdot \mathbf{F}) - (\mathbf{F} \cdot \nabla)\mathbf{G} + (\mathbf{G} \cdot \nabla)\mathbf{F}$

$\nabla \times (u\mathbf{F}) = (\nabla u) \times \mathbf{F} + u(\nabla \times \mathbf{F})$

$\nabla(\mathbf{F} \cdot \mathbf{G}) = \mathbf{F} \times (\nabla \times \mathbf{G}) - \mathbf{G} \times (\nabla \times \mathbf{F}) + (\mathbf{F} \cdot \nabla)\mathbf{G} + (\mathbf{G} \cdot \nabla)\mathbf{F}$

$\nabla \times (\nabla u) = 0$ $(\mathbf{curl}\,\mathbf{grad} = 0)$

$\nabla \cdot (\nabla \times \mathbf{F}) = 0$ $(\mathrm{div}\,\mathbf{curl} = 0)$

$$\Delta u(x,y,z) = \nabla^2 u(x,y,z) = \nabla \cdot \nabla u(x,y,z) = \mathrm{div}\,\mathbf{grad}\, u = \frac{\partial^2 u}{\partial x^2} + \frac{\partial u^2}{\partial y^2} + \frac{\partial^2 u}{\partial z^2} \qquad (\mathrm{div}\,\mathbf{grad} = \text{Laplacian})$$

$$\nabla \times (\nabla \times \mathbf{F}) = \nabla(\nabla \cdot \mathbf{F}) - \nabla^2 \mathbf{F} = \nabla(\nabla \cdot \mathbf{F}) - \Delta \mathbf{F} \qquad (\mathbf{curl}\,\mathbf{curl} = \mathbf{grad}\ \mathrm{div}\ \text{- Laplacian})$$

10.4 Integration by parts formulae in $\mathbb{R}$, $\mathbb{R}^2$, $\mathbb{R}^3$

$$\int_a^b f'(t)\,dt = f(b) - f(a) \qquad \textbf{(fundamental theorem)}$$

$$\int_\gamma \nabla u\,d\boldsymbol{\gamma} = \int_a^b \nabla u(\boldsymbol{\gamma})\,\boldsymbol{\gamma}'(t)\,dt = u(\boldsymbol{\gamma}(b)) - u(\boldsymbol{\gamma}(a)) \qquad (i)$$

$$\iint_\Omega \left(\frac{\partial F_2}{\partial x} - \frac{\partial F_1}{\partial y}\right) dx\,dy = \oint_{\partial\Omega} \mathbf{F}\cdot d\gamma = \oint_{\partial\Omega} (F_1(x,y)\,dx + F_2(x,y)\,dy) \qquad (ii)\ \textbf{(Green theorem)}$$

$$\iint_S \nabla\times\mathbf{F}\cdot\boldsymbol{\nu}\,d\sigma = \oint_{\partial S} \mathbf{F}\cdot d\gamma = \oint_{\partial S} (F_1(x,y,z)\,dx + F_2(x,y,z)\,dy + F_3(x,y,z)\,dz) \qquad (iii)\ \textbf{(Stokes theorem)}$$

$$\iiint_D \nabla\cdot\mathbf{F}\ dx\,dy\,dz = \iint_{\partial D} \mathbf{F}\cdot\boldsymbol{\nu}\,d\sigma \qquad (iv)\ \textbf{(divergence theorem)}$$

$$\iiint_D \nabla\times\mathbf{F}\,dx\,dy\,dz = -\iint_{\partial D} \mathbf{F}\times\boldsymbol{\nu}\,d\sigma \qquad (iv)\ \textbf{(curl theorem)}$$

$$\iiint_D \nabla u\,dx\,dy\,dz = \iint_{\partial D} u\,\boldsymbol{\nu}\,d\sigma \qquad (iv)\ \textbf{(gradient theorem)}$$

(i) where $\boldsymbol{\gamma}$ is the curve $\boldsymbol{\gamma}(t) = (x(t), y(t))$, $(a \le t \le b)$;
(ii) where $\partial\Omega$ is the boundary of the region $\Omega \subset \mathbb{R}^2$; $\partial\Omega$ is positively oriented and parameterized by $\boldsymbol{\gamma}(t) = (x(t), y(t))$;
(iii) where ∂S is the boundary of the surface $S \subset \mathbb{R}^3$; ∂S is positively oriented and parameterized by $\boldsymbol{\gamma}(t) = (x(t), y(t), z(t))$;
(iv) three dimensional formulation: ∂D is the boundary of the region $D \subset \mathbb{R}^3$, with external normal vector $\boldsymbol{\nu}$.

10.5 Integration by parts formulae in $\mathbb{R}^n$

Referring to the Euclidean space $\mathbb{R}^n$, $n \geq 2$, we consider:

- $\Omega \subset \mathbb{R}^n$ open set with smooth boundary $\partial\Omega$ and outer normal $\boldsymbol{\nu}$,
- u, v scalar fields defined in Ω and smooth up to the boundary $\partial\Omega$,
- $\mathbf{u}$, $\mathbf{v}$ vector fields defined in Ω and smooth up to the boundary $\partial\Omega$,
- $d\mathbf{x} = dx_1 dx_2 \dots dx_n$ volume element in Ω,
- $d\sigma(\mathbf{x})$ surface element in $\partial\Omega$.

The following (scalar-valued: 2-6; vector-valued: 1) identities hold:

1. $\displaystyle\int_\Omega \nabla u \, d\mathbf{x} = \int_{\partial\Omega} u \boldsymbol{\nu} \, d\sigma(\mathbf{x})$

2. $\displaystyle\int_\Omega \Delta u \, d\mathbf{x} = \int_{\partial\Omega} \nabla u \cdot \boldsymbol{\nu} \, d\sigma(\mathbf{x}) = \int_{\partial\Omega} \partial_\nu u \, d\sigma(\mathbf{x})$

3. $\displaystyle\int_\Omega \operatorname{div} \mathbf{u} \, d\mathbf{x} = \int_{\partial\Omega} \mathbf{u} \cdot \boldsymbol{\nu} \, d\sigma(\mathbf{x})$

4. $\displaystyle\int_\Omega u \operatorname{div} \mathbf{v} \, d\mathbf{x} = -\int_\Omega \nabla u \cdot \mathbf{v} + \int_{\partial\Omega} u \, \mathbf{v} \cdot \boldsymbol{\nu} \, d\sigma(\mathbf{x})$

5. $\displaystyle\int_\Omega u \, \Delta v \, d\mathbf{x} = -\int_\Omega \nabla u \, \nabla v \, d\mathbf{x} + \int_{\partial\Omega} u \, \partial_{\boldsymbol{\nu}} v \, d\sigma(\mathbf{x})$ (1st *Green id.*)

6. $\displaystyle\int_\Omega (u \, \Delta v - v \, \Delta u) \, d\mathbf{x} = \int_{\partial\Omega} (u \, \partial_{\boldsymbol{\nu}} v - v \, \partial_{\boldsymbol{\nu}} u) \, d\sigma(\mathbf{x})$ (2nd *Green id.*)

Referring to the Euclidean space $\mathbb{R}^3$ we consider:

- $S \subset \mathbb{R}^3$ oriented smooth surface, whose boundary is a closed and smooth simple curve γ,
- $\boldsymbol{\nu}$ unit vector normal to S, $\mathbf{t}$ unit vector tangent to γ, s.t. γ positively oriented with respect to S
- ds length element in γ,
- $d\sigma$ surface element in S,
- u, v scalar fields and $\mathbf{u}$ vector field, always smooth and defined in a neighborhood of S.

The following identities hold (**Stokes formulae**):

7. $\displaystyle\int_S (\mathbf{rot}\ \mathbf{u}) \cdot \boldsymbol{\nu} \, d\sigma(\mathbf{x}) = \int_\gamma \mathbf{u} \cdot \mathbf{t} \, ds$ (*curl flux = field circuitation*)

8. $\displaystyle\int_S \nabla u \times \boldsymbol{\nu} \, d\sigma(\mathbf{x}) = -\int_\gamma u \, \mathbf{t} \, ds$

9. $\displaystyle\int_\gamma u \, (\nabla v \cdot \mathbf{t}) \, ds == \int_\gamma u \frac{\partial v}{\partial s} \, ds = -\int_\gamma v \frac{\partial u}{\partial s} \, ds = -\int_\gamma v \, (\nabla u \cdot \mathbf{t}) \, ds$

10.6 Change of variables

10.6.1 Polar coordinates in $\mathbb{R}^2$

transformation: $x = \varrho \cos\vartheta, \quad y = \varrho \sin\vartheta, \qquad \varrho > 0,\ 0 \le \vartheta \le 2\pi$

position vector: $\mathbf{r} = \varrho\cos\vartheta\,\mathbf{i} + \varrho\sin\vartheta\,\mathbf{j}$

scale factor: $h_\varrho := \left|\dfrac{\partial \mathbf{r}}{\partial \varrho}\right| = 1, \quad h_\vartheta := \left|\dfrac{\partial \mathbf{r}}{\partial \vartheta}\right| = r$

local basis: $\mathbf{e}_\varrho = \cos\vartheta\,\mathbf{i} + \sin\vartheta\,\mathbf{j}, \quad \mathbf{e}_\vartheta = -\sin\vartheta\,\mathbf{i} + \cos\vartheta\,\mathbf{j}$

area element: $d\sigma = dxdy = \varrho\, d\varrho\, d\vartheta$

---

vector field: $\mathbf{v}(\varrho,\vartheta) = v_\varrho(\varrho,\vartheta)\mathbf{e}_\varrho + v_\vartheta(\varrho,\vartheta)\mathbf{e}_\vartheta$

divergence: $\nabla\cdot\mathbf{v} = \dfrac{1}{\varrho}\dfrac{\partial}{\partial\varrho}(\varrho F_\varrho) + \dfrac{1}{\varrho}\dfrac{\partial F_\vartheta}{\partial\vartheta}$

curl: $\nabla\times\mathbf{v} = \left[\dfrac{1}{\varrho}\dfrac{\partial}{\partial\varrho}(\varrho v_\vartheta) - \dfrac{1}{\varrho}\dfrac{\partial v_\varrho}{\partial\vartheta}\right]\mathbf{k}$

---

scalar field: $u(\varrho,\vartheta)$

gradient: $\nabla u = \dfrac{\partial u}{\partial\varrho}\mathbf{e}_\varrho + \dfrac{1}{\varrho}\dfrac{\partial u}{\partial\vartheta}\mathbf{e}_\vartheta$

Laplacian: $\Delta u = \dfrac{1}{\varrho}\dfrac{\partial}{\partial\varrho}\left(\varrho\dfrac{\partial u}{\partial\varrho}\right) + \dfrac{1}{\varrho^2}\dfrac{\partial^2 u}{\partial\vartheta^2} = \dfrac{\partial^2 u}{\partial\varrho^2} + \dfrac{1}{\varrho}\dfrac{\partial u}{\partial\varrho} + \dfrac{1}{\varrho^2}\dfrac{\partial^2 u}{\partial\vartheta^2}$

partial derivatives in polar form:
$$\begin{cases} \dfrac{\partial u}{\partial x} = \cos\vartheta\dfrac{\partial u}{\partial\varrho} - \dfrac{\sin\vartheta}{\varrho}\dfrac{\partial u}{\partial\vartheta} \\[2ex] \dfrac{\partial u}{\partial y} = \sin\vartheta\dfrac{\partial u}{\partial\varrho} + \dfrac{\cos\vartheta}{\varrho}\dfrac{\partial u}{\partial\vartheta} \end{cases}$$

energy density in polar form: $|\nabla u|^2 = \left|\dfrac{\partial u}{\partial x}\right|^2 + \left|\dfrac{\partial u}{\partial y}\right|^2 = \left|\dfrac{\partial u}{\partial\varrho}\right|^2 + \dfrac{1}{\varrho^2}\left|\dfrac{\partial u}{\partial\vartheta}\right|^2$

10.6.2 Spherical coordinates in $\mathbb{R}^3$

transformation: $x = \varrho\cos\vartheta\sin\varphi,\ y = \varrho\sin\vartheta\sin\varphi,\ z = \varrho\cos\varphi,$

$\varrho > 0,\ 0 \le \vartheta \le 2\pi,\ 0 \le \varphi \le \pi$, angles measured in radians, $\vartheta =$ geographic longitude, $\varphi = \pi/2 -$ geographic latitude

position vector: $\mathbf{x} = \varrho\sin\varphi\cos\vartheta\,\mathbf{i} + \varrho\sin\varphi\sin\vartheta\,\mathbf{j} + \varrho\cos\varphi\,\mathbf{k}$

scale factor: $h_\varrho := \left|\dfrac{\partial\mathbf{x}}{\partial\varrho}\right| = 1,\ h_\vartheta := \left|\dfrac{\partial\mathbf{x}}{\partial\vartheta}\right| = \varrho\sin\varphi,\ h_\varphi := \left|\dfrac{\partial\mathbf{x}}{\partial\varphi}\right| = \varrho$

local basis: $\mathbf{e}_\varrho = \cos\vartheta\sin\varphi\,\mathbf{i} + \sin\vartheta\sin\varphi\,\mathbf{j} + \cos\varphi\,\mathbf{k},$

$\mathbf{e}_\vartheta = -\sin\vartheta\mathbf{i} + \cos\vartheta\mathbf{j}$

$\mathbf{e}_\varphi = \cos\vartheta\cos\varphi\mathbf{i} + \sin\vartheta\cos\varphi\mathbf{j} - \sin\varphi\mathbf{k},$

volume element: $dV = dx\,dy\,dz = \varrho^2\ \sin\varphi\ d\varrho\ d\varphi\ d\vartheta$

---

vector field: $\mathbf{v}(\varrho,\varphi,\vartheta) = v_\varrho(\varrho,\varphi,\vartheta)\mathbf{e}_\varrho + v_\vartheta(\varrho,\varphi,\vartheta)\mathbf{e}_\vartheta + v_\varphi(\varrho,\varphi,\vartheta)\mathbf{e}_\varphi$

divergence: $$\nabla\cdot\mathbf{v} = \underbrace{\frac{1}{\varrho^2}\frac{\partial}{\partial\varrho}(\varrho^2 v_\varrho)}_{\text{radial part}} + \underbrace{\frac{1}{\varrho\sin\varphi}\frac{\partial v_\vartheta}{\partial\vartheta} + \frac{1}{\varrho\sin\varphi}\frac{\partial}{\partial\varphi}(\sin\varphi\,v_\varphi)}_{\text{spherical part}}$$

curl: $$\nabla\times\mathbf{v} = \left[\frac{1}{\varrho\sin\varphi}\frac{\partial}{\partial\varphi}(\sin\varphi\,v_\vartheta) - \frac{1}{\varrho\sin\varphi}\frac{\partial v_\varphi}{\partial\vartheta}\right]\mathbf{e}_\varrho +$$
$$+\left[\frac{1}{\varrho\sin\varphi}\frac{\partial v_\varrho}{\partial\vartheta} - \frac{1}{\varrho}\frac{\partial}{\partial\varrho}(\varrho v_\vartheta)\right]\mathbf{e}_\varphi + \left[\frac{1}{\varrho}\frac{\partial}{\partial\varrho}(\varrho F_\varphi) - \frac{1}{\varrho}\frac{\partial v_\varrho}{\partial\varphi}\right]\mathbf{e}_\vartheta$$

---

scalar field: $u(\varrho,\varphi,\vartheta)$

gradient: $$\nabla u = \frac{\partial u}{\partial\varrho}\mathbf{e}_\varrho + \frac{1}{\varrho\sin\varphi}\frac{\partial u}{\partial\vartheta}\mathbf{e}_\vartheta + \frac{1}{\varrho}\frac{\partial u}{\partial\varphi}\mathbf{e}_\varphi$$

Laplacian: $$\Delta u = \frac{1}{\varrho^2}\frac{\partial}{\partial\varrho}\left(\varrho^2\frac{\partial u}{\partial r}\right) + \frac{1}{\varrho^2\sin^2\varphi}\frac{\partial^2 u}{\partial\vartheta^2} + \frac{1}{\varrho^2\sin\varphi}\frac{\partial}{\partial\varphi}\left(\sin\varphi\frac{\partial u}{\partial\varphi}\right) =$$
$$= \underbrace{\frac{\partial^2 u}{\partial\varrho^2} + \frac{2}{\varrho}\frac{\partial u}{\partial\varrho}}_{\text{radial part}} + \underbrace{\frac{1}{\varrho^2\sin^2\varphi}\frac{\partial^2 u}{\partial\vartheta^2} + \frac{1}{\varrho^2\sin\varphi}\frac{\partial}{\partial\varphi}\left(\sin\varphi\frac{\partial u}{\partial\varphi}\right)}_{\text{spherical part}}$$

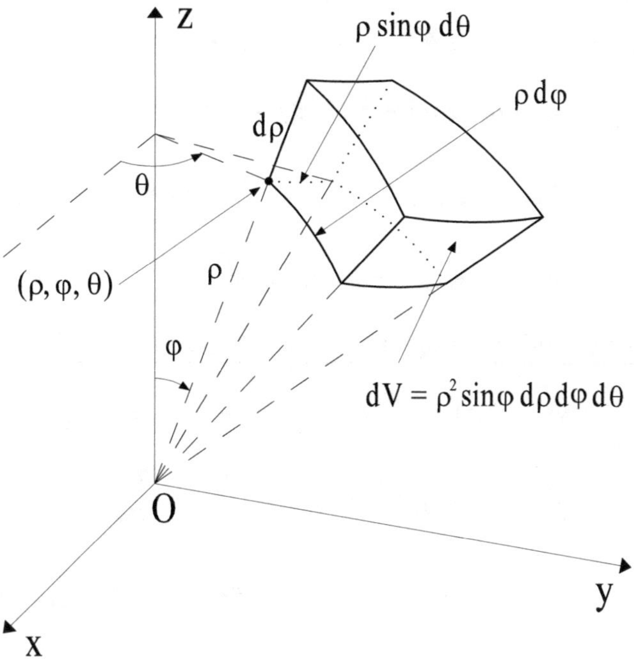

Figure 10.1 - Volume element dV in polar coordinates ($n = 3$).

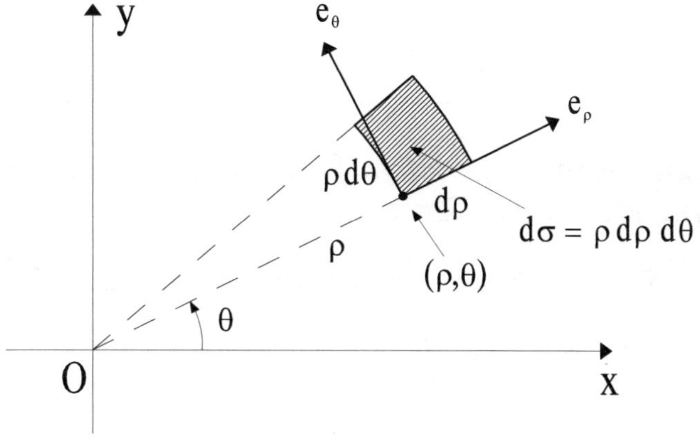

Figure 10.2 - Surface element dA in polar coordinates ($n = 2$).

10.6.3 Cylindrical coordinates in $\mathbb{R}^3$

transformation: $x = R\cos\vartheta,\ y = R\sin\vartheta,\ z = z,\ \ R > 0,\ 0\le\vartheta \le 2\pi,\ z \in \mathbb{R}$

position vector: $\mathbf{x} = R\cos\vartheta\,\mathbf{i} + R\sin\vartheta\,\mathbf{j} + z\,\mathbf{k}$

scale factor: $h_R = \left|\dfrac{\partial \mathbf{x}}{\partial R}\right| := 1,\ h_\vartheta := \left|\dfrac{\partial \mathbf{x}}{\partial \vartheta}\right| := R,\ h_z := \left|\dfrac{\partial \mathbf{x}}{\partial z}\right| = 1$

local basis: $\mathbf{e}_R = \cos\vartheta\,\mathbf{i} + \sin\vartheta\,\mathbf{j},\ \ \mathbf{e}_\vartheta = -\sin\vartheta\,\mathbf{i} + \cos\vartheta\,\mathbf{j},\ \ \mathbf{e}_z = \mathbf{k}$

volume element: $dV = dx\,dy\,dz = R\,dR\,d\vartheta\,dz$

---

vector field: $\mathbf{v}(R,\vartheta,z) = v_R(R,\vartheta,z)\mathbf{e}_R + v_\vartheta(R,\vartheta,z)\mathbf{e}_\vartheta + v_z(R,\vartheta,z)\mathbf{k}$

divergence: $\nabla\cdot\mathbf{v} = \dfrac{1}{R}\dfrac{\partial}{\partial R}(Rv_R) + \dfrac{1}{R}\dfrac{\partial v_\vartheta}{\partial \vartheta} + \dfrac{\partial v_z}{\partial z}$

curl:

$$\nabla\times\mathbf{v} = \left[\frac{1}{R}\frac{\partial v_z}{\partial\vartheta} - \frac{\partial v_\vartheta}{\partial z}\right]\mathbf{e}_R + \left[\frac{\partial v_R}{\partial z} - \frac{\partial v_z}{\partial R}\right]\mathbf{e}_\vartheta +$$

$$+\left[\frac{1}{R}\frac{\partial}{\partial R}(Rv_\vartheta) - \frac{1}{R}\frac{\partial v_R}{\partial\vartheta}\right]\mathbf{e}_z$$

---

scalar field: $u(R,\vartheta,z)$

gradient: $\nabla u = \dfrac{\partial u}{\partial R}\mathbf{e}_R + \dfrac{1}{R}\dfrac{\partial u}{\partial\vartheta}\mathbf{e}_\vartheta + \dfrac{\partial u}{\partial z}\mathbf{k}$

Laplacian: $\Delta u = \nabla^2 u = \dfrac{1}{R}\dfrac{\partial}{\partial R}\left(R\dfrac{\partial u}{\partial R}\right) + \dfrac{1}{R^2}\dfrac{\partial^2 u}{\partial\vartheta^2} + \dfrac{\partial^2 u}{\partial z^2}$

10.6.4 Laplacian operator for radially symmetric functions in $\mathbb{R}^n$

If $u \in \mathbb{C}^2(\mathbb{R}^n\setminus\{\mathbf{0}\}),\ n\ge 1,\ u\ \ (u(\mathbf{x}) = u(|\mathbf{x}|))$ and $\varrho = |\mathbf{x}|$. Then

$$\Delta u = \frac{1}{\varrho^{n-1}}\frac{\partial}{\partial\varrho}\left(\varrho^{n-1}\frac{\partial}{\partial\varrho}u\right) = \frac{\partial^2 u}{\partial\varrho^2} + \frac{n-1}{\varrho}\frac{\partial u}{\partial\varrho} \qquad \mathbb{R}^n\setminus\{\mathbf{0}\},$$

in particular:

$$\Delta u = \frac{1}{\varrho}\frac{\partial}{\partial\varrho}\left(\varrho\frac{\partial}{\partial\varrho}u\right) = \frac{\partial^2 u}{\partial\varrho^2} + \frac{1}{\varrho}\frac{\partial u}{\partial\varrho} \qquad \text{if } n = 2,$$

$$\Delta u = \frac{1}{\varrho^2}\frac{\partial}{\partial\varrho}\left(\varrho^2\frac{\partial}{\partial\varrho}u\right) = \frac{\partial^2 u}{\partial\varrho^2} + \frac{2}{\varrho}\frac{\partial u}{\partial\varrho} \qquad \text{if } n = 3.$$

10.7 Laplace transform (properties)

$\mathcal{L}$-transformable function: $f(t)$	Laplace transform: $\mathcal{L}\{f\}(s) = F(s) = \int_0^\infty e^{-st} f(t)\, dt$
$\alpha f + \beta g, \qquad \alpha, \beta \in \mathbb{C}$	$\alpha F + \beta G$
$f(at), \qquad a > 0$	$\frac{1}{a} F\left(\frac{s}{a}\right)$
$e^{\alpha t} f(t), \qquad \alpha \in \mathbb{C}$	$F(s - \alpha)$
$g(t) = f(t-\tau) = \begin{cases} 0 & t < T \\ f(t-\tau) & t \geq T \end{cases}$	$G(s) = e^{-sT} F(s)$
$\frac{f(t)}{t}$	$\int_s^\infty F(s)\, ds$
$t\, f(t)$	$-\frac{dF}{ds}$

$t^k f(t)$	$(-1)^k \dfrac{d^k F}{ds^k}$
$\dfrac{df}{dt}$ (*)	$sF(s) - f(0)$
$\dfrac{d^k f}{dt^k}$ (*)	$s^k F(s) - s^{k-1} f(0_+) - s^{k-2} \dfrac{df}{dt}(0_+) - \cdots - \dfrac{d^{k-1} f}{dt^{k-1}}(0_+)$
f' (*)	$s\,F(s)$
$f^{(k)}$ (*)	$s^k F(s)$
$g(t) = \int_0^t f(\tau)\, d\tau$	$G(s) = \dfrac{F(s)}{s}$

(*) In this Table $\dfrac{d^k f}{dt^k}$ denotes the k-th distributional derivative (or the classic one if it exists) in the open set $(0, +\infty)$; while $f^{(k)}$ denotes the k-th distributional derivative (or the classic one if it exists) in the open set $(-\infty, +\infty)$: $f' = df/dt + f(0_+)\delta\,, \ldots$

10.7.1 Laplace transforms of $\mathcal{L}$-transformable functions (spt $f \subset [0, +\infty)$, $e^{-\lambda t} f \in L^1(\mathbb{R})$)

$\mathcal{L}$-transformable function: f	Laplace transform: $F(s) = \int_0^{+\infty} e^{-st} f(t)\, dt$	convergence abscissa: λ_f
$H(t)$	$\frac{1}{s}$	0
t	$\frac{1}{s^2}$	0
t^n, $n \in \mathbb{N}$	$\frac{n!}{s^{n+1}}$	0
t^α, $\operatorname{Re} \alpha > -1$	$\frac{\Gamma(\alpha+1)}{s^{\alpha+1}}$	0
$e^{\alpha t}$, $\alpha \in \mathbb{C}$	$\frac{1}{s-\alpha}$	$\operatorname{Re} \alpha$

$\sin(\omega t), \quad \omega > 0$	$\dfrac{\omega}{s^2+\omega^2}$	0
$\cos(\omega t), \quad \omega > 0$	$\dfrac{s}{s^2+\omega^2}$	0
$\sinh(\omega t), \quad \omega \in \mathbb{R}$	$\dfrac{\omega}{s^2-\omega^2}$	$\lvert\omega\rvert$
$\cosh(\omega t), \quad \omega \in \mathbb{R}$	$\dfrac{s}{s^2-\omega^2}$	$\lvert\omega\rvert$
$\sin(\omega t+\phi), \quad \omega,\, \phi \in \mathbb{R}$	$\dfrac{s\sin\phi+\omega\cos\phi}{s^2+\omega^2}$	0
$\cos(\omega t+\phi), \quad \omega,\, \phi \in \mathbb{R}$	$\dfrac{s\cos\phi-\omega\sin\phi}{s^2+\omega^2}$	0
$e^{-\alpha t}\sin(\omega t), \quad \omega \in \mathbb{R},\, \alpha \in \mathbb{C}$	$\dfrac{\omega}{(s+\alpha)^2+\omega^2}$	$-\operatorname{Re}\alpha$

$e^{-\alpha t}\cos(\omega t), \qquad \omega \in \mathbb{R},\ \alpha \in \mathbb{C}$	$\dfrac{s+\alpha}{(s+\alpha)^2+\omega^2}$	$-\operatorname{Re}\alpha$
$\operatorname{Si} t = \displaystyle\int_0^t \frac{\sin\tau}{\tau}\,d\tau$	$\dfrac{1}{s}\operatorname{arccot} s$	0
e^{-t^2}	$\dfrac{\sqrt{\pi}}{2}\, e^{s^2/4}\operatorname{erfc}\left(\dfrac{s}{2}\right)$	$-\infty$
$\operatorname{erf}(t) = \dfrac{2}{\sqrt{\pi}}\displaystyle\int_0^t e^{-x^2}\,dx$	$\dfrac{1}{s}e^{s^2/4}\operatorname{erfc}\left(\dfrac{s}{2}\right)$	0
$\operatorname{erfc}(t) = 1 - \operatorname{erf}(t)$	$\dfrac{1}{s} - \dfrac{1}{s}e^{s^2/4}\operatorname{erfc}\left(\dfrac{s}{2}\right)$	0
$\operatorname{erf}(\sqrt{t})$	$\dfrac{1}{s\sqrt{s+1}}$	0
$\operatorname{erfc}\left(\dfrac{a}{2\sqrt{t}}\right) \qquad a > 0$	$\dfrac{e^{-a\sqrt{s}}}{s}$	0

$\frac{a}{2\sqrt{\pi t^3}}\, e^{-a^2/(4t)} \quad a > 0$	$e^{-a\sqrt{s}}$	0
$\frac{1}{\sqrt{\pi t}}\, e^{-a^2/(4t)} \quad a > 0$	$\frac{e^{-a\sqrt{s}}}{\sqrt{s}}$	0
$\ln t$	$\frac{\Gamma'(1) - \ln s}{s}$	0
$\frac{1 - e^{-t}}{t}$	$\ln\left(1 + \frac{1}{s}\right)$	0
$t^{1/n}$	$\Gamma\left(1 + \frac{1}{n}\right) s^{-(1+1/n)}$	0
$\mathrm{J}_0(t)$	$\frac{1}{\sqrt{s^2+1}}$	0
$\mathrm{J}_\mathrm{n}(t)$	$\left(\sqrt{s^2+1} - s\right)^n / \sqrt{s^2+1}$	0

$f(t+T)=f(t),\quad T>0, \forall t>0$	$\dfrac{1}{1-e^{-sT}}\displaystyle\int_0^T e^{-st} f(t)\, dt$	0

WARNING

Any function in the first column of the previous Table is implicitly thought as multiplied by the Heaviside function $H(t)$, disregarding its analytic extension to $\mathbb{R}^-$:
i.e. $\cos(\omega t)$ has to be considered as $\cos(\omega t)\, H(t)$ where H denotes the Heaviside function.

Moreover, any multi-valued function in the first column stands for the analytic branch which is real-valued on $\mathbb{R}^+$
i.e. : $t^\alpha = \exp(\alpha \ln t)\, H(t)$ where ln denotes the natural logarithm (or real valued logarithm referred to the base e);
$t^{1/n} = n$-th arithmetic root of t.

10.7.2 Laplace transforms of $\mathcal{L}$-transformable distributions (spt $f \subset [0,+\infty)$, $e^{-\lambda t} f \in \mathcal{S}'(\mathbb{R})$)

$\mathcal{L}$-transformable distribution: f		Laplace transform: $F(s)$ $F(s) = \int_{\mathbb{R}} e^{-(s-\lambda)t} \left(e^{-\lambda t f(t)}\right) dt\,,\ \ \mathrm{Re}\, s > \lambda > \lambda_f$	λ_f
$\delta(t)$		1	$-\infty$
$\delta'(t)$		s	$-\infty$
$\delta(t-\tau)$	$\tau > 0$	$e^{-\tau s}$	$-\infty$
$\delta^{(k)}(t)$	$k \in \mathbb{N}$	s^k	$-\infty$

10.8 Fourier transform

10.8.1 Fourier transforms in $L^1(\mathbb{R}) \cup L^2(\mathbb{R})$

Function $u = u(t) \qquad t \in \mathbb{R}$	Fourier transform $\mathcal{F}\{u\}(\omega) = \widehat{u}(\omega) = \int_{-\infty}^{+\infty} e^{-i\omega t}\, u(t)\, dt, \quad \omega \in \mathbb{R}$ the integral has to be considered as p.v. if $u \in L^2 \backslash L^1$
$\dfrac{1}{1+t^2}$	$\pi\, e^{-\lvert\omega\rvert}$
$e^{-\lvert t\rvert}$	$\dfrac{2}{1+\omega^2}$
$p(t) = \begin{cases} 1, & \lvert t\rvert < 1 \\ 0, & \lvert t\rvert > 1 \end{cases}$	$2\,\dfrac{\sin\omega}{\omega}$
$\dfrac{\sin t}{t} = \operatorname{Sinc} t$	$\pi\, p(\omega)$
$e^{-a t^2} \qquad a > 0,$	$\sqrt{\dfrac{\pi}{a}}\, e^{-\dfrac{\omega^2}{4a}}$
$p(t)\, \ln\lvert t\rvert$	$-2\,\dfrac{\operatorname{Si}(\omega)}{\omega} = -\dfrac{2}{\omega}\int_0^{\omega} \dfrac{\sin(\varphi)}{\varphi}\, d\varphi$
$J_0(t)$	$2\,\dfrac{p(\omega)}{\sqrt{1-\omega^2}}$
$\dfrac{\sin t}{t}\, e^{-\lvert t\rvert}$	$\arctan(\omega+1) - \arctan(\omega-1) = \arctan\left(\dfrac{2}{\omega^2}\right)$

$e^{-a\|t\|}$	$a > 0$	$\dfrac{2a}{a^2+\omega^2}$
$\dfrac{1}{a^2+t^2}$	$a > 0$	$\dfrac{\pi}{a}e^{-a\|\omega\|}$
$\dfrac{1}{(a^2+t^2)(b^2+t^2)}$	$, a > 0, b > 0, a \neq b$	$\dfrac{\pi(b^{-1}e^{-b\|\omega\|} - a^{-1}e^{-a\|\omega\|})}{a^2-b^2}$
$\dfrac{1}{(a^2+t^2)^2}$	$a > 0$	$\dfrac{\pi}{2a^3}(a\|\omega\|+1)e^{-a\|\omega\|}$
$\dfrac{1}{a^4+t^4}$	$a > 0$	$\dfrac{\pi}{a^3}e^{-a\|\omega\|/\sqrt{2}}\sin\left(\dfrac{\pi}{4}+\dfrac{a\|\omega\|}{\sqrt{2}}\right)$
$H(t)\,e^{-t}$		$\dfrac{1}{1+i\,\omega} = \dfrac{1-i\,\omega}{1+\omega^2}$
$H(t)\,t\,e^{-at}$	$a > 0$	$\dfrac{(a-i\omega)^2}{(a^2+\omega^2)^2}$

10.8.2 Fourier transforms in $\mathcal{S}'(\mathbb{R})$

Distribution $u = u(t), \quad t \in \mathbb{R}$	Fourier transform $\widehat{u}(\omega), \quad \omega \in \mathbb{R}$ $\int \widehat{u}\,\psi = \int \widehat{\psi}\,u \ \forall \psi \in \mathcal{S},\ \widehat{\psi}(\omega) = \int \psi(t) e^{-i\omega t} dt$
1	$2\pi\,\delta$
t^n	$2\pi\,(i)^n\,\delta^{(n)}$
δ	1
$\delta^{(n)}$	$(i\omega)^n$
$\delta(t - t_0)$	$e^{-i\,t_0\,\omega}$
$e^{i\,\omega_0\,t}$	$2\pi\,\delta(\omega - \omega_0)$
$\sin(\omega_0 t)$	$\pi\,i\ [\delta(\omega + \omega_0) - \delta(\omega - \omega_0)]$
$\cos(\omega_0 t)$	$\pi\ [\delta(\omega + \omega_0) + \delta(\omega - \omega_0)]$
$\text{p.v.}\,\dfrac{1}{t}$	$-\,\pi\,i\ \text{sign}(\omega)$
$\text{f.p.}\,\dfrac{1}{t^2} = -\dfrac{d}{dt}\,\text{p.v.}\,\dfrac{1}{t}$	$-\,\pi\,\lvert\omega\rvert$
$\text{f.p.}\,\dfrac{1}{t^3} = -\dfrac{1}{2}\dfrac{d}{dt}\,\text{f.p.}\,\dfrac{1}{t^2}$	$\dfrac{\pi}{2}\,i\,\omega^2\,\text{sign}(\omega)$

$\lvert t\rvert$	$-2\ \text{f.p.}\,\dfrac{1}{\omega^2}$
$t^+ = t\,H(t)$	$i\,\pi\,\delta' - \text{f.p.}\,\dfrac{1}{\omega^2}$
$\dfrac{1}{\lvert t\rvert^{1/2}}$	$\dfrac{\sqrt{2\pi}}{\lvert\omega\rvert^{1/2}}$
$\dfrac{1}{\lvert t\rvert^a}\,, \qquad 0<a<1$	$2^{1-a}\,\sqrt{\pi}\,\dfrac{\Gamma\big((1-a)/2\big)}{\Gamma(a/2)}\,\dfrac{1}{\lvert\omega\rvert^{1-a}}$
$\sin(\lvert t\rvert)$	$2\ \text{p.v.}\,\dfrac{1}{1-\omega^2}$
$\text{sign}(t)$	$-\,2i\ \text{p.v.}\,\dfrac{1}{\omega}$
$H(t) = \begin{cases} 0\,, & t<0 \\ 1\,, & t>0 \end{cases}$	$\pi\,\delta(\omega)\ -\ i\ \text{p.v.}\,\dfrac{1}{\omega}$
$\chi_{[-a,a]}\,, \qquad a>0$	$2\,\dfrac{\sin(a\,\omega)}{\omega}$
$\text{rect}(t) = \chi_{[-1/2,1/2]}$	$\text{Sinc}\left(\dfrac{\omega}{2}\right) = \dfrac{\sin(\omega/2)}{\omega/2}$
$\text{Sinc}(at)\,, \qquad a>0$	$\dfrac{\pi}{a}\,\chi_{[-1/a,1/a]}(\omega)$
$\text{sinc}(t) = \dfrac{\sin(\pi t)}{\pi t}$	$\chi_{[-\pi,\pi]}$

$\cos(\omega_0 t + \varphi)\,, \quad \omega, \varphi \in \mathbb{R}$	$\pi \left[e^{-i\varphi}\delta(\omega+\omega_0) + e^{i\varphi}\delta(\omega-\omega_0) \right]$
$\sin(\omega_0 t + \varphi)\,, \quad \omega, \varphi \in \mathbb{R}$	$i\,\pi \left[e^{-i\varphi}\delta(\omega+\omega_0) - e^{i\varphi}\delta(\omega-\omega_0) \right]$
$H(t)\cos(\omega_0 t)\,, \qquad \omega \in \mathbb{R}$	$\dfrac{\pi}{2}[\delta(\omega+\omega_0)+\delta(\omega-\omega_0)] + \dfrac{i\omega}{\omega_0^2-\omega^2}$
$H(t)\sin(\omega_0 t)\,, \qquad \omega \in \mathbb{R}$	$\dfrac{\pi}{2i}[\delta(\omega-\omega_0)-\delta(\omega+\omega_0)] + \dfrac{\omega^2}{\omega_0^2-\omega^2}$
$H(t)e^{-at}\,, \qquad a > 0$	$\dfrac{1}{a+i\omega}$
$H(t)e^{-at}\cos(\omega_0 t)\,, a > 0$	$\dfrac{(a+i\omega)}{\omega_0^2+(a+i\omega)^2}$
$H(t)e^{-at}\sin(\omega_0 t)\,, a > 0$	$\dfrac{\omega_0}{\omega_0^2+(a+i\omega)^2}$
$\displaystyle\int_{-\infty}^{t} u(\tau)\,d\tau, \;\; u \in L^1(\mathbb{R})$	$\pi \left(\displaystyle\int_{\mathbb{R}} u \right) \delta(\omega) - i\,\widehat{u}(\omega)\ \text{p.v.}\,\dfrac{1}{\omega}$
$\displaystyle\int_{0}^{t} u(\tau)\,d\tau, \quad u \in L^1(\mathbb{R})$	$\pi \left(\displaystyle\int_{0}^{+\infty} u - \int_{-\infty}^{0} u \right) \delta(\omega) - i\,\widehat{u}(\omega)\ \text{p.v.}\,\dfrac{1}{\omega}$

10.8.3 Fourier series and Fourier transform of a T periodic signal

$u \in \mathcal{S}'(\mathbb{R})\,,\quad u(t+T) = u(t)\,,\ T>0$	$\widehat{u}(\omega)\,,\quad \operatorname{spt}\widehat{u} \subset \{k\,2\pi/T,\ k \in \mathbb{Z}\}$
$u(t) = \sum_{k=-\infty}^{+\infty} \widehat{u}_k\, e^{i\,k\,\frac{2\pi}{T}\,t}$	$\widehat{u}(\omega) = 2\pi \sum_{k=-\infty}^{+\infty} \widehat{u}_k\ \delta\left(\omega - k\,\frac{2\pi}{T}\right)$

Where the coefficients $\widehat{u}_k$ are tempered: $\exists\, A, B :\ |\widehat{u}_k| \leq A(1+|k|)^B\ \forall k \in \mathbb{Z}$.

10.8.4 Fourier transform of a δ train (T spaced impulses)

$u \in \mathcal{S}'(\mathbb{R})\,,\quad \operatorname{spt} u \subset \{kT,\ k \in \mathbb{Z}\}\,,\ \ T>0$	$\widehat{u}\,,\ \ 2\pi/T$ periodic
$u(t) = \sum_{k=-\infty}^{+\infty} c_k\ \delta(t-kT)$	$\widehat{u}(\omega) = \sum_{k=-\infty}^{+\infty} c_k\, e^{-i\,k\,T\,\omega}$

Where the coefficients c_k are tempered: $\exists\, A, B :\ |c_k| \leq A(1+|k|)^B\ \forall k \in \mathbb{Z}$.

10.8.5 Fourier series and Fourier transforms of comb_T (T periodic Dirac comb)

$\mathrm{comb}_T(t) := \sum_{k=-\infty}^{+\infty} \delta(t-kT) = \frac{1}{T}\sum_{k=-\infty}^{+\infty} e^{i\,k\,\frac{2\pi}{T}\,t} = \frac{1}{T} + \frac{2}{T}\sum_{n=1}^{+\infty}\cos\left(n\frac{2\pi}{T}t\right),\ t \in \mathbb{R}$
$\widehat{\mathrm{comb}_T}(\omega) = \frac{2\pi}{T}\sum_{k=-\infty}^{+\infty} \delta\left(\omega - k\frac{2\pi}{T}\right) = \sum_{k=-\infty}^{+\infty} e^{i\,k\,T\,\omega} = 1 + 2\sum_{n=1}^{+\infty}\cos(nT\omega)\,,\ \omega \in \mathbb{R}$

10.8.6 Fourier transforms in $\mathcal{S}'(\mathbb{R}^n)$

Distribution $u = u(\mathbf{x})$, $\mathbf{x} \in \mathbb{R}^n$	Fourier transform $\widehat{u}(\boldsymbol{\xi})$, $\boldsymbol{\xi} \in \mathbb{R}^n$
$e^{-a\,\lvert\mathbf{x}\rvert^2}$, $a > 0,\ \mathbf{x} \in \mathbb{R}^n$	$\left(\frac{\pi}{a}\right)^{n/2} e^{-\lvert\boldsymbol{\xi}\rvert^2/(4a)}$, $\boldsymbol{\xi} \in \mathbb{R}^n$
$\frac{1}{\lvert\mathbf{x}\rvert^a}$, $n > a > 0$, $\mathbf{x} \in \mathbb{R}^n$	$\frac{C_{n,a}}{\lvert\boldsymbol{\xi}\rvert^{n-a}}$, $C_{n,a} = 2^{n-a}\pi^{n/2}\frac{\Gamma\left(\frac{n-a}{2}\right)}{\Gamma\left(\frac{a}{2}\right)}$
$\frac{1}{4\pi}\frac{e^{-a\lvert\mathbf{x}\rvert}}{\lvert\mathbf{x}\rvert}$, $a > 0,\ \mathbf{x} \in \mathbb{R}^3$	$\frac{1}{a^2 + \lvert\boldsymbol{\xi}\rvert^2}$, $\boldsymbol{\xi} \in \mathbb{R}^3$
$\frac{1}{(2\pi)^2}\lim_{R\to+\infty}\int_0^R \frac{J_0(\lvert\mathbf{x}\rvert\varrho)}{a^2+\varrho^2}\,\varrho\, d\varrho$, $a > 0, \mathbf{x} \in \mathbb{R}^2$	$\frac{1}{a^2 + \lvert\boldsymbol{\xi}\rvert^2}$, $\boldsymbol{\xi} \in \mathbb{R}^2$
$\delta(\mathbf{x})$, $\mathbf{x} \in \mathbb{R}^n$	1, $\boldsymbol{\xi} \in \mathbb{R}^n$
1, $\mathbf{x} \in \mathbb{R}^n$	$(2\pi)^n\,\delta(\boldsymbol{\xi})$, $\boldsymbol{\xi} \in \mathbb{R}^n$
$u(\mathbf{x}) = u_1(x_1)\,u_2(x_2)\cdots u_n(x_n)$, $\mathbf{x} \in \mathbb{R}^n$, $u_j \in \mathcal{S}'(\mathbb{R})$	$\widehat{u_1}(\xi_1)\,\widehat{u_2}(\xi_2)\cdots\widehat{u_n}(\xi_n)$, $\boldsymbol{\xi} \in \mathbb{R}^n$

10.8.7 *Fourier transforms of radially symmetric distributions in* $\mathcal{S}'(\mathbb{R}^n) \cap L^1_{loc}(\mathbb{R}^n)$

Distribution $u = u(\mathbf{x}) = \psi(\lvert\mathbf{x}\rvert)\,, \quad \mathbf{x} \in \mathbb{R}^n$	Fourier transform $\widehat{u}(\boldsymbol{\xi}), \quad \boldsymbol{\xi} \in \mathbb{R}^n$
$n = 1, \quad \mathbf{x} \in \mathbb{R}$	$2 \displaystyle\int_0^{+\infty} \cos(\xi\varrho)\, \psi(\varrho)\, d\varrho$
$n = 2, \quad \mathbf{x} \in \mathbb{R}^2$	$2\pi \displaystyle\int_0^{+\infty} J_0(\lvert\boldsymbol{\xi}\rvert\varrho)\, \psi(\varrho)\, \varrho\, d\varrho$
$n = 3, \quad \mathbf{x} \in \mathbb{R}^3$	$4\pi \displaystyle\int_0^{+\infty} \frac{\sin(\lvert\boldsymbol{\xi}\rvert\varrho)}{\lvert\boldsymbol{\xi}\rvert}\, \psi(\varrho)\, \varrho\, d\varrho$
$n \geq 1, \quad \mathbf{x} \in \mathbb{R}^n$	$(2\pi)^{n/2} \displaystyle\int_0^{+\infty} J_{\frac{n-2}{2}}(\sin(\lvert\boldsymbol{\xi}\rvert\varrho)) \left(\frac{\varrho}{\lvert\boldsymbol{\xi}\rvert}\right)^{\frac{n-2}{2}} \psi(\varrho)\, \varrho\, d\varrho$

In the previous table the integral $\displaystyle\int_0^{\infty} \ldots\, d\varrho$ stands for $\displaystyle\lim_{R\to+\infty} \int_0^{R} \ldots\, d\varrho$ in cases when $u \notin L^1(\mathbb{R}^n)$, hence in such cases:

$$\widehat{u}(\boldsymbol{\xi}) = \text{p.v.} \int_{\mathbb{R}^n} e^{-i\boldsymbol{\xi}\cdot\mathbf{x}} u(\mathbf{x})\, d\mathbf{x} = \lim_{R\to+\infty} \int_{\lvert\mathbf{x}\rvert<R} e^{-i\boldsymbol{\xi}\cdot\mathbf{x}} u(\mathbf{x})\, d\mathbf{x}\,.$$

10.8.8 Fourier transforms of homogeneous distributions in $\mathbb{R}^n$

$\mathbf{x} \in \mathbb{R}^n,\ \boldsymbol{\xi} \in \mathbb{R}^n$	$u(\mathbf{x})$	$\widehat{u}(\boldsymbol{\xi})$
$n = 1,\quad x \in \mathbb{R},\ \xi \in \mathbb{R}$	$\dfrac{1}{\sqrt{\lvert x\rvert}}$	$\dfrac{\sqrt{2\pi}}{\sqrt{\lvert \xi\rvert}}$
$n = 2,\quad \mathbf{x} \in \mathbb{R}^2,\ \boldsymbol{\xi} \in \mathbb{R}^2$	$\dfrac{1}{\lvert\mathbf{x}\rvert}$	$\dfrac{2\pi}{\lvert\boldsymbol{\xi}\rvert}$
$n = 3,\quad \mathbf{x} \in \mathbb{R}^3,\ \boldsymbol{\xi} \in \mathbb{R}^3$	$\dfrac{1}{\lvert\mathbf{x}\rvert^2}$	$\dfrac{2\pi^2}{\lvert\boldsymbol{\xi}\rvert}$
$n = 3,\quad \mathbf{x} \in \mathbb{R}^3,\ \boldsymbol{\xi} \in \mathbb{R}^3$	$\dfrac{1}{\lvert\mathbf{x}\rvert}$	$\dfrac{4\pi}{\lvert\boldsymbol{\xi}\rvert^2}$
$n \geq 3,\quad \mathbf{x} \in \mathbb{R}^n,\ \boldsymbol{\xi} \in \mathbb{R}^n$	$\dfrac{1}{\lvert\mathbf{x}\rvert^{n-2}}$	$\dfrac{4\,\pi^{n/2}}{\Gamma\left(\frac{n-2}{2}\right)}\,\dfrac{1}{\lvert\boldsymbol{\xi}\rvert^2}$
$n = 1,\quad x \in \mathbb{R},\ \xi \in \mathbb{R}$	x^2	$-\,2\pi\,\delta''(\xi)$
$n \geq 1,\quad \mathbf{x} \in \mathbb{R}^n,\ \boldsymbol{\xi} \in \mathbb{R}^n$	$\lvert\mathbf{x}\rvert^2$	$-\,(2\pi)^n\,\Delta_{\boldsymbol{\xi}}\,\delta(\boldsymbol{\xi})$
$n \geq 1,\quad \mathbf{x} \in \mathbb{R}^n,\ \boldsymbol{\xi} \in \mathbb{R}^n$	$\lvert\mathbf{x}\rvert^4$	$(2\pi)^n\,\Delta_{\boldsymbol{\xi}}^2\,\delta(\boldsymbol{\xi})$
$n \geq 1,\quad \mathbf{x} \in \mathbb{R}^n,\ \boldsymbol{\xi} \in \mathbb{R}^n$	$\lvert\mathbf{x}\rvert$	$(2\pi)^{n/2}\,(-\Delta_{\boldsymbol{\xi}})^{1/2}\,\delta(\boldsymbol{\xi})$
$n \geq 1,\quad \mathbf{x} \in \mathbb{R}^n,\ \boldsymbol{\xi} \in \mathbb{R}^n$	$-\,\Delta_{\mathbf{x}}\,\delta(\mathbf{x})$	$\lvert\boldsymbol{\xi}\rvert^2$

References

1. Amerio L., (1986): Analisi Matematica 3, UTET (4a Ed.), Torino. vol.3.1 ISBN:9788802039817; vol.3.2 ISBN:9788802037486.
2. Brezis H., (1986): Analisi funzionale, Ed. Liguori, Napoli, ISBN 88-207-1501-5.
3. Cartan H., (1997): Théorie élémentaire des fonctions analytiques, Hermann (6th Ed.), Paris, ISBN 9782705652159.
4. Evans L.C., (1998): Partial differential equations, Ed. American Mathematical Society, ISBN 9780821807729.
5. Ferrero A., Gazzola F., Zanotti M., (2010): Elementi di Analisi Superiore per la Fisica e l'Ingegneria, Esculapio (2nd Ed.), Bologna, ISBN 978-88-7488-353-0.
6. Gazzola F., Tomarelli F., Zanotti M. (2023): Analisi Complessa, Trasformate, Equazioni Differenziali, Quarta Edizione. Esculapio. ISBN 978-88-9385-344-6.
7. Gilardi G., (1994): Analisi Tre, McGraw-Hill, Milano, ISBN: 88-386-0659-5
8. John F., (1971): Partial differential equations, Springer, New York, ISBN 9780387900216.
9. Oldham K., Myland J. & Spanier J., (2000): An Atlas of functions, (2a Ed.), Springer, ISBN 978-0-387-48806-6
10. Quarteroni A., (2008): Modellistica Numerica per Problemi Differenziali, Springer Unitext 39 (4a Ed.), Milano, ISBN 9788847008427.
11. Quarteroni A., Sacco R., Saleri F. (2007): Numerical Mathematics. 2nd ed., Texts in Applied Mathematics, 12 (3a Ed.), Springer-Verlag, Berlin. ISBN: 978-3-540-34658-6 (2007).
12. Rauch J., (1991): Partial Differential Equations, Springer, New-York, ISBN 0387974725.
13. Roux D., (1984): Istituzioni di analisi superiore, Ed. Masson, Milano, ISBN 88-214-0561-3.
14. Rudin W., (2006): Real and Complex Analysis, McGraw-Hill, ISBN 978-0070619876.
15. Rudin W., (1991): Functional Analysis, Ed. McGraw-Hill, Singapore, ISBN 978-0070542365.
16. Salinelli E., Tomarelli F. (2014): Discrete Dynamical Models, Springer Unitext, 76 (2014), ISBN 978-3-319-02290-1 (eBook: ISBN 978-3-319-02291-8).
17. Salsa S., (2010): Equazioni a Derivate Parziali, Springer Unitext 45 (2a Ed.), Milano, ISBN 978-88-470-1645-3.
18. Salsa S., Vegni F., Zaretti A., Zunino P. (2009): Invito alle Equazioni a Derivate Parziali, Springer Unitext 42 (2a Ed.), Milano, ISBN 978-88-470-1179-3.
19. Smoller J., (1983): Shock waves and reaction-diffusion equations, Springer-Verlag, New-York, ISBN 978-0387907529.
20. Soardi P.M., (1998): Appunti sulle Ondine, Quaderni U.M.I., Pitagora, Bologna, ISBN 9788837110260.

21. Spiegel M.R., (1976): Trasformate di Laplace, Collana Schaum, Etas Libri, Milano, ISBN 8845300250.
22. Strang G., (1986): Introduction to Applied Mathematics, Wellesley-Cambridge Press, ISBN 0-9614088-0-4.
23. Strang G., Nguyen T., (1996): Wavelets and Filter Banks, Wellesley - Cambridge Press, ISBN 0961408871.
24. Tomarelli F., (1993): Esercizi di metodi matematici per l'ingegneria, CittàStudi, Milano, ISBN: 8825172583.
25. Tomarelli F. (2023): Mathematical Analysis Tools for Engineering, Esculapio. ISBN 978-88-9385-346-0.
26. Wegert E., (2012): Visual Complex Functions, Birkhäuser, ISBN 978-3-0348-0179-9.
27. Wolfram S., (2003): The Mathematica Book, Fifth Edition, Wolfram Media, ISBN 1-57955-022-3.

Index

Made in the USA
Monee, IL
05 November 2026

67812671R00216